Modern Sample Preparation Approaches for Separation Science

Modern Sample Preparation Approaches for Separation Science

Special Issue Editor
Nuno Neng

MDPI • Basel • Beijing • Wuhan • Barcelona • Belgrade

Special Issue Editor
Nuno Neng
Universidade de Lisboa
Portugal

Editorial Office
MDPI
St. Alban-Anlage 66
4052 Basel, Switzerland

This is a reprint of articles from the Special Issue published online in the open access journal *Molecules* (ISSN 1420-3049) from 2018 to 2019 (available at: https://www.mdpi.com/journal/molecules/special issues/modern sample separation).

For citation purposes, cite each article independently as indicated on the article page online and as indicated below:

LastName, A.A.; LastName, B.B.; LastName, C.C. Article Title. *Journal Name* **Year**, *Article Number*, Page Range.

ISBN 978-3-03921-411-2 (Pbk)
ISBN 978-3-03921-412-9 (PDF)

© 2019 by the authors. Articles in this book are Open Access and distributed under the Creative Commons Attribution (CC BY) license, which allows users to download, copy and build upon published articles, as long as the author and publisher are properly credited, which ensures maximum dissemination and a wider impact of our publications.

The book as a whole is distributed by MDPI under the terms and conditions of the Creative Commons license CC BY-NC-ND.

Contents

About the Special Issue Editor vii

Preface to "Modern Sample Preparation Approaches for Separation Science" ix

Ruiqin Zhu, Ying Dong, Xiangyang Cai and Chuixiu Huang
Determination of Barbiturates in Biological Specimens by Flat Membrane-Based Liquid-Phase Microextraction and Liquid Chromatography-Mass Spectrometry
Reprinted from: *molecules* **2019**, *24*, 1494, doi:10.3390/molecules24081494 1

Thomas Ribette, Bertrand Leroux, Balkis Eddhif, Audrey Allavena, Marc David, Robert Sternberg, Pauline Poinot and Claude Geffroy-Rodier
Primary Step Towards In Situ Detection of Chemical Biomarkers in the UNIVERSE via Liquid-Based Analytical System: Development of an Automated Online Trapping/Liquid Chromatography System
Reprinted from: *molecules* **2019**, *24*, 1429, doi:10.3390/molecules24071429 13

Rui Zhang, Zhen-Chao Tan, Ke-Cheng Huang, Yan Wen, Xiang-Ying Li, Jun-Long Zhao and Cheng-Lan Liu
A Vortex-Assisted Dispersive Liquid-Liquid Microextraction Followed by UPLC-MS/MS for Simultaneous Determination of Pesticides and Aflatoxins in Herbal Tea
Reprinted from: *molecules* **2019**, *24*, 1029, doi:10.3390/molecules24061029 26

Wenbin Chen, Xijuan Tu, Dehui Wu, Zhaosheng Gao, Siyuan Wu and Shaokang Huang
Comparison of the Partition Efficiencies of Multiple Phenolic Compounds Contained in Propolis in Different Modes of Acetonitrile–Water-Based Homogenous Liquid–Liquid Extraction
Reprinted from: *molecules* **2019**, *24*, 442, doi:10.3390/molecules24030422 38

Ying Xue, Xian-Shun Xu, Li Yong, Bin Hu, Xing-De Li, Shi-Hong Zhong, Yi Li, Jing Xie and Lin-Sen Qing
Optimization of Vortex-Assisted Dispersive Liquid-Liquid Microextraction for the Simultaneous Quantitation of Eleven Non-Anthocyanin Polyphenols in Commercial Blueberry Using the Multi-Objective Response Surface Methodology and Desirability Function Approach
Reprinted from: *molecules* **2018**, *23*, 2921, doi:10.3390/molecules23112921 49

Ola Svahn and Erland Björklund
High Flow-Rate Sample Loading in Large Volume Whole Water Organic Trace Analysis Using Positive Pressure and Finely Ground Sand as a SPE-Column In-Line Filter
Reprinted from: *molecules* **2019**, *24*, 1426, doi:10.3390/molecules24071426 63

Luyi Jiang, Jie Wang, Huan Zhang, Caijing Liu, Yiping Tang and Chu Chu
New Vortex-Synchronized Matrix Solid-Phase Dispersion Method for Simultaneous Determination of Four Anthraquinones in Cassiae Semen
Reprinted from: *molecules* **2019**, *24*, 1312, doi:10.3390/molecules24071312 78

Wenbang Li, Fangling Wu, Yongwei Dai, Jing Zhang, Bichen Ni and Jiabin Wang
Poly (Octadecyl Methacrylate-Co-Trimethylolpropane Trimethacrylate) Monolithic Column for Hydrophobic in-Tube Solid-Phase Microextraction of Chlorophenoxy Acid Herbicides
Reprinted from: *molecules* **2019**, *24*, 1678, doi:10.3390/molecules24091678 92

Luiz G. M. Beloti, Luis F. C. Miranda and Maria Eugênia C. Queiroz
Butyl Methacrylate-Co-Ethylene Glycol Dimethacrylate Monolith for Online in-Tube SPME-UHPLC-MS/MS to Determine Chlopromazine, Clozapine, Quetiapine, Olanzapine, and Their Metabolites in Plasma Samples
Reprinted from: *molecules* 2019, *24*, 310, doi:10.3390/molecules24020310 . . . . **105**

Huifang Liu, Geun Su Noh, Yange Luan, Zhen Qiao, Bonhan Koo, Yoon Ok Jang and Yong Shin
A Sample Preparation Technique Using Biocompatible Composites for Biomedical Applications
Reprinted from: *molecules* 2019, *24*, 1321, doi:10.3390/molecules24071321 . . . . **118**

Naiara M. F. M. Sampaio, Natara D. B. Castilhos, Bruno C. da Silva, Izabel C. Riegel-Vidotti and Bruno J. G. Silva
Evaluation of Polyvinyl Alcohol/Pectin-Based Hydrogel Disks as Extraction Phase for Determination of Steroidal Hormones in Aqueous Samples by GC-MS/MS
Reprinted from: *molecules* 2019, *24*, 40, doi:10.3390/molecules24010040 . . . . **128**

Ana R. M. Silva, Nuno R. Neng and José M. F. Nogueira
Multi-Spheres Adsorptive Microextraction (MSAμE)—Application of a Novel Analytical Approach for Monitoring Chemical Anthropogenic Markers in Environmental Water Matrices
Reprinted from: *molecules* 2019, *24*, 931, doi:10.3390/molecules24050931 . . . . **141**

Anna Klimek-Turek, Kamila Jaglińska, Magdalena Imbierowicz and Tadeusz Henryk Dzido
Solvent Front Position Extraction with Semi-Automatic Device as a Powerful Sample Preparation Procedure Prior to Quantitative Instrumental Analysis
Reprinted from: *molecules* 2019, *24*, 1358, doi:10.3390/molecules24071358 . . . . **151**

Fumiki Takahashi, Masaru Kobayashi, Atsushi Kobayashi, Kanya Kobayashi and Hideki Asamura
High-Frequency Heating Extraction Method for Sensitive Drug Analysis in Human Nails
Reprinted from: *molecules* 2018, *23*, 3231, doi:10.3390/molecules23123231 . . . . **164**

Xijuan Tu and Wenbin Chen
A Review on the Recent Progress in Matrix Solid Phase Dispersion
Reprinted from: *molecules* 2018, *23*, 2767, doi:10.3390/molecules23112767 . . . . **175**

Yan-hong Liu, Bo Wan and Ding-shuai Xue
Sample Digestion and Combined Preconcentration Methods for the Determination of Ultra-Low Gold Levels in Rocks
Reprinted from: *molecules* 2019, *24*, 1778, doi:10.3390/molecules24091778 . . . . **188**

Ivan Liakh, Alicja Pakiet, Tomasz Sledzinski and Adriana Mika
Modern Methods of Sample Preparation for the Analysis of Oxylipins in Biological Samples
Reprinted from: *molecules* 2019, *24*, 1639, doi:10.3390/molecules24081639 . . . . **208**

Yuan Zhang, Wei-e Zhou, Jia-qing Yan, Min Liu, Yu Zhou, Xin Shen, Ying-lin Ma, Xue-song Feng, Jun Yang and Guo-hui Li
A Review of the Extraction and Determination Methods of Thirteen Essential Vitamins to the Human Body: An Update from 2010
Reprinted from: *molecules* 2018, *24*, 1484, doi:10.3390/molecules23061484 . . . . **246**

About the Special Issue Editor

Nuno Neng graduated in Chemistry from Faculdade de Ciências da Universidade de Lisboa (FCUL) in 2005 and completed a Master's degree in Biomedical Inorganic Chemistry—Applications in Diagnostic and Therapy in 2007 at the same institution. In 2012, N.R. Neng obtained a PhD degree in Analytical Chemistry by the FCUL and gained a post-doctoral fellowship from Fundação para a Ciência e a Tecnologia. In 2017, N.R. Neng was invited auxiliary professor and more recently as a junior researcher at FCUL. N.R. Neng specializes in experimental techniques in the area of separation science and technology with prominence to the development, optimization and validation of new analytical strategies for the trace analysis of chemical compounds. In this context, his research field falls upon areas such as environment, food, forensic, and clinical trials, among others.

Preface to "Modern Sample Preparation Approaches for Separation Science"

My thoughts for this book are to provide a set of high-quality studies and ideas of modern sample preparation techniques for analytical separation. Sample preparation is an essential step in most analytical methods for environmental and biomedical analysis, since the target analytes are often not detected in their in-situ forms, or the results are distorted by interfering species. In the past decade, modern sample preparation techniques have aimed to comply with green analytical chemistry principles, leading to simplification, miniaturization, easy manipulation of the analytical devices, low costs, strong reduction or absence of toxic organic solvents, and low sample volume requirements.

The Special Issue gathered in this printed book was able to gather 18 peer-reviewed manuscripts, with 14 original research articles and 4 reviews. The manuscripts were submitted by research groups from different countries that fit the aims and scope of this Special Issue.

I would like to thank all contributors, authors, and colleagues who chose to publish their works here, as well as the reviewers who dedicated their time, effort, and expertise to evaluating the submissions and assuring the high quality of the published work. I would also like to thank the publisher MDPI and the editorial staff of the journal for their constant and professional support as well as for their invitation to edit this Special Issue. Furthermore, I would like to thank all readers, and I hope that the content of this book will offer new perspectives and ideas to initiate and continue further research in sample preparation.

Nuno Neng
Special Issue Editor

Article

Determination of Barbiturates in Biological Specimens by Flat Membrane-Based Liquid-Phase Microextraction and Liquid Chromatography-Mass Spectrometry

Ruiqin Zhu [†], Ying Dong [†], Xiangyang Cai and Chuixiu Huang *

Department of Forensic Medicine, Huazhong University of Science and Technology, 13 Hangkong Road, Wuhan 430030, China; zhuruiqin@hust.edu.cn (R.Z.); yingdong@hust.edu.cn (Y.D.); xiangyangc@hust.edu.cn (X.C.)
* Correspondence: chuixiuh@hust.edu.cn; Tel.: +86-027-8369-2042
† These authors contributed equally to this work.

Academic Editor: Nuno Neng
Received: 19 March 2019; Accepted: 16 April 2019; Published: 16 April 2019

Abstract: The wide abuse of barbiturates has aroused extensive public concern. Therefore, the determination of such drugs is becoming essential in therapeutic drug monitoring and forensic science. Herein, a simple, efficient, and inexpensive sample preparation technique, namely, flat membrane-based liquid-phase microextraction (FM-LPME) followed by liquid chromatography-mass spectrometry (LC-MS), was used to determine barbiturates in biological specimens. Factors that may influence the efficiency including organic extraction solvent, pH, and composition of donor and acceptor phases, extraction time, and salt addition to the sample (donor phase) were investigated and optimized. Under the optimized extraction conditions, the linear ranges of the proposed FM-LPME/LC-MS method (with correlation coefficient factors ≥ 0.99) were 7.5–750 ng mL^{-1} for whole blood, 5.0–500 ng mL^{-1} for urine, and 25–2500 ng g^{-1} for liver. Repeatability between 5.0 and 13.7% was obtained and the limit of detection (LOD) values ranged from 1.5 to 3.1 ng mL^{-1}, from 0.6 to 3.6 ng mL^{-1}, and from 5.2 to 10.0 ng g^{-1} for whole blood, urine, and liver samples, respectively. This method was successfully applied for the analysis of barbiturates in blood and liver from rats treated with these drugs, and excellent sample cleanup was achieved.

Keywords: membrane-based microextraction; barbiturates; simultaneous determination; whole blood; urine; liver

1. Introduction

Barbiturates, which typically act as central nervous system depressants, are principally used as anxiolytics, hypnotics, and anticonvulsants in medical practice [1–3]. Depending on the dosage, barbiturates can produce a wide spectrum of effects [4–6]. For example, they can cause relaxation and sleepiness at a relatively low dose but depress the respiratory system at a high dose. Moreover, they have a potential risk of physical and psychological addiction and may result in serious adverse effects [5]. Due to the addictive properties, the use of barbiturates as sedative/hypnotics has largely been superseded by the benzodiazepine group [6]. Nowadays, the wide abuse of such drugs has aroused extensive public concern. Therefore, the determination of barbiturates in biological specimens is not only essential in therapeutic drug monitoring to investigate poisoning but also important in new formulation development, as well as in forensic science [7].

In recent decades, the analysis of barbiturates has attracted extensive attention worldwide, and several methods such as ultraviolet–visible spectroscopy (UV-Vis) [8], capillary electrophoresis

(CE) [9–12], liquid chromatography (LC) [13,14], liquid chromatography–mass spectrometry (LC-MS) [15], and gas chromatography–mass spectrometry (GC-MS) [16–20] have been reported for the determination of barbiturates in biological specimens. Traditional technologies such as UV-Vis spectroscopy is still widely used in forensic science for its convenience but it lacks specificity and sensitivity. CE method features high resolving power, low solvent consumption, and simple pretreatment, however, its major drawback is the inherent low concentration sensitivity. GC-MS was commonly applied for the analysis of drugs because it can achieve low limits of detection, however, the sensitivity of GC-MS for barbiturates is not high enough because it requires derivatization. With the recent advances of instruments, LC-MS has become an efficient analytical method to determine barbiturates in drug monitoring due to its high sensitivity and specificity [21].

However, it is still difficult to determine the target analyte concentrations at low levels in biological specimens without sample preparation in view of the limited sample volumes and complex sample matrices [8,22–24]. Hence, appropriate sample preparation is necessary and of great significance in the whole analysis process. With the development of extraction techniques, miniaturized techniques such as solid-phase microextraction (SPME) and liquid-phase microextraction (LPME) have already been the current trend in sample cleanup [25–31]. Compared to conventional liquid-liquid extraction (LLE) and solid-phase extraction (SPE), SPME and LPME offer short extraction time and high extraction efficiency without using large volumes of organic solvent. For example, barbiturates in human whole blood, urine, and hair were extracted by SPME and detected by GC-MS [19,32]. However, the construction of a suitable SPME setup often involves a tedious and complicated procedure. Moreover, the fiber used in the SPME suffers from high cost, short life, and the possibility of carryover [33].

On the other hand, LPME, which is considered as a "green" extraction technique, has attracted widespread attention for its good purification capability, economical efficiency, and easy operation [18,28–31,34]. Until now, LPME has already been applied in the purification and enrichment steps in different biological samples [8,16,18,23,34]. For example, Zarei et al. adopted a dispersive liquid–liquid microextraction (DLLME) technique combined with spectrophotometric analysis for the determination of trace amounts of barbituric acid in human serum [8]. Hollow fiber–liquid phase microextraction (HF-LPME) was also developed to isolate barbiturates in hair [18], blood [16], and liver samples [35], and coupled with GC-MS, satisfactory results can be reached.

Taking advantage of LPME, in this work, we aim to develop a method for the simultaneous quantification of barbital, phenobarbital, and pentobarbital in urine, blood, and liver tissue (the structures of the three barbiturates can be found in Figure 1). In our previous work [29], flat membrane-based liquid-phase microextraction (FM-LPME) was applied for the extraction of acidic drugs from human plasma. Compared to hollow fibers, the flat membrane device can accommodate a larger amount of acceptor phase to promote high efficiency. In addition, the FM-LPME setup is more convenient and easier to manipulate. Here, LC-MS, which has a lower limit of detection and no requirement for pretreatment of derivatization compared to GC-MS, is applied to detect the target analytes. To the best of our knowledge, this is the first report on the analysis of barbiturates in biological samples using FM-LPME coupled with LC-MS.

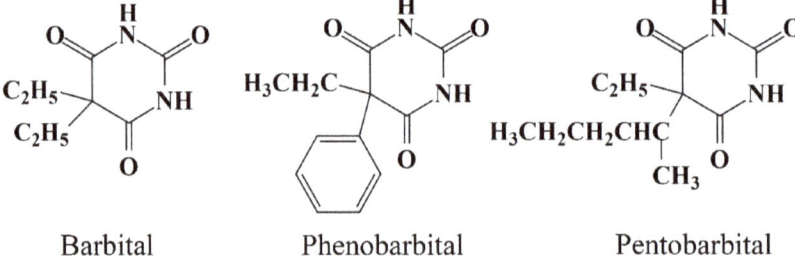

Figure 1. Structure diagram of barbital, phenobarbital, and pentobarbital.

2. Results and Discussion

2.1. Optimization of the Extraction Conditions

In order to optimize LPME of barbiturates from biological specimens, we systematically studied the analytical factors such as solvent type, donor and acceptor phase type, extraction time, and salt addition that may potentially affect the sample extraction efficiency. In this study, the extraction efficiency is defined as extraction recovery, and the recovery for each analyte was calculated by the following equation:

$$\text{Recovery} = C_A \times \frac{V_A}{C_D^0 \times V_D} \times 100\%$$

where C_A represents the concentration of the analytes in the acceptor solution after extraction, and C_D^0 is the initial concentration of the analytes in the sample solution, while V_A and V_D are the volumes of the acceptor and sample solution, respectively.

2.1.1. Selection of the Organic Extraction Solvent

The type of organic solvent is a key parameter in LPME. The organic solvent is impregnated in the pores of the membranes, constructing the supported liquid membrane (SLM). Therefore, the ideal organic solvent should be compatible with the membrane, immiscible in the donor and acceptor phase, and have good stability over the extraction process and excellent affinity for the target analyte [36]. On the basis of these considerations, we tested five types of organic solvents including 2-octanone, 2-nonanone, 2-undecanone, 1-octanol, and dihexyl ether (DHE). As shown in Figure 2a, 2-nonanone provided the highest extraction recovery for barbiturates among the tested organic solvents. Thus, 2-nonanone was selected and used in the rest of the experiments.

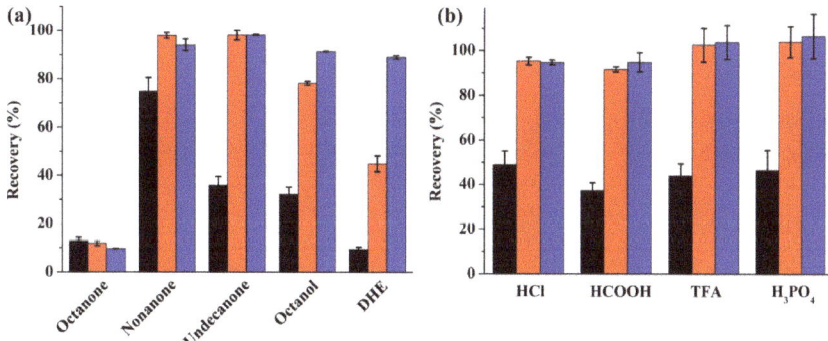

Figure 2. Extraction efficiency for barbiturates (black, red, and blue columns represent barbital, phenobarbital, and pentobarbital, respectively) using different organic solvents (**a**) and donor phases (**b**). Extraction conditions: extraction time: 60 min; stirring speed: 1000 rpm; acceptor phase: 100 μL 20 mM Na$_3$PO$_4$; (**a**) donor phase: 800 μL 10 mM HCl and 4 μL different organic solvent; (**b**) organic solvent: 4 μL 2-nonanone and 800 μL donor phase.

2.1.2. Optimization of the Donor Phase

Barbiturates are acidic drugs with a pKa at about 7. In order to get high extraction efficiency, the donor phase should be acidified so that the analytes can be deionized and consequently transfer from the donor phase into the organic phase. In this study, four different acids including hydrochloric acid, formic acid, trifluoroacetic acid (TFA), and phosphoric acid were tested at a concentration of 10 mM. The results in Figure 2b display no obvious difference using the four acids. The pH-value of all the above-selected donor phase background electrolytes are below 3, meaning that the targets can be completely deionized in all of the tested donor phases. Therefore, recoveries of all three barbiturates were similar to the tested donor phases. Here, hydrochloric acid was selected for the subsequent experiments because it provided the highest recovery for barbital and it is one of the most commonly used background electrolytes in LPME.

2.1.3. Optimization of the Acceptor Phase

Since barbiturates are acidic analytes, the acceptor phase should be basic to ionize them in order to prevent the analytes from re-entering into the organic phase [37]. In this study, the pH of the acceptor phase was adjusted in the range of 8–12 using sodium hydroxide. According to the results in Figure 3a, the highest recovery was obtained at a pH value of 12, and higher pH (≥ 13) resulted in an M-shaped peak for barbital and phenobarbital. As a result, further studies were conducted with an acceptor solution pH 12.

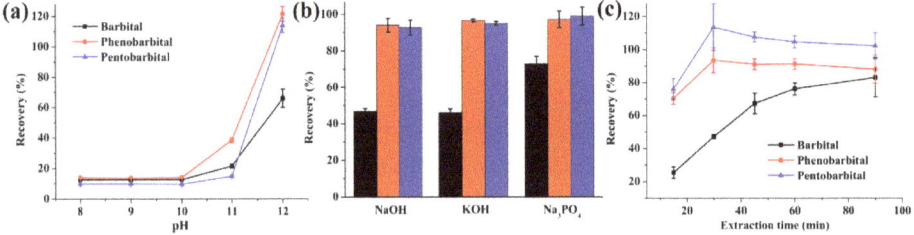

Figure 3. Effect of pH (**a**), composition of acceptor phase (**b**), extraction time (**c**) on the recovery of barbiturates. In (**b**), the black, red, and blue columns represent barbital, phenobarbital, and pentobarbital, respectively. Extraction conditions: stirring speed: 1000 rpm; donor phase: 800 μL 10 mM HCl; 4 μL 2-nonanone (**a**) 100 μL acceptor phase with different pH and 60 min extraction time; (**b**) 100 μL different acceptor phases with pH 12 and 60 min extraction time (**c**) 100 μL trisodium phosphate with pH 12 and different extraction time.

For the determination of the composition of the acceptor phase, three different basic chemicals including sodium hydroxide, potassium hydroxide, and trisodium phosphate were tested. As observed in Figure 3b, trisodium phosphate gave the best recovery for barbiturates. Trisodium sodium as the acceptor solution obtained the best extraction efficiency, possibly due to its better buffering capacity, thus, facilitating pH gradient transport of the targets.

2.1.4. Effect of Stirring Rate and Extraction Time

Normally, stirring of the sample solution can accelerate the extraction because it facilitates the analyte diffusion from the donor phase to the interface of the SLM [37], reducing the time required to reach thermodynamic equilibrium. The effect of sample agitation was tested using a stirring speed between 250 and 1000 rpm. It was observed that with an increase in the stirring rate, the barbiturate extraction efficiencies were improved. Therefore, 1000 rpm was chosen as the optimal stirring speed in subsequent experiments.

The mass transfer kinetics of LPME is passive diffusion, so it takes time for the analytes to reach equilibrium within the three phases. As a consequence, the extraction time can influence the distribution of the analyte between the sample, SLM, and acceptor phase. Therefore, the influence of extraction time (from 15 to 90 min) on the recovery of barbiturates was investigated (Figure 3c). It was clearly shown that with an increase in the extraction time up to 30 min, the recovery increased rapidly for all three barbiturates but decreased slightly thereafter for pentobarbital and phenobarbital because the system reached equilibrium. For barbital, the recovery kept increasing with extending the extraction time. It has been reported that the validation data are not affected by the extraction time under non-equilibrium conditions using LPME [38]. From the view of practical application, we selected 60 min as the extraction time for the following experiments.

2.2. FM-LPME of Barbiturates from Biological Specimens

Subsequently, the optimized extraction procedures were performed on biological specimens including whole blood, urine, and liver. The addition of salt in the blood sample may increase the recovery in microextraction procedures, especially for the more polar analytes because of the salting-out effect [38]. Moreover, the diffusion of analytes might be reduced due to the interaction of the analyte molecules with the added ions [39]. For that purpose, the salt influence was tested with the addition of NaCl at concentrations between 0.5 and 20% (w/v) in the extraction solution in whole blood samples. As shown in Figure 4a, with the addition of salt, the extraction efficiency decreased initially and then increased steadily until the salt concentration reached 12.5%. As a result, a concentration of 12.5% of salt was added to the whole blood samples to improve the extraction efficiency. Due to the viscosity of the biological samples, appropriate dilution has an apparent effect on recovery improvement. As a result, the whole blood, urine, and homogenized liver were diluted with HCl solution by different times before the extraction procedure, which is depicted in detail in the Materials and Method part. The extraction efficiencies for the biological samples and water sample are shown in Figure 4b. Compared with the water sample, lower recoveries were obtained from the biological samples for all three barbiturates. It has already been proven that barbiturates tend to bind to proteins in biological samples, and protein adsorption to the membrane surface also leads to a lower mass transport rate during the extraction [9]. A distinct decrease could be observed in the recovery of pentobarbital from the blood and liver samples, which might have resulted from the stronger protein binding ability of pentobarbital [40]. Premised on these considerations, the extraction efficiencies from the biological samples could be regarded as satisfying.

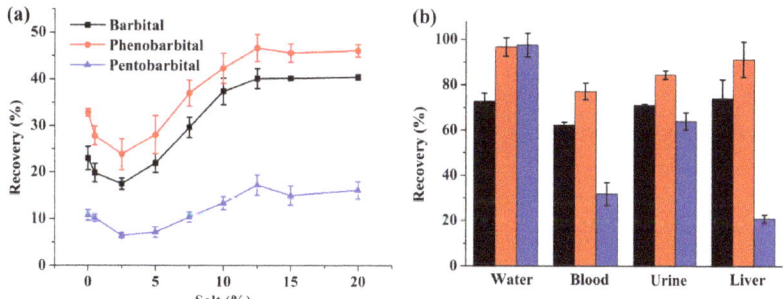

Figure 4. (**a**) The effect of salt addition on the extraction of barbiturates from the whole blood sample; and (**b**) extraction efficiencies of barbiturates from different samples applying FM-LPME. Black, red, and blue columns represent barbital, phenobarbital, and pentobarbital, respectively. Extraction conditions: extraction time: 60 min; stirring speed: 1000 rpm; acceptor phase: 100 µL 20 mM Na_3PO_4; organic solvent: 4 µL 2-nonanone.

2.3. Method Evaluation

To evaluate the analytical performance of the proposed method, figures of merit of this method including linear range, limit of detection and quantification (LOD and LOQ), and repeatability were studied for the extraction of barbiturates from biological samples under the optimum conditions and the results are illustrated in Table 1.

Table 1. Validation of the proposed method for determination of the three barbiturate drugs in biological specimens.

Matrices	Analytes	Linearity (ng mL^{-1})	LOD (ng mL^{-1})	LOQ (ng mL^{-1})	Repeatability (%)
	Barbital	15–750	2.3	7.7	10
Blood	Phenobarbital	7.5–750	1.5	5.0	6
	Pentobarbital	15–750	3.1	10.2	8
	Barbital	20–500	3.6	12.0	11
Urine	Phenobarbital	5–500	1.2	4.0	5
	Pentobarbital	5–500	0.6	2.0	5
	Barbital	50–2500	10.0	33.3	9
Liver [1]	Phenobarbital	25–2500	5.2	17.3	11
	Pentobarbital	25–2500	7.4	24.7	14

[1] The concentration unit for liver is ng g^{-1}.

The calibration curves show good linearity for all analytes in the ranges as shown in Table 1 with correlation coefficient factors all greater than 0.99. The LODs for whole blood were 1.5–3.1 ng mL^{-1}, for urine were 0.6–3.6 ng mL^{-1}, and for liver were 5.2–10.0 ng g^{-1}. It was reported that the therapeutic blood levels of the barbiturates were several µg ml^{-1} to several 10 µg mL^{-1} [41]. Therefore, the proposed method is sensitive enough to meet the therapeutic levels.

The repeatability of the proposed method was evaluated by analyzing the biological specimens spiked with barbiturates ($n = 5$) at a concentration of 50 ng mL^{-1}. Repeatability results are expressed as relative standard deviation (RSD %). The RSD values were below 20% in all cases. The assay results demonstrate that our present method can provide good repeatability for complex biological specimens.

2.4. Application

Due to the importance of determining barbiturates in biological specimens, the optimized and evaluated method was applied to determine the concentration of barbiturates in whole blood and liver samples from two rats treated with barbiturates. Two male Sprague-Dawley (SD) rats weighing 250 g and 300 g were gavaged with a barbiturate mixture at a dose of 50 mg kg^{-1} [42]. Then, two hours later, their whole blood and liver were collected for analysis, and the results are summarized in Table 2.

Table 2. Concentrations of barbiturates in whole blood and liver found in two actual cases.

	Analytes	Blood	Liver
		Drug Concentration [1] (µg mL^{-1})	Drug Concentration [1] (µg g^{-1})
	Barbital	51.1 ± 4.9	17.2 ± 3.0
Rat 1	Phenobarbital	50.3 ± 4.1	18.8 ± 3.2
	Pentobarbital	36.0 ± 2.4	52.6 ± 8.2
	Barbital	44.9 ± 3.0	19.6 ± 1.5
Rat 2	Phenobarbital	44.8 ± 6.7	20.0 ± 1.6
	Pentobarbital	34.2 ± 3.7	58.9 ± 4.4

[1] The concentrations were calculated based on the dilution times.

2.5. Comparison of the Proposed Method with Other Reported Methods

Table 3 summarizes different methods reported in the literature for determining barbital, phenobarbital, and pentobarbital in biological samples. As can be observed from the table, compared with other reported methods, our method provided almost the lowest LOD value for all three barbiturates.

Table 3. Summary of reported methods for determining barbiturates in biological specimens.

Sample	Analytes	Extraction	Detection	Linear Range (ng mL^{-1})	LOD (ng mL^{-1})	Ref
Urine	Phenobarbital Barbital	SPE	CE	2–500	0.5–5.0	[11]
Blood	Pentobarbital	SPME	GC-MS	200–40000	50	[32]
Serum	Barbital Phenobarbital	LLE	CE-UV	2900–43290	830–1390	[12]
Liver	Pentobarbital Phenobarbital	HF-LPME	GC-MS	1000–10,000 (ng g^{-1})	500 (ng g^{-1})	[35]
Blood	Pentobarbital Phenobarbital	HF-LPME	GC-MS	1000–10,000	1000	[16]
Blood	Barbital Phenobarbital Pentobarbital	LLE	LC-MS	2–2000	0.2–0.5	[43]
Blood Urine Liver	Barbital Phenobarbital Pentobarbital	FM-LPME	LC-MS	7.5–750 [1] 5–500 [2] 25–2500 [3] (ng g^{-1})	1.5–3.1 [1] 0.6–3.6 [2] 5.2–10.0 [3] (ng g^{-1})	Our work

[1] For blood. [2] For urine. [3] For liver.

3. Materials and Methods

3.1. Chemicals and Materials

Barbital was obtained from Shenyang Trial Three Biochemical Technology Development Co., Ltd. (Shenyang, China). Phenobarbital and pentobarbital sodium were purchased from Shanghai Chemical Reagent Factory (Shanghai, China). Formic acid, acetic acid, 2-octanone, 2-nonanone, 2-undecanone, 1-octanol, dihexyl ether (DHE), trifluoroacetic acid (TFA), and diclofenac sodium were all purchased from Aladdin Chemical Reagent Co. (Shanghai, China). Hydrochloric acid (HCl), sodium hydroxide (NaOH), potassium hydroxide (KOH), and trisodium phosphate (Na_3PO_4) were supplied by Sinopharm Chemical Reagent Co., Ltd. (Shanghai, China). Methanol was from Tedia (Fairfield, OH, USA). Formic acid, methanol, and acetic acid were of chromatographic purity grade while other chemicals were of analytical grade. Milli-Q water purification system (Mollsheim, France) was employed to produce deionized water.

Accurel PP 1E (R/P) flat membrane (polypropylene membrane, average thickness of 100 μm) was from Membrana (Wuppertal, Germany). The standard 1000 μL pipette tips were from Kirgen (Shanghai, China). The Eppendorf safe lock 2.0 mL PP tubes were obtained from Eppendorf AG (Hamburg, Germany).

3.2. FM-LPME Setup and Extraction Procedures

The setup diagram of FM-LPME is shown in Figure 5. It comprised two aqueous phases—acceptor phase and donor phase—isolated by SLM, which was prepared by immobilization of some organic solvent into the pores of the Accurel PP 1E (R/P) flat membrane. The fabrication of this FM-LPME configuration was described in our previous work [44]. The container of acceptor solution was a wide end-closed 1000 μL pipet tip sealed with a piece of flat membrane. The narrow end of the pipet tip was cut off for easy operation. The donor compartment was a 2 mL Eppendorf PP tube. Afterward, the acceptor compartment was inserted into the sample compartment. The LPME process was initiated by starting the MIC-100 constant temperature mixer (Hangzhou MIULAB Instrument Co. Ltd., Hangzhou, China) with a speed of 1000 rpm. After the default extraction time, the extraction was terminated by manually turning off the agitator. Immediately, the acceptor solution after LPME was collected individually and subsequently analyzed by HPLC-UV or LC-MS.

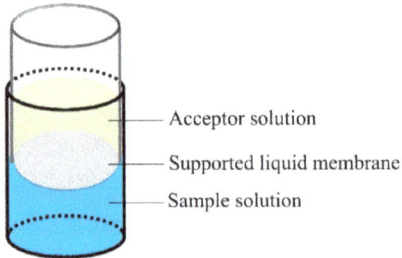

Figure 5. Schematic representation of flat membrane-based liquid-phase microextraction (FM-LPME) setup.

3.3. Sample Preparation

3.3.1. Water Samples

The stock solution of barbiturates was prepared by dissolving the drugs in methanol with a concentration of 1 mg mL^{-1} and stored at 4 °C in the dark. The working solution (water sample) was obtained by diluting the stock solutions to a concentration of 5 µg mL^{-1} with 10 mM HCl solution.

3.3.2. Biological Samples

Drug-free whole blood, urine, and liver were used for the optimization of the LPME conditions and the validation of the proposed analytical method. For the validation of FM-LPME-LC-MS, diclofenac sodium (50 ng mL^{-1}) was used as the internal standard (IS). The drug-free whole blood and urine samples were collected from the volunteers who had not been exposed to barbiturates. Liver specimens were obtained from Tongji Medicolegal Expertise Center in Hubei.

The whole blood sample was diluted three times using 10 mM HCl solution containing NaCl, three barbiturates at desired concentration, and IS (if applicable). The urine sample was diluted 1:1 with 10 mM HCl containing the three barbiturates at the desired concentration and IS (if applicable).

One gram liver tissue was weighed and homogenized with Freezer Mixer (Shanghai Jingxin Industrial Development Co. Ltd., China) after overnight lyophilization. The homogenization of lyophilized liver tissue was conducted for 30 min in a 5 mL grinding jar containing 3 mL of 10 mM HCl and four grinding beads at 60 rpm. The homogenized sample was transferred to a 5 mL volumetric flask and diluted to 5 mL (with the liver concentration of 0.2 g mL^{-1}) using 10 mM HCl. Prior to FM-LPME, the mixture was 1:1 (v/v) diluted with 10 mM HCl containing barbiturates and IS (if applicable) and equilibrated overnight at 4 °C in the dark.

3.3.3. Method Evaluation

In order to assess the practical applicability of the proposed method, extractions were performed under the optimal conditions for diluted whole blood, urine, and liver tissue spiked with model analytes at concentrations of 1, 2, 5, 10, 20, 50, 100, 200, and 500 ng mL^{-1} (n = 4 for each concentration). Moreover, diclofenac sodium with a concentration of 50 ng mL^{-1} was used as the internal standard (IS). The acceptor solution after FM-LPME was then analyzed by LC-MS. The peak area ratio between analyte and IS versus analyte concentration was plotted to construct the calibration curve. LOD and LOQ were determined to estimate the sensitivity of the method and calculated as the concentration of the inject sample to yield a signal-to-noise ratio of three and ten, respectively.

To assess repeatability of the method, replicate analysis (n = 5) of biological samples spiked at ng mL^{-1} of each analyte was performed. Repeatability results are expressed as RSD from the replicate theoretical value.

3.3.4. Application Experiment

Two male SD rats weighing 250 g and 300 g were first fed with a mixture of the three barbiturates at a dose of 50 mg kg^{-1} by gavage. Two hours later, the whole blood and liver tissue of the rats were collected individually by open chest cardiac puncture and laparotomy, respectively. The whole blood was diluted 100 times with 10 mM HCl solution containing 12.5% NaCl and 50 ng mL^{-1} of IS was used as the donor phase of FM-LPME. The liver samples for FM-LPME were prepared as described above. The concentration of the barbiturates in the liver was out of the linearity range for FM-LPME/LC-MS. Therefore, the liver samples for FM-LPME were finally diluted with 10 mM HCl to a concentration of approximately 2.5 mg mL^{-1}. The diluted liver samples were mixed 1:1 with 10 mM HCl solution containing IS and equilibrated overnight at 4 °C in the dark before FM-LPME. The acceptor phases were collected and analyzed by LC-MS. All procedures related to animals were in accordance with the international, national, and/or institutional guidelines and were approved by the Institutional Animal Care and Use Committee of Tongji Medical College of Huazhong University of Science and Technology in January 10, 2018 ([2018] IACUC Number: 2127).

3.4. HPLC-UV Analysis

An Ultimate 3000 system equipped with a pump (LPG-3400RS), an autosampler (WPS-3000RS), a column oven (TCC-3000RC), and a VWD-3400RS UV/Vis detector (all from Thermo Scientific, Waltham, MA, USA) was used for the chromatographic separation, and the UV-Vis detector was operated at 214 nm. Data were collected and processed by Chromeleon software 7.2 SR5 (Thermo Scientific, Waltham, MA, USA). Separation was carried out on a Hypersil GOLD C18 column (100 mm × 2.1 mm, 5 μm) (Thermo Scientific, Waltham, MA, USA) at 45 °C, with an injection volume of 10 μL. Mobile phase A was 20 mM formic acid containing 5% of methanol (v/v), and mobile phase B was methanol containing 5% of 20 mM formic acid (v/v). Mobile phase B was increased from 20% to 80% within 3.5 min at a flow rate of 0.8 mL min^{-1}; afterward, it was decreased to 20% within 0.1 min, and this condition was kept for 2.5 min for equilibration.

3.5. LC-MS Analysis

Analysis of barbiturates was performed using an Ultimate 3000 UPLC system interfaced with a TSQ Quantum Access MAX triple quadrupole Mass Spectrometry (Thermo Scientific, Waltham, MA, USA). Chromeleon client software (Thermo Scientific, Waltham, MA, USA) was used for LC control, and Xcalibur software (Thermo Scientific, Waltham, MA, USA) was used to control the MS, data acquisition, and data processing. Chromatographic separation was conducted at 45 °C on an Accucore C18 column (2.1 mm × 100 mm, 2.6 μm) (Thermo Scientific, Waltham, MA, USA) with gradient elution. Water containing 0.5% of acetic acid (v/v) was used as mobile phase A, and methanol was used as mobile phase B. Mobile phase B started from 20% for 0.5 min, and then increased to 95% in 1.5 min and 95% was kept for 2 min. At the end, mobile phase B was decreased to 20% within 0.1 min, which was kept for 1.9 min for equilibration. The flow rate was set to 0.4 mL min^{-1} and the injection volume was 10 μL. Mass spectrometry was performed with an ESI source in the negative-ionization mode with a sheath gas of 40 Arb and aux gas of 10 Arb. The capillary temperature was set at 320 °C and the vaporizer temperature was set at 350 °C. The spray voltage was 3.2 kV. The parameters for the quantification selected reaction monitoring (SRM) transitions are presented in Table 4.

Table 4. Mass spectrometry parameters for three barbiturates and internal standard (IS).

Analyte	Parent (m/z)	Product (m/z)	Collision energy (eV)	Tube lens (V)	Retention Time (min)
Barbital	183.0	42.4 / 140.0	80 / 13	52	1.54
Phenobarbital	231.0	42.4 / 188.0	17 / 10	57	3.33
Pentobarbital	225.0	42.4 / 182.0	53 / 18	48	3.78
IS	294.0	214.0 / 249.9	24 / 14	74	4.22

4. Conclusions

In this study, a simple, efficient, and inexpensive sample preparation technique, namely, FM-LPME coupled with LC-MS, was used to determine barbiturates in biological specimens. Compared to hollow fibers, the flat membrane device is easier for operation and can house a larger amount of acceptor solution, which might lead to high recovery. To our best knowledge, this is the first report on the simultaneous measurement of three barbiturate drugs in human biological specimens using FM-LPME/LC-MS analysis. Three barbiturate drugs could be rapidly and efficiently extracted and simultaneously determined even at trace concentration. This established method has the potential to be used in therapeutic drug monitoring and clinical toxicology, as well as forensic toxicology.

Author Contributions: Conceptualization, C.H.; methodology, R.Z. and C.H.; experimental work, R.Z., Y.D. and X.C.; analysis of the results, R.Z., Y.D. and C.H.; writing—original draft preparation, R.Z., Y.D. and C.H.; writing—review and editing, R.Z., Y.D., X.C. and C.H.

Funding: This research was funded by the Fundamental Research Funds for the Central Universities in China (Grant number 2017KFYXJJ021).

Conflicts of Interest: The authors declare no conflict of interest.

References

1. Swarbrick, J. *Encyclopedia of Pharmaceutical Technology*, 3rd ed.; CRC Press: Boca Raton, FL, USA, 2013.
2. Yasiry, Z.; Shorvon, S.D. How phenobarbital revolutionized epilepsy therapy: The story of phenobarbital therapy in epilepsy in the last 100 years. *Epilepsia* **2012**, *53*, 26–39. [CrossRef] [PubMed]
3. López-Muñoz, F.; Ucha-Udabe, R.; Alamo, C. The history of barbiturates a century after their clinical introduction. *Neuropsychiatr. Dis. Treat.* **2005**, *1*, 329. [PubMed]
4. Ito, T.; Suzuki, T.; Wellman, S.E.; Ho, K. Pharmacology of barbiturate tolerance/dependence: GABAA receptors and molecular aspects. *Life Sci.* **1996**, *59*, 169–195. [CrossRef]
5. Vlasses, P.H.; Rocci, M.L., Jr.; Koffer, H.; Ferguson, R.K. Combined phenytoin and phenobarbital overdose. *Drug Intell. Clin. Pharm.* **1982**, *16*, 487–488. [CrossRef] [PubMed]
6. Fritch, D.; Blum, K.; Nonnemacher, S.; Kardos, K.; Buchhalter, A.R.; Cone, E.J. Barbiturate detection in oral fluid, plasma, and urine. *Ther. Drug Monit.* **2011**, *33*, 72–79. [CrossRef]
7. Coupey, S.M. Barbiturates. *Pediatr. Rev.* **1997**, *18*, 260–264; quiz 265. [CrossRef]
8. Zarei, A.R.; Gholamian, F. Development of a dispersive liquid–liquid microextraction method for spectrophotometric determination of barbituric acid in pharmaceutical formulation and biological samples. *Anal. Biochem.* **2011**, *412*, 224–228. [CrossRef]
9. Li, S.; Weber, S.G. Determination of barbiturates by solid-phase microextraction and capillary electrophoresis. *Anal. Chem.* **1997**, *69*, 1217–1222. [CrossRef]
10. Jiang, T.-F.; Wang, Y.-H.; Lv, Z.-H.; Yue, M.-E. Direct determination of barbiturates in urine by capillary electrophoresis using a capillary coated dynamically with polycationic polymers. *Chromatographia* **2007**, *65*, 611–615. [CrossRef]

11. Botello, I.; Borrull, F.; Calull, M.; Aguilar, C.; Somsen, G.W.; de Jong, G.J. In-line solid-phase extraction–capillary electrophoresis coupled with mass spectrometry for determination of drugs of abuse in human urine. *Anal. Bioanal. Chem.* **2012**, *403*, 777–784. [CrossRef]
12. Ohyama, K.; Wada, M.; Lord, G.A.; Ohba, Y.; Fujishita, O.; Nakashima, K.; Lim, C.K.; Kuroda, N. Capillary electrochromatographic analysis of barbiturates in serum. *Electrophoresis* **2010**, *25*, 594–599. [CrossRef]
13. Tanaka, E.; Terada, M.; Tanno, K.; Misawa, S.; Wakasugi, C. Forensic analysis of 10 barbiturates in human biological samples using a new reversed-phase chromatographic column packed with 2-micrometre porous microspherical silica-gel. *Forensic Sci. Int.* **1997**, *85*, 73–82. [CrossRef]
14. Capella-Peiró, M.E.; Gil-Agustí, M.; Martinavarro-Domínguez, A.; Esteve-Romero, J. Determination in serum of some barbiturates using micellar liquid chromatography with direct injection. *Anal. Biochem.* **2002**, *309*, 261–268. [CrossRef]
15. La Marca, G.; Malvagia, S.; Filippi, L.; Luceri, F.; Moneti, G.; Guerrini, R. A new rapid micromethod for the assay of phenobarbital from dried blood spots by LC-tandem mass spectrometry. *Epilepsia* **2009**, *50*, 2658–2662. [CrossRef]
16. Menck, R.A.; De Lima, D.S.; Seulin, S.C.; Leyton, V.; Pasqualucci, C.A.; Muñoz, D.R.; Osselton, M.D.; Yonamine, M. Hollow-fiber liquid-phase microextraction and gas chromatography-mass spectrometry of barbiturates in whole blood samples. *J. Sep. Sci.* **2012**, *35*, 3361–3368. [CrossRef]
17. Hall, B.J.; Brodbelt, J.S. Determination of barbiturates by solid-phase microextraction (SPME) and ion trap gas chromatography–mass spectrometry. *J. Chromatogr. A* **1997**, *777*, 275–282. [CrossRef]
18. Roveri, F.L.; Paranhos, B.A.P.B.; Yonamine, M. Determination of phenobarbital in hair matrix by liquid phase microextraction (LPME) and gas chromatography–mass spectrometry (GC–MS). *Forensic Sci. Int.* **2016**, *265*, 75–80. [CrossRef]
19. Frison, G.; Favretto, D.; Tedeschi, L.; Ferrara, S.D. Detection of thiopental and pentobarbital in head and pubic hair in a case of drug-facilitated sexual assault. *Forensic Sci. Int.* **2003**, *133*, 171–174. [CrossRef]
20. Johnson, L.L.; Garg, U. Quantitation of amobarbital, butalbital, pentobarbital, phenobarbital, and secobarbital in urine, serum, and plasma using gas chromatography-mass spectrometry (GC-MS). In *Clinical Applications of Mass Spectrometry, Methods in Molecular Biology (Methods and Protocols)*; Garg, U., Hammett-Stabler, C., Eds.; Humana Press: New York, NY, USA, 2010; pp. 65–74.
21. Deveaux, M.; Cheze, M.; Pépin, G. The role of liquid chromatography-tandem mass spectrometry (LC-MS/MS) to test blood and urine samples for the toxicological investigation of drug-facilitated crimes. *Ther. Drug Monit.* **2008**, *30*, 225–228. [CrossRef]
22. Baciu, T.; Borrull, F.; Aguilar, C.; Calull, M. Recent trends in analytical methods and separation techniques for drugs of abuse in hair. *Anal. Chim. Acta* **2015**, *856*, 1–26. [CrossRef]
23. Zhang, Y.; Zhou, W.-E.; Yan, J.-Q.; Liu, M.; Zhou, Y.; Shen, X.; Ma, Y.-L.; Feng, X.-S.; Yang, J.; Li, G.-H. A review of the extraction and determination methods of thirteen essential vitamins to the human body: An update from 2010. *Molecules* **2018**, *23*, 1484. [CrossRef] [PubMed]
24. Tu, X.; Chen, W. A Review on the Recent Progress in Matrix Solid Phase Dispersion. *Molecules* **2018**, *23*, 2767. [CrossRef]
25. Kataoka, H.; Lord, H.L.; Pawliszyn, J. Applications of solid-phase microextraction in food analysis. *J. Chromatogr. A* **2000**, *880*, 35–62. [CrossRef]
26. Souza-Silva, E.A.; Jiang, R.; Rodriguez-Lafuente, A.; Gionfriddo, E.; Pawliszyn, J. A critical review of the state of the art of solid-phase microextraction of complex matrices I. Environmental analysis. *TrAC Trends Anal. Chem.* **2015**, *71*, 224–235. [CrossRef]
27. Snow, N.H. Solid-phase micro-extraction of drugs from biological matrices. *J. Chromatogr. A* **2000**, *885*, 445–455. [CrossRef]
28. Lee, J.; Lee, H.K.; Rasmussen, K.E.; Pedersen-Bjergaard, S. Environmental and bioanalytical applications of hollow fiber membrane liquid-phase microextraction: A review. *Anal. Chim. Acta* **2008**, *624*, 253–268. [CrossRef]
29. Huang, C.; Seip, K.F.; Gjelstad, A.; Shen, X.; Pedersen-Bjergaard, S. Combination of electromembrane extraction and liquid-phase microextraction in a single step: Simultaneous group separation of acidic and basic drugs. *Anal. Chem.* **2015**, *87*, 6951–6957. [CrossRef]

30. Bello-López, M.Á.; Ramos-Payán, M.; Ocaña-González, J.A.; Fernández-Torres, R.; Callejón-Mochón, M. Analytical applications of hollow fiber liquid phase microextraction (HF-LPME): A review. *Anal. Lett.* **2012**, *45*, 804–830. [CrossRef]
31. Yan, Y.; Chen, X.; Hu, S.; Bai, X. Applications of liquid-phase microextraction techniques in natural product analysis: A review. *J. Chromatogr. A* **2014**, *1368*, 1–17. [CrossRef]
32. Iwai, M.; Hattori, H.; Arinobu, T.; Ishii, A.; Kumazawa, T.; Noguchi, H.; Noguchi, H.; Suzuki, O.; Seno, H. Simultaneous determination of barbiturates in human biological fluids by direct immersion solid-phase microextraction and gas chromatography–mass spectrometry. *J. Chromatogr. B* **2004**, *806*, 65–73. [CrossRef]
33. Jun, X.; Jie, C.; Man, H.; Bin, H. Simultaneous quantification of amphetamines, caffeine and ketamine in urine by hollow fiber liquid phase microextraction combined with gas chromatography-flame ionization detector. *Talanta* **2010**, *82*, 969–975.
34. Overstreet, D.H.; Mathe, A.A.; Nicolau, G.; Feighner, J.P.; Jimenez-Vasquez, P.A.; Hlavka, J.; Morrison, J.; Abajian, H. Liquid-phase microextraction of protein-bound drugs under non-equilibrium conditions. *Analyst* **2002**, *127*, 608–613.
35. Menck, R.A.; de Oliveira, C.D.R.; de Lima, D.S.; Goes, L.E.; Leyton, V.; Pasqualucci, C.A.; Munoz, D.R.; Yonamine, M. Hollow fiber–liquid phase microextraction of barbiturates in liver samples. *Forensic Toxicol.* **2013**, *31*, 31–36. [CrossRef]
36. Pedersen-Bjergaard, S.; Rasmussen, K.E. Bioanalysis of drugs by liquid-phase microextraction coupled to separation techniques. *J. Chromatogr. B* **2005**, *817*, 3–12. [CrossRef] [PubMed]
37. Hadjmohammadi, M.; Ghambari, H. Three-phase hollow fiber liquid phase microextraction of warfarin from human plasma and its determination by high-performance liquid chromatography. *J. Pharm. Biomed. Anal.* **2012**, *61*, 44–49. [CrossRef]
38. Oliveira, A.F.F.; de Figueiredo, E.C.; dos Santos-Neto, Á.J. Analysis of fluoxetine and norfluoxetine in human plasma by liquid-phase microextraction and injection port derivatization GC–MS. *J. Pharm. Biomed. Anal.* **2013**, *73*, 53–58. [CrossRef]
39. Shen, G.; Lee, H.K. Hollow fiber-protected liquid-phase microextraction of triazine herbicides. *Anal. Chem.* **2002**, *74*, 648–654. [CrossRef]
40. Clarke, E.G.C.; Moffat, A.C.; Osselton, M.D.; Widdop, B. *Clarke's Analysis of Drugs and Poisons: In Pharmaceuticals, Body Fluids and Postmortem Material*; Pharmaceutical Press: London, UK, 2004.
41. Winek, C.L.; Wahba, W.W.; Winek, C.L., Jr.; Balzer, T.W. Drug and chemical blood-level data 2001. *Forensic Sci. Int.* **2001**, *122*, 107–123. [CrossRef]
42. Chemical Toxicity Database. Available online: http://www.drugfuture.com/toxic/index.html (accessed on 10 March 2019).
43. Zhang, X.; Lin, Z.; Li, J.; Huang, Z.; Rao, Y.; Liang, H.; Yan, J.; Zheng, F. Rapid determination of nine barbiturates in human whole blood by liquid chromatography-tandem mass spectrometry. *Drug Test. Analysis* **2016**, *9*, 588–595. [CrossRef] [PubMed]
44. Huang, C.; Eibak, L.E.E.; Gjelstad, A.; Shen, X.; Trones, R.; Jensen, H.; Pedersen-Bjergaard, S. Development of a flat membrane based device for electromembrane extraction: A new approach for exhaustive extraction of basic drugs from human plasma. *J. Chromatogr. A* **2014**, *1326*, 7–12. [CrossRef]

Sample Availability: Not available.

© 2019 by the authors. Licensee MDPI, Basel, Switzerland. This article is an open access article distributed under the terms and conditions of the Creative Commons Attribution (CC BY) license (http://creativecommons.org/licenses/by/4.0/).

Article

Primary Step Towards In Situ Detection of Chemical Biomarkers in the UNIVERSE via Liquid-Based Analytical System: Development of an Automated Online Trapping/Liquid Chromatography System

Thomas Ribette [1], Bertrand Leroux [1], Balkis Eddhif [1], Audrey Allavena [1], Marc David [2], Robert Sternberg [2], Pauline Poinot [1] and Claude Geffroy-Rodier [1],*

[1] Institut de Chimie des Milieux et Matériaux de Poitiers (IC2MP), Université de Poitiers, UMR CNRS 7285, Equipe Eau Géochimie Santé, 4 rue Michel Brunet, 86076 Poitiers, France; thomas.ribette@univ-poitiers.fr (T.R.); Bertrand.leroux@univ-poitiers.fr (B.L.); balkis.eddhif@univ-poitiers.fr (B.E.); audrey.allavena@univ-poitiers.fr (A.A.); pauline.poinot@univ-poitiers.fr (P.P.)

[2] Laboratoire Interuniversitaire des Systèmes Atmosphériques (LISA), Université Paris Est Creteil, UMR CNRS 7583, 61 avenue du General de Gaulle, 94010 Créteil, France; marc.david@lisa.u-pec.fr (M.D.); robert.sternberg@lisa.u-pec.fr (R.S.)

* Correspondence: claude.geffroy@univ-poitiers.fr; Tel.: +335-4945-3590

Academic Editor: Nuno Neng
Received: 28 March 2019; Accepted: 8 April 2019; Published: 11 April 2019

Abstract: The search for biomarkers in our solar system is a fundamental challenge for the space research community. It encompasses major difficulties linked to their very low concentration levels, their ambiguous origins (biotic or abiotic), as well as their diversity and complexity. Even if, in 40 years' time, great improvements in sample pre-treatment, chromatographic separation and mass spectrometry detection have been achieved, there is still a need for new in situ scientific instrumentation. This work presents an original liquid chromatographic system with a trapping unit dedicated to the one-pot detection of a large set of non-volatile extra-terrestrial compounds. It is composed of two units, monitored by a single pump. The first unit is an online trapping unit able to trap polar, apolar, monomeric and polymeric organics. The second unit is an online analytical unit with a high-resolution Q-Orbitrap mass spectrometer. The designed single pump system was as efficient as a laboratory dual-trap LC system for the analysis of amino acids, nucleobases and oligopeptides. The overall setup significantly improves sensitivity, providing limits of detection ranging from ppb to ppt levels, thus meeting with in situ enquiries.

Keywords: space instrumentation; liquid chromatography; oligopeptides; trapping system

1. Introduction

The search for traces of past or present life in the solar system arouses the curiosity of many scientists. Detection of organic biomarkers has become a key challenge in planetary exploration, in order to understand whether they played a role in the origin of life on Earth [1,2]. Taking Earth-based techniques to develop a spatial instrument suite with multiple capabilities is, however, a real challenge. In space, even simple chemical analyses involve complex sample handling inside a probe's instrument systems and require a sophisticated design with mass and power constraints as major factors. Within this framework, the technique of choice which was, and is still, employed for landed missions dedicated to the quest of life traces is pyrolysis gas chromatography mass spectrometry (Py-GC-MS) [3]. Based on a compound's volatility, it was designed to detect low-to intermediate-molecular-weight organic biomarkers with improved sensitivity, leading to the in situ detection of organic molecules, as demonstrated by the Rosetta mission [4–7]. From Viking to the next ExoMars 2020 mission, this approach has been greatly improved with the use of online chemical derivatization agents, multi-column chromatographs and integrated traps (tenax, carbosieve, glass bead) [3,4,8–12]. These onboard instruments enable the detection of organic molecules, such as polycyclic aromatic hydrocarbons (PAHs) [13], amino acids [14,15] and sugars [16,17]. While such molecules have already been found in the interstellar medium and in many meteorites, their detection on extra-terrestrial planets remains a difficult task, and their assignment to a definite biological origin is still questioned, as they can be abiotically produced. To find additional data leading to unambiguous features of extant or extinct life and/or of prebiotic chemistry beyond Earth, researchers now pay attention to molecular biological polymers [18]. As a consequence, future onboard instruments must be able to detect and quantify all these compounds in one pot. As solid, liquid and aerosol samples are anticipated, the future instrument platform should involve versatile sample analysis. Gas and/or vapor samples can be analyzed directly by mass spectrometers or gas chromatograph-mass spectrometers [19,20], which have already been successfully used in previous planetary [11,21] and cometary missions [10,22,23]. Liquid (lake, icy regolith/cryovolcanic meltwater) and solid samples, undergoing melting, extraction or solubilization into a liquid mobile phase before analysis, could be analyzed either by gas chromatography (GC) or liquid chromatography (LC). In recent years, micro-fluidic systems involving sandwich and/or competitive immunoassays [24,25], microchip capillary electrophoreses [26–28] or nanopore-based analysis [29] have been designed [30]. A wide range of molecular-sized compounds, from amino acids and nucleobases to oligopeptides and oligonucleotides, would then likely be detected. Although these methods have already demonstrated real benefits in terms of sensitivity (1 µM to 0.1 nM) [31,32], some problems remain unresolved, even on Earth. Firstly, only a few molecules can be analyzed simultaneously, compared to multidimensional methods used in laboratories, e.g., two-dimensional gel electrophoresis (2D-PAGE) or two-dimensional liquid chromatography (2D-LC) [28]. Another major hurdle is the very small sample volume that can be analyzed in one run (from 1 nL to 10 µL [31]), which reduces the method's sensitivity and sample representativity. In addition, only a pure extract can be injected in these micro-fluidic systems; as a result, they require a previous complex and multi-step off-line sample preparation [25,27,33–35]. Thanks to remarkable improvements in multidimensional systems and UPLC stationary phases, an online analytical platform allowing both the purification and analysis of a various set of compounds could now be considered for exobiological studies.

The aim of the present study is to develop a liquid setup able to concentrate and separate a wide range of potential extra-terrestrial peptide-like molecules. The developed configuration, in addition to its intrinsic qualities, such as the concentration and great versatility despite the drastic operating conditions, would have to facilitate a simple and fast separation suitable for in situ analysis. Placed online with a spatialized detector [36,37], the generic unit should potentiate the one-pot detection of diverse molecules with increasing complexity and present at nanomolar or picomolar levels.

MS detectors, coupled to liquid chromatography, have already allowed major advances in organics characterization of meteorite, tholin and comet analogues [38–40]. Thanks to a Q-Orbitrap High Resolution Mass Spectrometer (HRMS) that allows a comprehensive assessment of the data obtained, several analytical features were assessed to define the best trapping and separation conditions. Finally, the approach, combined with a simplified sample preparation protocol and spatialized detector, could potentially be validated for space life-search experiments.

2. Materials and Methods

2.1. Chemicals and Solutions

Several monomers and polymers considered to be strong biosignatures of life (amino acids, nucleic acids and oligopeptides) were used at different stages of optimization, as they are distinct in terms of polarities, chemical structures and molecular masses, and such biomarkers were used to select generic trapping parameters.

A commercial mix of five oligopeptides (Sigma-Aldrich, Steinheim, Germany) contained a dipeptide (glycine-tyrosine (gly-tyr)), a tripeptide (valine-tyrosine-valine (val-tyr-val)) and three oligopeptides (leu-enkephalin, met-enkephalin and angiotensin II), each at 0.5 mg. Nucleic acids such as cytosine (>99%), uracil (>99%) and thymine (>99%) were purchased from Sigma-Aldrich. Amino acids (alanine, glycine, valine, leucine, isoleucine, proline, methionine, arginine, cysteine, threonine, serine, aspartic acid, glutamic acid, histidine, lysine, phenylalanine, tyrosine) and other oligopeptides were supplied by Sigma-Aldrich (St. Louis, MO, USA). LC mobile phases were prepared with HPLC grade acetonitrile and ultrapure grade formic acid, purchased from Sigma-Aldrich. Purified water was generated by a Purelab Flex purifier system (Veolia, Paris, France).

2.2. Preparation of Standard Solutions

Two stock solutions were prepared with oligopeptides solubilized in high purity water, amino acids in HCl 1 M and nucleobases in NaOH 0.1 M The first solution contained the peptides (0.5 µg/mL) in high-purity water, the second was composed of the amino acids and nucleobases (1.5×10^{-6} M). The standard solution containing the oligopeptides was used to prepare a working solution at 0.01 µg·mL^{-1}, which was used to select the optimal trapping parameters of the trapping-LC setup in comparison with the 1D-LC method. This solution was further mixed and diluted with the standard solutions, containing amino acids and bases, to prepare the different series of calibration solutions to validate the optimized system.

2.3. 1D-LC Setup

The analysis was performed with a Wadose LC isocratic pump, interfaced with a Q-Exactive Hybrid Quadrupole-Orbitrap mass spectrometer equipped with an ESI source (Thermo Fisher Scientific, Waltham, MA, USA). The MS functions were controlled by the Xcalibur data system (Thermo Fisher Scientific), whereas injection and HPLC solvent elution were monitored and controlled by our software developed on LabVIEW. The analytical column was a semi-polar Hypersil Gold aQ (50 × 1 mm, 1.9 µm, 175 Å, Thermo Fisher Scientific). The mobile phase consisted of acetonitrile (ACN)–0.1% formic acid and water–0.1% formic acid. Elution was performed with 10% and 20% ACN at a constant flow rate of 110 µL·min^{-1}. Experiments were conducted at 40 °C.

2.4. Trapping-LC Setup

The analysis was performed with the same instrumentation. The pump enabled the loading of the sample on the trapping setup, followed by the backflush and the analytical separation of analytes. Two different trapping columns, a semi-polar Hypersil Gold aQ (20 × 2.1 mm, 12 µm, 175 Å; Thermo Fisher Scientific) and a polar Hypercarb (20 × 2.1 mm, 7 µm, 175 Å; Thermo Fisher Scientific), were used.

One thousand microliters of the sample was injected in the preparative loop. The compounds were transferred to the trapping columns at 500 µL·min^{-1} for 180 s. Non-retained compounds (e.g., matrix interferences, salt) were flushed to waste. Once the loading completed, trapped analytes were backflushed at 110 µL·min^{-1} until all targets were eluted on the analytical column. During this time, trapping and analytical columns were connected in series and eluted by means of the 1D-LC mobile phase. The valve scheme is described in Section 3.3. The automation was performed by LabVIEW®software (version 2016, National Instrument Corporation, Austin, TX, USA) that controlled the pump, valves and column oven.

The system was compared to a laboratory dual-trap system with two quaternary Accela LC pumps (600 and 1250) working together and interfaced with the same Q-Exactive Hybrid Quadrupole-Orbitrap Mass Spectrometer. The Accela 600 pump provided the loading of the sample on a trapping-LC unit, while the 1250 pump controlled the backflush and the analytical separation of analytes. Before injection, samples were stored at 4 °C using a Stack cooler CW (CTC Analytics AG, Zwingen, Switzerland). MS

2.5. Mass Spectrometry

The analysis was carried out on a Q-Exactive mass spectrometer. Mass detection was performed in positive ion mode. The electrospray voltage was set at 4.0 Kv. The capillary and heater temperatures were 275 °C and 300 °C, respectively. The sheath, sweep and auxiliary gas (nitrogen) flow rates were set at 35, 10 and 20, respectively (arbitrary units).

MS analyses were performed by either full scan or targeted selected ion monitoring mode (tSIM). The full scan mode was employed when standard solutions were analyzed. Mass spectra were acquired at 70,000 resolution, AGC target 10^6 and max IT 200 ms. Compounds were analyzed in the range of 300–2000 m/z when solutions of oligopeptides were analyzed (i.e., solutions used for the selection of optimal multidimensional parameters), and in the range of 75–1100 m/z when solutions contained amino acids and nucleobases (i.e., calibration solutions).

tSIM MS offered superior sensitivity when complex samples were analyzed. It was then used to determine the recovery of compounds with resolution at 17500, AGC target at 10^5, max IT at 200 ms and MSX count at 4. Detection of organics was set within 2 ppm of the theoretical mass.

Leu-Leu m/z 245.18657 $[M + H]^+$; Leu-Leu-Leu m/z 358.27003 $[M + H]^+$, Phe-Phe m/z 312.36 $[M + H]^+$; Met-Met-Met m/z 412.13985 $[M + H]^+$; Gly-Tyr, m/z 239.10263 $[M + H]^+$; Val-Tyr-Val, m/z 380.21747 $[M + H]^+$; Gly-Gly-Gly m/z 190,08281 $[M + H]^+$; Ala-Ala-Ala-Ala, m/z 303.16685 $[M + H]^+$ Leu-enkephalin, m/z 556.27658 $[M + H]^+$; Met-enkephalin, m/z 574.23300 $[M + H]^+$; Angiotensin II, m/z 523.77453 $[M + 2H]^{2+}$; Guanine, m/z 152.0567 $[M + H]^+$; Thymine, m/z 127.05669 $[M + H]^+$; Cytosine, m/z 112.05054 $[M + H]^+$; Uracil, m/z 113.03512 $[M + H]^+$.

3. Results and Discussion

3.1. UPLC General Features

In this rationale, peptides were used as molecular targets. To overcome the lack of extra-terrestrial peptide standards, amino-acids polymers differing in terms of molecular weight and polarity were considered [33]. Oligopeptides with alanine, glycine and leucine were considered as valuable targets, since their building blocks are the main amino acids in the acid hydrolysates of meteorites, tholins and comets [22,41,42]. The sensitivity of the system was thus investigated based on concentrations of meteorite compounds. Amino acid and nucleobase concentrations present in carbonaceous meteorites range from ppb to ppm levels (ng·g^{-1} to µg·g^{-1} of meteorite) [15,43]. Assuming a similar range in the universe, the limit of detection of any technique used in situ has to be at least at the ppb level. For a 1 g sample of liquid, melted or extracted in 1 mL of solvent, a detection limit of ng·mL^{-1} is mandatory.

To develop a simple but efficient liquid chromatographic system for space experimentation, a single isocratic pump was used to perform the trapping and separation of both compounds. For that purpose, a minimum length of flexible stainless steel capillaries together with Viper Fingertight Fitting system were selected to provide virtually dead-volume-free plumbing, minimizing extra-column dispersion.

Stationary phases were chosen according to their relevance for peptide-like compound analysis. Alone and in series, short columns (20–150 mm) with stationary phases exhibiting different polarities were previously evaluated in gradient mode for the analysis of laboratory cometary analogues [33]. Briefly, in isocratic mode, elution on Reverse Phase C_{18} Hypersil Gold aQ, with acetonitrile as the organic solvent, allowed the study of complex mixtures with high peptide retention. To decrease the mobile phase consumption and increase the sensitivity, a low-diameter (1 mm) Hypersil Gold aQ column was selected as the analytical column. To ensure the high solubility of the oligopeptides in the mobile phase, with the lower energy consumption of the column oven, the temperature was set at 40 °C.

Automation and control (oven, pump and valves) were performed by an interface programmed on LabVIEW.

3.2. 1D-LC Configuration

For direct injection, best separation was achieved with a 10/90 ACN/H$_2$O mobile phase, regarding intensities and retention times of the different oligopeptides (Figure 1, Table 1).

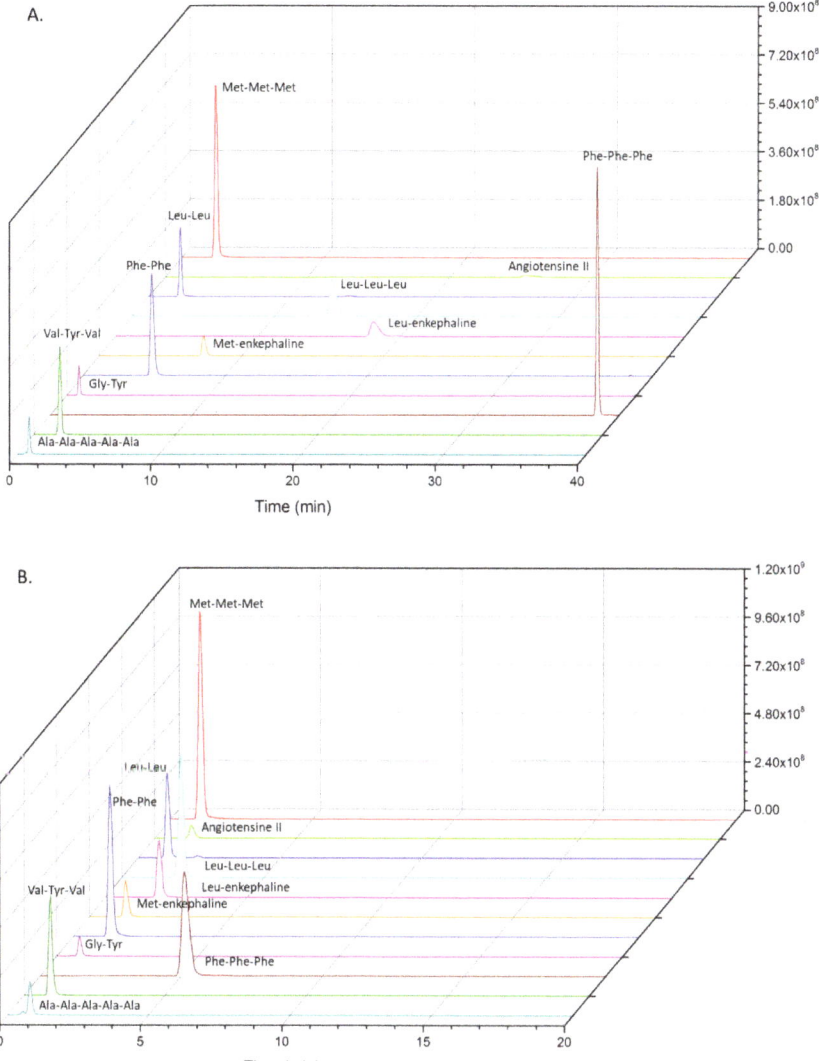

Figure 1. Direct injection extracted ion chromatogram: 20 µL of peptides solution (0.5 µg/mL) eluting on Hypersil Gold aQ (50 mm × 1 mm, 40 °C) with (**A**) 10/90 acetonitrile (ACN)/H$_2$O and (**B**) 20/80 ACN/H$_2$O as the mobile phase.

Table 1. Retention times (min) of oligopeptides eluted on Hypersil Gold aQ (50 mm × 1 mm, 40 °C) with 10/90 ACN/H$_2$O and 20/80 ACN/H$_2$O as the mobile phase.

Compounds	10% ACN	20% ACN
Ala-Ala-Ala-Ala	0.80	0.79
Gly-Tyr	0.85	0.80
Val-Tyr-Val	1.80	0.93
Leu-Leu	2.22	1.02
Met-Met-Met	2.39	1.01
Met-enkephalin	7.39	1.30
Phe-Phe	4,87	1.33
Leu-Leu-Leu	14.07	2.07
Leu-enkephalin	18.13	1.90
Angiotensin II	25.33	1.29
Phe-Phe-Phe	38.44	5.12

Increasing the volume of injection would be a way to improve sensitivity as a slight amount of complex and highly diluted sample is expected to be available [44]. An online liquid-trapping system would then be necessary to enable a large volume injection, to clean up samples (highly aqueous, salts-containing, etc.) and to selectively trap molecules of interest.

3.3. Trapping-LC Configuration

Regarding space constraints, trapping must be performed under an unusual configuration with a single pump for trapping and elution.

Various trapping factors, such as column stationary phases, loading and backflush parameters, strongly influence compound recovery and cleanup efficiency [45–47]. Stationary phases of the trapping columns were previously selected to characterize high-molecular-weight compounds in a cometary ice analogue. Briefly, Hypersil Gold aQ allowed the retention of semi-polar and apolar peptides, while more polar and low-mass compounds were refocused at the head of a Hypercarb column. By serially coupling both columns and setting a loading flow rate of 500 µL·min^{-1} for 120 s and a backflush of 240 s, this dual-trap setup led to the best retention of all standards [33].

To adapt this system to in situ analysis, the laboratory loading pump was suppressed and a switching valve was added (Figure 2). The designed system was thus composed of two trapping columns coupled to the analytical dimension. In that configuration of a single pump, the backflush step corresponded to elution on the analytical column. The only parameter to be optimized was then the nature of the mobile phase. To evaluate the system, trimethionine was chosen as an internal standard.

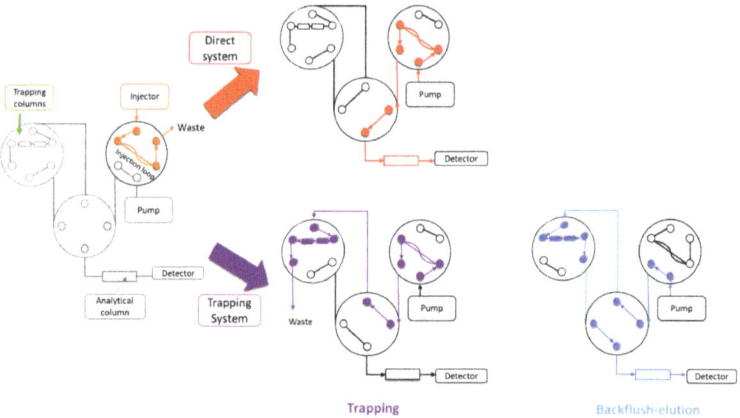

Figure 2. Schematic representation of the direct and trapping-LC setups.

Peaks tailing and broadening of the highest molecular weight oligopeptides with 10% ACN were not suitable for the elution of non-targeted oligopeptides. Backflush and elution with 20% ACN gave, on the contrary, a real benefit in terms of separation, as less coelution occurred for the studied peptides compared to the direct injection 1D-LC configuration (Figures 1 and 3). On the whole, peaks were well-defined. The delay of 180s in elution was particularly interesting for polar and/or very-low-molecular-weight compounds, which were no longer eluted at the death retention time. Backflush with 20% ACN gave also the best recoveries, except for phenylalanine tripeptide (loss of 22%, Figure 4).

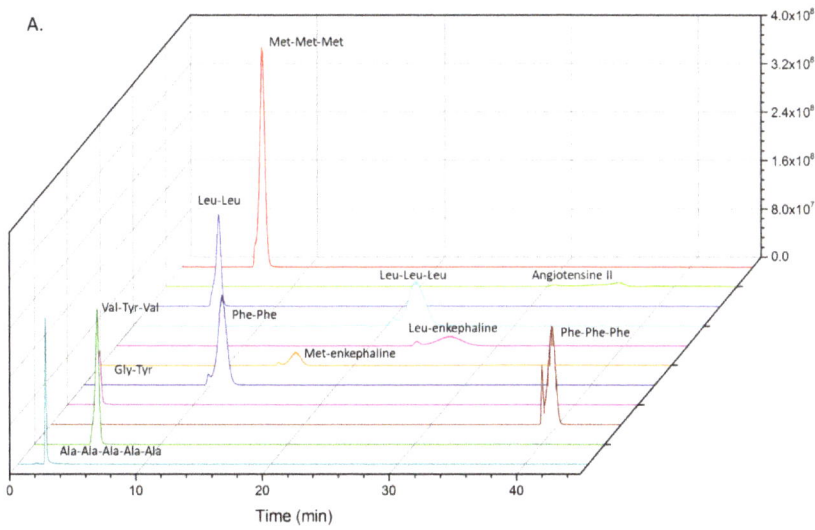

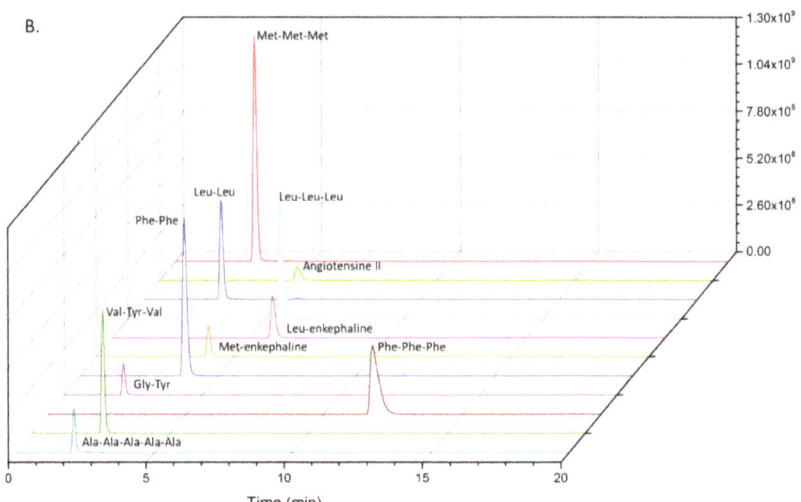

Figure 3. Extracted ion chromatogram of trapped peptides (0.5 µg/mL) on Hypersil Gold aQ (50 mm × 1 mm, 40 °C). Backflush-elution with (**A**) 10 % ACN and (**B**) 20% ACN.

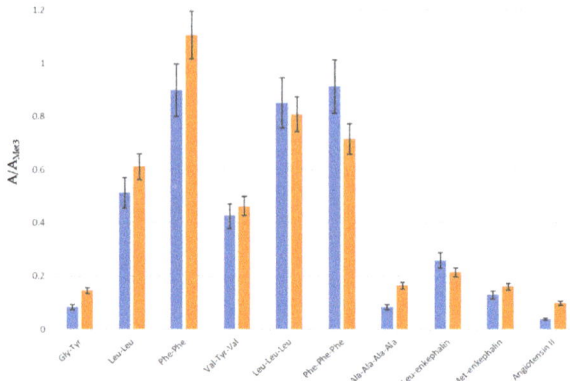

Figure 4. Recoveries of trapped peptides from a standard sample 1.5×10^{-6} M (0.5 µg/mL) after backflush and isocratic elution with 10% (in blue) or 20% ACN (in orange) on Hypersil Gold. Results are given as the ratio of compound area over internal standard one (A/A$_{Met3}$).

3.4. Interest of the Trapping-LC Setup for In Situ Experiments

Under space constraints, time and solvent consumption have to be considered. In our configuration, if LC was chosen to be part of the on-board instrumentation, it would analyze samples in less than 20 min with 2 mL of a single mobile phase. These features comply with in situ conditions and constitute a good basis for future improvements.

The retention capability of our designed system was compared to direct injection without trapping. The performance of the system was evaluated by injecting the same number of oligopeptides in direct (20 µL, 0.5 µg/mL) and trapping configurations (1000 µL, 10 ng/mL). As illustrated by Figure 5, there was no major difference in peptide retention and detection. Both distributions of oligopeptides were similar. The trapping was, however, not efficient for all the peptides, as alanine one was not retained.

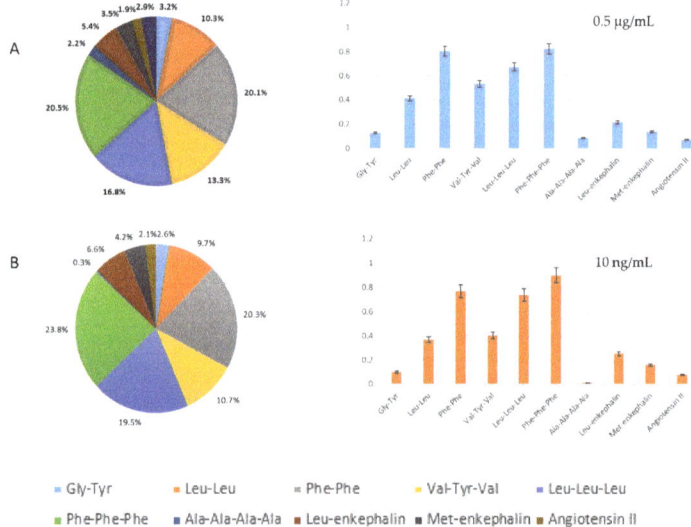

Figure 5. Repartition and recoveries of each peptide (**A**) in direct injection configuration 0.5 µg/mL and (**B**) in trapping configuration 10 ng/mL. Results are given as the ratio of compound area over internal standard one (A/A$_{Met3}$).

The trapping-LC system led, however, to significantly higher signal intensities as similar responses were obtained with a 50-fold lower concentration in trapping configuration.

Regarding targets for future space exploration missions, amino acid and nucleobase trapping was then evaluated. Contrary to nucleobases, in the optimized peptide trapping conditions, no amino acid was retained, except for phenylalanine and tyrosine (Figure 6).

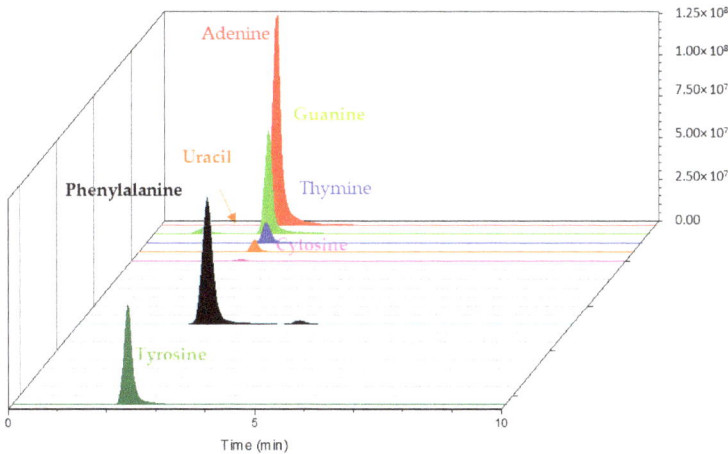

Figure 6. Extracted ion chromatogram of trapped amino acids and nucleobases on Hypersil Gold aQ (50 mm × 1 mm, 40 °C) from a 1 mL water sample. Backflush-elution with 20% ACN.

Analyses were then performed with a laboratory dual-trap system. Figure 7 shows the responses of the two trapping systems for retained amino acids, bases and oligopeptides. For all the targets, similar retention was obtained but with a higher standard deviation for the single-pump system (up to 26% for Phe-Phe).

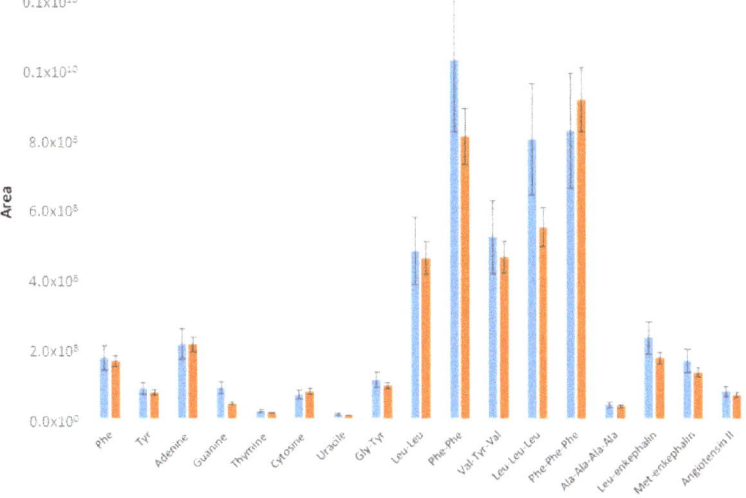

Figure 7. Recoveries of trapped molecules from a standard sample 1.5×10^{-6} M (0.5 µg/mL), with the trapping-LC system in blue and a laboratory dual-trap LC system after backflush and isocratic elution with 20% ACN on the Hypersil Gold column.

To further exemplify the sensitivity of the system when coupled to a mass spectrometer, the recovery of some targeted compounds was calculated using calibration curves (Table 2). Linearity ranged from 0.25 to 10 ng·mL^{-1}.

Table 2. Performance hallmarks of the designed trapping-LC system.

Compounds	R^2	RSD% (n = 18 [a])	LOD [b] ng·mL^{-1}	LOQ [c] ng·mL^{-1}
Gly-Tyr	0.912	11.1	0.16	0.54
Val-Tyr-Val	0.983	10.3	0.11	0.36
Leu-Enkephalin	0.987	6.3	0.18	0.60
Angiotensin II	0.986	30.2	2.77	9.24
Phenylalanine	0.978	9.0	4.03	13.42
Tryptophan	0.976	16.6	3.08	10.27
Uracil	0.956	7.2	1.95	6.15
Cytosine	0.863 *	15.0	2.22	7.41
Thymine	0.965	10.3	1.85	6.18

[a] Average relative standard deviation (six calibration points; three replicates per calibrant; n = 18). [b] Detection limit: 3.3 × (residual standard deviation/slope). [c] Quantification limit: 10 × residual standard deviation/slope).
* Coefficient of determination and subsequent values (Relative Standard Deviation RSD, Limit of Detection LOD, Limit of Quantification LOQ) were below 0.9. Regression models were not validated.

Retained peptides, nucleobases and amino acids were detected at the ng·mL^{-1} level. This clearly demonstrates the effectiveness of this online trapping approach when highly diluted and complex samples are analyzed. This trapping unit, coupled to a liquid chromatography system, would then enlarge the set of data about potential exobiological molecules without denaturing them. Despite its ability to retain different organics in terms of polarity, chemical structure and molecular weight in a single run, this broad approach should also be able to raise the signal of highly diluted compounds. This is fundamental for liquid in situ experiments, since compound extraction would previously have to be reduced to an extreme simplicity with large volumes of final extracts (in the order of milliliters), and thus with a low recovery achievement.

4. Conclusions

In situ detection of biomarkers in the solar system has become an appealing project, partly guiding past, present and future space missions. Up to now, in situ instrumentation was mainly designed to detect and determine concentrations of volatile organic compounds or derivatives. In this work, we present the first trapping unit for extra-terrestrial peptide-like compounds. The screening of several parameters showed that a trapping unit placed in series with an analytical column significantly enhanced the range of potential compounds to be analyzed. Through this LC setup, we do not pretend to separate all individual compounds in the sample as MD-LC systems do. Nevertheless, by avoiding mobile phase changes and reducing the system's dead volume due to long tubing and viper connections, the chromatographic dilution of a compound's band, as well as the tailing and broadening of peaks, are limited. As a result, the detection of very low concentrations of analytes is facilitated. Under space conditions, this system could present several advantages, since it would (1) elude chemical derivatization of non-volatile and polar compounds, as is necessary for current on-board GC-MS instruments, (2) limit the misinterpretation of chromatograms, (3) enlarge the range of potential biomarkers targeted and (4) reduce the complexity of offline sample preparation protocol used with microfluidic systems, without decreasing a compound's signal intensity. It could then represent a powerful tool for exobiological studies.

Author Contributions: C.G.R. and P.P. conceived and designed the experiments; T.R., A.A. and B.E. performed the experiments and analyzed the data; B.L. performed the automation. C.G.R. and P.P. wrote the paper. M.D. and R.S. discussed and commented on the manuscript. All authors read and approved the final manuscript.

Acknowledgments: The authors acknowledge S. Riffaut for the graphical abstract. Nouvelle Aquitaine Region and F. Courtade from CNES for financial support.

Conflicts of Interest: The authors declare no conflict of interest.

References

1. Irvine, W.M. Extraterrestrial organic matter: A review. *Orig. Life Evol. Biosph.* **1998**, *28*, 365–383. [CrossRef] [PubMed]
2. Simoneit, B.R.; Summons, R.E.; Jahnke, L.L. Biomarkers as Tracers for Life on Early Earth and Mars. *Orig. Life Evol. Biosph.* **1998**, *28*, 475–483. [CrossRef]
3. Sternberg, R.; Raulin, F.; Szopa, C.; Buch, A.; Vidal-Madjar, C. Reference Module in Chemistry, Molecular Sciences and Chemical Engineering. In *Encyclopedia of Separation Science*; Reedijk J: Philadelphia, PA, USA, 2007; pp. 1–13.
4. Goesmann, F.; Raulin, F.; Bredehöft, J.H.; Cabane, M.; Ehrenfreund, P.; MacDermott, A.J.; McKenna-Lawlor, S.; Meierhenrich, U.J.; Muñoz Caro, G.M.; Szopa, C.; et al. COSAC prepares for sampling and in situ analysis of cometary matter from comet 67P/Churyumov–Gerasimenko. *Planet. Space Sci.* **2014**, *103*, 318–330. [CrossRef]
5. Geffroy-Rodier, C.; Grasset, L.; Sternberg, R.; Buch, A.; Amblès, A. Thermochemolysis in search for organics in extraterrestrial environments. *J. Anal. Appl. Pyrolysis* **2009**, *85*, 454–459. [CrossRef]
6. Freissinet, C.; Buch, A.; Sternberg, R.; Szopa, C.; Geffroy-Rodier, C.; Jelinek, C.; Stambouli, M. Search for evidence of life in space: Analysis of enantiomeric organic molecules by N,N-dimethylformamide dimethylacetal derivative dependant Gas Chromatography–Mass Spectrometry. *J. Chromatogr. A* **2010**, *1217*, 731–740. [CrossRef]
7. Gibney, E. Philae's 64 hours of science. *Nat. News* **2014**, *515*, 319–320. [CrossRef]
8. Geffroy-Rodier, C.; Buch, A.; Sternberg, R.; Papot, S. Gas chromatography–mass spectrometry of hexafluoroacetone derivatives: First time utilization of a gaseous phase derivatizing agent for analysis of extraterrestrial amino acids. *J. Chromatogr. A* **2012**, *1245*, 158–166. [CrossRef]
9. Rodier, C.; Vandenabeele-Trambouze, O.; Sternberg, R.; Coscia, D.; Coll, P.; Szopa, C.; Raulin, F.; Vidal-Madjar, C.; Cabane, M.; Israel, G.; et al. Detection of martian amino acids by chemical derivatization coupled to gas chromatography: In situ and laboratory analysis. *Adv. Space Res.* **2001**, *27*, 195–199. [CrossRef]
10. Sternberg, R.; Szopa, C.; Coscia, D.; Zubrzycki, S.; Raulin, F.; Vidal-Madjar, C.; Niemann, H.; Israel, G. Gas chromatography in space exploration: Capillary and micropacked columns for in situ analysis of Titan's atmosphere. *J. Chromatogr. A* **1999**, *846*, 307–315. [CrossRef]
11. Mahaffy, P.R.; Webster, C.R.; Cabane, M.; Conrad, P.G.; Coll, P.; Atreya, S.K.; Arvey, R.; Barciniak, M.; Benna, M.; Bleacher, L.; et al. The Sample Analysis at Mars Investigation and Instrument Suite. *Space Sci. Rev.* **2012**, *170*, 401–478. [CrossRef]
12. Miller, K.E.; Kotrc, B.; Summons, R.E.; Belmahdi, I.; Buch, A.; Eigenbrode, J.L.; Freissinet, C.; Glavin, D.P.; Szopa, C. Evaluation of the Tenax trap in the Sample Analysis at Mars instrument suite on the Curiosity rover as a potential hydrocarbon source for chlorinated organics detected in Gale Crater: SAM TRAP AS SOURCE OF ORGANICS ON MARS. *J. Geophys. Res. Planets* **2015**, *120*, 1446–1459. [CrossRef]
13. Becker, L.; Glavin, D.P.; Bada, J.L. Polycyclic aromatic hydrocarbons (PAHs) in Antarctic Martian meteorites, carbonaceous chondrites, and polar ice. *Geochim. Cosmochim. Acta* **1997**, *61*, 475–481. [CrossRef]
14. Cronin, J.R.; Pizzarello, S. Amino acids in meteorites. *Adv. Space Res.* **1983**, *3*, 5–18. [CrossRef]
15. Pizzarello, S.; Cooper, G.W.; Flynn, G.J. The nature and distribution of the organic material in carbonaceous chondrites and interplanetary dust particles. *Meteor. Early Sol. Syst. II* **2006**, *1*, 625–651.
16. Cooper, G.; Kimmich, N.; Belisle, W.; Sarinana, J.; Brabham, K.; Garrel, L. Carbonaceous meteorites as a source of sugar-related organic compounds for the early Earth. *Nature* **2001**, *414*, 879–883. [CrossRef]
17. Cronin, J.R.; Chang, S. *Chemistry of Life's Origins*; Greenberg, J.M., Pirronello, V., Mendoza-Gomez, C., Eds.; Springer Science & Business Media: Dordrecht, The Netherlands, 1993; pp. 209–258.
18. Fernández-Calvo, P.; Näke, C.; Rivas, L.A.; García-Villadangos, M.; Gómez-Elvira, J.; Parro, V. A multi-array competitive immunoassay for the detection of broad-range molecular size organic compounds relevant for astrobiology. *Planet. Space Sci.* **2006**, *54*, 1612–1621. [CrossRef]
19. Postberg, F.; Khawaja, N.; Abel, B.; Choblet, G.; Glein, C.R.; Gudipati, M.S.; Henderson, B.L.; Hsu, H.-W.; Kempf, S.; Klenner, F.; et al. Macromolecular organic compounds from the depths of Enceladus. *Nature* **2018**, *558*, 564–568. [CrossRef]

20. Palmer, P.T.; Limero, T.F. Mass spectrometry in the U.S. space program: Past, present, and future. *J. Am. Soc. Mass Spectrom.* **2001**, *12*, 656–675. [CrossRef]
21. Freissinet, C.; Glavin, D.P.; Mahaffy, P.R.; Miller, K.E.; Eigenbrode, J.L.; Summons, R.E.; Brunner, A.E.; Buch, A.; Szopa, C.; Archer, P.D., Jr. *Organic Molecules in the Sheepbed Mudstone, Gale Crater, Mars*; NASA: Washington, DC, USA, 2015.
22. Altwegg, K.; Balsiger, H.; Bar-Nun, A.; Berthelier, J.-J.; Bieler, A.; Bochsler, P.; Briois, C.; Calmonte, U.; Combi, M.R.; Cottin, H.; et al. Prebiotic chemicals—Amino acid and phosphorus—In the coma of comet 67P/Churyumov-Gerasimenko. *Sci. Adv.* **2016**, *2*, e1600285. [CrossRef]
23. Goesmann, F.; Rosenbauer, H.; Bredehöft, J.H.; Cabane, M.; Ehrenfreund, P.; Gautier, T.; Giri, C.; Kruger, H.; Le Roy, L.; MacDermott, A.J.; et al. Organic compounds on comet 67P/Churyumov-Gerasimenko revealed by COSAC mass spectrometry. *Science* **2015**, *349*, aab0689. [CrossRef]
24. Parro, V.; Rodríguez-Manfredi, J.A.; Briones, C.; Compostizo, C.; Herrero, P.L.; Vez, E.; Sebastián, E.; Moreno-Paz, M.; Garcia-Villadangos, M.; Fernández-Calvo, P.; et al. Instrument development to search for biomarkers on Mars: Terrestrial acidophile, iron-powered chemolithoautotrophic communities as model systems. *Planet. Space Sci.* **2005**, *53*, 729–737. [CrossRef]
25. Sims, M.R.; Cullen, D.C.; Rix, C.S.; Buckley, A.; Derveni, M.; Evans, D.; Miguel García-Con, L.; Rhodes, A.; Rato, C.C.; Stefinovic, M.; et al. Development status of the life marker chip instrument for ExoMars. *Planet. Space Sci.* **2012**, *72*, 129–137. [CrossRef]
26. Bada, J.L.; Ehrenfreund, P.; Grunthaner, F.; Blaney, D.; Coleman, M.; Farrington, A.; Yen, A.; Mathies, R.; Amudson, R.; Quinn, R.; et al. Urey: Mars Organic and Oxidant Detector. *Space Sci. Rev.* **2008**, *135*, 269–279. [CrossRef]
27. Cable, M.L.; Stockton, A.M.; Mora, M.F.; Willis, P.A. Low-Temperature Microchip Nonaqueous Capillary Electrophoresis of Aliphatic Primary Amines: Applications to Titan Chemistry. *Anal. Chem.* **2013**, *85*, 1124–1131. [CrossRef]
28. Kim, J.; Jensen, E.C.; Stockton, A.M.; Mathies, R.A. Universal Microfluidic Automaton for Autonomous Sample Processing: Application to the Mars Organic Analyzer. *Anal. Chem.* **2013**, *85*, 7682–7688. [CrossRef]
29. Schulze-Makuch, D.; Head, J.N.; Houtkooper, J.M.; Knoblauch, M.; Furfaro, R.; Fink, W.; Fairen, A.G.; Vali, H.; Sears, S.K.; Daly, M.; et al. The Biological Oxidant and Life Detection (BOLD) mission: A proposal for a mission to Mars. *Planet. Space Sci.* **2012**, *67*, 57–69. [CrossRef]
30. Poinot, P.; Geffroy-Rodier, C. Searching for organic compounds in the Universe. *TrAC Trends Anal. Chem.* **2015**, *65*, 1–12. [CrossRef]
31. Skelley, A.M.; Scherer, J.R.; Aubrey, A.D.; Grover, W.H.; Ivester, R.H.; Ehrenfreund, P.; Grunthaner, F.J.; Bada, J.L.; Mathies, R.A. Development and evaluation of a microdevice for amino acid biomarker detection and analysis on Mars. *Proc. Natl. Acad. Sci. USA* **2005**, *102*, 1041–1046. [CrossRef]
32. Parro, V.; de Diego-Castilla, G.; Graciela de Diego-Castilla, J.A.; Rivas, L.A.; Blanco-López, Y.; Sebastián, E.; Romeral, J.; Compostizo, C.; Herrero, P.L.; García-Marín, A.; et al. SOLID3: A multiplex antibody microarray-based optical sensor instrument for in situ life detection in planetary exploration. *Astrobiology* **2011**, *11*, 15–28. [CrossRef]
33. Eddhif, B.; Allavena, A.; Liu, S.; Ribette, T.; Abou Mrad, N.; Chiavassa, T.; d'Hendecourt, L.L.S.; Sternberg, R.; Danger, G.; Geffroy-Rodier, C.; et al. Development of liquid chromatography high resolution mass spectrometry strategies for the screening of complex organic matter: Application to astrophysical simulated materials. *Talanta* **2018**, *179*, 238–245. [CrossRef]
34. Court, R.W.; Rix, C.S.; Sims, M.R.; Cullen, D.C.; Sephton, M.A. Extraction of polar and nonpolar biomarkers from the martian soil using aqueous surfactant solutions. *Planet. Space Sci.* **2012**, *67*, 109–118. [CrossRef]
35. Luong, D.; Court, R.W.; Sims, M.R.; Cullen, D.C.; Sephton, M.A. Extracting organic matter on Mars: A comparison of methods involving subcritical water, surfactant solutions and organic solvents. *Planet. Space Sci.* **2014**, *99*, 19–27. [CrossRef]
36. Briois, C.; Thissen, R.; Thirkell, L.; Aradj, K.; Bouabdellah, A.; Boukrara, A.; Carrasco, N.; Chalumeau, G.; Chapelon, O.; Colin, F.; et al. Orbitrap mass analyser for in situ characterisation of planetary environments: Performance evaluation of a laboratory prototype. *Planet. Space Sci.* **2016**, *131*, 33–45. [CrossRef]
37. Yokota, S. Isotope Mass Spectrometry in the Solar System Exploration. *Mass Spectrom.* **2018**, *7*, S0076. [CrossRef]

38. Vuitton, V.; Bonnet, J.-Y.; Frisari, M.; Thissen, R.; Quirico, E.; Dutuit, O.; Schmitt, B.; Le Roy, L.; Fray, N.; Cottin, H.; et al. Very high resolution mass spectrometry of HCN polymers and tholins. *Faraday Discuss.* **2010**, *147*, 495–508. [CrossRef]
39. Callahan, M.P.; Martin, M.G.; Burton, A.S.; Glavin, D.P.; Dworkin, J.P. Amino acid analysis in micrograms of meteorite sample by nanoliquid chromatography–high-resolution mass spectrometry. *J. Chromatogr. A* **2014**, *1332*, 30–34. [CrossRef]
40. Yamashita, Y.; Naraoka, H. Two homologous series of alkylpyridines in the Murchison meteorite. *Geochem. J.* **2014**, *48*, 519–525. [CrossRef]
41. Khare, B.N.; Sagan, C.; Ogino, H.; Nagy, B.; Er, C.; Schram, K.H.; Arakawa, E.T. Amino acids derived from Titan Tholins. *Icarus* **1986**, *68*, 176–184. [CrossRef]
42. Martins, Z.; Alexander, C.M.O.; Orzechowska, G.E.; Fogel, M.L.; Ehrenfreund, P. Indigenous amino acids in primitive CR meteorites. *Meteorit. Planet. Sci.* **2007**, *42*, 2125–2136. [CrossRef]
43. Callahan, M.P.; Smith, K.E.; Cleaves, H.J.; Ruzicka, J.; Stern, J.C.; Glavin, D.P.; House, C.H.; Dworkin, J.P. Carbonaceous meteorites contain a wide range of extraterrestrial nucleobases. *Proc. Natl. Acad. Sci. USA* **2011**, *108*, 13995–13998. [CrossRef]
44. Koppen, V.; Jones, R.; Bockx, M.; Cuyckens, F. High volume injections of biological samples for sensitive metabolite profiling and quantitation. *J. Chromatogr. A* **2014**, *1372*, 102–109. [CrossRef]
45. Nestola, M.; Friedrich, R.; Bluhme, P.; Schmidt, T.C. Universal Route to Polycyclic Aromatic Hydrocarbon Analysis in Foodstuff: Two-Dimensional Heart-Cut Liquid Chromatography–Gas Chromatography–Mass Spectrometry. *Anal. Chem.* **2015**, *87*, 6195–6203. [CrossRef]
46. Motoyama, A.; Venable, J.D.; Ruse, C.I.; Yates, J.R. Automated Ultra-High-Pressure Multidimensional Protein Identification Technology (UHP-MudPIT) for Improved Peptide Identification of Proteomic Samples. *Anal. Chem.* **2006**, *78*, 5109–5118. [CrossRef]
47. Magdeldin, S.; Moresco, J.J.; Yamamoto, T.; Yates, J.R. Off-Line Multidimensional Liquid Chromatography and Auto Sampling Result in Sample Loss in LC/LC–MS/MS. *J. Proteome Res.* **2014**, *13*, 3826–3836. [CrossRef]

Sample Availability: Samples of the compounds are not available from the authors.

© 2019 by the authors. Licensee MDPI, Basel, Switzerland. This article is an open access article distributed under the terms and conditions of the Creative Commons Attribution (CC BY) license (http://creativecommons.org/licenses/by/4.0/).

Article

A Vortex-Assisted Dispersive Liquid-Liquid Microextraction Followed by UPLC-MS/MS for Simultaneous Determination of Pesticides and Aflatoxins in Herbal Tea

Rui Zhang [1], Zhen-Chao Tan [1], Ke-Cheng Huang [2], Yan Wen [1], Xiang-Ying Li [1], Jun-Long Zhao [1] and Cheng-Lan Liu [1,*]

1. Key Laboratory of Natural Pesticide and Chemical Biology, Ministry of Agriculture & Key Laboratory of Bio-Pesticide Innovation and Application of Guangdong Province, South China Agricultural University, Wushan Road 483, Guangzhou 510642, China; 13570447655@163.com (R.Z.); tanzhenchao@outlook.com (Z.-C.T.); 18131378639@163.com (Y.W.); lxykity521@163.com (X.-Y.L.); 17602038236@163.com (J.-L.Z.)
2. Shenzhen Noposion Agrochemical Co. Ltd., Shenzhen 510640, Guangdong, China; kechenghuang@163.com
* Correspondence: liuchenglan@scau.edu.cn; Tel.: +86-20-85284925

Received: 18 February 2019; Accepted: 13 March 2019; Published: 15 March 2019

Abstract: A method for detecting the organophosphorus pesticides residue and aflatoxins in China herbal tea has been developed by UPLC-MS/MS coupled with vortex-assisted dispersive liquid-liquid microextraction (DLLME). The extraction conditions for vortex-assisted DLLME extraction were optimized using single-factor experiments and response surface design. The optimum conditions for the experiment were the pH 5.1, 347 µL of chloroform (extraction solvent) and 1614 µL of acetonitrile (dispersive solvent). Under the optimum conditions, the targets were good linearity in the range of 0.1 µg/L–25 µg/L and the correlation coefficient above 0.9998. The mean recoveries of all analytes were in the ranged from 70.06%–115.65% with RSDs below 8.54%. The detection limits were in the range of 0.001 µg/L–0.01µg/L. The proposed method is a fast and effective sample preparation with good enrichment and extraction efficiency, which can simultaneously detect pesticides and aflatoxins in China herbal tea.

Keywords: vortex-assisted dispersive liquid-liquid microextraction; China herbal tea; pesticides residue; aflatoxins; UPLC-MS/MS

1. Introduction

Herbal tea is a kind of soup made with natural Chinese herbal medicine as raw materials according to the local climate, water and soil characteristics with the unique cooking methods by residents in the southern coastal areas of China [1–4], and guided by the theory of traditional Chinese medicine (TCM) health in the process of longer-term disease prevention and health care. It has the functions of clearing away heat and detoxifying, stimulating thirst, and preventing diseases [5–7] and as a social and recreational pastime [8–11]. It is also a widely used traditional health drink that has been widely circulated for generations.

Currently, there are many brands of Chinese herbal tea, mainly Wang Laoji, Jia Duo Bao, and Huang Zhenglong [12]. The herbal tea is composed of a variety of medicinal materials, which are susceptible to pests and diseases during growth and storage. It is necessary for using chemical pesticides during the cultivation of Chinese herbal medicines plants. Therefore, pesticides residues are inevitable in Chinese herbal materials [13–15]. Ultimately, the pesticide residues maybe also be detected in herbal tea. The organophosphorus pesticides (OPPs) were intensively applied at large-scale spraying on crops and sometimes they are detected in agricultural products. Although their residual time is short, OPPs can cause many acute and chronic neurotoxic diseases [16,17]. On the other hand, the herbal medicines plants are often infected some toxigenic fungi, such as *Aspergillus*, *Penicillium* and *Fusarium*. These fungi can produce mycotoxins during herbal plants growth and storage under suitable environmental conditions [18,19]. Aflatoxins (AFs) are secondary toxic metabolites mainly produced by *Aspergillus flavus* and *A. parasiticus*. The four main aflatoxins universally contaminated with food are AFB1, AFB2, AFG1, and AFG2, which were classified as class I human carcinogens by the International Agency for Research on Cancer in 1993 [20]. AFs not only harms the health of consumers but also causes the loss of economic benefits of Chinese herbal medicines. Hence, it is necessary to establish a rapid and effective method for detecting the OPPs and aflatoxins in the herbal tea.

Many studies have reported methods for the detection of trace pesticides, which some pesticides have endocrine activity (EDCs) [21] in Chinese herbal medicines: liquid chromatography (LC) and gas chromatography (GC) [22,23], gas chromatography-mass spectrometry (GC-MS) [24] and liquid chromatography-mass spectrometry (tandem) mass spectrometry (HPLC-MS or HPLC-MS/MS) [25]. The methods for detecting mycotoxins such as aflatoxins in traditional Chinese medicine have enzyme-linked immunosorbent assay (ELISA) [26], HPLC [27] and LC-MS/MS [28]. At the same time, pre-treatment procedure such as extraction and concentration are crucial to improving sensitivity and selectivity of the analytical methods owing to the presence of trace amounts of OPPs and aflatoxins and the complexity of real samples. The most frequently sample pre-treatment methods are solid phase extraction (SPE) [29] and liquid-liquid extraction (LLE) [30]. However, LLE has some disadvantage such as using a large volume of organic solvents and time-consuming. SPE needs to use expensive SPE cartridges. At present, a novel method named dispersive-liquid-liquid microextraction (DLLME) has been widely used to treat samples for pesticides residues, mycotoxins and plant ingredients analysis [31–33]. The DLLME have many advantages such as rapidity, simplicity, low cost, low solvent usage and high enrichment factor. However, most of the above-mentioned methods are only used for pesticides residues or mycotoxins [34–37].

Our study was to establish a vortex-assisted DLLME combined with UPLC-MS/MS method for detecting eight OPPs (dichlorvos, phoxim, Chlorpyrifos-methyl, chlorpyrifos, tolcofos-methyl, ediphenphos, ethion, and profenofos) and four aflatoxins (AFB1, AFB2, AFG1 and AFG2) in herbal tea. The important parameters of the DLLME procedure were optimized by single factor experiment and response surface design. The developed method was validated and applied to analyze the real herbal tea samples. To the best of our knowledge, it is the first report that a DLLME combined with UPLC-MS/MS method has been developed to simultaneously determine the pesticides residues and aflatoxins in herbal tea.

2. Materials and Methods

2.1. Chemicals and Standards

The dichlorvos (purity 98.0%), phoxim (97.0%), chlorpyrifos-methyl (99.7%), chlorpyrifos (99.8%), tolcofos-methyl (98.3%), ediphenphos (97.9%), ethion (99.1%), profenofos (99.0%), and aflatoxin B1 (99.0%), aflatoxin B2 (99.0%), aflatoxin G1 (99.0%) and aflatoxin G2 (99.0%) were bought from Dr. Ehrenstorfer (Augsburg, Germany). Acetonitrile (ACN, HPLC grade) and methanol (MeOH, HPLC grade) were purchased from Shanghai Anpel Scientific Instrument Corporation (Shanghai, China). Analytical-grade carbon tetrachloride (CCl_4), chlorobenzene (C_6H_5Cl), chloroform ($CHCl_3$) and dichloromethane (CH_2Cl_2), 1.1.2.2-tetrachloroethane ($C_2H_2Cl_4$) were purchased from Tianjin Dongtian zheng Chemical Co. (Tianjin, China). Sodium chloride (NaCl) and hydrochloric acid (HCl) were purchased from Guangzhou Qian Hui Instrument Co., Ltd. (Guangzhou, China). Ultrapure water (UNIQUE-R20 purification system with UV + UF optional accessories, Research, Xiamen, China) was used in our work. A 0.22 mm cellulose membrane filter (Sterlitech, Kent, WA, USA) was used to filter the stock standard solution and herbal tea samples.

The stock solutions of eight target pesticides standards and four aflatoxins standards were prepared with acetone at 1000 mg/L and 100 mg/L, respectively, and stored in an amber glass vial at −20 °C. The working solutions were prepared by diluting the stock solution with acetonitrile.

2.2. Instruments and Equipment

The target analytes were determined with an ultra-performance liquid chromatography-tandem mass spectrometry (Waters TS-Q, Milford, MA, American). Separations were performed in an Acquity UPLC BEH C_{18} column (1.7 µm, 2.1 × 50 mm, Waters) under the condition of 40 °C. The mobile phase A was 2% (v/v) formic acid and the mobile phase B was methanol, at a flow rate of 0.3 mL/min. The injection volume was 5 µL. The elution solution was put into practice as follows: 0 min, 3% B; 0.5 min, 30% B; 6.5 min, 95% B; 7.5 min, 95% B; 9 min, 30% B and 10 min, 30% B.

The mass spectrometric analysis was carried out in the positive spray ionization mode and multiple reaction monitoring mode. Dry gas and atomizer are both nitrogen (N_2). The optimal spray voltage was at 1.0 KV. Source and desolvation temperatures were 150 °C and 400 °C, respectively. The gas flow was at 650 L/h and the collision gas flow was at 0.25 mL/min. The achieved MS/MS parameters were generalized in Table 1.

Table 1. The MS/MS parameters of the aimed pesticides and aflatoxins.

Analytes	Adduct On	Retention Time	Precursor Ion (m/z)	Product Ion (m/z)	Collision Energy/eV	Cone Voltage/V
dichlorvos	$[M + H]^+$	2.89	221	79/109	34/22	30
phoxim	$[M + H]^+$	5.41	299	129/153	13/7	30
Chlorpyrifos-methyl	$[M + H]^+$	5.61	321.8	125/289.9	20/16	30
chlorpyrifos	$[M + H]^+$	6.18	349.9	97/198	32/20	30
tolcofos-methyl	$[M + H]^+$	5.47	263.9	79/109	36/22	30
ediphenphos	$[M + H]^+$	5.24	311	109/111	32/26	30
ethion	$[M + H]^+$	6.09	385	199.1/143	10/20	30
profenofos	$[M + H]^+$	5.89	372.9	127.9/302.6	40/20	30
AFB1	$[M + H]^+$	2.87	313.2	241.1/285.1	36/24	40
AFB2	$[M + H]^+$	2.64	315.2	259.1/287.1	30/26	40
AFG1	$[M + H]^+$	2.43	329.2	243.1/283.1	30/30	40
AFG2	$[M + H]^+$	2.18	331.2	243.1/257.1	25/25	35

2.3. Sample Preparation

Several herbal tea samples were collected from local supermarkets in Guangzhou, China. The samples were filtered using 0.22 mm cellulose membrane filters in order to remove some solid residues, and the filtered herbal tea samples were adjusted to pH 5.1 with 0.1 M of hydrochloric acid (HCl) or sodium hydroxide (NaOH).

2.4. Optimization of the Vortex Assisted DLLME Process

Chloroform (347 µL) (as extraction solvent) was added to 1614 µL acetonitrile (as dispersive solvent). The mixture was then injected into a 15 mL conical centrifuge tube that contained 5 mL herbal tea sample (pH 5.1). The tube was shaken for 60 s with a vortex mixer. A cloudy, turbid solution was rapidly obtained in the tube. Then the tube was centrifuged for 5 min at 3800 rpm. The upper aqueous phase was removed and the $CHCl_3$ phase was quantitatively moved to a new centrifuge tube using a micro-syringe and evaporated to dryness under a stream of nitrogen at 45 °C. The evaporation residues reconstituted with 200 µL of acetonitrile. Finally, 5 µL was injected into the UPLC-MS/MS system for analysis.

2.5. Experimental Design and Data Analysis

In this study, the central composite design (CCD) was selected to optimize the three main factors that influenced the recovery efficiency (A: the sample pH; B: the volume of acetonitrile and C: the volume of CHCl$_3$). The response value (Y) was the mean recoveries of twelve aimed compounds. According to the design, each of the three factors (A, B and C) was studied at five levels (Table 2). For each of the three studied variables, low and high set points were constructed for an orthogonal design (Table 2). The CCD design consisted of six replicates of the central points and twenty combinations. The resulting of twenty combinations, in which 5 mL of deionized water added into 0.01 mg/L of twelve analytes, were randomly performed. Every combination was done with three replicates and the obtained twelve analytes of mean recoveries were used as the response by statistical software. The relationship between the response and the three variables were expressed as the following quadratic polynomial equation:

$$Y = b0 + b1A + b2B + b3C + b1b1A2 + b2b2B2 + b3b3C2 + b1b2AB + b1b3AC + b2b3BC \quad (1)$$

where Y is the response; A, B, C were the independent variables; b0 was the model intercept coefficient; b1, b2 and b3 were the linear coefficients; b1b1, b2b2 and b3b3 were the quadratic coefficients.

Table 2. The experimental range and levels of the variables in the central composite design (CCD).

variable	Parameter	Variable Levels				
		$-\alpha$(low)	-1	0	1	$+\alpha$(high)
A	pH	4	4.4	5	5.6	6
B	Volume of CHCl$_3$ (µL)	310	326	350	374	400
C	Volume of ACN (µL)	1500	1541	1600	1659	1700

3. Results and Discussion

3.1. DLLME Procedure Optimization

The optimal DLLME conditions were determined for twelve targets in herbal tea samples, and the chloroform and acetonitrile be used as the extractant and dispersant, respectively. The influence of different parameters on extraction efficiency such as the extraction solvent types and volume, the dispersant types and volume, the aqueous phase pH were carefully investigated. In these experiments, a blank herbal tea sample spiked with 12 analytes (10 µg/L) was applied to assess the performance of the pre-treatment method and were calculated using the following equations:

$$EF = \frac{C_{sed}}{C_0} \quad (2)$$

$$ER\% = \frac{C_{sed}V_{sed}}{C_0V_{aq}} \times 100 = EF \times \frac{V_{sed}}{V_{aq}} \times 100 \quad (3)$$

where EF is the enrichment factor; C_{sed} is the concentration of target in the sedimentary phase; C_0 is the initial concentration of target in the aqueous phase. ER% is the extraction recovery, V_{sed} is the volume of deposition phase, V_{aq} is the volume of the aqueous phase.

3.2. Extraction Solvent Selection

The extraction solvent plays a vital role in the extraction that affects the efficiency of microextraction specific condition. The appropriate extraction solvents must have a higher density than water, high partition coefficient, and poorly soluble in water. According to previous reports [38,39], we selected five halogenated hydrocarbons with a density higher than 1 g/mL, including carbon tetrachloride (CCl$_4$), chloroform (CHCl$_3$), dichloromethane (CH$_2$Cl$_2$), chlorobenzene (C$_6$H$_5$Cl) and

1.1.2.2-tetrachloroethane ($C_2H_2Cl_4$). 5 mL of herbal tea (pH = 4, all the target analytes at 0.01 mg/L) was extracted by the mixture of 400 μL extraction solvent (carbon tetrachloride, chloroform, dichloromethane, chlorobenzene or 1.1.2.2-terachloroethane) and 800 μL of acetonitrile (dispersive solvent). The extraction efficiency was assessed by comparing the recovery rates of each compound. The results revealed that the extraction recoveries using $CHCl_3$ as extraction solvent were higher than that of the other chlorinated solvent (shown as Figure 1a). The extraction recoveries of twelve compounds are 64.81%, 72.62%, 67.54%, 72.98%, 69.94%, 81.76%, 56.96%, 62.21%, 80.69%, 75.64%, 73.10% and 67.57%, respectively. Therefore, chloroform was selected as the extraction solvent.

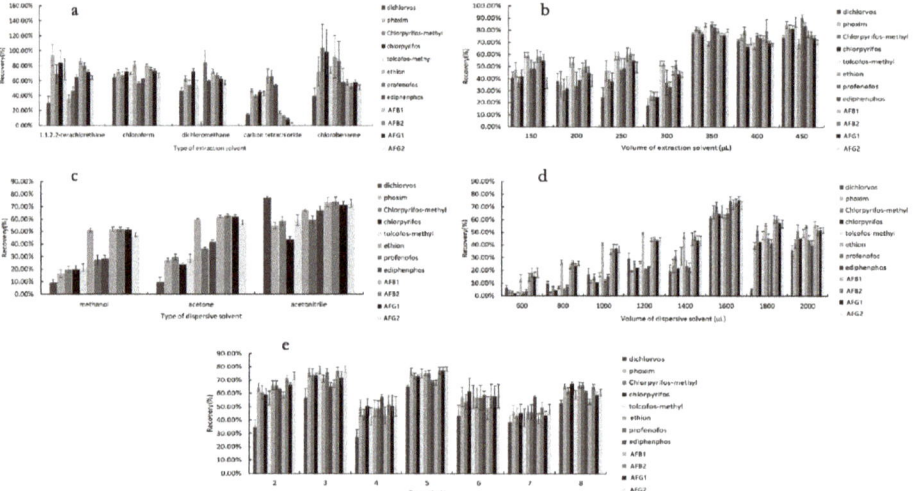

Figure 1. Optimized parameters of the DLLME procedure: (**a**) type of extraction solvent (**b**) volume of extraction solvent (μL) (**c**) type of dispersive solvent (**d**) volume of dispersive solvent (μL) and (**e**) pH. Extraction conditions: volume of chloroform, 350 μL; volume of acetonitrile,1600 μL; sample pH, 5; vortex-shaken time, 1 min; centrifuging for 5 min at 3800 rpm [a: sample pH 4, 400 μL of extraction solvent (carbon tetrachloride, chloroform, dichloromethane, chlorobenzene or 1.1.2.2-tetrachloroethane) and 800 μL of acetonitrile ; b: sample pH: 4, $CHCl_3$ (150, 200, 250, 300, 350, 400 and 450 μL) and 800 μL of acetonitrile; c: sample pH: 4, 350 μL of chloroform and 800 μL of dispersive solvent (acetonitrile, acetone and methanol); d: sample pH: 4, 350 μL of chloroform and different volume of acetonitrile (600, 800, 1000, 1200, 1400, 1600, 1800 and 2000 μL); e: sample pH range of 2–8, 350 μL of chloroform and 1600 μL of acetonitrile].

3.3. Effect of the Extraction Solvent Volume (Chloroform)

The volume of extraction solvent is an important factor in the extraction recovery. Normally, a low volume of extraction solvent can achieve higher enrichment [40]. Increasing the volume of the extraction solvent would improve the extraction efficiency. However, they might decrease the enrichment factor [41]. In the second experimental step, we observed different volumes of $CHCl_3$ (150, 200, 250, 300, 350, 400 and 450 μL) on the extraction efficiency at the same DLLME conditions. The results were shown in Figure 1b. The extraction recoveries for all analytes increased with increasing volume from 150 to 350 μL. However, the extraction recoveries of all analytes achieved a constant level under the volume above 350 μL. When the volume of chloroform was lower than 150 μL, it is difficult to withdraw the sedimentary phase. Thus, 350 μL was chosen as the optimal volume.

3.4. Effect of Dispersive Solvent

The dispersive solvent also obviously affect the extraction efficiency of DLLME. The role of the dispersive solvent converts the extraction solvent into droplets in an aqueous sample to help the analytes transfer into the organic phase. The dispersive solvents not only has a good solubility with the extraction solvent but also is miscible with water [42,43]. In the DLLME method, acetonitrile, acetone and methanol are usually chosen as the dispersive solvents. In our experiment, the effects of three dispersants (acetonitrile, acetone and methanol) on the extraction recovery were studied under the condition of 350 µL chloroform. The results showed that acetonitrile provided the highest extraction recoveries (Figure 1c). Thus, acetonitrile was chosen as the dispersive solvent.

3.5. Effect of Dispersive Solvent Volume (Acetonitrile)

The dispersive solvent volume directly affects the concentration of the extraction solvent in the aqueous phase, then affects the volume of deposition phase and extraction efficiency [44]. In order to get the optimal volume of acetonitrile, multiple experiments were carried out under different volumes of acetonitrile (600, 800, 1000, 1200, 1400, 1600, 1800, and 2000 µL) with 350 µL $CHCl_3$. As results in Figure1d, the extraction recoveries increased with increasing volume of acetonitrile. However, the extraction recoveries decreased when the volume of acetonitrile was higher than 1600 µL. Therefore, 1600 µL was selected for the next study.

3.6. Effect of pH

The pH of the aqueous sample not only affects the presence of the target compounds (such as an ionic or neutral form) but also can change the distribution ratio of targets between the organic phase and the aqueous phase [45,46]. What is more, different herbal tea samples with different pH values may affect DLLME extraction efficiency. So, the effect of pH on the DLLME procedure was investigated by adding 0.1M HCl or NaOH into the herbal tea sample within the pH range of 2–8. As shown as Figure1e, the extraction recoveries for all analytes increased with the pH increased from 2 to 5. However, the extraction recoveries of all analytes presented a slowly declining trend with increasing pH (from 5 to 8). Furthermore, the standard error of each target was lower when the pH value was 5. Therefore, pH 5 was selected for the following experiments.

3.7. Experimental Design

The CCD was used to select the optimal experimental conditions and maximize recoveries. Three variables were studied by using the CCD at the center point at five levels with six replicates. The levels of each factor, high and low set points were established in orthogonal design (Table 2). The average recoveries of twelve targets were used as the parameters for the response surface curve, the polynomial regression analysis was performed on the response values in the experiment and the quadratic regression equation was obtained:

$$R = 1.01 + 0.051A - 0.035B + 0.014C - 0.038AB + 0.019AC + 0.068BC - 0.13A^2 - 0.90B^2 - 0.12C^2 \quad (4)$$

where R is average recoveries of twelve analytes as a function of A (pH), B (volume of acetonitrile) and C (volume of $CHCl_3$).

The method used ANOVA to evaluate the significance level and of each factor and interaction term, the larger the F value, the smaller the *p*-value, the more reliable the regression model obtained. As shown in Table 3, the model reached a very significant level with a *p*-value of less than 0.0001 and an F value of 77.74. The closer the decision coefficient (R^2) of the model to 1, the closer the predicted value and the real values are 104.53% and 101.69%, respectively. The decision coefficient (R^2) and the modified decision coefficient ($AdjR^2$) of the model are 98.59% and 97.32%, respectively, and the coefficient of variation (CV) is 4.49%, that indicated a high correlation between the experimental and predicted values. The lack-fit-*p*-value of 0.0005 indicates that the model is susceptible to interference from non-experimental factor. The above test data showed that the model can reliably predict and analyze the mean recoveries of 12 targets by DLLME method. Figure 2 depicts the outline and 3D surface map of the mean recovery versus variable pair. The response surface map was applied to determine the extraction amount of eight pesticides and four mycotoxins on the interaction variables A–C. It could be seen from Figure 2a, the 3D map of the response surface model indicates that A (pH) and B (VE: volume of extraction solvent) have strong interactions. The target analytes get optimal extraction volume between 342 μL–350 μL and the most suitable pH range of 5.0–5.3. The A (pH) and C (VD: volume of dispersive solvent) have an impact on the mean recoveries of target extractants (shown in Figure 2b). When the B (VE: volume of extraction solvent) was fixed value, the extraction rate increased in the range of 1541 μL–1600 μL for acetonitrile. However, the extraction rate decreased at the volume of acetonitrile more than 1600 μL. Figure 2c shows that the maximum recoveries were achieved at the volume of chloroform (B) and acetonitrile in the range of 342 μL–350 μL and 1600 μL–1629 μL, respectively, when A (pH) was a fixed amount. The optimal conditions for this model are pH 5.1, 347 μL of extraction solvent and 1614 μL of dispersive solvent.

Table 3. Analysis of variance (ANOVA) for response surface quadratic model (12 analytes).

Source	Sum of Squares	d.f [a]	Mean Square	F-Value [b]	p-Value [c]	Prof > F
Model	0.86	9	0.095	77.74	<0.0001	significant
A-pH	0.035	1	0.035	28.73	0.0003	
B-VE	0.016	1	0.016	13.33	0.0045	
C-VD	0.28	1	0.28	226.20	<0.0001	
AB	0.011	1	0.011	9.31	0.0122	
AC	3.306×10^{-3}	1	3.306×10^{-3}	2.48	0.14666	
BC	0.037	1	0.037	30.27	0.0003	
A^2	0.25	1	0.25	203.26	<0.0001	
B^2	0.12	1	0.12	94.77	<0.0001	
C^2	0.20	1	0.20	164.98	<0.0001	
Redisual	0.012	10	1.205×10^{-3}			
Lack of fit	0.012	5	2.393×10^{-3}	38.76	0.0005	significant
Pure Error	3.082×10^{-4}	5	6.165×10^{-5}			
Cor Total	0.87	19				

[a] Degrees of freedom. [b] Test for comparing model variance with residual (error) variance. [c] Probability of seeing the observed F-value if the null hypothesis is true.

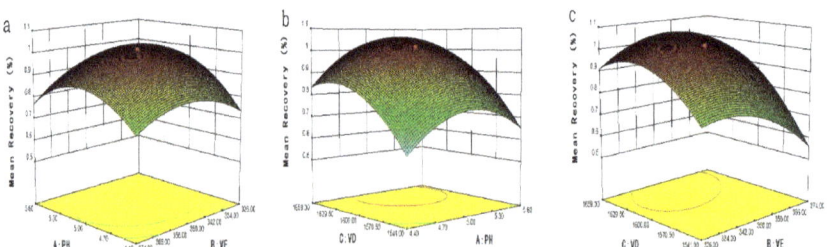

Figure 2. Response using the central composite design obtained by plotting: (A) pH; (B) VE: volume of extraction solvents; (C) VD: Volume of dispersive solvent, and (f) NaCl percentage vs. volume of dispersive solvent.

3.8. Method Evaluation

For evaluating the suitability of the method for simultaneous analyze the organophosphorous pesticides and aflatoxins in herbal tea, the linear range, the detection limits (LODs), the quantification limits (LOQs), precision and accuracy of the method were investigated under optimal conditions.

Since the herbal tea contains sugar, pigments and other impurities, which may affect the chromatographic signal of the targets and cause matrix effects. The magnitude of matrix effects can be assessed by comparing the matrix-matching calibration curve with the solvent standard curve. If the slope is 0~±20%, it indicates that the sample has a weak matrix effect; if the slope is between ±20%~±50%, there is a medium matrix effect; if the slope exceeds ±50%, which shows a strong matrix effect [46]. Therefore, the standard solutions were prepared by using a blank matrix extract without the target analytes to eliminate and compensate for the matrix effect. The matrix effect formula was evaluated as follows:

$$\text{ME (\%)} = \left(\frac{\text{slope of the calibration curve in the matrix}}{\text{slope of the calibration curve in the solvent}} - 1 \right) \times 100 \tag{5}$$

According to Table 3, there was a weak matrix effect for all the targets, indicated that the matrix does not significantly interfere with the chromatographic signals of the targets. Due to the weak matrix effect, seven different concentrations of pesticides and mycotoxins (0.1–25 µg/L) were applied in acetonitrile. The 12 targets had a good linearity in the linear range (0.1 µg/L–25 µg/L) and the correlation coefficient (R^2) higher than 0.9980. (Table 4). The LODs were ranged from 0.001 µg/L–0.01 µg/L at the signal-to-noise ratio of 3 (S/N = 3). The LOQs were used as the lowest added concentration in herbal tea samples and the LOQs of the aimed pesticides and aflatoxins were 1 µg/L and 0.2 µg/L, respectively.

Table 4. Calibration data of the DLLME procedure for pesticide and mycotoxins in Wang Lo Kat and Jia Duo Bao samples.

Samples	Analytes	Linearity (µg·L^{-1})	S(S$_a$) [a]	R^2(R$_a^2$) [b]	Ratio (%)	Matrix Effect
Wang Laoji	dichlorvos	0.1–25	519,146(485,390)	0.9992(0.9994)	6.50	Mild
	phoxim	0.1–25	192,788(190,297)	0.9995(0.9995)	1.30	Mild
	Chlorpyrifos-methyl	0.1–25	89,333(85,254)	0.9996(0.9994)	4.57	Mild
	chlorpyrifos	0.1–25	183,791(172,781)	0.9997(0.9995)	5.99	Mild
	tolcofos-methyl	0.1–25	39,699 (38,327)	0.9990(0.9996)	3.46	Mild
	ediphenphos	0.1–25	750,665(761,053)	0.9993(0.9991)	1.38	Mild
	profenofos	0.1–25	160,427(157,929)	0.9990(0.9998)	3.72	Mild
	ethion	0.1–25	387,602(383,630)	0.9996(0.9995)	1.03	Mild
	AFB1	0.1–25	213,455(207,545)	0.9998(0.9998)	2.77	Mild
	AFB2	0.1–25	7042.5(7133.2)	0.9995(0.9995)	1.27	Mild
	AFG1	0.1–25	155,448(135,238)	0.9994(0.9996)	13.00	Mild
	AFG2	0.1–25	55,640(51,872)	0.9995(0.9995)	6.77	Mild
Jia Duo Bao	dichlorvos	0.1–25	519,146(470,095)	0.9992(0.9989)	9.45	Mild
	phoxim	0.1–25	192,788(185,570)	0.9995(0.9987)	3.74	Mild
	Chlorpyrifos-methyl	0.1–25	89,333(81,727)	0.9991(0.9989)	8.51	Mild
	chlorpyrifos	0.1–25	183,791(162,751)	0.9995(0.9991)	11.45	Mild
	tolcofos-methyl	0.1–25	39,699(36,143)	0.9996(0.9999)	8.96	Mild
	ediphenphos	0.1–25	750,665(760,249)	0.999(0.9991)	1.28	Mild
	profenofos	0.1–25	164,027(152,981)	0.9997(0.9998)	6.73	Mild
	ethion	0.1–25	387,602(370,279)	0.9996(0.9996)	4.35	Mild
	AFB1	0.1–25	213,455(189,305)	0.9998(0.9990)	11.31	Mild
	AFB2	0.1–25	7133.2(7457.9)	0.9995(0.9996)	4.55	Mild
	AFG1	0.1–25	155,448(136,653)	0.9994(0.9998)	12.09	Mild
	AFG2	0.1–25	55,640(50,823)	0.9995(0.9998)	8.66	Mild

[a] S and R^2, slope and determination coefficient of the calibration curves obtained from ACN solution. [b] S_a and R_a^2, slope and determination coefficient of the calibration curves obtained from matrix matched standard solutions.

To validate the vortex-assisted DLLME method, four levels of concentrations of the added recoveries experiments (n = 5) were carried out on two different kinds of herbal tea samples (Wang Laoji and Jia Duo Bao). The results of the recoveries and precision were indicated in Table 5. From the results as shown in Table 5, the average recoveries of all targets were between 70.06% and 115.65%, and the RSDs were low at 8.54%, in agreement with the established performance criteria [47]. When the addition of 500, 100 and 10 µg/L in the blank sample of herbal tea, the response value exceeded the linear range, and the dilution was 500, 100 and 10 times before injection.

Table 5. Recoveries of the OPPs and aflatoxins from herbal tea samples using optimized vortexed-assisted DLLME ($n = 5$).

Analytes	Spiked Level µg/L	Wang Laoji		Jia Duo Bao	
		Recovery (%)	RSD (%)	Recovery (%)	RSD (%)
dichlorvos	500	75.34	8.14	72.48	7.77
	100	70.12	5.57	71.61	6.83
	10	70.27	2.28	72.45	1.53
	1	70.06	2.95	70.44	3.86
phoxim	500	97.81	5.12	90.97	7.29
	100	79.78	5.33	74.50	4.85
	10	78.35	1.30	88.06	7.49
	1	71.19	4.43	70.21	4.54
Chlorpyrifos-methyl	500	82.39	5.17	89.05	8.26
	100	72.65	7.31	83.22	3.69
	10	77.15	3.14	79.53	7.45
	1	77.05	8.50	70.32	4.05
chlorpyrifos	500	84.84	3.00	83.11	5.92
	100	72.94	4.98	78.19	5.90
	10	74.63	2.39	86.64	5.03
	1	74.17	4.45	71.68	2.52
Tolcofos-methyl	500	75.70	8.28	104.34	5.03
	100	73.64	7.68	79.59	6.41
	10	76.23	5.92	76.23	5.92
	1	73.29	5.50	86.65	5.11
ediphenphos	500	71.04	3.63	72.28	2.51
	100	70.64	6.23	72.39	4.99
	10	71.80	6.96	70.44	4.69
	1	95.25	4.63	107.73	2.12
ethion	500	101.12	5.35	106.38	2.29
	100	78.08	4.68	82.26	4.05
	10	94.12	2.17	86.59	2.86
	1	115.65	2.40	114.09	2.21
profenofos	500	102.36	3.40	102.55	6.62
	100	81.89	5.28	78.78	3.29
	10	99.49	4.00	91.24	2.73
	1	81.88	5.29	83.59	4.72
AFB1	10	94.59	5.25	92.14	7.62
	1	83.28	3.83	77.44	2.22
	0.2	72.68	2.88	71.04	1.90
AFB2	10	101.19	5.22	97.07	5.12
	1	74.87	8.41	72.61	8.54
	0.2	73.56	4.76	75.81	3.42
AFG1	10	99.39	1.25	91.03	2.33
	1	70.67	4.07	70.67	2.32
	0.2	72.36	7.53	70.36	7.84
AFG2	10	102.99	2.62	92.15	2.46
	1	72.59	6.17	70.68	3.24
	0.2	70.10	2.70	72.40	1.89

3.9. Application of the Developed DLLME Method to Real Herbal Tea Samples

The established method was used to detect the target analytes (eight organophosphorus pesticides and four aflatoxins) in 10 batches of herb tea samples (five of Wang Laoji herbal tea beverages, five of Jia Duo Bao herbal tea beverages), which were purchased from local supermarkets on 2018 in Guangzhou, China. We found that four aflatoxins were not detected in all herbal tea samples. Traces of phoxim and chlorpyrifos were detected in one batch with concentrations of 0.16×10^{-3} mg/kg and 0.10×10^{-3} mg/kg, respectively. However, the concentration of two pesticides did not exceed the maximum residue limit (MRL) of phoxim (0.01 mg/kg) and chlorpyrifos (0.02 mg/kg) in Chinese herbal medicines formulated by the European Union.

4. Conclusions

In our study, a novel vortex-assisted DLLME method combined with UPLC-MS/MS was developed for the detection of organophosphorus pesticides and aflatoxins in herbal tea beverages. The response surface method based on the central composite design was used to optimize the important parameters affecting the extraction recoveries, so as to determine the interaction between variables and obtain the optimal experimental combination. The optimized extraction conditions were pH 5.1, 347 μL of chloroform and 1614 μL of acetonitrile. Under the optimum conditions, the linear range was 0.1–25 μg/L for all targets and the correlation coefficient (R^2) > 0.998. The LOD ranged from 0.001 μg/L–0.01 μg/L (S/N > 3) and the quantitative limits were 1 μg/L for pesticides and 0.2 μg/L for mycotoxins. The fortified recoveries were 70.06%–115.65%, and the relative standard deviation is low at 8.54% (n = 5). The developed method has the advantages of simple and rapid operation, high concentration factor and environment friendliness. The method can simultaneously analyze and detect organophosphorus pesticides and aflatoxins in different kinds of herbal tea and liquid samples. As far as we know, this is the first time that vortex-assisted DLLME combined with UPLC-MS/MS is used to simultaneously detect the pesticides and aflatoxins in herbal tea.

Author Contributions: C.-L.L. and K.-C.H. conceived the study. R.Z. designed the experiments, analyzed the data, and wrote the manuscript. Z.-C.T., Y.W., X.-Y.L. and J.-L.Z. participated in the experiments, discussed and commented on the manuscript. All authors read and approved the final manuscript.

Funding: This work was funded by the Project of Science and Technology of Guangdong Province (No. 2016A040403102).

Conflicts of Interest: The authors declare no conflict of interest.

References

1. Jin, B.; Liu, Y.; Xie, J.; Luo, B.; Long, C. Ethnobotanical survey of plant species for herbal tea in a Yao autonomous county (Jianghua, China): Results of a 2-year study of traditional medicinal markets on the Dragon Boat Festival. *J. Ethnobiol. Ethnomed.* **2018**, *14*, 58–78. [CrossRef] [PubMed]
2. Fu, Y.; Yang, J.; Cunningham, A.B.; Towns, A.M.; Zhang, Y.; Yang, H.Y.; Li, J.W.; Yang, X.F. A billion cups: The diversity, traditional uses, safety issues and potential of Chinese herbal teas. *J. Ethnopharmacol.* **2018**, *222*, 217–228. [CrossRef]
3. Liu, Y.; Ahmed, S.; Long, C. Ethnobotanical survey of cooling herbal drinks from southern China. *J. Ethnobiol. Ethnomed.* **2013**, *9*, 82–89. [CrossRef] [PubMed]
4. De Santayana, M.P.; Blanco, E.; Morales, R. Plants known as té in Spain: An ethno-pharmaco-botanical review. *J. Ethnopharmacol.* **2005**, *98*, 1–19. [CrossRef]
5. Han, B.Q.; Peng, Y.; Xiao, P.G. Systematic Research on Chinese Non-Camellia Tea. *Mod. Chin. Med.* **2013**, *15*, 259–269.
6. Li, S.; Li, S.K.; Li, H.B.; Xu, X.R.; Deng, G.F.; Xu, D.P. *Antioxidant Capacities of Herbal Infusions*; South China Sea Institute of Oceanology, Chinese Academy of Sciences: Guangzhou, China, 2014; pp. 41–51.
7. Ranasinghe, S.; Ansumana, R.; Lamin, J.M.; Bockarie, A.S.; Bangura, U.; Jacob, A.G.; Stenger, D.A.; Jacobsen, K.H. Herbs and herbal combinations used to treat suspected malaria in Bo, Sierra Leone. *J. Ethnopharmacol.* **2015**, *166*, 200–204. [CrossRef] [PubMed]
8. Bussmann, R.W.; Paniagua-Zambrana, N.; Sifuentes, R.Y.C.; Velazco, Y.A.P.; Mandujano, J. Health in a Pot-The Ethnobotany of Emolientes and Emolienteros in Peru. *Econ. Bot.* **2015**, *69*, 83–88. [CrossRef]
9. Obón, C.; Rivera, D.; Alcaraz, F.; Attieh, L. Beverage and culture. "Zhourat", a multivariate analysis of the globalization of an herbal tea from the Middle East. *Appetite* **2014**, *79*, 1–10. [CrossRef] [PubMed]
10. Pardo, D.S.M.; Blanco, E.R. Plants known as té in Spain: an ethno-pharmaco-botanical review. *J. Ethnopharmacol.* **2005**, *98*, 1–19. [CrossRef]
11. Chi, L.; Li, Z.; Dong, S.; He, P.; Wang, Q.; Fang, Y. Simultaneous determination of flavonoids and phenolic acids in Chinese herbal tea by beta-cyclodextrin based capillary zone electrophoresis. *Microchim. Acta* **2009**, *167*, 179–185. [CrossRef]

12. Hajjo, R.M.; Battah, F.U.A.A. Multiresidue pesticide analysis of the medicinal plant *Origanum syriacum*. *Food Addit. Contam.* **2007**, *24*, 274–279. [CrossRef] [PubMed]
13. Leung, K.S.Y.; Chan, K.; Chan, C.L.; Lu, G.H. Systematic evaluation of organochlorine pesticide residues in Chinese materia medica. *Phytother. Res.* **2010**, *19*, 514–518. [CrossRef] [PubMed]
14. Li, J.; Li, D.; Wu, J.; Qin, J.; Hu, J.; Huang, W.; Wang, Z.; Xiao, W.; Wang, Y. Simultaneous determination of 35 ultra-trace level organophosphorus pesticide residues in Sanjie Zhentong capsules of traditional Chinese medicine using ultra high performance liquid chromatography with tandem mass spectrometry. *J. Sep. Sci.* **2017**, *40*, 999–1009. [CrossRef] [PubMed]
15. Jokanović, M.; Kosanović, M. Neurotoxic effects in patients poisoned with organophosphorus pesticides. *Environ. Toxicol. Pharmacol.* **2010**, *29*, 195–201. [CrossRef]
16. Wei, J.; Hu, J.; Cao, J.L.; Wan, J.B.; He, C.W.; Hu, Y.J.; Hu, H.; Li, P. Sensitive detection of organophosphorus pesticides in medicinal plants using ultrasound-assisted dispersive liquid–liquid microextraction combined with sweeping micellar electrokinetic chromatography. *J. Agric. Food Chem.* **2016**, *64*, 932–940. [CrossRef] [PubMed]
17. Sun, S.; Yao, K.; Zhao, S.; Zheng, P.; Wang, S.; Zeng, Y.; Liang, D.; Ke, Y.; Jiang, H. Determination of aflatoxin and zearalenone analogs in edible and medicinal herbs using a group-specific immunoaffinity column coupled to ultra-high-performance liquid chromatography with tandem mass spectrometry. *J. Chromatogr. B* **2018**, *1092*, 228–236. [CrossRef]
18. International Agency for Research on Cancer (IARC). IARC Monographs on the Evaluation of Carcinogenic Risks to Humans. In *Working Group Evaluation Carcinogenic Risks Humans I*; IARC: Lyon, France, 2010; Volume 96, pp. 27–338.
19. International Agency for Research on Cancer (IARC). Aflatoxins: B1, B2, G1, G2, M1, 245–395. In *IARC Monographs on the Evaluation of Carcinogenic Risks of Chemicals to Humans*; IARC: Lyon, France, 2010; Volume 56.
20. Locatelli, M.; Sciascia, F.; Cifelli, R.; Malatesta, L.; Bruni, P.; Croce, F. Analytical methods for the endocrine disruptor compounds determination in environmental water samples. *J. Chromotogr. A* **2016**, *1434*, 1–18. [CrossRef]
21. Chen, L.; Xing, W. Simple one-step preconcentration and cleanup with a micellar system for high performance liquid chromatography determination of pyrethroids in traditional Chinese medicine. *Anal. Methods* **2015**, *7*, 1691–1700. [CrossRef]
22. Guo, Q.; Lv, X.; Tan, L.; Yu, B.Y. Simultaneous determination of 26 pesticide residues in 5 Chinese medicinal materials using solid-phase extraction and GC-ECD Method. *Chin. J. Nat. Med.* **2009**, *7*, 210–216. [CrossRef]
23. Rutkowska, E.; łozowicka, B.; Kaczyński, P. Three approaches to minimize matrix effects in residue analysis of multiclass pesticides in dried complex matrices using gas chromatography tandem mass spectrometry. *Food Chem.* **2019**, *279*, 20–29. [CrossRef]
24. Qin, Y.; Chen, L.; Yang, X.; Li, S.; Wang, Y.; Tang, Y.; Liu, C. Multi-residue method for determination of selected neonicotinoid insecticides in traditional Chinese medicine using modified dispersive solid-phase extraction combined with ultra-performance liquid chromatography tandem mass spectrometry. *Anal. Sci.* **2015**, *31*, 823–830. [CrossRef]
25. Burmistrova, N.A.; Rusanova, T.Y.; Yurasov, N.A.; De Saeger, S.; Goryacheva, I. Simultaneous determination of several mycotoxins by rapid immunofiltration assay. *J. Anal. Chem.* **2014**, *69*, 525–534. [CrossRef]
26. Mohamadi, S.A.; Nikpooyan, H. Determination of aflatoxin M1 in milk by high-performance liquid chromatography in Mashhad (North East of Iran). *Toxicol. Ind. Health* **2012**, *29*, 334.
27. Han, Z.; Zheng, Y.; Luan, L.; Cai, Z.; Ren, Y.; Ren, Y. An ultra-high-performance liquid chromatography-tandem mass spectrometry method for simultaneous determination of aflatoxins B1, B2, G1, G2, M1 and M2 in traditional Chinese medicines. *Anal. Chim. Acta* **2010**, *664*, 165–171. [CrossRef]
28. Boulanouar, S.; Combès, A.; Mezzache, S.; Pichon, V. Synthesis and application of molecularly imprinted silica for the selective extraction of some polar organophosphorus pesticides from almond oil. *Anal. Chim. Acta* **2018**, *1018*, 35–44. [CrossRef]
29. Notardonato, I.; Russo, M.V.; Vitali, M.; Protano, C.; Avino, P. Analytical method validation for determining organophosphorus pesticides in baby foods by a modified liquid–liquid microextraction method and gas chromatography–ion trap/mass spectrometry analysis. *Food Anal. Method* **2019**, *12*, 41–50. [CrossRef]

30. Prosen, H. Applications of Liquid-Phase Microextraction in the Sample Preparation of Environmental Solid Samples. *Molecules* **2014**, *19*, 6776–6808. [CrossRef]
31. Kissoudi, M.; Samanidou, V. Recent advances in applications of ionic liquids in miniaturized microextraction techniques. *Molecules* **2018**, *23*, 1437. [CrossRef]
32. Diuzheva, A.; Carradori, S.; Andruch, V.; Locatelli, M.; De Luca, E.; Tiecco, M.; Germani, R.; Menghini, L.; Nocentini, A.; Gratteri, P.; et al. Use of innovative (Micro)Extraction techniques to CharacteriseHarpagophytum procumbens root and its commercial food supplements. *Phytochem. Anal.* **2018**, *29*, 233–241. [CrossRef]
33. Moyakao, K.; Santaladchaiyakit, Y.; Srijaranai, S.; Vichapong, J. Preconcentration of trace neonicotinoid insecticide residues using vortex-assisted dispersive micro solid-phase extraction with montmorillonite as an efficient sorbent. *Molecules* **2018**, *23*, 883. [CrossRef]
34. De Souza, G.; Mithöfer, A.; Daolio, C.; Schneider, B.; Rodrigues-Filho, E. Identification of Alternaria alternata mycotoxins by LC-SPE-NMR and their cytotoxic effects to soybean (Glycine max) cell suspension culture. *Molecules* **2013**, *18*, 2528–2538. [CrossRef]
35. Ho, Y.; Tsoi, Y.; Leung, K.S. Highly sensitive and selective organophosphate screening in twelve commodities of fruits, vegetables and herbal medicines by dispersive liquid–liquid microextraction. *Anal. Chim. Acta* **2013**, *775*, 58–66. [CrossRef] [PubMed]
36. Chen, F.; Luan, C.; Wang, L.; Wang, S.; Shao, L. Simultaneous determination of six mycotoxins in peanut by high-performance liquid chromatography with a fluorescence detector. *J. Sci. Food Agric.* **2017**, *97*, 1805–1810. [CrossRef] [PubMed]
37. Yan, H.; Wang, H. Recent development and applications of dispersive liquid–liquid microextraction. *J. Chromatogr. A* **2013**, *1295*, 1–15. [CrossRef]
38. Herrera-Herrera, A.V.; Asensio-Ramos, M.; Hernández-Borges, J.; Rodríguez-Delgado, M. Dispersive liquid-liquid microextraction for determination of organic analytes. *TrAC Trends Anal. Chem.* **2010**, *29*, 728–751. [CrossRef]
39. Chen, B.; Wu, F.; Wu, W.; Jin, B.; Xie, L.; Feng, W.; Ouyang, G. Determination of 27 pesticides in wine by dispersive liquid–liquid microextraction and gas chromatography–mass spectrometry. *Microchem. J.* **2016**, *126*, 415–422. [CrossRef]
40. Rodríguez-Cabo, T.; Rodríguez, I.; Ramil, M.; Cela, R. Dispersive liquid–liquid microextraction using non-chlorinated, lighter than water solvents for gas chromatography–mass spectrometry determination of fungicides in wine. *J. Chromotogr. A* **2011**, *1218*, 6603–6611. [CrossRef] [PubMed]
41. Zgoła-Grześkowiak, A.; Grześkowiak, T. Dispersive liquid-liquid microextraction. *TrAC Trends Anal. Chem.* **2011**, *30*, 1382–1399. [CrossRef]
42. Chaiyamate, P.; Seebunrueng, K.; Srijaranai, S. Vortex-assisted low density solvent and surfactant based dispersive liquid-liquid microextraction for sensitive spectrophotometric determination of cobalt. *RSC Adv.* **2018**, *8*, 7243–7251. [CrossRef]
43. Lai, X.; Ruan, C.; Li, R.; Liu, C. Application of ionic liquid-based dispersive liquid–liquid microextraction for the analysis of ochratoxin A in rice wines. *Food Chem.* **2014**, *161*, 317–322. [CrossRef]
44. Zhang, H.; Shi, Y. Temperature-assisted ionic liquid dispersive liquid–liquid microextraction combined with high performance liquid chromatography for the determination of anthraquinones in Radix et Rhizoma Rhei samples. *Talanta* **2010**, *82*, 1010–1016. [CrossRef]
45. Hendriks, G.; Uges, D.R.A.; Franke, J.P. Reconsideration of sample pH adjustment in bioanalytical liquid–liquid extraction of ionisable compounds. *J. Chromatogr. B* **2007**, *853*, 234–241. [CrossRef] [PubMed]
46. Fotopoulou, A. Matrix effect in gas chromatographic determination of insecticides and fungicides in vegetables. *Int. J. Environ. Chem.* **2004**, *84*, 15–27.
47. Commission, E. Commission Regulation (EC) No 401/2006 of 23 February 2006 laying down the methods of sampling and analysis for the official control of the levels of mycotoxins in foodstuffs. *J. Eur. Union* **2006**, *70*, 12–34.

Sample Availability: Samples of the compounds are not available from the authors.

© 2019 by the authors. Licensee MDPI, Basel, Switzerland. This article is an open access article distributed under the terms and conditions of the Creative Commons Attribution (CC BY) license (http://creativecommons.org/licenses/by/4.0/).

Article

Comparison of the Partition Efficiencies of Multiple Phenolic Compounds Contained in Propolis in Different Modes of Acetonitrile–Water-Based Homogenous Liquid–Liquid Extraction

Wenbin Chen [1,2,*], Xijuan Tu [1,2], Dehui Wu [1], Zhaosheng Gao [1], Siyuan Wu [1] and Shaokang Huang [1]

1. College of Bee Science, Fujian Agriculture and Forestry University, Fuzhou 350002, China; xjtu@fafu.edu.cn (X.T.); dehui2580@163.com (D.W.); gzs100@126.com (Z.G.); wusiyuan2018@126.com (S.W.); skhuang@fafu.edu.cn (S.H.)
2. MOE Engineering Research Center of Bee Products Processing and Application, Fujian Agriculture and Forestry University, Fuzhou 350002, China
* Correspondence: wbchen@fafu.edu.cn; Tel.: +86-591-83789482

Academic Editor: Nuno Neng
Received: 12 December 2018; Accepted: 23 January 2019; Published: 26 January 2019

Abstract: Homogeneous liquid–liquid extraction (HLLE) has attracted considerable interest in the sample preparation of multi-analyte analysis. In this study, HLLEs of multiple phenolic compounds in propolis, a polyphenol-enriched resinous substance collected by honeybees, were performed for improving the understanding of the differences in partition efficiencies in four acetonitrile–water-based HLLE methods, including salting-out assisted liquid–liquid extraction (SALLE), sugaring-out assisted liquid–liquid extraction (SULLE), hydrophobic-solvent assisted liquid–liquid extraction (HSLLE), and subzero-temperature assisted liquid–liquid extraction (STLLE). Phenolic compounds were separated in reversed-phase HPLC, and the partition efficiencies in different experimental conditions were evaluated. Results showed that less-polar phenolic compounds (kaempferol and caffeic acid phenethyl ester) were highly efficiently partitioned into the upper acetonitrile (ACN) phase in all four HLLE methods. For more-polar phenolic compounds (caffeic acid, *p*-coumaric acid, isoferulic acid, dimethoxycinnamic acid, and cinnamic acid), increasing the concentration of ACN in the ACN–H_2O mixture could dramatically improve the partition efficiency. Moreover, results indicated that NaCl-based SALLE, HSLLE, and STLLE with ACN concentrations of 50:50 (ACN:H_2O, v/v) could be used for the selective extraction of low-polarity phenolic compounds. $MgSO_4$-based SALLE in the 50:50 ACN–H_2O mixture (ACN:H_2O, v/v) and the NaCl-based SALLE, SULLE, and STLLE with ACN concentrations of 70:30 (ACN:H_2O, v/v) could be used as general extraction methods for multiple phenolic compounds.

Keywords: salting-out assisted liquid–liquid extraction; sugaring-out assisted liquid–liquid extraction; hydrophobic-solvent assisted liquid–liquid extraction; subzero-temperature assisted liquid–liquid extraction; phenolic compounds

1. Introduction

Increasing demands on monitoring a large number of target compounds have promoted the development of multi-analyte analytical methods. For instance, to assess a broad spectrum of possible metabolites, metabolomics requires multi-analyte methods to analyze the entire metabolome [1,2]. As another example, improper usage and the cross-contamination of chemicals in agricultural practice may lead to multi-residues of contaminants in agricultural products. Thus, multi-analyte methods have been developed to monitor the unknown chemical treatment history and protect the health of consumers [3]. Additionally, fingerprint profiles based on multi-analyte analysis of phytochemical compounds or volatile fractions in foods have been applied in foodomics for the issue of food quality [4].

Modern analytical instruments, especially the chromatography tandem mass spectrometry techniques, are capable of analyzing a large number of target compounds in a single analysis [5]. However, sample preparation procedure is still the crucial variable of multi-analyte analysis in achieving complete and accurate information [6]. Conventional liquid–liquid extraction has been widely used for the extraction of multiple low-polar compounds from aqueous sample solutions. Nevertheless, its applications in multi-analyte analysis are limited by the low extraction efficiency towards high-polarity compounds. Recently, homogeneous liquid–liquid extraction (HLLE) methods have been developed and extensively applied in multi-analyte analysis, due to their effective extraction of target compounds with a wide range of polarities [7–9]. In addition, HLLEs are receiving increasing interest from researchers because of the reduction of reagent consumption, extraction time, and the cost of analysis [7].

In acetonitrile–water-based HLLE, the acetonitrile (ACN) is mixed with water to form a homogenous solution for the extraction. Then, the ACN phase is triggered to partition from the aqueous solution with the addition of phase separation agents such as salts [8], sugars [10], hydrophobic solvents [11], or the cooling performance [12]. For example, Valente et al. reported the capabilities of salting-out assisted liquid–liquid extraction (SALLE) in phytochemical analysis [13]. This HLLE technique was demonstrated to be simple, of low cost, and versatile for the identification of various volatile and non-volatile compounds in fennel seeds (*Foeniculum vulgare* Mill). In addition, sugars can also trigger the phase separation in ACN–water mixtures, to develop the sugaring-out assisted liquid–liquid extraction (SULLE) method. Compared with SALLE, SULLE showed the advantage of being environment friendly [14]. Multi-analyte analysis of drugs in honey and plasma by using SULLE have been reported [15,16]. Recently, Liu et al. reported a similar phase separation phenomenon of ACN–water mixtures, induced by hydrophobic solvents [11]. This hydrophobic-solvent assisted liquid–liquid extraction (HSLLE) has been used for the profiling of endogenous phytohormones in plants [17]. Additionally, Yoshida et al. reported that ACN was separated from the aqueous solution at a subzero temperature ($-20\,^\circ$C) [12]. This technique, subzero-temperature assisted liquid–liquid extraction (STLLE), can avoid the residues of phase separation agents in the ACN phase, compared with the other three HLLE methods. This simple HLLE method has been applied for the determination of anthraquinone derivatives in sticky traditional Chinese medicines [18]. More recently, microextraction methods based on HLLE have been developed for the extraction and preconcentration of multiple contaminants in foodstuffs [19,20].

In multi-analyte analysis, understanding the distribution of compounds in the extractive is valuable for the design of sample preparation protocol to achieve a wide extraction of multi-analytes and minimize the co-extraction of interferences. Phenolic compounds play an important role in human diets, resulting from their nutritional significance and potentially beneficial health effects [21]. Despite HLLE having been widely used in multi-analyte analysis, little is known about the partition efficiencies of phenolic compounds in different acetonitrile–water-based HLLE methods, which are fundamentally important for the design of sample preparation protocol for analytical purposes. Propolis, a resinous substance collected by honeybees, is rich in polyphenols [22,23]. The reported methods for the extraction of phenolic compounds in propolis include maceration extraction [24], ultrasonic-assisted extraction [25], and microwave-assisted extraction [26]. In the present work, propolis is used as a model to systematically compare the partition efficiencies of phenolic compounds in four typical acetonitrile–water-based HLLE methods and discuss more details between the partition efficiencies and polarities of target phenolic compounds. To the best of our knowledge, this is the first report on the investigation of HLLE in propolis.

2. Results and Discussion

2.1. Salting-Out Assisted Liquid–Liquid Extraction

Seven typical phenolic compounds observed in propolis, including caffeic acid, *p*-coumaric acid, isoferulic acid, dimethoxycinnamic acid, cinnamic acid, kaempferol, and caffeic acid phenethyl ester (CAPE), were well-separated in reversed-phase HPLC, as shown in Figure 1. The chromatogram of standards is shown in Figure S1. Chromatographic peaks were identified based on the UV absorption spectra and the retention time compared to the standards. Effects of salt concentration on the partitioning of the phenolic compounds in the ACN–H_2O mixture (50:50, v/v) are shown in Figure 2. To illustrate the correlation between partition efficiency and polarity, extraction yields (EYs) in the upper phase are plotted against the LogD (distribution coefficient) value of the estimated compounds, which were collected from ChemSpider [27] and shown in Table 1. Since the extraction solution was in the neutral pH, the value of LogD in pH 7.4 was selected. For NaCl-based SALLE, shown in Figure 2a, EYs of phenolic compounds increased as the salt concentration increased from 25 to 125 g/L. The detailed trends are shown in Figure S2a. When the concentration of NaCl was 25 g/L, EYs of the phenolic compounds were between 29.5% and 82.8%. Then, the EYs were dramatically raised to the range between 43.6% and 95.8% under the NaCl concentration of 50 g/L. As the salt concentration further increased to 125 g/L, slight growth in EYs was observed. Additionally, trends of increasing EYs with the increase of LogD values were observed at each salt concentration. The maximum EYs were found in the least polar compound (CAPE), while the minimum EYs were observed in the most polar compound (caffeic acid).

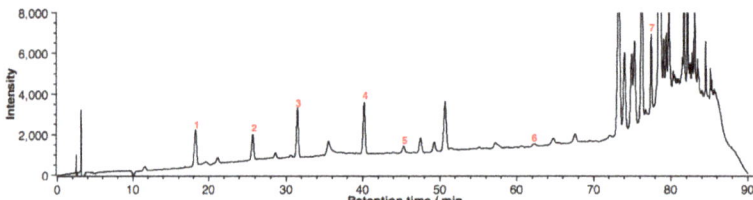

Figure 1. Representative HPLC-DAD chromatogram (λ = 280 nm) of phenolic compounds in propolis. 1, caffeic acid; 2, *p*-coumaric acid; 3, isoferulic acid; 4, dimethoxycinnamic acid; 5, cinnamic acid; 6, kaempferol; 7, caffeic acid phenethyl ester.

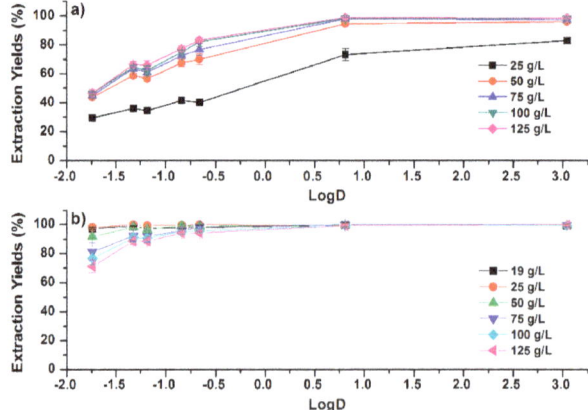

Figure 2. Correlations between extraction yields (EYs) and the LogD (pH 7.4) values of phenolic compounds under different addition amounts of (**a**) NaCl and (**b**) $MgSO_4$ in acetonitrile–water (ACN–H_2O) mixture (50:50, v/v). Error bars present the standard deviation (n = 3).

Table 1. LogD (pH 7.4) values of the phenolic compounds.

Compounds	LogD (pH 7.4) [a]
Caffeic acid	−1.74
p-Coumaric acid	−1.32
Isoferulic acid	−1.18
Dimethoxycinnamic acid	−0.84
Cinnamic acid	−0.66
Kaempferol	0.81
Caffeic acid phenethyl ester	3.04

[a] Data were collected from [27].

Compared with NaCl, lower salt concentrations of MgSO$_4$ are required to trigger the phase separation. In addition, higher EYs are achieved for all the estimated phenolic compounds in MgSO$_4$-based SALLE. As shown in Figure 2b, EYs of the phenolic compounds were all larger than 97% when the concentration of MgSO$_4$ was 19 g/L in ACN–H$_2$O mixture (50:50, v/v). This higher partition efficiency of MgSO$_4$ compared with NaCl was similar with reports on the SALLE of dicarbonyl [28] and fatty acid [29] compounds, which can be attributed to the larger phase ratio in MgSO$_4$-based SALLE than that in NaCl-based SALLE. As shown in Figure 2, EYs of more-polar phenolic compounds with LogD < 0 (caffeic acid, p-coumaric acid, isoferulic acid, dimethoxycinnamic acid, and cinnamic acid) in NaCl-based SALLE are lower than those in MgSO$_4$. However, EYs of the less-polar phenolic compounds with logD > 0 (kaempferol and CAPE) in NaCl-based SALLE are similar with those in MgSO$_4$-based SALLE. This means that though the phase ratio is lower in NaCl-based SALLE [28], EYs of less-polar compounds are comparable to the high phase ratio conditions of MgSO$_4$-based SALLE. Additionally, when the concentration of MgSO$_4$ was increased to 125 g/L, the volume of the upper phase was decreased. This decrease of phase ratio with the increase in MgSO$_4$ concentration is consistent with the results reported by Valente et al. [28]. Consequently, the decrease in the EY of the most polar compound (caffeic acid) was from 98.0% to 71.2%, while for the least polar compound (CAPE), EY varied between 99.3% and 99.8% (Figure S2b). These observations indicate that the influence of the phase ratio on partition efficiency is more significant when the polarity of the estimated compounds is higher. It also implies that EYs of polar compounds may be improved by increasing the phase ratio.

Increasing the initial concentration of ACN in ACN–H$_2$O mixtures can lead to the increase of phase ratio [29]. Consequently, the improvement of EYs for more-polar compounds were observed. For NaCl-based SALLE, as shown in Figure 3a, when the ACN concentration in the ACN–H$_2$O mixture was increased to 70:30 (ACN:H$_2$O, v/v), the EYs of phenolic compounds were all higher than 81.8%. In addition, the influence of initial ACN concentration on EYs is more significant for more-polar phenolic compounds than for less-polar phenolic compounds. For instance, as the initial concentration of ACN increased from 40:60 to 70:30 (ACN:H$_2$O, v/v), EYs of caffeic acid were increased from 21.0% to 81.8%, while for CAPE, EYs were increased from 88.4% to 99.9%. In MgSO$_4$-based SALLE, as shown in Figure 3b, the influence of the initial concentration of ACN on EYs is much lower than that in NaCl-based SALLE. As the concentration of ACN increased from 40:60 to 70:30, EYs of caffeic acid and CAPE in MgSO$_4$-based SALLE varied from 79.5% to 81.2% and from 98.1% to 99.8%, respectively.

From these above experiments, it becomes clear that MgSO$_4$-based SALLE with low salt concentration may be used as a general HLLE method for compounds with a wide range of polarities because of the high EYs achieved (>98%) for all the investigated compounds with LogD ≥ −1.74. NaCl-based SALLE with an ACN concentration of 70:30 in the ACN–H$_2$O mixture (ACN:H$_2$O, v/v) may also be a general extraction method, as the EYs were not less than 90% for investigated compounds with LogD ≥ −1.18. If the multi-analyte compounds are less polar, NaCl-based SALLE with a salt concentration of 50 g/L and an ACN concentration of 50:50 would be a suitable choice. In this condition, EYs were larger than 94% for the investigated compounds with LogD ≥ 0.81. These EYs values are comparable with MgSO$_4$-based SALLE. Furthermore, the volume of the upper phase is lower than that in MgSO$_4$-based SALLE. This could be helpful for enhancing the sensitivity of analysis, and the lower EYs towards polar compounds in this condition would reduce the co-extraction of high-polarity interferences.

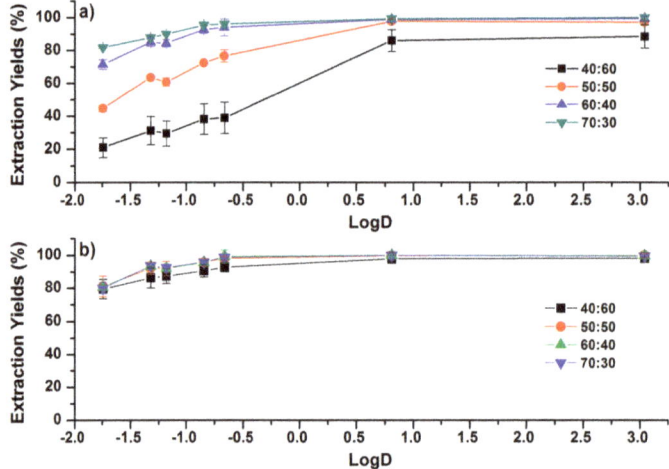

Figure 3. Correlations between EYs and the LogD (pH 7.4) values of phenolic compounds under different concentration of acetonitrile (ACN) in the ACN–H$_2$O mixture (ACN:H$_2$O, v/v) for (**a**) NaCl- and (**b**) MgSO$_4$-based salting-out assisted liquid–liquid extraction (SALLE). The addition amounts of salts were 75 g/L. Error bars present the standard deviation (n = 3).

2.2. Sugaring-Out Assisted Liquid–Liquid Extraction

Sugars, including glucose, fructose, and sucrose etc., have been reported to trigger the phase separation of ACN–H$_2$O mixtures [14]. Glucose was chosen in this study because it has been demonstrated to be the better phase separation agent than other sugars [10]. Effects of glucose concentration on the partition efficiencies of phenolic compounds in ACN–H$_2$O mixtures (50:50, v/v) are shown in Figure 4a. Results indicated that EYs were increased as more glucose was introduced. The required amount of glucose to trigger the phase separation is larger than the required amount of NaCl and MgSO$_4$. Increasing the concentration of glucose from 125 to 225 g/L led to the increase of EYs, and the obtained maximum values of EYs ranged from 45.6% (caffeic acid) to 86.1% (CAPE). The detailed trends are shown in Figure S3. In addition, EYs of the phenolic compounds displayed the trends of increasing as the value of LogD increased at each glucose concentration, which are similar with the trends in NaCl-based SALLE.

Effects of the initial concentration of ACN on EYs are shown in Figure 4b. Increasing the concentration of ACN in ACN–H$_2$O mixtures (ACN:H$_2$O, v/v) from 50:50 to 70:30 significantly increased the EYs of phenolic compounds, and the increment appeared to decrease as the polarity of the compounds became less polar. For instance, EYs of the most polar compound (cafferic acid) were increased from 45.5% to 89.6%, while EYs of the least polar compound (CAPE) were increased from 85.8% to 99.7%. When the concentration of ACN was 70:30 (ACN:H$_2$O, v/v), EYs were in the range of 90% (cafferic acid) to 99.7% (CAPE). The obtained EYs for more-polar compounds in SULLE are higher than NaCl-based SALLE, and are comparable with MgSO$_4$-based SALLE. This could be attributed to the high phase ratio under the condition of high ACN concentration in SULLE [29].

It is important to note that SULLE may work as a general method under the ACN concentration of 70:30 (ACN:H$_2$O, v/v) with a glucose concentration of 200 g/L. In this condition, investigated phenolic compounds with LogD ≥ -1.74 could be partitioned into the upper phase, with EYs not less than

90%. SULLE in the high ACN concentration may be used as an alternative method for MgSO$_4$-based SALLE, as the volume of the upper phase and the obtained EYs are comparable with MgSO$_4$.

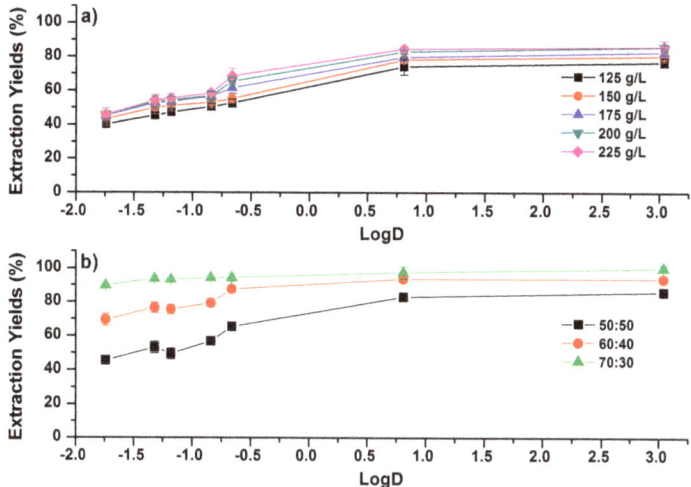

Figure 4. Correlations between EYs and the LogD (pH 7.4) values of phenolic compounds (**a**) under different addition amounts of glucose in ACN–H$_2$O mixtures (50:50, v/v); (**b**) under different concentrations of ACN in the ACN–H$_2$O mixture (ACN:H$_2$O, v/v) with the addition of 200 g/L glucose. Error bars present the standard deviation ($n = 3$).

2.3. Hydrophobic-Solvent Assisted Liquid–Liquid Extraction

In HSLLE, two typical hydrophobic solvents, dichloromethane (DCM) and chloroform, were studied. The volume of DCM required to trigger the phase separation is larger than that of chloroform. The investigated volumes for the HSLLE were in the range of 200 to 500 μL and 60 to 300 μL for DCM and chloroform, respectively. As shown in Figure 5a, EYs of the more-polar phenolic compounds were increased with the introduction of more DCM into the ACN–H$_2$O mixture (50:50, v/v), whereas EYs of the less-polar phenolic compounds were slightly varied, with values larger than 92%. Furthermore, the influence of solvent volume on EYs was more significant for chloroform than for DCM. As the volume of chloroform increased from 60 to 300 μL, EYs of cafferic acid were increased from 11.0% to 29.9%, and EYs of CAPE were increased from 29.6% to 95.0%, as shown in Figure 5b. This observation might be attributed to the significant increase of the upper phase volume when more chloroform is introduced [11]. It is interesting to find that EYs of more-polar phenolic compounds are much lower than those of less-polar phenolic compounds under the introduction of DCM or chloroform. This means that HSLLE may be used for the selective extraction of less-polar phenolic compounds.

Increasing the initial concentration of ACN results in the increase of EYs in both DCM and chloroform HSLLEs, as shown in Figure 6. The increments of EYs were dramatic for more-polar phenolic compounds, but the increment reduced as the polarity of the compounds decreased. As the concentration of ACN (ACN:H$_2$O, v/v) increased from 30:70 to 70:30 in DCM-based HSLLE, as shown in Figure 6a, EYs of caffeic acid and CAPE increased from 25.0% to 77.8% and from 91.0% to 99.0%, respectively. In chloroform-based HSLLE, shown in Figure 6b, EYs of caffeic acid and CAPE increased from 30.0% to 71.4% and 95.0% to 99.2%, respectively, when the concentration of ACN (ACN:H$_2$O, v/v) increased from 50:50 to 70:30. Therefore, EYs of the more-polar phenolic compounds display more sensitivity to the increase of the ACN concentration than less-polar phenolic compounds, which has also been found in the above results of SALLE and SULLE.

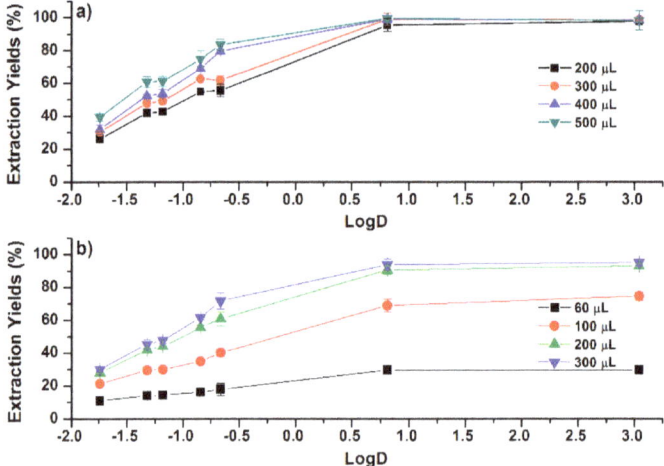

Figure 5. Correlations between EYs and the LogD (pH 7.4) values of phenolic compounds under different addition volumes of (**a**) dichloromethane (DCM) and (**b**) chloroform in ACN–H$_2$O mixtures (50:50, v/v). Error bars present the standard deviation (n = 3).

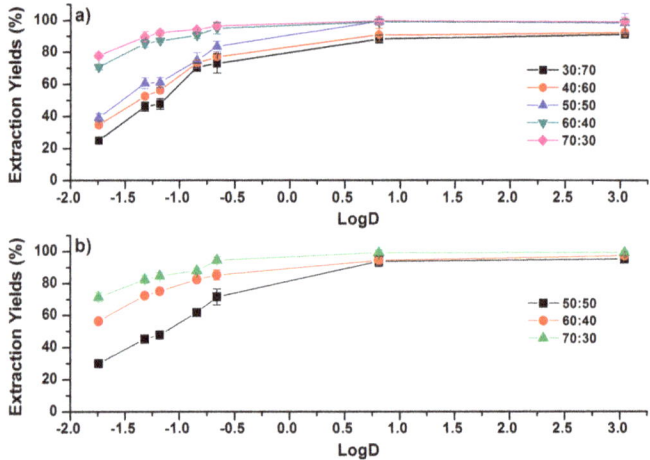

Figure 6. Correlations between EYs and the LogD (pH 7.4) values of phenolic compounds under different concentration of ACN in the ACN–H$_2$O mixture (ACN:H$_2$O, v/v) for (**a**) DCM and (**b**) chloroform. The addition volumes of DCM and chloroform were 25% and 15% of the initial volume of ACN in the ACN–H$_2$O mixture, respectively. Error bars present the standard deviation (n = 3).

Compared with SULLE and SALLE, HSLLE shows the better selective extraction of less-polar phenolic compounds. With the addition of 200 µL DCM and chloroform, EYs of the investigated compounds with LogD > 0.81 in the ACN–H$_2$O mixture (50:50, v/v) were higher than 95% and 90%, respectively. Furthermore, EYs of polar compounds are much lower, and thus the co-extraction of high-polarity interference compounds might be dramatically reduced.

2.4. Subzero-Temperature Assisted Liquid–Liquid Extraction

In STLLE, the extraction solution of ACN–H$_2$O mixture was cooled at a low temperature (-20 °C) to induce the phase separation. The cooling time significantly influences the partition performance [12].

As shown in Figure 7a, EYs of the phenolic compounds increased with the extending of cooling time. The phase separation began at the cooling time of 30 min. The volume of the upper phase was increased as the cooling time extended, and the lower aqueous phase was nearly frozen at 60 min. Consequently, EYs were increased with the extending of cooling time, and reached the plateau at 60 min. In addition, increments of EYs in less-polar phenolic compounds were larger than those of more-polar phenolic compounds. For example, EYs of cafferic acid and CAPE were increased from 30.4% to 44.8% and 62.8% to 97.0%, respectively. Moreover, STLLE showed the selective partitioning of less-polar phenolic compounds. When the phase separation was performed in the ACN–H$_2$O mixture (50:50, v/v) under 60 min cooling, EYs of more-polar compounds were under 61%, while the EYs of investigated compounds with LogD $\geq$ 0.81 were higher than 96%.

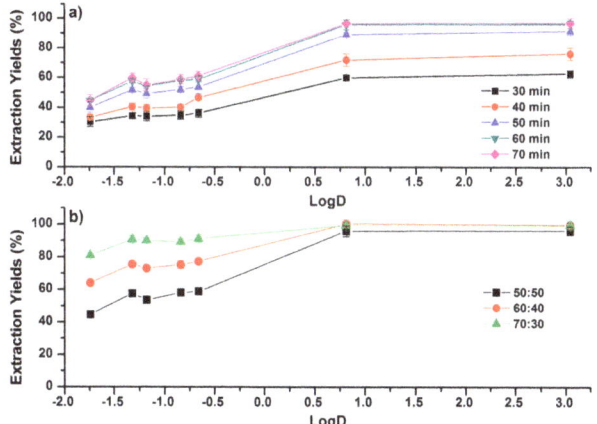

Figure 7. Correlations between EYs and the LogD (pH 7.4) values of phenolic compounds (**a**) under different cooling times in ACN–H$_2$O mixture (50:50, v/v); (**b**) under different concentrations of ACN in the ACN–H$_2$O mixture (ACN:H$_2$O, v/v) with a cooling time of 60 min. Error bars present the standard deviation (n = 3).

Increasing the initial concentration of ACN in the ACN–H$_2$O mixture also results in the increase of EYs in STLLE, as shown in Figure 7b. The influences of ACN concentration on EYs are more significant in more-polar phenolic compounds than in less-polar phenolic compounds. As the concentration of ACN increased from 50:50 to 70:30 (ACN:H$_2$O, v/v), EYs of caffeic acid and CAPE increased from 44.7% to 80.8% and from 96.1% to 99.3%, respectively. Additionally, EYs of the investigated compounds were all increased to be larger than 80.8% when the initial concentration of ACN was 70:30 (ACN:H$_2$O, v/v). This observation implies that STLLE, at high initial concentrations of ACN, may be applied for the extraction of wide-polarity multi-analyte compounds with high EYs. The drawback of STLLE might be the relatively long time for the complete phase separation, but it shows the advantages of having a simple procedure, easy collection of the upper phase, and the elimination of additional phase separation agents.

3. Materials and Methods

3.1. Materials

Methanol and ACN with HPLC grades were obtained from Merck (Darmstadt, Germany). Standards of caffeic acid, *p*-coumaric acid, isoferulic acid, dimethoxycinnamic acid, cinnamic acid, kaempferol, and caffeic acid phenethyl ester (CAPE) were purchased from Aladdin (Shanghai, China). Anhydrous magnesium sulfate, sodium chloride, glucose, anhydrous ethanol, dichloromethane (DCM), and chloroform were all of analytical grade and obtained from Sinopharm Chemical Reagent Co., Ltd

(Shanghai, China). Ultrapure water (18.2 MΩ) was used throughout this article. Raw propolis was collected from Hubei, China.

3.2. Homogeneous Liquid–Liquid Extraction

Raw propolis was purified according to the reported method to remove the insoluble subjects and wax compounds [30]. Raw propolis was frozen and ground prior to the extraction. The ground samples were extracted by maceration for 7 days at room temperature, with 10 mL anhydrous ethanol for every 3 g of raw propolis. The insoluble subjects were separated by filtration. The filtrates were frozen overnight and filtered again to remove the wax compounds. Solvent was then evaporated on a water bath at 50 °C to obtain dry extracts of propolis. Purified propolis (100 mg) was weighted into the 100 mL volumetric flask and diluted to the volume with different ACN–water mixtures, and sonicated for 20 min. For SALLE, SULLE, and HSLLE, this propolis solution (4 mL) was added with different amounts of phase separation agents (NaCl and $MgSO_4$ for SALLE, glucose for SULLE, and DCM and chloroform for HSLLE) and then vortexed for 1 min. The mixed solution was centrifuged at 6000 rpm for 5 min to make a clear phase separation. The upper phase was collected by a micro-syringe and then diluted to 25 mL with anhydrous ethanol. For STLLE, the propolis solution (4 mL) was cooled at −20 °C for different times. The upper phase was repeatedly collected and transferred into a 25 mL volumetric flask by micro-syringe (100 μL), and then diluted to 25 mL with anhydrous ethanol. This dilution procedure was performed to make all the collected upper phases into the same volume. These final extractive solutions were analyzed by HPLC. Another propolis solution (4 mL) without phase separation was diluted to 25 mL with anhydrous ethanol. This control solution was also analyzed by HPLC and used for the calculation of extraction yields. All experiments were triplicates.

3.3. HPLC Analysis

Phenolic compounds were analyzed based on the reversed-phase HPLC method reported by Zhang et al. [31]. The HPLC system (Shimadzu, Kyoto, Japan) was composed of LC-20AT pumps, a SIL-20AC autosampler, a CTO-20AC column oven, and a SPD-M20A photodiode array detector (190~800 nm). A Wonda Cract (Shimadzu-GL) C18 column (5 μm, 4.6 × 150 mm) was used for the separation. The mobile phase A was 0.1% aqueous acetic acid solution (v/v) and the mobile phase B was methanol. Gradient elution was as follows: 15–40% B at 0–30 min, 40–55% B at 30–65 min, 55–62% B at 65–70 min, 62–100% B at 70–80 min, 100–15% at 80–85 min, and stayed at 15% for 5 min. The flow rate was 0.8 mL/min, the injection volume was 10 μL, and the column temperature was 35 °C. Standards were used for the identification of chromatographic peaks. The partition efficiency was compared by extraction yields (EYs, %) = (peak areas of target compounds in the HLLE extractive solution/peak areas of target compounds in the control solution) × 100. A chromatogram at 280 nm was used for the calculation of peak area.

4. Conclusions

In summary, we present the first report of HLLE in propolis. Partitioning of seven phenolic compounds in the ACN–H_2O mixture, triggered by four HLLE methods, were investigated. The partition efficiencies were found to be correlated with the polarity of the target compounds. The less-polar phenolic compounds (kaempferol and caffeic acid phenethyl ester) could be highly efficiently partitioned into the upper ACN phase in all four investigated HLLE methods. For more-polar phenolic compounds (caffeic acid, *p*-coumaric acid, isoferulic acid, dimethoxycinnamic acid, and cinnamic acid), increasing the initial concentration of ACN in the ACN–H_2O mixture is suggested for archiving higher EYs. This study has also suggested that $MgSO_4$-based SALLE under the ACN concentration of 50:50 (ACN:H_2O, v/v), together with NaCl-based SALLE, SULLE, and STLLE under the ACN concentration of 70:30 (ACN:H_2O, v/v) might be used as general HLLE methods for the extraction of multiple phenolic compounds with a wide range of polarities. Additionally, NaCl-based SALLE, HSLLE, and STLLE with an ACN concentration of 50:50 (ACN:H_2O, v/v) might be used for the selective extraction of low-polarity phenolic compounds. These observations show a better understanding of the partitioning of multiple phenolic compounds in HLLE methods, and would be valuable for the development of sample preparation protocol in phytoanalysis.

Supplementary Materials: The following are available online: Figure S1: Representative HPLC-DAD chromatogram (λ = 280 nm) of phenolic standards. 1, caffeic acid; 2, *p*-coumaric acid; 3, isoferulic acid; 4, dimethoxycinnamic acid; 5, cinnamic acid; 6, kaempferol; 7, caffeic acid phenethyl ester. Figure S2: Extraction yields of investigated phenolic compounds under different addition amounts of (**a**) NaCl and (**b**) MgSO$_4$ in ACN–H$_2$O mixture (50:50, v/v). Error bars present the standard deviation (n = 3). Figure S3: Extraction yields of investigated phenolic compounds under different addition amounts of glucose in ACN–H$_2$O mixture (50:50, v/v). Error bars present the standard deviation (n = 3).

Author Contributions: W.C. conceived the study. X.T. and W.C. designed the experiments, analyzed the data, and wrote the manuscript. D.W., Z.G., and S.W. participated in the experiments. S.H. discussed and commented on the manuscript. All authors read and approved the final manuscript.

Funding: This research was funded by Natural Science Foundation of China, grant number 31201861 and 51202030.

Conflicts of Interest: The authors declare no conflict of interest.

References

1. Bijttebier, S.; der Auwera, A.V.; Foubert, K.; Voorspoels, S.; Pieters, L.; Apers, S. Bridging the gap between comprehensive extraction protocols in plant metabolomics studies and method validation. *Anal. Chim. Acta* **2016**, *935*, 136–150. [CrossRef] [PubMed]
2. Naz, S.; dos Santos, D.C.M.; Garcia, A.; Barbas, C. Analytical protocols based on LC-MS; GC-MS and CE-MS for nontargeted metabolomics of biological tissues. *Bioanalysis* **2014**, *6*, 1657–1677. [CrossRef] [PubMed]
3. Villaverde, J.J.; Sevilla-Morán, B.; López-Goti, C.; Alonso-Prados, J.L.; Sandín-España, P. Trends in analysis of pesticide residues to fulfil the European Regulation (EC) No. 1107/2009. *Trends Anal. Chem.* **2016**, *80*, 568–580. [CrossRef]
4. Herrero, M.; Simó, C.; García-Cañas, V.; Ibáñez, E.; Cifuentes, A. Foodomics: MS-based strategies in modern food science and nutrition. *Mass Spectrom. Rev.* **2012**, *31*, 49–69. [CrossRef] [PubMed]
5. Zhu, M.Z.; Chen, G.L.; Wu, J.L.; Li, N.; Liu, Z.H.; Guo, M.Q. Recent development in mass spectrometry and its hyphenated techniques for the analysis of medicinal plants. *Phytochem. Anal.* **2018**, *29*, 365–374. [CrossRef] [PubMed]
6. Causon, T.J.; Hann, S. Review of sample preparation strategies for MS-based metabolomic studies in industrial biotechnology. *Anal. Chim. Acta* **2016**, *938*, 18–32. [CrossRef] [PubMed]
7. Anthemidis, A.N.; Ioannou, K.-I.G. Recent developments in homogeneous and dispersive liquid-liquid extraction for inorganic elements determination. A review. *Talanta* **2009**, *80*, 413–421. [CrossRef]
8. Valente, I.M.; Rodrigues, J.A. Recent advances in salt-assisted LLE for analyzing biological samples. *Bioanalysis* **2015**, *7*, 2187–2193. [CrossRef]
9. Farajzadeh, M.A.; Mohebbi, A.; Feriduni, B. Development of a simple and efficient pretreatment technique named pH-dependent continuous homogenous liquid–liquid extraction. *Anal. Methods* **2016**, *8*, 5676–5683. [CrossRef]
10. Wang, B.; Ezejias, T.; Feng, H.; Blaschek, H. Sugaring-out: A novel phase separation and extraction system. *Chem. Eng. Sci.* **2008**, *63*, 2595–2600. [CrossRef]
11. Liu, G.Z.; Zhou, N.Y.; Zhang, M.S.; Li, S.J.; Tian, Q.Q.; Chen, J.T.; Chen, B.; Wu, Y.N.; Yao, S.Z. Hydrophobic solvent induced phase transition extraction to extract drugs from plasma for high performance liquid chromatography-mass spectrometric analysis. *J. Chromatogr. A* **2010**, *1217*, 243–249. [CrossRef] [PubMed]
12. Yoshida, M.; Akane, A. Subzero-temperature liquid-liquid extraction of benzodiazepines for high performance liquid chromatography. *Anal. Chem.* **1999**, *71*, 1918–1921. [CrossRef] [PubMed]
13. Valente, I.M.; Moreira, M.M.; Neves, P.; da Fé, T.; Gonçalves, L.M.; Almeida, P.J.; Rodrigues, J.A. An insight on salting-out assisted liquid-liquid extraction for phytoanalysis. *Phytochem. Anal.* **2017**, *28*, 297–304. [CrossRef] [PubMed]
14. Wang, B.; Feng, H.; Ezeji, T.; Blaschek, H. Sugaring-out separation of acetonitrile from its aqueous solution. *Chem. Eng. Technol.* **2008**, *31*, 1869–1874. [CrossRef]

15. Tsai, W.H.; Chuang, H.Y.; Chen, H.H.; Wu, Y.W.; Cheng, S.H.; Huang, T.C. Application of sugaring-out extraction for the determination of sulfonamides in honey by high-performance liquid chromatography with fluorescence detection. *J. Chromatogr. A* **2010**, *1217*, 7812–7815. [CrossRef] [PubMed]
16. Zhang, J.; Myasein, F.; Wu, H.Q.; El-Shourbagy, T.A. Sugaring-out assisted liquid/liquid extraction with acetonitrile for bioanalysis using liquid chromatography-mass spectrometry. *Microchem. J.* **2013**, *108*, 198–202. [CrossRef]
17. Cai, B.D.; Ye, E.C.; Yuan, B.F.; Feng, Y.Q. Sequential solvent induced phase transition extraction for profiling of endogenous phytohormones in plants by liquid chromatography-mass spectrometry. *J. Chromatogr. B* **2015**, *1004*, 23–29. [CrossRef]
18. Zhang, H.Y.; Li, S.S.; Liu, X.Z.; Yuan, F.F.; Liang, Y.H.; Shi, Z.H. Determination of five anthraquinone derivatives in sticky traditional Chinese patent medicines by subzero-temperature liquid-liquid extraction combined with high-performance liquid chromatography. *J. Liq. Chromatogr. Relat. Technol.* **2015**, *38*, 584–590. [CrossRef]
19. Torbati, M.; Farajzadeh, M.A.; Torbati, M.; Nabil, A.A.A.; Mohebbi, A.; Mogaddam, M.R.A. Development of salt and pH–induced solidified floating organic droplets homogeneous liquid–liquid microextraction for extraction of ten pyrethroid insecticides in fresh fruits and fruit juices followed by gas chromatography mass spectrometry. *Talanta* **2018**, *176*, 565–572. [CrossRef]
20. Farajzadeh, M.A.; Mohebbi, A.; Mogadam, M.R.A.; Davaran, M.; Norouzi, M. Development of salt-induced homogenous liquid-liquid microextraction based on iso-propanol/sodium sulfate system for extraction of some pesticides in fruit juices. *Food Anal. Methods* **2018**, *11*, 2497–2507. [CrossRef]
21. Bravo, L. Polyphenols: Chemistry, dietary sources, metabolism, and nutritional significance. *Nutr. Rev.* **1998**, *56*, 317–333. [CrossRef] [PubMed]
22. Salatino, A.; Fernandes-Silva, C.C.; Gighi, A.A.; Salatino, M.L.F. Propolis research and the chemistry of plant products. *Nat. Prod. Rep.* **2011**, *28*, 925–936. [CrossRef] [PubMed]
23. Falcão, S.I.; Vale, N.; Gomes, P.; Domingues, M.R.M.; Freire, C.; Cardoso, S.M.; Vilas-Boas, M. Phenolic profiling of Portuguese propolis by LC–MS spectrometry: Uncommon propolis rich in flavonoid glycosides. *Phytochem. Anal.* **2013**, *24*, 309–318. [CrossRef] [PubMed]
24. Papotti, G.; Bertelli, D.; Bortolotti, L.; Plessi, M. Chemical and functional characterization of Italian propolis obtained by different harvesting methods. *J. Agric. Food Chem.* **2012**, *60*, 2852–2862. [CrossRef] [PubMed]
25. Sun, C.L.; Wu, Z.S.; Wang, Z.Y.; Zhang, H.C. Effect of ethanol/water solvents on phenolic profiles and antioxidant properties of Beijing propolis extracts. *Evid-Based Compl. Alt.* **2015**, *2015*, 595393. [CrossRef] [PubMed]
26. Hamzah, N.; Leo, C.P. Microwave-assisted extraction of *Trigona* propolis: The effects of processing parameters. *Int. J. Food Eng.* **2015**, *11*, 861–870. [CrossRef]
27. ChemSpider. Available online: http://www.chemspider.com (accessed on 3 January 2018).
28. Valente, I.M.; Goncalves, L.M.; Rodrigues, J.A. Another glimpse over the salting-out assisted liquid-liquid extraction in acetonitrile/water mixtures. *J. Chromatogr. A* **2013**, *1308*, 58–62. [CrossRef]
29. Tu, X.J.; Sun, F.Y.; Wu, S.Y.; Liu, W.Y.; Gao, Z.S.; Huang, S.K.; Chen, W.B. Comparison of salting-out and sugaring-out liquid-liquid extraction methods for the partition of 10-hydroxy-2-decenoic acid in royal jelly and their co-extracted protein content. *J. Chromatogr. B* **2018**, *1073*, 90–95. [CrossRef]
30. Sawaya, A.C.H.F.; da Silva Cunha, I.B.; Marcucci, M.C.; Aidar, D.S.; Silva, E.C.A.; Carvalho, C.A.L.; Eberlin, M.N. Electrospray ionization mass spectrometry fingerprinting of propolis of native Brazilian stingless bees. *Apidologie* **2007**, *38*, 93–103. [CrossRef]
31. Zhang, C.P.; Huang, S.; Wei, W.T.; Ping, S.; Shen, X.G.; Li, Y.J.; Hu, F.L. Development of high-performance liquid chromatographic for quality and authenticity control of Chinese propolis. *J. Food Sci.* **2014**, *79*, C1315–C1322.

Sample Availability: Not available.

© 2019 by the authors. Licensee MDPI, Basel, Switzerland. This article is an open access article distributed under the terms and conditions of the Creative Commons Attribution (CC BY) license (http://creativecommons.org/licenses/by/4.0/).

Article

Optimization of Vortex-Assisted Dispersive Liquid-Liquid Microextraction for the Simultaneous Quantitation of Eleven Non-Anthocyanin Polyphenols in Commercial Blueberry Using the Multi-Objective Response Surface Methodology and Desirability Function Approach

Ying Xue [1,2,3], Xian-Shun Xu [3], Li Yong [3], Bin Hu [3], Xing-De Li [2], Shi-Hong Zhong [1], Yi Li [1], Jing Xie [1,*] and Lin-Sen Qing [1,2,*]

1. School of Pharmacy, Collaborative Innovation Center of Sichuan for Elderly Care and Health, Chengdu Medical College, Chengdu 610500, China; xuecher0221@sina.com (Y.X.); saraca1980@126.com (S.-H.Z.); lychengdu@aliyun.com (Y.L.)
2. Chengdu Institute of Biology, Chinese Academy of Sciences, Chengdu 610041, China; li_xingde1975@163.com
3. Sichuan Provincial Center for Disease Control and Prevention, Chengdu 610041, China; xuxianshun@163.com (X.-S.X.); yongch121@163.com (L.Y.); hubin1258@163.com (B.H.)
* Correspondence: aggie-xj@163.com (J.X.); qingls@cib.ac.cn (L.-S.Q.); Tel.: +86-28-62308658 (J.X.); +86-28-82890640 (L.-S.Q.)

Academic Editor: Nuno Neng
Received: 29 September 2018; Accepted: 7 November 2018; Published: 9 November 2018

Abstract: In the present study, 11 non-anthocyanin polyphenols, gallic acid, protocatechuate, vanillic acid, syringic acid, ferulic acid, quercetin, catechin, epicatechin, epigallocatechin gallate, gallocatechin gallate and epicatechin gallate—were firstly screened and identified from blueberries using an ultra performance liquid chromatography–time of flight mass spectrography (UPLC-TOF/MS) method. Then, a sample preparation method was developed based on vortex-assisted dispersive liquid-liquid microextraction. The microextraction conditions, including the amount of ethyl acetate, the amount of acetonitrile and the solution pH, were optimized through the multi-objective response surface methodology and desirability function approach. Finally, an ultra performance liquid chromatography–triple quadrupole mass spectrography (UPLC-QqQ/MS) method was developed to determine the 11 non-anthocyanin polyphenols in 25 commercial blueberry samples from Sichuan province and Chongqing city. The results show that this new method with high accuracy, good precision and simple operation characteristics, can be used to determine non-anthocyanin polyphenols in blueberries and is expected to be used in the analysis of other fruits and vegetables.

Keywords: blueberry; non-anthocyanin polyphenol; vortex-assisted dispersive liquid-liquid microextraction; response surface methodology; desirability function approach

1. Introduction

Blueberry, edible as a fresh fruit, can also be processed into various foods such as blueberry-pizza, blueberry-cake, jam and beverage and so forth. Increasing evidence shows that blueberry and food rich in blueberry have significant benefits to human health [1–4]. It is generally believed that these benefits derive from rich polyphenols in blueberry [5,6]. At present, the research on blueberry polyphenols is mainly focused on anthocyanins but a few studies have focused on non-anthocyanin polyphenols [7–11], especially in commercial samples. Non-anthocyanin polyphenols are important components in commercial blueberry fruits and contribute to their characteristic flavor. Therefore, in order to fully evaluate the chemical constituents in commercial blueberries, it is necessary to comprehensively study the composition and content of non-anthocyanin polyphenols.

At present, the methods of sample preparation for the determination of polyphenols in food are mainly liquid-liquid extraction (LLE) and solid-phase extraction (SPE) [12,13]. However, LLE and SPE have some inherent disadvantages. For example, LLE requires a large number of toxic solvents and is time-consuming and the SPE materials are expensive. In recent years, the development of dispersive liquid-liquid microextraction (DLLME) can make up the deficiencies of LLE and SPE [14,15]. The principle of DLLME is to form a ternary solvent system using extraction solvent (organic solvent), dispersant (hydrophilic solvent) and sample aqueous solution. The huge contact area allows the emulsification equilibrium and a large number of small droplets to be rapidly formed. Subsequently, the droplets are polymerized into individual organic phases by centrifugation, thereby achieving extraction/enrichment of the target analyte. This method has such advantages as less solvent consumption, simple operation, high enrichment factor and time-saving and so forth. In order to speed up the mass transfer process and shorten the equilibration time, some auxiliary emulsification methods such as vortex-assisted [14], ultrasound-assisted [16], microwave-assisted [17] and air-assisted [18] were also applied to improve the performance of DLLME.

When using the DLLME method for sample preparation, each parameter should be optimized to achieve high-efficient extraction. Inevitably, it is a time-consuming and tedious work to optimize each variable one by one and examine the interaction between variables. Response surface methodology (RSM) can optimize multivariable parameters by reasonable experimental design [19]. Furthermore, combined with the desirability function (DF) approach, RSM could comprehensively optimize multiple parameters to effectively predict the overall best experimental conditions for multiple target analytes simultaneously [20].

In this study, 11 non-anthocyanin polyphenols from blueberry were identified by UPLC-TOF/MS and subsequently determined by UPLC-QqQ/MS using the vortex-assisted dispersive liquid-liquid microextraction (VA-DLLME) method for sample preparation. The optimization based on RSM and DF was performed on VA-DLLME variables including the amount of extraction solvent, the amount of dispersant and pH of the sample solution. Subsequently, UPLC-QqQ/MS was used to determine the content of 11 non-anthocyanin polyphenols in 25 commercial blueberry samples from Sichuan province and Chongqing city. As far as we know, this study is the first assay to describe a non-anthocyanin polyphenols determination method through UPLC-QqQ/MS based on RSM and DF experimental design and optimization, as well as its application in blueberry sample analysis.

2. Results and Discussion

2.1. Compounds Confirmation by UPLC-TOF/MS

Time-of-flight mass spectrometry (TOF/MS) has the advantages of high-quality accuracy, high sensitivity and high resolution. It can determine the exact mass of the compound, enabling qualitative structure identification in a complex matrix [21,22]. In this study, UPLC-TOF was used to scan polyphenolic compounds in negative ion mode. After referring to the literature on polyphenolic compounds in *Vaccinium* spp. and confirmation by authentic standards under the same chromatographic conditions, 11 non-anthocyanin polyphenols were identified from blueberries as GA (gallic acid, m/z 169.0143, 0.6 ppm $[M - H]^-$), PC (protocatechuate, m/z 153.0193, 0.1 ppm $[M - H]^-$), VA (vanillic acid, m/z 167.0350, -0.1 ppm $[M - H]^-$), SA (syringic acid, m/z 197.0456, 0.1 ppm $[M - H]^-$), FA (ferulic acid, m/z 193.0505, -0.5 ppm $[M - H]^-$), Que (quercetin, m/z 301.0352, -0.5 ppm $[M - H]^-$), C (catechin, m/z 289.0716, -0.7 ppm $[M - H]^-$), EC (epicatechin, m/z 289.0716, -0.7 ppm $[M - H]^-$), ECG (epigallocatechin gallate, m/z 441.0827, 0 ppm $[M - H]^-$),

GCG (gallocatechin gallate, m/z 457.0773, -0.8 ppm $[M - H]^-$) and EGCG (epicatechin gallate, m/z 457.0773, -0.7 ppm $[M - H]^-$), respectively.

2.2. Optimization of UPLC-QqQ/MS Conditions

All 11 polyphenols are acidic compounds. Therefore, the acid mobile phase could increase the separating degree, symmetry factor and the number of theoretical plates [23,24]. Considering the ion suppression induced by a high concentration of acid, 0.1% formic acid was finally added into the mobile phase. The performance of 11 polyphenols was compared on three columns of BEH C18 (2.1 mm × 100 mm, 1.7 μm), Shimadzu XR-ODS (2.0 mm × 100 mm, 1.8 μm) and Agilent ZORBAX SB-Aq (3.0 mm × 100 mm, 1.8 μm). The ZORBAX SB-Aq column exhibits a better separation of two pairs of isomers (EC and C, EGCG and ECG) and more symmetrical peak shapes for all of the 11 non-anthocyanin polyphenols can be obtained on it. The MS detection uses negative ion mode for acidic compounds. By optimizing the collision energy (CE) for collision-induced dissociation, two stable product ions with high sensitivity are selected for multiple reaction monitoring (Table S1). The representative mass spectra of blueberry samples are shown in Figure 1.

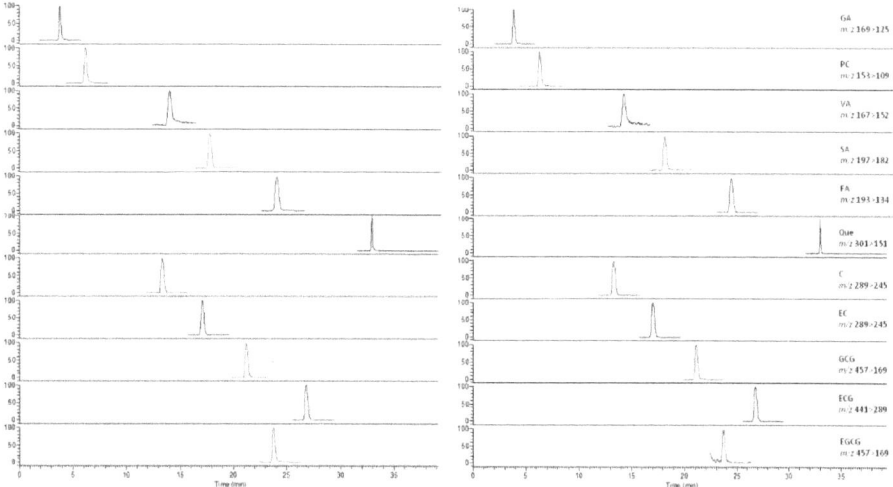

Figure 1. UPLC-MS total ion count chromatograms of 11 non-anthocyanin polyphenols ((**Left**), standard; (**Right**), commercial blueberry sample).

2.3. Optimization of DLLME Condition by Experimental Design

2.3.1. Single Factor Experimental Analysis

Since the extraction conditions have a crucial influence on the extraction results of VA-DLLME, it is essential to define the range of experimental variation and establish corresponding control methods. Therefore, it is necessary to conduct single-factor experiments to determine the variables, including the amount of EtOAc, the amount of ACN and the pH of the solution, which had significant influences on the response. In the present study, the recoveries of 11 non-anthocyanin polyphenols were assessed by means of changing a single variable and fixing two variables. The results are shown in Figure S1. Due to structural differences, the recoveries of the 11 non-anthocyanin polyphenols were different but the optimal values were relatively consistent: when the amount of EtOAc was about 1000 μL, the amount of ACN was about 500 μL and the pH was about 4, there was a high recovery rate. Based on this, the variables and their levels were fixed as shown in Table S2 for the following multi-factor optimization study.

2.3.2. Multi-Response Design and Analysis

Based on the results of single factor experiments, the optimization of VA-DLLME extraction conditions was performed based on Box-Behnken design with three factors and three levels. The experimental design and results are shown in Table 1. Among 11 non-anthocyanin polyphenols, syringic acid (SA) was randomly selected as an example to elucidate the results and the process of analysis. The quadratic regression model of SA was plotted based on response (Y_{SA}) versus variables (A, the amount of EtOAc; B, the amount of ACN; and C, the pH of the solution) as below:

$$Y_{SA} = 94.79 + 15.56A - 5.71B - 7.93C + 3.07AB - 1.95BC - 14.47A^2 - 10.33B^2 - 41.16C^2$$

The analysis of variance (ANOVA), goodness-of-fit and adequacy of the regression model are summarized in Table 2. The p-value in the F-test was used to evaluate the significance level of each variable. The model was considered statistically significant when $p < 0.05$ and was considered highly significant when $p < 0.01$. ANOVA showed that the model's $F = 826.42$ ($p < 0.0001$), indicating that the experimental error was minimal and the model had highly statistical significance and could be applied to data analysis. Lack of fit $F = 3.11$ ($p = 0.1511$) was not statistically significant, indicating that the experimental error has no significant effect on the model. The results of ANOVA also showed the statistical significance of items A, B and C, interaction items AB and BC and quadratic items A^2, B^2 and C^2 respectively, indicating that the effect of each variable on the response was not only linear but also interactive. The model verified by ANOVA could reflect the relationship between the SA recovery and the amount of EtOAc, the amount of ACN and the pH of the solution. The regression equation could be used to predict and optimize the VA-DLLME process of SA.

Response surfaces (three-dimensional) were plotted by Design Expert software (8.0.6) to illustrate the interactions of variables for the maximum response. The corresponding 3D response surfaces are shown in Figure 2. Each figure showed the effects of two factors on the SA recovery while the third factor was kept at zero level.

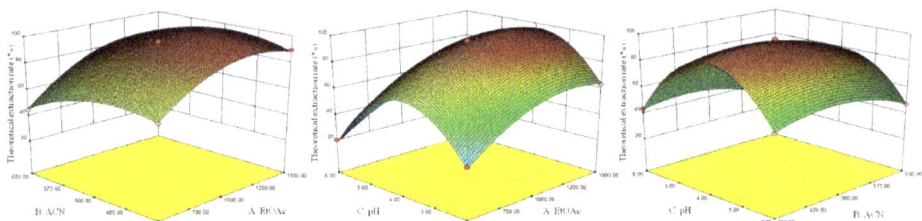

Figure 2. Response surface plots of syringic acid by response (%) versus significant variables (A, the amount of EtOAc; B, the amount of ACN and C, the pH of the solution).

Using the same method, the quadratic regression models of the remaining 10 non-anthocyanin polyphenols were plotted based on response (Y_i) versus variables (A, the amount of EtOAc; B, the amount of ACN and C, the pH of the solution) as below.

$Y_{GA} = 93.70 + 10.26A - 12.82C + 6.86AC - 9.86A^2 - 6.91B^2 - 33.64C^2$

$Y_{PC} = 91.76 + 9.82A - 13.48C + 2.24BC - 12.92A^2 - 34.65C^2$

$Y_{VA} = 94.65 + 5.13A - 4.58B - 5.81C + 6.52AC - 14.46A^2 - 2.93B^2 - 25.52C^2$

$Y_{FA} = 93.29 + 7.62A + 2.00B - 12.38C + 3.00AC + 2.25BC - 4.89A^2 - 3.64B^2 - 27.89C^2$

$Y_{Que} = 88.24 + 7.98A + 1.32B + 4.32AC + 6.64BC - 11.13A^2 - 7.83B^2 - 49.88C^2$

$Y_C = 91.26 + 6.54A - 11.98C + 5.09AC - 1.59BC - 11.83A^2 - 2.11B^2 - 25.07C^2$

$Y_{EC} = 95.87 + 8.06A - 2.40B - 9.85C + 4.52AC + 1.96BC - 16.80A^2 - 4.88B^2 - 28.30C^2$

$Y_{GCG} = 93.31 + 7.23A + 1.76B + 10.22C + 4.16BC - 9.71A^2 - 6.89B^2 - 50.18C^2$

$Y_{ECG} = 92.83 + 7.91A + 1.15B + 10.00C - 2.24AB - 5.43AC + 6.54BC - 7.25A^2 - 8.84B^2 - 43.04C^2$

$Y_{EGCG} = 92.76 + 7.12A - 1.94B + 5.93C + 2.20AB + 1.33AC - 8.85A^2 - 5.77B^2 - 41.79C^2$

Table 1. Box-Behnken design matrix and observed responses values of the extraction recovery.

Run	EtOAc/mL	ACN/mL	pH	GA/%	PC/%	VA/%	SA/%	FA/%	Que/%	C/%	EC/%	GCG/%	ECG/%	EGCG/%
1	500	650	4	70.56	66.70	67.73	44.58	77.88	62.34	71.07	62.45	71.01	72.23	66.87
2	1000	650	2	63.73	64.10	68.01	47.23	73.88	25.08	77.61	66.82	23.63	25.57	37.92
3	1000	350	6	38.45	42.72	65.54	43.26	45.13	22.69	53.74	54.63	40.53	43.26	53.66
4	1000	500	4	91.75	93.05	96.68	95.44	91.30	87.13	90.61	97.70	91.98	93.42	93.92
5	1000	350	2	69.15	73.06	76.00	54.35	74.38	35.72	74.52	75.87	29.92	36.35	40.61
6	1000	500	4	92.85	90.09	94.55	93.56	92.40	88.45	92.18	94.57	93.35	92.35	92.89
7	1000	500	4	93.89	91.31	93.12	94.36	94.19	89.80	91.35	94.9	92.69	94.28	91.39
8	1500	350	4	85.30	85.26	87.14	89.24	89.13	75.66	84.24	83.72	80.78	85.73	84.99
9	500	500	6	19.45	17.97	37.20	17.34	37.50	15.05	30.76	27.49	36.20	50.06	39.59
10	1500	500	2	67.24	68.27	59.09	63.46	77.50	30.77	67.8	64.98	28.39	45.88	41.97
11	500	500	2	57.39	48.19	61.86	31.58	68.25	23.44	64.9	58.60	16.47	19.21	30.40
12	500	350	4	65.32	71.02	76.52	62.54	76.38	60.26	70.5	69.12	67.64	65.44	75.16
13	1000	650	6	41.28	42.72	55.22	28.35	53.63	38.61	50.48	53.42	50.88	58.65	48.58
14	1000	500	4	94.12	92.46	93.19	95.89	93.29	86.46	90.49	95.39	94.69	91.24	92.39
15	1500	650	4	86.83	86.54	77.63	83.57	95.63	78.87	83.5	81.46	87.41	83.56	85.51
16	1000	500	4	95.89	91.9	95.69	94.68	95.25	89.35	91.68	96.77	93.84	92.87	93.2
17	1500	500	6	56.24	42.33	60.5	44.25	58.75	39.65	54.02	51.97	52.59	55.01	56.49

Table 2. ANOVA for dependent variable: the recovery of SA.

Source	Sum of Squares	df	Mean Square	F-Value	p-Value	Prob > F
Model	11,823.31	9	1313.7	826.42	<0.0001	significant
A-EtOAc	1936.91	1	1936.91	1218.47	<0.0001	
B-ACN	260.6	1	260.6	163.94	<0.0001	
C-pH	502.76	1	502.76	316.28	<0.0001	
AB	37.76	1	37.76	23.75	0.0018	
AC	6.18	1	6.18	3.88	0.0894	
BC	15.17	1	15.17	9.54	0.0176	
A^2	881.82	1	881.82	554.73	<0.0001	
B^2	449.45	1	449.45	282.74	<0.0001	
C^2	7132.12	1	7132.12	4486.67	<0.0001	
Residual	11.13	7	1.59			
Lack of fit	7.79	3	2.6	3.11	0.1511	not significant
Pure error	3.34	4	0.84			
Cor total	11,834.44	16				

2.3.3. Optimization Analysis

The desirability function (DF) is a way to optimize multiple responses at the same time. The overall desirability (D) was used to search for optimal extraction conditions by the geometric mean of the individual desirability of each response (d_i). The use of the geometric mean ensures that the selected condition could satisfy at least the minimum desired value of all responses and D increases as most responses come close to their desired value. For example, if the selected extraction conditions favored some recoveries with high d but other recoveries with low d, the overall D would still be very low. Similarly, the condition having equal but low d, yields a low D as well. In this case, the scenario is not desirable. In optimizing the extraction conditions, it will be preferable to maximize the recovery of each polyphenol as high as possible, towards high D, so that the selected condition ensures the overall optimal performance.

The higher its sample recovery rate, the better it is; and when multiple components are measured at the same time, the recovery rate of some compounds could be compromised. Therefore, we define $Y_{i\text{-max}}$ in Equation (2) at 100% (theoretical maximum recovery rate) and $Y_{i\text{-min}}$ at 80%. In this way, the d_i value of each compound could be calculated according to Equation (2) and the D value of 11 polyphenols could be calculated according to Equation (3), subsequently. With D as a new response, its optimal value could be obtained through regression analysis. In the present study, the best conditions for achieving the highest D value (0.723) were 1184.52 µL EtOAc, 492.9 µL ACN and solution pH of 3.92. The corresponding theoretical maximum of 11 compounds were GA 96.48%, PC 94.15%, VA 94.88%, SA 99.02%, FA 95.73%, Que 89.44%, C 92.44%, EC 96.93%, GCG 94.03%, ECG 94.34% and EGCG 93.89%, respectively.

2.3.4. Verification of the Predictive Model

In order to verify the applicability of the model equations, six verification experiments ($n = 6$) were performed under optimized conditions. Considering the operability of actual production, the optimal conditions were slightly modified as follows: 1185 µL EtOAc, 493 µL ACN and solution pH of 3.92. Under these conditions, the recoveries of 11 compounds were GA 93.56 ± 1.39%, PC 91.71 ± 1.02%, VA 93.61 ± 1.52%, SA 94.75 ± 0.79%, FA 92.26 ± 1.52%, Que 88.29 ± 1.32%, C 91.13 ± 0.72%, EC 95.76 ± 1.16%, GCG 93.35 ± 0.92% ECG 92.67 ± 1.08% and EGCG 92.71 ± 0.84%. The consistency between the predicted values and experimental values was assessed by Bland-Altman analysis [25]. Using 95% as the confidence interval, the limits of agreement for measurement difference were [−0.0029, 0.0421]. The Bland-Altman plot is depicted in Figure S2, which shows that the experimental values are close to the predicted values. The result confirmed that the models proposed by a combination of RSM and DF methods can be used to optimize the VA-DLLME extraction process.

2.4. Analytical Performance of VA-DLLME-UPLC-QqQ/MS

The method validation data are summarized in Table 3, including the calibration curve, linear range, LOD, LOQ, precision, repeatability and recovery. The results indicated that the VA-DLLME-UPLC-QqQ/MS method was satisfactory for the determination of 11 polyphenols in blueberry samples. Besides, the analytical figures of merit were compared with those of several other quantitative methods reported for non-anthocyanin polyphenols determination as shown in Table S3.

Table 3. The results of method validation.

Analyte	Regression Equation (y = ax + b, r²)	Linear Range (ng/mL)	LOD * (ng)	LOQ * (ng)	Precision (RSD, n = 6)		Repeatability (n = 6)		Recovery (n = 6)	
					Intra-Day (%)	Inter-Day (%)	Mean (μg/kg)	RSD (%)	Mean (%)	RSD (%)
GA	y = 327102x + 293214, 0.9962	1.05–21.0	0.01	0.03	0.59	1.18	9.085	1.27	93.56	1.39
PC	y = 75436x + 19495, 0.9993	0.98–19.6	0.02	0.06	0.96	1.11	10.053	1.34	91.71	1.02
VA	y = 3045x − 254, 0.9998	0.99–19.8	0.3	0.9	2.79	2.92	6.266	1.87	93.61	1.52
SA	y = 3500x − 1177, 0.9997	0.95–19.0	0.1	0.3	1.86	0.95	5.093	0.95	94.75	0.79
FA	y = 67479x − 1274, 0.9999	0.96–19.2	0.1	0.3	1.12	2.95	12.917	0.38	92.26	1.52
Que	y = 442647x − 283610, 0.9986	1.04–20.8	0.01	0.03	1.18	1.88	354.675	0.70	88.29	1.32
C	y = 43132x − 8789, 0.9999	0.96–19.2	0.1	0.3	0.58	1.49	8.295	1.11	91.13	0.72
EC	y = 87001x + 31241, 0.9989	0.92–18.4	0.1	0.3	1.32	1.87	15.177	0.95	95.76	1.16
GCG	y = 49998x − 10512, 0.9997	0.96–19.2	0.1	0.3	1.13	2.12	0.634	2.46	93.35	0.92
ECG	y = 117448x − 15171, 0.9997	1.08–21.6	0.1	0.3	1.36	2.48	9.528	1.88	92.67	1.08
EGCG	Y = 7895x − 1558, 0.9998	0.98–19.6	0.1	0.3	1.04	2.35	0.827	1.31	92.71	0.84

* LOD and LOQ were calculated at the signal-to-noise ratio (S/N) 3 and 10 by a gradual dilution of standard solution, respectively.

2.5. Sample Analysis

Twenty-five batches of commercially available blueberry were collected from Sichuan province and Chongqing city. The contents of eleven polyphenols were determined by the present VA-DLLME-UPLC-QqQ/MS method. As shown in Table 4, The difference of GA content was the largest, which was 6.09 times and the difference of VA content was the least, which was 2.46 times. The content of these 11 polyphenols was 208.408–552.326 µg/kg. It was noted that a small amount of EGCG was detected in the four fresh samples picked from local blueberry orchards but samples from shops and supermarkets were not measured. It is estimated that because EGCG is unstable, it decomposes and isomerizes to other catechins during storage [26].

Table 4. The contents of 11 polyphenols in blueberry samples (μg/kg) (Mean ± SD, $n = 3$).

No	Source	GA	PC	VA	SA	FA	Que	C	EC	GCG	ECG	EGCG	Total
1	orchard, Deyang	9.21 ± 0.15	10.10 ± 0.11	5.33 ± 0.02	5.09 ± 0.10	12.92 ± 0.09	357.54 ± 2.12	8.33 ± 0.11	15.24 ± 0.09	0.66 ± 0.06	9.79 ± 0.07	0.85 ± 0.03	439.096
2	orchard, Chengdu	19.02 ± 0.19	10.87 ± 0.14	7.13 ± 0.06	2.67 ± 0.12	10.09 ± 0.04	332.24 ± 2.38	7.09 ± 0.12	21.12 ± 0.21	1.22 ± 0.09	9.25 ± 0.07	1.72 ± 0.11	422.403
3	orchard, Chengdu	10.55 ± 0.15	4.83 ± 0.10	3.81 ± 0.11	3.25 ± 0.10	6.48 ± 0.11	160.99 ± 1.16	6.48 ± 0.20	37.27 ± 0.20	1.03 ± 0.08	6.44 ± 0.13	1.15 ± 0.05	247.286
4	orchard, Chengdu	18.37 ± 0.18	3.27 ± 0.13	10.112 ± 0.10	2.27 ± 0.14	6.77 ± 0.12	283.86 ± 2.07	8.65 ± 0.20	18.72 ± 0.19	0.88 ± 0.06	11.53 ± 0.10	0.73 ± 0.04	367.165
5	fruit shop, Leshan	13.03 ± 0.16	3.74 ± 0.06	4.39 ± 0.13	5.65 ± 0.10	16.59 ± 0.14	319.60 ± 2.20	8.58 ± 0.12	19.58 ± 0.13	0.85 ± 0.01	13.41 ± 0.09	n.d.	405.423
6	fruit shop, Chengdu	13.92 ± 0.09	5.98 ± 0.10	8.51 ± 0.18	1.86 ± 0.11	6.63 ± 0.08	439.85 ± 2.55	3.77 ± 0.12	36.04 ± 0.06	1.66 ± 0.08	6.14 ± 0.07	n.d.	524.376
7	fruit shop, Chengdu	17.13 ± 0.18	4.74 ± 0.10	10.09 ± 0.15	2.81 ± 0.14	10.50 ± 0.13	390.38 ± 2.61	10.81 ± 0.14	22.16 ± 0.13	1.37 ± 0.08	8.35 ± 0.11	n.d.	478.338
8	fruit shop, Chengdu	19.57 ± 0.13	5.63 ± 0.18	5.71 ± 0.15	3.12 ± 0.12	13.71 ± 0.12	137.62 ± 1.84	9.08 ± 0.08	13.56 ± 0.110	0.67 ± 0.05	8.63 ± 0.13	n.d.	217.294
9	fruit shop, Yibing	12.24 ± 0.21	5.72 ± 0.15	8.2 ± 0.13	4.69 ± 0.09	15.30 ± 0.15	430.71 ± 2.63	3.95 ± 0.11	15.26 ± 0.12	0.57 ± 0.05	9.22 ± 0.10	n.d.	505.877
10	fruit shop, Deyang	11.02 ± 0.11	4.77 ± 0.16	8.74 ± 0.15	5.83 ± 0.12	16.40 ± 0.10	231.83 ± 2.07	9.23 ± 0.14	26.41 ± 0.11	1.56 ± 0.07	6.54 ± 0.12	n.d.	322.336
11	fruit shop, Suining	13.46 ± 0.12	5.44 ± 0.09	5.67 ± 0.12	1.89 ± 0.11	15.70 ± 0.09	220.23 ± 2.59	8.89 ± 0.10	17.34 ± 0.10	1.28 ± 0.08	13.01 ± 0.10	n.d.	302.876
12	fruit shop, Neijiang	14.41 ± 0.16	8.22 ± 0.10	5.80 ± 0.18	3.51 ± 0.12	8.14 ± 0.07	134.57 ± 1.67	6.30 ± 0.10	21.69 ± 0.15	1.54 ± 0.06	4.22 ± 0.10	n.d.	208.408
13	fruit shop, Mianyang	11.91 ± 0.08	8.98 ± 0.16	7.85 ± 0.15	2.35 ± 0.09	14.45 ± 0.134	394.96 ± 2.56	7.41 ± 0.15	19.89 ± 0.08	0.80 ± 0.03	12.83 ± 0.13	n.d.	481.433
14	fruit shop, Nanchong	19.52 ± 0.12	9.97 ± 0.11	8.65 ± 0.18	3.73 ± 0.10	10.23 ± 0.10	376.28 ± 2.00	11.15 ± 0.08	33.17 ± 0.10	0.91 ± 0.05	9.86 ± 0.14	n.d.	483.473
15	fruit shop, Langzhong	7.80 ± 0.14	5.58 ± 0.07	7.67 ± 0.13	4.68 ± 0.08	14.08 ± 0.11	180.87 ± 1.54	8.71 ± 0.10	28.82 ± 0.14	1.10 ± 0.08	6.03 ± 0.04	n.d.	265.320
16	fruit shop, Chongqing	14.97 ± 0.12	7.88 ± 0.13	6.0C ± 0.15	4.69 ± 0.10	5.89 ± 0.10	357.77 ± 2.23	4.32 ± 0.12	23.25 ± 0.11	1.06 ± 0.07	9.86 ± 0.11	n.d.	435.707
17	fruit shop, Chongqing	18.02 ± 0.08	6.91 ± 0.10	13.26 ± 0.10	3.06 ± 0.11	13.93 ± 0.09	229.77 ± 1.84	8.35 ± 0.11	16.92 ± 0.10	1.27 ± 0.05	4.02 ± 0.09	n.d.	312.492
18	supermarket, Deyang	3.25 ± 0.17	9.25 ± 0.12	8.8c ± 0.09	2.71 ± 0.13	12.60 ± 0.13	302.55 ± 1.83	10.30 ± 0.18	23.57 ± 0.14	0.80 ± 0.04	13.03 ± 0.07	n.d.	386.919
19	supermarket, Chengdu	19.82 ± 0.11	7.12 ± 0.09	8.21 ± 0.18	4.59 ± 0.08	15.74 ± 0.14	450.23 ± 3.27	5.74 ± 0.90	28.95 ± 0.14	1.16 ± 0.07	10.76 ± 0.09	n.d.	552.326
20	supermarket, Chengdu	14.35 ± 0.12	5.98 ± 0.09	8.27 ± 0.10	3.87 ± 0.14	10.19 ± 0.08	397.70 ± 2.14	8.61 ± 0.11	18.29 ± 0.06	1.07 ± 0.07	13.03 ± 0.13	n.d.	481.339
21	supermarket, Chengdu	16.68 ± 0.22	8.77 ± 0.16	7.65 ± 0.10	5.16 ± 0.06	8.50 ± 0.10	405.34 ± 1.85	4.81 ± 0.10	28.00 ± 0.12	1.61 ± 0.02	4.88 ± 0.13	n.d.	491.432
22	supermarket, Chengdu	12.95 ± 0.18	7.37 ± 0.12	4.17 ± 0.07	2.75 ± 0.13	7.12 ± 0.09	387.03 ± 2.31	6.55 ± 0.11	18.13 ± 0.11	0.73 ± 0.03	7.97 ± 0.10	n.d.	454.761
23	supermarket, Mianyang	7.66 ± 0.12	5.51 ± 0.09	7.06 ± 0.14	3.70 ± 0.08	10.80 ± 0.11	261.42 ± 2.10	10.34 ± 0.11	20.37 ± 0.12	0.88 ± 0.05	11.31 ± 0.11	n.d.	339.054
24	supermarket, Chongqing	11.96 ± 0.16	7.16 ± 0.12	10.1 ± 0.11	4.70 ± 0.16	7.54 ± 0.11	223.72 ± 1.74	4.29 ± 0.11	32.41 ± 0.13	1.07 ± 0.07	11.52 ± 0.11	n.d.	314.477
25	supermarket, Chongqing	13.92 ± 0.16	7.26 ± 0.15	8.09 ± 0.05	4.19 ± 0.19	11.51 ± 0.10	319.40 ± 2.21	8.06 ± 0.09	25.40 ± 0.11	1.14 ± 0.04	9.42 ± 0.12	n.d.	408.386

57

3. Materials and Methods

3.1. Chemical and Apparatus

Eleven authentic reference standards (Figure S3)—gallic acid (GA), protocatechuate (PC), vanillic acid (VA), syringic acid (SA), ferulic acid (FA), quercetin (Que), catechin (C), epicatechin (EC), gallocatechin gallate (GCG), epicatechin gallate (ECG) and epigallocatechin gallate (EGCG)—were purchased from Chengdu Push Bio-technology Co., Ltd. (Chengdu, China). Ultra-pure water was prepared by the Milli-Q water purification system (Millipore, Bedford, MA, USA). HPLC grade acetonitrile (Fisher, Fair Lawn, NJ, USA) was used for UPLC-MS analysis. Other chemicals and reagents of analytical grade were purchased from Chengdu Kelong Chemical Reagent Plant (Chengdu, China).

3.2. Preparation of Standard Solution

Eleven stock solutions were prepared by dissolving each authentic reference standard into a 10 mL volumetric flask with methanol, respectively. The exact concentrations were as follows: GA 10.5 mg/mL; PC 9.8 mg/mL; VA 9.9 mg/mL; SA 9.5 mg/mL; FA 9.6 mg/mL; Que 10.4 mg/mL; C 9.6 mg/mL; EC 9.2 mg/ ML; GCG 9.6 mg/mL; EGC 10.8 mg/mL; EGCG 9.8 mg/mL.

The mixed-standard stock solution was prepared by adding 1 mL of each stock solution to a 100 mL volumetric flask and subsequently setting the volume with methanol. Work solutions I-V were obtained by dilution of the mixed-standard stock solution with methanol to the final concentrations of about 1 ng/mL, 2 ng/mL, 5 ng/mL, 10 ng/mL and 20 ng/mL. All the solutions were kept in a refrigerator at 4 °C before use.

3.3. Sample Preparation by the DLLME Procedure

Accurately add 1 g blueberry homogenate to a 100 mL volumetric flask; perform ultrasonic extraction with water for 30 min; centrifuge and add 1 mL of supernatant to a 4 mL centrifuge tube; adjust to pH 2–6 using a pH meter by dilute hydrochloric acid; add 500–1500 µL EtOAc and 350–650 µL ACN, then vortex for 30 s. After centrifugation, the upper organic phase was transferred. The extraction was repeated once, combined and the solvent was removed by a Termovap Sample Concentrator. The resulting residue was re-dissolved in 1 mL of methanol and filtered through a 0.45 µm filter for UPLC-MS analysis.

3.4. UPLC-MS Analysis

The Shimadzu UPLC system consisted of two pumps (LC-30AD), an auto-sampler (SIL-30AC), a column oven (CTO-30aHE) and an online degasser (DGU-20A^{5R}). Chromatographic separation was performed on an Agilent ZORBAX SB-Aq analytical column (3.0 × 100 mm, 1.8 µm) using formic acid solution (0.1%, solvent A) and acetonitrile (solvent B). The gradient elution conditions were as follows: 0~5 min to maintain 5% B and 5~30 min linear increase from 5% B up to 20% B. The column temperature was kept at 40 °C. Two microliters of sample solution was injected into the UPLC system by the auto-sampler.

Qualitative scanning and identification of polyphenols were accomplished with a 4600 Q-TOF mass spectrometer (AB Sciex, Concord, CA, USA) equipped with an electrospray ionization source. Q-TOF/MS analysis was performed in negative mode as follows: ion source gas 1 (GS1) (N_2) 50 psi, ion source gas 1 (GS1) 50 psi, curtain gas 35 psi, temperature 500 °C, ionspray voltage floating −4500 V. The mass range was set at m/z 100–1000. The system was operated under Analyst 1.6 and Peak 2.0 (AB Sciex, Concord, CA, USA) and used APCI negative calibration solution to calibrate the instrument's mass accuracy in real time.

Quantitative analysis of polyphenols was accomplished by a triple quadrupole mass spectrometer (Thermo, TSQ VANTAGE, Waltham, MA USA) equipped with an electrospray ionization source. The MS spectra were acquired in negative ion mode. The quantitative analysis of the target analytes was performed under multi-reaction monitoring (MRM) mode. The mass spectrometry detector (MSD) parameters were as follows: capillary temperature 350 °C, vaporizer temperature 350 °C, sheath gas pressure 40 Arb, aux gas pressure 10 Arb, ion sweep gas pressure 2 Arb, spray voltage −2000 V, discharge current 4.0 μA. Data collection and processing were conducted with Thermo Xcalibur Workstation (Version 2.2, Thermo).

3.5. Experimental Design

3.5.1. Single Factor Experimental Design

The most important factors affecting the VA-DLLME assay were the amount of EtOAc, the amount of ACN and the pH of the solution. In the present study, the recoveries of 11 polyphenols were assessed to determine their inflection points by means of changing a single variable and fixing two variables.

3.5.2. Box-Behnken Design

Box-Behnken design (BBD) is a class of (nearly) rotatable second-order design based on three-level incomplete factorial design [27]. On the basis of single-factor experiments, the VA-DLLME conditions were further optimized by BBD of three factors and three levels. Three independent variables—the amount of EtOAc, the mount of ACN and solution pH—were designated as A, B and C, respectively. Furthermore, the range of values was based on the results of the above single factor experiments. The response variables were the recoveries of 11 compounds (Y_1–Y_{11}). There was a total of 17 runs based on BBD design with four center points performed in random order.

3.6. Statistical Analysis and Optimization

The parameter and analysis of variance (ANOVA) of the response equations were executed by Design Expert Software (Version 8.0.6). The fitted second-order polynomial model of the response value and independent variables is shown in Equation (1). The significance test of the model was performed and the influence of each parameter on the extraction rates of 11 compounds was analyzed and response surfaces (three-dimensional) were drawn. The statistical significance of variance of each response was tested by ANOVA, where insignificant items were removed from the model and only the data were re-fit to significant parameters ($p < 0.05$) and the surface was constructed. By $p_{lack\text{-}of\text{-}fit}$, the coefficient of determination (R^2) and their adjusted R^2 are used to assess the adequacy and quality of the fit.

$$Y_n(x) = b_0 + \sum_{i=1}^{3} b_i X_i + \sum_{i=1}^{3} b_{ii} X_i^2 + \sum_{i=1}^{2}\sum_{j=i+1}^{3} b_{ij} X_i X_j \tag{1}$$

where Y_n is the predicted response, b_0 is the intercept parameter, b_i, b_{ii} and b_{ij} are the regression coefficients of linear, quadratic and interaction effects, respectively and X_i and X_j are the independent variables.

The desirability function (DF) is a way to optimize multiple responses at the same time. In this study, DF was used to convert the eleven independent responses into one desirable value (D). First of all, each response (recovery value of 11 compounds, Y_i) is firstly converted into a dimensionless desirable value d_i according to Equation (2) and then integrated into D by Equation (3). The D value increases as the desired response value increases with the range of 0 to 1. $D = 0$ indicates that the response value has seriously deviated from the target value and $D = 1$ indicates that the response value is consistent with the target value. Finally, according to the optimal variable parameter suggested by DF, the experiments with triplicates were conducted and checked to ensure the consistency of the predicted value and the actual value.

$$d_i = \begin{cases} 0 & Y_i \leq Y_{i-\min} \\ \left[\dfrac{Y_i - Y_{i-\min}}{Y_{i-\max} - Y_{i-\min}}\right]^r & Y_{i-\min} < Y_i < Y_{i-\max} \\ 1 & Y_i \geq Y_{i-\max} \end{cases} \tag{2}$$

$$D = \sqrt[\Sigma w_i]{d_1^{w_1} \times d_2^{w_2} \times d_3^{w_3} \times d_i^{w_i}} \qquad (3)$$

where d_i and D are the individual response desirability and overall desirability, respectively; Y_i, $Y_{i\text{-min}}$ and $Y_{i\text{-max}}$ are the actual value of response i, the minimum acceptable value for response i and the maximum acceptable value for response i, respectively; r and w_i are the relative weight of the response i and both equal 1 in this work.

4. Conclusions

In this study, eleven non-anthocyanin polyphenols were firstly screened and identified from blueberries using UPLC-TOF/MS. Then, a vortex-assisted dispersive liquid–liquid microextraction method was developed for sample preparation. With the optimization of the multi-objective response surface methodology and desirability function approach, the best extraction condition was as follows: 1185 µL ethyl acetate, 493 µL acetonitrile and solution pH of 3.92. Finally, the content of 11 polyphenols in 25 batches of blueberry samples from different regions was determined by UPLC-QqQ/MS. This work was designed to be complementary to the work of others and provides a basis for the comprehensive evaluation of active ingredients in blueberries. These results can thus be utilized for extraction standardizations and quality control protocols of blueberry.

Supplementary Materials: The following are available online.

Author Contributions: Y.X. performed the experiments; X.-S.X., L.Y., B.H. and X.-D.L. collected samples, and analyzed the data; S.-H.Z. and Y.L. participated in the experiment; J.X. and L.-S.Q. designed the experiments, and wrote the manuscript.

Funding: This work was supported by the Chinese Academy of Sciences (CAS) 'Light of West China' Program, the Application Development and Achievement Transformation Cultivation Project (CYCG17-04), and the Collaborative Innovation Center of Sichuan for Elderly Care and Health (YLZBZ1820), Chengdu Medical College.

Conflicts of Interest: The authors declare no conflict of interest.

References

1. Giacalone, M.; Di Sacco, F.; Traupe, I.; Topini, R.; Forfori, F.; Giunta, F. Antioxidant and neuroprotective properties of blueberry polyphenols: A critical review. *Nutr. Neurosci.* **2011**, *14*, 119–125. [CrossRef] [PubMed]
2. Neto, C.C. Cranberry and blueberry: Evidence for protective effects against cancer and vascular diseases. *Mol. Nutr. Food Res.* **2007**, *51*, 652–664. [CrossRef] [PubMed]
3. Shi, M.; Loftus, H.; McAinch, A.J.; Su, X.Q. Blueberry as a source of bioactive compounds for the treatment of obesity, type 2 diabetes and chronic inflammation. *J. Funct. Foods* **2017**, *30*, 16–29. [CrossRef]
4. Zhu, Y.; Sun, J.; Lu, W.; Wang, X.; Wang, X.; Han, Z.; Qiu, C. Effects of blueberry supplementation on blood pressure: A systematic review and meta-analysis of randomized clinical trials. *J. Hum. Hypertens.* **2017**, *31*, 165–171. [CrossRef] [PubMed]
5. Correa-Betanzo, J.; Allen-Vercoe, E.; McDonald, J.; Schroeter, K.; Corredig, M.; Paliyath, G. Stability and biological activity of wild blueberry (*Vaccinium angustifolium*) polyphenols during simulated in vitro gastrointestinal digestion. *Food Chem.* **2014**, *165*, 522–531. [CrossRef] [PubMed]
6. Louis, X.L.; Thandapilly, S.J.; Kalt, W.; Vinqvist-Tymchuk, M.; Aloud, B.M.; Raj, P.; Yu, L.; Le, H.; Netticadan, T. Blueberry polyphenols prevent cardiomyocyte death by preventing calpain activation and oxidative stress. *Food Funct.* **2014**, *5*, 1785–1794. [CrossRef] [PubMed]
7. Harris, C.S.; Burt, A.J.; Saleem, A.; Le, P.M.; Martineau, L.C.; Haddad, P.S.; Bennett, S.A.L.; Arnason, J.T. A single HPLC-PAD-APCI/MS method for the quantitative comparison of phenolic compounds found in leaf, stem, root and fruit extracts of Vaccinium angustifolium. *Phytochem. Anal.* **2007**, *18*, 161–169. [CrossRef] [PubMed]
8. Li, J.; Yuan, C.; Pan, L.; Benatrehina, P.A.; Chai, H.; Keller, W.J.; Naman, C.B.; Kinghorn, A.D. Bioassay-guided isolation of antioxidant and cytoprotective constituents from a Maqui Berry (*Aristotelia chilensis*) dietary supplement ingredient as markers for qualitative and quantitative analysis. *J. Agric. Food Chem.* **2017**, *65*, 8634–8642. [CrossRef] [PubMed]

9. Wang, Y.F.; Jennifer, J.-C.; Ajay, P.S.; Vorsa, N. Characterization and quantification of flavonoids and organic acids over fruit development in American cranberry (*Vaccinium macrocarpon*) cultivars using HPLC and APCI-MS/MS. *Plant Sci.* **2017**, *262*, 91–102. [CrossRef] [PubMed]
10. Ayaz, F.A.; Hayirlioglu-Ayaz, S.; Gruz, J.; Novak, O.; Strnad, M. Separation, characterization, and quantitation of phenolic acids in a little-known blueberry (*Vaccinium arctostaphylos* L.) fruit by HPLC-MS. *J. Agric. Food Chem.* **2005**, *53*, 8116–8122. [CrossRef] [PubMed]
11. Su, X.; Zhang, J.; Wang, H.; Xu, J.; He, J.; Liu, L.; Zhang, T.; Chen, R.; Kang, J. Phenolic Acid Profiling, Antioxidant, and Anti-Inflammatory Activities, and miRNA Regulation in the Polyphenols of 16 Blueberry Samples from China. *Molecules* **2017**, *22*, 312. [CrossRef] [PubMed]
12. De Souza Dias, F.; Silva, M.F.; David, J.M. Determination of quercetin, gallic acid, resveratrol, catechin and malvidin in brazilian wines elaborated in the Vale do São Francisco using liquid-liquid extraction assisted by ultrasound and GC-MS. *Food Anal. Methods* **2013**, *6*, 963–968. [CrossRef]
13. Mülek, M.; Högger, P. Highly sensitive analysis of polyphenols and their metabolites in human blood cells using dispersive SPE extraction and LC-MS/MS. *Anal. Bioanal. Chem.* **2015**, *407*, 1885–1899. [CrossRef] [PubMed]
14. Shalash, M.; Makahleh, A.; Salhimi, S.M.; Saad, B. Vortex-assisted liquid-liquid liquid microextraction followed by high performance liquid chromatography for the simultaneous determination of fourteen phenolic acids in honey, iced tea and canned coffee drinks. *Talanta* **2017**, *174*, 428–435. [CrossRef] [PubMed]
15. Faraji, H.; Helalizadeh, M.; Kordi, M.R. Overcoming the challenges of conventional dispersive liquid-liquid microextraction: Analysis of THMs in chlorinated swimming pools. *Anal. Bioanal. Chem.* **2018**, *410*, 605–614. [CrossRef] [PubMed]
16. Rezaee, M.; Khalilian, F. Application of ultrasound-assisted extraction followed by solid-phase extraction followed by dispersive liquid-liquid microextraction for the determination of chloramphenicol in chicken meat. *Food Anal. Methods* **2018**, *11*, 759–767. [CrossRef]
17. Wang, K.; Xie, X.; Zhang, Y.; Huang, Y.; Zhou, S.; Zhang, W.; Lin, Y.; Fan, H. Combination of microwave-assisted extraction and ultrasonic-assisted dispersive liquid-liquid microextraction for separation and enrichment of pyrethroids residues in Litchi fruit prior to HPLC determination. *Food Chem.* **2018**, *240*, 1233–1242. [CrossRef] [PubMed]
18. Wang, L.; Huang, T.; Cao, H.X.; Yuan, Q.X.; Liang, Z.P.; Liang, G.X. Application of air-assisted liquid-liquid microextraction for determination of some fluoroquinolones in milk powder and egg samples: Comparison with conventional dispersive liquid-liquid microextraction. *Food Anal. Methods* **2016**, *9*, 2223–2230. [CrossRef]
19. Rahmani, M.; Ghasemi, E.; Sasani, M. Application of response surface methodology for air assisted-dispersive liquid-liquid microextraction of deoxynivalenol in rice samples prior to HPLC-DAD analysis and comparison with solid phase extraction cleanup. *Talanta* **2017**, *165*, 27–32. [CrossRef] [PubMed]
20. Matias-Guiu, P.; Rodríguez-Bencomo, J.J.; Pérez-Correa, J.R.; López, F. Aroma profile design of wine spirits: Multi-objective optimization using response surface methodology. *Food Chem.* **2018**, *245*, 1087–1097. [CrossRef] [PubMed]
21. Chen, C.; Xue, Y.; Li, Q.-M.; Wu, Y.; Liang, J.; Qing, L.-S. Neutral loss scan-based strategy for integrated identification of amorfrutin derivatives, new peroxisome proliferator-activated receptor gamma agonists, from Amorpha Fruticosa by UPLC-QqQ-MS/MS and UPLC-Q-TOF-MS. *J. Am. Soc. Mass Spectrom.* **2018**, *29*, 685–693. [CrossRef] [PubMed]
22. Sun, W.-X.; Zhang, Z.-Z.; Xie, J.; He, Y.; Cheng, Y.; Ding, L.-S.; Luo, P.; Qing, L.-S. Determination of a astragaloside IV derivative LS-102 in plasma by ultra-performance liquid chromatography-tandem mass spectrometry in dog plasma and its application in a pharmacokinetic study. *Phytomedicine* **2019**, *53*, 243–251. [CrossRef]
23. Xie, J.; Li, J.; Liang, J.; Luo, P.; Qing, L.-S.; Ding, L.-S. Determination of contents of catechins in oolong teas by quantitative analysis of multi-components via a single marker (QAMS) method. *Food Anal. Methods* **2017**, *10*, 363–368. [CrossRef]
24. Qing, L.-S.; Xue, Y.; Zhang, J.-G.; Zhang, Z.-F.; Liang, J.; Jiang, Y.; Liu, Y.-M.; Liao, X. Identification of flavonoid glycosides in Rosa chinensis flowers by liquid chromatography–tandem mass spectrometry in combination with ^{13}C nuclear magnetic resonance. *J. Chromatogr. A* **2012**, *1249*, 130–137. [CrossRef] [PubMed]

25. Wang, Q.-L.; Li, J.; Li, X.-D.; Tao, W.-J.; Ding, L.-S.; Luo, P.; Qing, L.-S. An efficient direct competitive nano-ELISA for residual BSA determination in vaccines. *Anal. Bioanal. Chem.* **2017**, *409*, 4607–4614. [CrossRef] [PubMed]
26. Zhu, Q.Y.; Zhang, A.; Tsang, D.; Huang, Y.; Chen, Z. Stability of Green Tea Catechins. *J. Agric. Food Chem.* **1997**, *45*, 4624–4628. [CrossRef]
27. Ferreira, S.L.C.; Bruns, R.E.; Ferreira, H.S.; Matos, G.D.; David, J.M.; Brandão, G.C.; da Silva, E.G.P.; Portugal, L.A.; dos Reis, P.S.; Souza, A.S.; et al. Box-Behnken design: An alternative for the optimization of analytical methods. *Anal. Chim. Acta* **2007**, *597*, 179–186. [CrossRef] [PubMed]

Sample Availability: Samples of the all compounds are available from the authors.

© 2018 by the authors. Licensee MDPI, Basel, Switzerland. This article is an open access article distributed under the terms and conditions of the Creative Commons Attribution (CC BY) license (http://creativecommons.org/licenses/by/4.0/).

Article

High Flow-Rate Sample Loading in Large Volume Whole Water Organic Trace Analysis Using Positive Pressure and Finely Ground Sand as a SPE-Column In-Line Filter

Ola Svahn * and Erland Björklund

Department of Environmental Science and Bioscience, Faculty of Natural Science, Kristianstad University, SE-291 39 Kristianstad, Sweden; erland.bjorklund@hkr.se
* Correspondence: ola.svahn@hkr.se

Academic Editor: Nuno Neng
Received: 15 March 2019; Accepted: 9 April 2019; Published: 11 April 2019

Abstract: By using an innovative, positive pressure sample loading technique in combination with an in-line filter of finely ground sand the bottleneck of solid phase extraction (SPE) can be reduced. Recently published work by us has shown the proof of concept of the technique. In this work, emphasis is put on the SPE flow rate and method validation for 26 compounds of emerging environmental concern, mainly from the 1st and 2nd EU Watch List, with various physicochemical properties. The mean absolute recoveries in % and relative standard deviations (RSD) in % for the investigated compounds from spiked pure water samples at the three investigated flow rates of 10, 20, and 40 mL/min were 63.2% (3.2%), 66.9% (3.3%), and 69.0% (4.0%), respectively. All three flow rates produced highly repeatable results, and this allowed a flow rate increase of up to 40 mL/min for a 200 mg, 6 mL, reversed phase SPE cartridge without compromising the recoveries. This figure is more than four times the maximum flow rate recommended by manufacturers. It was indicated that some compounds, especially pronounced for the investigated macrolide molecules, might suffer when long contact times with the sample glass bottle occurs. A reduced contact time somewhat decreases this complication. A very good repeatability also held true for experiments on both spiked matrix-rich pond water (high and low concentrations) and recipient waters (river and wastewater) applying 40 mL/min. This work has shown that, for a large number of compounds of widely differing physicochemical properties, there is a generous flow rate window from 10 to 40 mL/min where sample loading can be conducted. A sample volume of 0.5 L, which at the recommended maximum flow rate speed of 10 mL/min, would previously take 50 min, can now be processed in 12 min using a flow rate of 40 mL/min. This saves 38 min per processed sample. This low-cost technology allows the sample to be transferred to the SPE-column, closer to the sample location and by the person taking the sample. This further means that only the sample cartridge would need to be sent to the laboratory, instead of the whole water sample, like today's procedure.

Keywords: environmental analysis; whole water; trace analysis; SPE; large volume; in-line filter; sand; flow rate; pharmaceuticals; hormones; pesticides

---

1. Introduction

In environmental organic trace analysis, there is most often a need to enrich the compounds of interest from a large sample volume, where the use of solid phase extraction (SPE) is more or less the standard technique. Extraction and clean-up protocols related to the SPE methodology commonly constitute the bottleneck of the analytical procedure and are estimated to account for 75% of the analysis time [1]. SPE-column manufacturers recommend maximum flow rates in reverse phase SPE

(200 mg HLB, Hydrophilic Lipophilic Balanced) and 6 mL from 1 to 10 mL/min, to avoid component breakthrough [2–4]. A sample volume of 0.5 L will therefore take between 50 to 500 min to process if such flow rates are applied. These figures are under ideal conditions, without the presence of a possible flow-interfering matrix. The most common commercial technique for loading samples onto SPE columns utilizes a vacuum manifold. However, as the sample volume increases, the matrix content transferred to the SPE column also increases. To prevent clogging, and/if focusing on soluble compounds, samples are often pre-filtered, and in some methods even filtered twice [5]. Even when filtering the samples, the vacuum manifold technique might become tediously slow, especially at the end of a sample load. This problem often remains in spite of reducing the pressure near to the minimum allowed, as recommended by the manufacturer. Often, as is specifically expressed on the 1st and 2nd EU Watch List, substances shall be monitored in whole water samples [6,7]. The reason behind this is that some pharmaceuticals are hydrophobic and lipophilic, and therefore might be associated with particulate matter. As a consequence, for a meaningful monitoring of pharmaceuticals in the aquatic environment, the particulate phase should also be considered. Even so, the majority of all published monitoring studies mainly focus on the water phase [8].

We recently published a short communication presenting a technique, which addresses this obstacle, using positive pressure by compressed air and finely ground sand as an SPE in-line filter [9]. This design facilitates loading larger sample volumes onto SPE columns, as needed in organic trace analysis of environmental surface waters.

The schematic arrangement of both the sample loader using compressed air, combined with sand as an in-line filter, and the drying procedure of the SPE cartridge containing analytes, matrix, SPE-polymer, and sand is outlined in Figure 1.

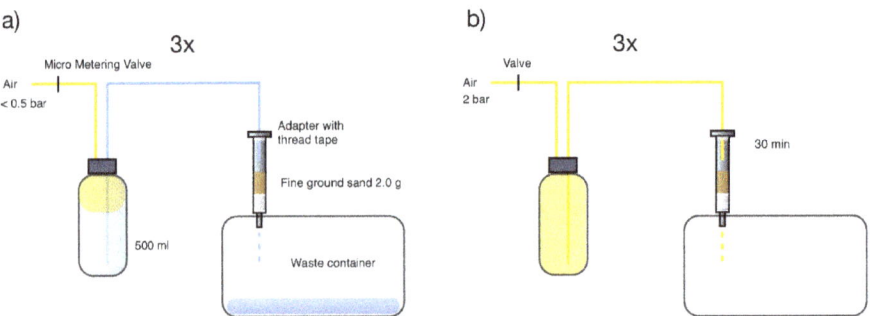

Figure 1. Schematic arrangement of (**a**) the sample loader using compressed air and sand as an in-line filter, and (**b**) the drying procedure of the solid phase extraction (SPE) columns containing analytes, matrix, SPE-polymer, and sand. Adapted from reference [9].

In this work, we investigate the possibility of increasing the sample loading flow rate with the purpose of increasing laboratory sample throughput without compromising repeatability and robustness. Flow rate investigations are crucial in determining if this low-cost technique could be applied outside the fully equipped laboratory. We have limited this study to 26 compounds of emerging concern, including all the compounds on the 1st and 2nd EU Watch List with an indicative analytical method of SPE LC-MS-MS (Liquid Chromatography - Tandem Mass Spectrometry) [6,7]. This plethora of compounds possess widely differing physicochemical properties, which will reveal if different properties affect break-through on the hydrophilic lipophilic balanced SPE cartridge at high flow rates.

Three separate experiments were carried out, and the general construction is outlined in Figure 2. The first experiment was performed in order to determine the maximum flow rate in terms of recovery and repeatability. An increased flow rate raises the need to compare if the produced means are significantly different, or if they are indistinguishable, within the limits of the experimental error.

Based on manufacturers recommendations, it was expected that a low flow rate should be less error prone compared to a higher flow rate, and that recoveries would be higher at lower flow rates. Following the manufacturers recommendation, the lowest experimental flow rate value was set to their maximum recommendation at 10 mL/min [4].

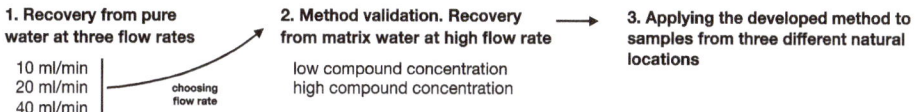

Figure 2. Flow chart illustrating the three main experiments.

Next, after determining the flow rate and letting this be the methods foundation, the technique was applied to an artificial matrix water spiked at two concentration levels in order to validate the method for the investigated compounds. Finally, in experiment number three, we investigated the robustness and repeatability of the method by applying it to determine the occurrence and concentrations of the 26 selected compounds in natural river samples and wastewater from three different recipient locations in the county of Scania, Sweden.

2. Results

2.1. Absolute and Relative Recoveries for the Three Experiments

2.1.1. Recovery from Pure Water at Three Flow Rates

The results of the absolute recoveries of the three investigated flow rates, 10, 20 and 40 mL/min, are presented in Table 1. Mean absolute recovery was found to be 64.2%, 68.2%, and 70.7%, for the three flow rates, respectively, with relative standard deviations (RSDs) of 3.3%, 3.5%, and 4.3%. A total of 17 of the 35 investigated compounds (26 + 9 isotope labeled standards) showed an absolute recovery above 75%, while 8 fell below 50% at 40 mL/min. The corresponding figures for the two other investigated flow rates were 16 above 75% and 8 below 50% at 10 mL/min, and 16 above 75% and 7 below 50% at 20 mL/min. It is also worth pointing out that recoveries are particularly low for amoxicillin, azithromycin, and triallate, with never more than 10% at any condition.

One-way ANOVA was applied to test whether the difference between the flow rate mean results were too great to be explained by a random error. The one-way ANOVA analysis clearly showed that the absolute recovery for 6 (grey marked) of the 35 compounds are significantly different (F-crit. value at 5.14, $p = 0.05$), Table 1. In these cases, there was always a higher mean recovery at the highest flow rate of 40 mL/min. The two antibiotic macrolides azithromycin and erythromycin, represent good examples of this behavior. Three of the significantly different mean values are identified among the group of isotope labeled standards (IS); ciprofloxacin_D8, clarithromycin_D3, and sulfamethoxazole_C6. Finally, the sunscreen ingredient 2-ethylhexyl 4-methoxycinnamate also showed a higher recovery at the highest flow rate. In this respect, it could be noted that adsorption of analytes to sample containers of different types has been investigated in trace analysis and found to be significant for some compounds [10].

Each pair of compound-IS combination used to calculate the relative recovery for the three experiments are shown in the second to last column of Table 1. Using the IS method, 4, 5, and 4 compounds out of 26 showed a recovery lower than 50% (grey marked) at 10, 20, and 40 mL/min. The mean relative recovery was found to be 83.1%, 76.7%, and 74.1% for the three flow rates, respectively, with an RSD of 6.6%, 3.3%, and 4.5%. There was a trend of generally higher relative calculated concentrations for the two lower flow rates compared to 40 mL/min. The IS method does not compensate to any greater extent for the low absolute recovery previously noted for amoxicillin and azithromycin.

Table 1. Absolute and relative recoveries of 26 compounds and 9 isotope labeled standards (IS) spiked in 0.5 liter pure water at a high concentration level applying three different flow rates; A = 10 mL/min, B = 20 mL/min, and C = 40 mL/min. Each experiment (A, B, and C) ran as triplicate. One-way ANOVA was performed on the absolute recovery results at $p = 0.05$, $F_{crit} = 5.1$. Absolute recovery for 6 compounds (grey marked) exceeded $F_{crit} = 5.1$. Compounds showing relative recovery lower than 50% are marked in light grey. The absolute amounts added ranged from 10–1000 ng depending on the compound (see Table 2, high, column 10). *Compound is run in both acid and basic UPLC (Ultra-Performance Liquid Chromatography) method. The 9 IS are indicated in italic.

Compound	Absolute Recovery						ANOVA, F	Relative Recovery						IS standard	Method
	A (%)	RSD (%)	B (%)	RSD (%)	C (%)	RSD (%)		A (%)	RSD (%)	B (%)	RSD (%)	C (%)	RSD (%)		
Acetamiprid	97.4	2.4	100.1	1.3	98.4	0.8	1.9	107.9	9.4	101.2	2.5	94.6	3.3	Carbamazepine_D10	Basic
Amoxicillin	3.3	0.3	3.2	0.1	3.1	0.4	0.4	15.7	2.7	10.7	0.7	8.3	1.8	Ciprofloxacin_D8	Acid
Azithromycin	3.5	1.7	4.0	0.4	9.6	0.8	28.1	19.9	11.8	15.0	2.4	27.8	3.5	Ciprofloxacin_D8	Acid
Carbamazepine	93.2	3.6	93.4	2.5	92.0	0.7	0.3	103.0	5.5	94.5	4.7	88.5	1.8	Carbamazepine_D10	Basic
Carbamazepine_D10	90.7	8.1	98.9	3.5	104.0	2.8	4.7								Basic/Acid*
Ciprofloxacin	22.7	4.1	26.9	3.3	32.2	6.8	2.7	105.2	8.0	89.9	2.7	84.1	11.6	Ciprofloxacin_D8	Acid
Ciprofloxacin_D8	21.4	2.2	29.8	3.2	37.9	3.1	24.6								Acid
Citalopram	53.5	3.8	57.2	6.6	55.4	5.4	0.3	51.9	1.4	51.5	6.1	48.2	5.9	Carbamazepine_D10	Acid
Clarithromycin	55.5	4.3	61.1	1.7	63.2	5.2	3.0	101.4	8.0	94.0	3.3	89.1	2.6	Clarithromycin_D3	Basic
Clarithromycin_D3	54.7	0.7	65.0	2.9	70.9	4.8	19.2								Basic
Clothianidin	101.9	1.6	101.5	4.1	102.3	1.8	0.1	99.1	6.9	91.4	5.5	89.0	2.3	Carbamazepine_D10	Acid
Diclofenac	80.8	2.9	84.7	7.1	85.9	5.9	0.7	104.0	6.1	97.2	2.4	91.1	2.5	Diclofenac_C6	Basic/Acid*
Diclofenac_C6	77.9	6.9	87.1	5.3	94.4	8.3	4.3								Basic
Doxycycline	77.4	1.8	83.6	3.6	76.6	9.2	1.3	75.3	4.2	75.3	2.0	66.8	9.7	Ciprofloxacin_D8	Acid
Erythromycin	21.2	0.1	27.8	1.4	37.7	3.4	45.5	38.7	0.5	42.7	0.7	53.2	4.0	Clarithromycin_D10	Basic
17-Beta-estradiol (E2)	57.6	2.5	62.6	7.2	63.1	3.4	1.2	101.5	6.7	95.7	4.9	90.1	1.7	17-Beta-estradiol (E2)_D5	Neutral
17-Beta-estradiol (E2)_D5	56.9	6.0	65.3	5.2	70.0	5.0	4.5								Neutral
Estrone (E1)	57.0	2.9	62.3	8.0	61.6	4.3	0.8	102.3	7.1	96.1	3.6	89.3	1.7	Estrone (E1)_D4	Neutral
Estrone (E1)_D4	56.0	6.4	64.8	6.8	69.0	6.0	3.2								Neutral
17-Alpha-ethinylestradiol (EE2)	52.2	2.9	59.1	8.7	57.5	5.4	1.1	93.6	6.0	91.1	5.2	83.5	3.1	Estrone (E1)_D4	Basic
Fluconazole	96.4	0.7	96.3	1.1	97.7	0.8	2.2	106.9	9.3	97.4	3.4	93.9	2.6	Carbamazepine_D10	Basic
Imidacloprid	97.3	0.2	98.0	1.2	97.9	1.7	0.3	107.8	10.4	99.2	2.8	94.2	2.0	Carbamazepine_D10	Basic
Metaflumizone	64.1	12.3	69.9	5.2	89.2	16.5	3.5	84.1	9.3	80.0	2.1	98.0	13.8	Diclofenac_C6	Acid
Methiocarb	68.2	4.0	69.8	5.6	69.4	3.1	0.1	102.0	6.0	94.0	5.5	87.4	3.6	Methiocarb_D3	Basic
Methiocarb_D3	67.1	6.5	74.3	4.2	79.5	5.3	3.9								Basic
Metoprolol	88.8	2.3	90.9	1.5	89.6	3.4	0.5	98.8	4.0	93.4	4.2	89.8	6.0	Metoprolol_D7	Basic
Metoprolol_D7	90.0	3.5	97.5	4.8	99.9	6.1	3.3								Basic
2-Ethylhexyl 4-methoxycinnamate	16.5	4.9	16.2	0.6	29.0	7.5	6.0	15.9	4.1	14.6	0.4	25.3	7.1	Carbamazepine_D10	Acid
Oxadiazone	48.8	3.9	53.3	4.1	52.9	8.8	0.5	47.3	1.3	48.0	3.9	46.1	8.7	Carbamazepine_D10	Acid
Oxazepame	85.7	2.5	87.0	4.1	86.7	3.4	0.1	110.4	7.0	100.0	3.1	92.1	4.3	Carbamazepine_D10	Basic
Sulfamethoxazole	96.2	1.8	95.5	0.4	93.5	3.2	1.2	102.1	6.8	94.0	3.4	88.0	3.3	Sulfamethoxazole_C6	Acid
Sulfamethoxazole_C6	94.4	4.8	101.6	3.8	106.3	0.8	8.5								Acid
Thiamethoxam	100.2	1.3	101.4	0.7	100.8	1.1	1.0	111.0	9.2	102.6	3.1	97.0	3.6	Carbamazepine_D10	Basic
Triallate	9.9	0.7	9.8	0.9	10.4	2.4	0.1	55.7	11.4	36.8	4.9	29.5	2.4	Carbamazepine_D10	Acid
Trimethoprim	88.3	0.9	87.2	0.2	86.1	1.4	3.9	97.9	9.6	88.2	3.0	82.9	3.5	Carbamazepine_D10	Basic
Mean:	64.2	3.3	68.2	3.5	70.7	4.3	F-crit 5.14	83.1	6.6	76.7	3.3	74.1	4.5		

2.1.2. Method Validation. Recovery Evaluation at High and Low Concentrations in Matrix-Rich Pond Water at the Highest Flow Rate of 40 mL/min

Electrospray ionization processes are known for both ion suppression and ion enhancement, with the former being more common [11]. A diluted humic-rich pond water provided the necessary matrix effect for this to be expressed. The carbon content was found to be 2.0 mg/L in the 1:5 diluted pond water. The absolute recovery results from the former experiment made us confident in picking and proceeding with the highest flow rate of 40 mL/min as the foundation for further method validation. It is noteworthy that not a single value, no matter what chemical functional groups, pointed in the direction of a higher recovery at a lower flow.

The results for this second set of experiments are presented in Table 2, ranked from the compound with the highest absolute recovery in the pond water to the compound with the lowest absolute recovery. IS compound recoveries are ranked as a separate group at the bottom of Table 2. The absolute amounts of compounds spiked to the 0.5 L sample volume can be found in the last two columns of Table 2. The high value is the same as that applied in the previous set of experiments (Table 1), and the low value represents a spiking level 10 times below that amount. As an example, carbamazepine has a high concentration of 20 ng/L (10 ng/0.5 L) and a low concentration of 2 ng/L (1 ng/0.5 L).

In the matrix-rich pond water, the mean absolute recovery at the high concentration dropped to 54.7% as compared to 70.7% for pure water (Table 1, absolute recovery C, column 6), but still showed a low mean RSD at 3.6%. In column 4 (Aq-matrix) in Table 2 the difference between absolute recovery in pure water (Table 1, absolute recovery C, column 6) and the pond water (Table 2, high, column 2) are presented for each individual compound. The majority of the compounds expressed a decrease in recovery (indicated as positive values). The decrease ranged from 2–63%, with fluconazole showing the largest difference between pure spiked water and matrix-rich pond water. However, six compounds showed higher absolute recovery in the presence of matrix (grey marked). The increased recovery results in the presence of matrix are particularly pronounced in three cases, namely triallate and the two antibiotics, azithromycin and doxycycline. A moderate enhancement could be noted for erythromycin, while clarithromycin and 2-ethylhexyl 4-methoxycinnamate had very minor increases. One extra experiment; processing un-spiked diluted pond water through the entire sample preparation chain, followed by post-spiking of the compounds to the obtained extract, showed that ion enhancement was indeed the case for these compounds, as blank pond water, processed and analyzed, revealed no traces of any of the investigated compounds.

It was also verified by this experiment that fluconazole suffers from true ion suppression, as post-spiking of the pond water showed a 59% difference in absolute recovery (data not shown) as compared to the pre-spiked pond water with a very similar decrease of 63% (Table 2, Aq-matrix, column 4). The average difference in absolute recovery for the 20 compounds showing ion suppression was 26.4%. Additionally, all 9 IS also suffered from ion suppression with an average difference of 32.9%.

Relative recoveries at low and high concentrations, using the IS calibration method that compensates for matrix interferences, are presented in column 5–8 in Table 2. Sometimes the compensation becomes too high as is the case for azithromycin, doxycycline, triallate and 2-ethylhexyl 4-methoxycinnamate, marked in light grey in Table 2 (high, column 7). A re-calculation using sulfamethoxazole_C6 as an IS (it was the IS with highest absolute recovery), gave a lower and more correct recovery situation for triallate and 2-ethylhexyl 4-methoxycinnamate, but this could not fully adjust for the matrix effect attributed to the two antibiotics, doxycycline and azithromycin. However, both antibiotics were paired with sulfamethoxazole_C6 (final method) before moving into the third experiment below. Native IS would be appropriate to consider regarding these two compounds to achieve better relative recoveries.

Table 2. Absolute and relative recoveries of 26 compounds spiked in 0.5 liter diluted matrix-rich pond water (1:5) at high and low concentration levels applying a flow rate of 40 mL/min. The experiment ran as a triplicate. The absolute amount (ng) of compounds spiked to the sample at low and high levels are shown in column 9 and 10. The absolute recovery is calculated for the high-level values as in the previous experiment (Table 1). Compounds are ranked based on absolute recovery, highest to lowest. The 9 IS are indicated in italic. Six compounds showed higher absolute recovery in the presence of matrix and two of them exceeded 100%, grey marked. Four compounds showed excessive relative recoveries, light grey.

Compound	Absolute Recovery		Aq-Matrix			Relative Recovery			
	High (%)	RSD (%)	Difference	Low (%)	RSD (%)	High (%)	RSD (%)	Low (ng)	High (ng)
Doxycycline	198.3	36.7	−121.7	30.4	2.3	436.7	97.8	5.0	50.0
Azithromycin	107.0	2.7	−97.4	87.1	0.7	382.4	16.6	5.0	50.0
Clothianidin	86.7	2.8	2.3	98.9	2.2	125.8	3.7	1.0	10.0
Sulfamethoxazole	82.3	2.4	11.2	116.7	0.4	119.3	4.0	1.0	10.0
Clarithromycin	65.9	3.6	−2.7	78.3	0.5	105.8	3.8	1.0	10.0
Carbamazepine	64.5	0.7	27.5	122.4	0.4	127.9	6.2	1.0	10.0
Metoprolol	63.4	1.6	26.2	94.4	0.5	100.7	2.9	1.0	10.0
Erythromycin	62.6	3.6	−24.9	112.7	0.2	100.6	4.9	1.0	10.0
Trimethoprim	60.8	1.4	25.4	125.9	0.0	120.4	5.2	1.0	10.0
Acetamiprid	60.7	1.0	37.6	119.8	0.3	120.4	4.9	1.0	10.0
Imidacloprid	60.0	2.5	38.0	116.1	0.7	118.9	1.8	1.0	10.0
Thiamethoxam	58.0	1.3	42.8	105.4	0.2	115.0	3.9	1.0	10.0
Oxazepame	56.9	1.6	29.8	132.8	0.2	112.8	3.8	1.0	10.0
Triallate	50.9	3.9	−40.5	92.5	0.6	182.1	17.2	1.0	10.0
Methiocarb	50.5	2.5	18.9	91.0	0.2	102.0	6.6	1.0	10.0
Oxadiazone	50.3	2.4	2.6	91.9	0.6	110.3	10.5	2.0	20.0
Diclofenac	42.3	3.0	43.6	115.4	1.3	105.1	2.3	1.0	10.0
2-Ethylhexyl 4-methoxycinnamate	38.9	9.5	−9.9	119.3	1.7	224.5	59.3	5.0	50.0
Metaflumizone	37.8	4.4	51.4	66.2	0.2	61.9	8.2	5.0	50.0
Citalopram	36.7	1.8	18.6	68.3	0.1	80.7	8.0	1.0	10.0
17-Beta-estradiol (E2)	34.8	1.1	28.3	102.3	0.8	105.6	7.1	50.0	500.0
Fluconazole	34.3	1.2	63.3	65.0	0.1	68.0	1.6	1.0	10.0
17-Alpha-ethinylestradiol (EE2)	31.6	1.0	25.9	91.9	0.9	100.4	5.5	100.0	1000.0
Ciprofloxacin	30.1	3.3	0.12	26.6	0.6	108.4	13.5	5.0	50.0
Estrone (E1)	28.8	1.2	32.8	88.8	0.3	91.7	2.7	10.0	100.0
Amoxicillin	1.3	0.3	1.9	137.6	3.1	44.6	9.8	10.0	100.0
Sulfamethoxazole_C6	68.5	4.0	37.8						
Metoprolol_D7	62.8	2.9	37.1						
Clarithromycin_D3	62.1	4.7	8.8						
Carbamazepine_D10	50.4	2.8	53.6						
Methiocarb_D3	49.5	3.2	29.9						
Diclofenac_C6	40.2	2.1	54.3						
17-Beta-estradiol (E2)_D5	33.0	2.8	37.0						
Estrone (E1)_D4	31.4	2.2	37.6						
Ciprofloxacin_D8	27.4	0.5	0.01						
Mean:	54.7	3.6							

2.1.3. Applying the Developed Method to Recipient Samples from Three Different Natural Locations

Results from recipient samples taken from three natural locations in the southern parts of Sweden are found in Table 3. In short, Degeberga represents a sample taken from the Segesholmsån River (average flow 0.6 m^3/s) downstream of Degeberga village. Degeberga wastewater treatment plant (WWTP) treats wastewater from a population of ca. 1000 people. St Olof represents a sample taken in the Rörums Södra Å River (average flow roughly 0.4 m^3/s) downstream of St Olof village that has a WWTP treating wastewater from a population of ca. 600 people. Both samples were expected to contain measurable concentrations of a number of compounds occurring in the developed method, though at a low concentration level due to small WWTPs with a great extent of dilution of the wastewater in the river. Pynten is the outlet point of basically undiluted wastewater from the city of Kristianstad's WWTP into the Hammarsjön Lake. The water has only been transported in a 1500 m long artificial channel from the WWTP to the lake, with only minor inflow of additional surface water. Kristianstad WWTP treats wastewater from ca. 52,000 people. Therefore, at the sampling point Pynten, relatively high concentrations of many of the compounds were expected. A map showing the sampling locations within the county of Scania (Sweden) can be found in Figure S1.

Given the fact that the location near the major city of Kristianstad showed the highest concentrations, the results were ranked based on these concentrations with the compound identified and quantified at the lowest concentration first, as shown in Table 3 (Pynten, column 12).

The IS standards are put alphabetically at the lower part of Table 3 (Degeberga, column 2; St Olof, column 4; Pynten, column 6). The location at Degeberga had a carbon content of 2.6 mg/L in the water and showed the highest absolute recovery values for all IS, except for clarithromycin_D3. Sulphamethoxazole_C6 was almost recovered to 100% in Degeberga. The St Olof water, with a carbon content of 6.2 mg/L, had the second highest absolute recovery values, and last came the water sample from Pynten (Kristianstad), which had a carbon content of 2.4 mg/L. All three sample locations showed low RSDs for the IS in general. However slightly higher RSDs could be noted for Degeberga compared to the other two locations. The matrix composition at Pynten (Kristianstad) caused a decrease in flow rate at the end of the sample loading procedure when approximately 100 mL of samples were left to process. Raising the pressure to the maximum of 1.5 bar could not compensate for the increased back pressure in the SPE column, and the sample finished with a flow rate of 27 mL/min instead of the starting value set at 40 mL/min.

A total of 11 out of 26 investigated compounds were identified in Degeberga and St Olof, while 15 compounds were found in Pynten (Kristianstad), out of which five compounds displayed concentrations above 100 ng/L—sulfamethoxazole, carbamazepine, diclofenac, metoprolol and oxazepam (marked in grey). RSD values were very low for these real-world recipient samples at all sampling sites. With very few exceptions, RSDs never exceeded 15%. It could be noted that at Degeberga the RSDs deviated a bit more as compared to the results at the other two locations. The two antibiotics, sulfamethoxazole and trimethoprim, were present at all three locations, as were citalopram, erythromycin, estrone, and fluconazole.

Table 3. Results for the 26 investigated compounds in recipient samples at three different natural locations: Degeberga, St Olof and Pynten. Each recipient water sample was previously processed in triplicates applying a flow rate of 40 mL/min. The results are presented based on findings in Pynten (close to the city of Kristianstad), ranging from the highest to lowest concentration. Five compounds displayed concentrations above 100 ng/L, light grey. The sampling locations can be seen in Figure S1.

Compound	Absolute Recovery						Relative Recovery					
	Degeberga (%)	RSD (%)	St Olof (%)	RSD (%)	Pynten (%)	RSD (%)	Degeberga (%)	RSD (%)	St Olof (%)	RSD (%)	Pynten (%)	RSD (%)
Acetamiprid							nd	-	nd	-	0.2	13.9
Thiamethoxam							nd	-	nd	-	0.9	23.9
Estrone (E1)							0.1	3.6	0.7	2.2	2.2	5.5
Imidacloprid							0.2	12.8	nd	-	3.0	2.3
Clarithromycin							nd	-	0.04	2.9	12.6	4.9
Azithromycin							nd	-	nd	-	15.8	1.4
Fluconazole							1.1	13.0	0.5	9.5	29.7	3.8
Citalopram							2.7	17.0	3.4	3.4	58.0	7.5
Erythromycin							1.2	11.9	0.7	9.7	62.7	2.5
Trimethoprim							0.2	3.6	3.6	1.7	75.2	4.9
Sulfamethoxazole							2.5	12.3	4.8	3.5	124.0	4.4
Carbamazepine							79.1	12.7	0.6	10.3	224.3	3.2
Diclofenac							30.3	13.1	20.7	1.8	374.5	3.5
Metoprolol							2.5	10.7	33.6	3.0	422.5	5.1
Oxazepame							20.8	14.8	14.1	6.8	473.0	4.1
Amoxicillin							nd	-	nd	-	nd	-
Ciprofloxacin							nd	-	nd	-	nd	-
Clothianidin							nd	-	nd	-	nd	-
Doxycycline							nd	-	nd	-	nd	-
17-Beta-estradiol (E2)							nd	-	nd	-	nd	-
17-Alpha-ethinylestradiol (EE2)							nd	-	nd	-	nd	-
Metaflumizone							nd	-	nd	-	nd	-
Methiocarb							nd	-	nd	-	nd	-
2-Ethylhexyl 4-methoxycinnamate							nd	-	nd	-	nd	-
Oxadiazone							nd	-	nd	-	nd	-
Triallate							nd	-	nd	-	nd	-
Carbamazepine-D10	63.6	8.1	51.9	0.8	33.8	1.1						
Ciprofloxacin_D8	7.8	0.6	4.3	2.1	3.1	1.1						
Clarithromycin_D3	51.5	5.6	47.8	1.8	60.7	2.3						
Diclofenac_C6	51.3	8.0	40.2	3.3	30.1	1.1						
17-Beta-estradiol (E2)_D5	42.2	4.6	39.9	2.2	38.6	2.0						
Estrone (E1)_D4	40.5	4.7	39.3	3.1	32.1	1.1						
Methiocarb_D3	62.5	8.0	57.2	4.8	48.0	1.1						
Metoprolol_D7	64.9	7.2	55.7	0.8	41.5	1.8						
Sulfamethoxazole_C6	99.7	12.8	70.9	3.6	45.7	1.8						

3. Discussion

As mentioned in the introduction, commercial literature speaks of recommended and applied flow rates in the development of SPE methods intended for polar and semi-polar organic trace analysis [2–4]. This is also true for scientific publications [12]. These flow rate figures are also excluded in the scientific literature [13,14], hence they are not to be considered a crucial factor when SPE methodology is under targeted evaluation [15]. The recommendation is generally not to exceed 10 mL/min, and sometimes recommendations are even less [3]. Only in rare occasions has the flow rate exceeded 10 mL/min as exemplified by Öllers and co-workers, who set the flow rate to 15 mL/min using Waters Oasis HLB SPE cartridges (60 mg, 3 mL) for quantification of neutral and acidic pharmaceuticals and pesticides [16].

We were surprised by the results in the recovery experiments of spiked pure water samples as presented in Table 1. Severe breakthroughs were expected at the higher flow rates since the recommended flow rates by several manufacturers exceeded four times at the highest flow rate of 40 mL/min [2–4]. On the contrary, the ANOVA analysis showed very clearly that there were no significant differences in absolute recoveries between the investigated flow rates for a vast majority of the compounds. The most likely explanation at this point is probably the fact that flow rates previously have been considered a factor not to impinge upon.

The significantly higher absolute recovery at 40 mL/min for six of the compounds may be interpreted as a result of decreased compound adsorption to the inner wall of the glass bottle as the contact time decreased at the higher flow rate. It is known that untreated glassware contains silicate and silanol groups that can act as ion-exchange and nucleophilic centers [10]. The low recoveries of ciprofloxacin, both as a mother compound and as IS, may be explained by the fact that ciprofloxacin can bind to exchangeable cations (which are bound to negatively charged surfaces) via the β-keto acid structure in the molecule [17].

The trend of generally higher calculated relative recoveries for the lowest flow rate compared with the higher flow rates can be explained by the difference in absolute response expressed by the labeled compounds (Table 1). One could also consider shifting the matchmaking of compound-IS pairs, already at this early point in method development, based on the relative recoveries. For example, citalopram could be better off with an IS that shows less absolute recovery, instead of using carbamazepine_D10 as an IS.

The noticed general drop in absolute recovery for the spiking experiment in pond water was as expected, due to matrix effects [11]. The low mean RSD of 3.6% at the high-level concentration, must therefore be considered satisfying given the complexity of both matrix and analysis method (Table 2). Doxycycline was an exception, with an RSD of 36.7%. Azithromycin and doxycycline, originally linked to the low recovery of ciprofloxacin, behaved differently in presence of the matrix. Turning to the relative recoveries at both low and high concentration levels (Table 2, low, column 6 and high, column 7), it is clear that for a majority of the compounds these are very good, meaning that in most cases the choice of IS works well (Table 1, IS standard, column 15) in the presence of matrix. For a few compounds, the relative recoveries are somewhat too high or low though. Some of the more striking examples of high relative recoveries at the high concentration level are azithromycin (382.4%) and doxycycline (436.7%), both originally linked to the low recovery of the IS ciprofloxacin. Two other examples of high relative recoveries at the high concentration level are triallate (182.1%) and 2-ethylhexyl 4 methoxycinnamate (224.5%), both combined with carbamazepine_D8, for which recombination of IS could be enough. The large relative recoveries of doxycycline and azithromycin would most likely be avoided by allowing them to have their own labeled IS. At the other end of the scale, there is also amoxicillin, which has a relatively low recovery of 44.6%, that possibly could gain from its own labeled IS.

There are many scientific publications dealing with analyzing these types of waters, so the individual compound results in absolute figures will not be discussed in detail. We can conclude, though, that the produced concentration levels (Table 3) and findings are in line with what other researchers have found in recipients affected by STP effluents [18]. Petrovic argued that for a multimethod, recovery outside of the required range of 70%–130% is acceptable if the precision is good and RSD is 20% [8]. All three investigated waters procedures had RSDs < 20%, regardless of the concentration of the reported compound. In Pynten (Kristianstad), where the concentrations were relatively high, RSDs were very low, and in parity with those observed for spiked matrix-rich pond water. This is expected since there is always an error margin in the analysis, which will have a greater impact on a low concentration number. The concentration results from Degeberga showed somewhat higher RSDs than St Olof, despite the concentrations being in the same order of magnitude. When examining the individual samples from Degeberga, it became clear that sample 1 had a somewhat deviating internal standard response (data not shown). This could probably be caused by a somewhat increased amount of labeled standard added to this sample, and as a gentle reminder of the importance of humbleness towards organic trace analysis.

4. Materials and Methods

As this work's primary focus was to test the hypothesis that flow rate can be increased without compromising analytical recovery in environmental analysis, detailed method descriptions will only be given for the SPE loading procedure. In Figure 3, a schematic layout is presented to put the sample loading procedure in the analytical chain in context, and to give a brief summary of the UPLC MS/MS system set up. As time and volume are key parameters in environmental chemistry analysis, these variables are listed for the individual analytical steps. The analytical chain is relatively straight forward between elution (C) and injection (F). The required part of the final analysis is comprised of a unique UPLC MS/MS setup. In short, the UPLC MS/MS (G) analysis was built up around three individual chromatographic methods; acid (A), basic (B), and neutral (N). During method development, each compound was evaluated and optimized without prejudices regarding historical residence in terms of chromatographic conditions and ESI (Electrospray ionization) mode. This unique strategy embraces the full potential of a multi-component UPLC-ESI-MS/MS method adapted to cover emerging contaminants of many different polarities, minimizing the elements of compromise in the performance of the final analytical separation and detection. Both sample preparation and final analysis is described in detail (chemicals, preparations, instrumentation, and quantification) in [19–21].

	A) Sample loading	B) Drying	C) Eluting	D) Evaporation	E) Reconstitution	F) Injection	G) Analysis
Volumes	500 ml	-	6 ml	-	1 ml	1 ul, 10 ul	-
Time (min)	12.5 25 50	20	2	22	1	-	6.5 (A) 6.5 (B) 8 (N)

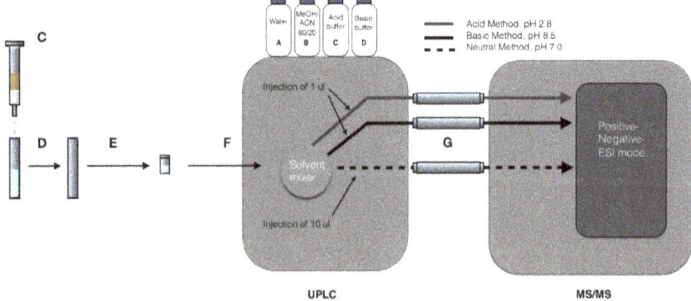

Figure 3. Schematic layout of the entire analytical procedure to put the sample preparation steps of the analytical chain in context, and to give a brief summary of the UPLC MS/MS system setup.

In Table 1 and column 16, each compound's original method belonging is noted as acid, basic or neutral. A fraction (1 µL or 10 µL) of each prepared sample vial volume (F) was injected once in each of the three methods. First, the basic method was applied to all samples, then the acid method, and finally the neutral method. One sample faced a total analysis time of 6.5 + 6.5 + 8 min, which included washing and equilibration between individual injections. Every method has its own dedicated UPLC column, as shown in Figure 3.

4.1. Chemicals

We used fine ground sand, extra pure, low iron, 40x100 mesh, from Fischer Scientific (Gothenburg, Sweden). Methanol HPLC-grade was from Sigma–Aldrich (Steinheim, Germany). Ultra-pure water (18.2 MOhm) was obtained from an OPTIMA water purification system (Elga Ltd., High Wycombe, Buckinghamshire, UK). SPE cartridges HLB Oasis (200 mg, 6 mL) were purchased from Waters (Gothenburg, Sweden).

4.2. Instrumentation

4.2.1. Sample Loader

Clean pressurized air was produced by an Atlas Copco SF2 Oil-free air system (Atlas Copco Airpower n.v. B, Wilrijk, Belgium). A flow meter measured the flow rate (Global Quality Analytical Resources, Scantec, Sweden). Two pressure regulators were installed and set to operate at a maximum 1.5 bar. VICI-Cap GL45, equipped with two ports, were used as sample bottle lids (VICI Jour, Schenkon, Switzerland). A Micro Metering Valve Assy 1/8 (IDEX Health & Scientific, Oak Harbor, WA, USA) controlled the flow rate. PTFE tubing (1/8", 2.40 mm ID), PTFE Ferrules 1/8", Nut, PPS (flangeless) 1/8" connector and SPE-adapters were provided by Scantec, Partille, Sweden. DURAN® laboratory bottle, pressure plus+ is recommended (500 mL), Maintz, Germany. A wastewater container and gasket PTFE-tape were purchased at Biltema, Kristianstad, Sweden. SPE-column.

4.2.2. Sample Loader Construction

The sample loader parts described above were assembled according to the schematics in Figure 1. Two PTFE-tubings of 50 cm each were connected to the two holes in the lid (Figure 1a). One of the tubes is dedicated for air supply and the other tubing will transfer the sample to the SPE-column. The end of this tubing is expanded, hindering it from popping out as pressure is applied, by screwing a screw a couple of turns into the tubing. A total of 2.0 g of finely ground sand was put on top of the SPE column frit. This sand will efficiently hinder larger particles from clogging the SPE inlet frit, enabling efficient whole water analysis. Gasket tape was applied around the SPE- adapter to prevent leakage, and then the adapter, joint with the VICI-lid, and the SPE-column were assembled according to Figure 1a.

4.3. Methods

The three separate experiments (4.3.1, 4.3.2, and 4.3.3) shared the same sample preparation, sample loader construction, sample loader operation, and drying procedure. The samples from these experiments were eluted, reconstituted, and analyzed in the same fashion: 500 g of sample water were weighed in to a 500 mL sample bottle (Figure 1), and 100 µL FoA (10%) and 100 µL EDTA (sat. sol.) was added together with 30 µL of IS mixture. All experiments were carried out in triplicates.

4.3.1. Recovery from Pure Water at Three Flow Rates

The investigated flow rates were 10, 20, and 40 mL/min. Samples were prepared according to 4.3. A total of 100 µL of compound mixture was added at high level (Table 2).

4.3.2. Method Validation. Recovery from Matrix Containing Pond Water at High Flow Rate

The flow rate was set to 40 mL/min. The pond water used was diluted 1:5 in pure water and was estimated to have a humic carbon content of 2.0 mg/L. The carbon content was determined using the UV_{254} method and ERM-CA 100 as reference (carbon content of 20 mg/L).

Samples were prepared according to 4.3. A total of 100 µL of compound mixture was added at low and high levels (Table 2).

4.3.3. Applying the Method to Recipient Samples from Three Different Natural Locations

Samples were collected from three different locations in Scania, Sweden with the following coordinates; St Olof: 55°38′25.3″N 14°8′3.3E, Degeberga: 55°49′47.4″N 14°7′12.0″E, and Kristanstad: 56°0′50.3″N 14°11′9.8″E. Coordinates are given in the format of the World Geodetic System 1984 (WGS 84). Each sample set triplicate was prepared according to 4.3. An overview of the sampling locations is seen in Figure S1.

4.3.4. Sample Loader Operation

IMPORTANT: Always use two pressure regulators to ensure a safe operation, keeping pressure within the safety margins as described by the glass manufacturer [22]. The bottle used must always: (1) be visually inspected to check that it is in good condition, and (2) as a safety measure placed in a shatterproof container.

Pressure was set to <0.5 bar. Each of the three trials were executed in the exact same way: flow was measured and set to the desired rate by connecting the flow meter to each SPE outlet, and pressure was adjusted by gently increasing the pressure and fine-tuning by adjusting the Micro Metering Valve (Figure 1a). The flow was monitored at regular intervals during the sample loading procedure, and pressure was adjusted if needed.

4.3.5. Drying Procedure, Eluting, Reconstituting, and Final Analysis

After each round, pressure was relieved and the SPE-columns brought to drying, according to Figure 1b. A flow of pure air at 2 bar for 30 min dried the columns. The columns were eluted into glass tubes with 6 mL MeOH, and then evaporated at 40 °C for 22 min in a TurboVap, Biotage. The samples were reconstituted in 100 µL MeO, 5 µL EDTA, and 895 µL H_2O, then analyzed according to the UPLC MS/MS methods described in [19–21].

4.4. Calculations

4.4.1. Absolute Recovery

Absolute recovery results were calculated as:

$$\text{Absolute recovery (\%)} = 100 \cdot (a/b) \tag{1}$$

where a is the MS/MS response of a given compound and b is the MS/MS mean response, generated from four injections from two individual, high-level standards.

4.4.2. Relative Recovery

The relative recovery results were calculated as described in the comprehensive analytical methodology by the United States Environmental Protection Agency; EPA Method1694 (U.S. Environmental Protection Agency 2007) [20]. The relative response (RR) (labeled to native) vs. concentration in the calibration solutions was computed over the calibration range according to (2):

$$RR = (A_n C_l)/(A_l C_n) \tag{2}$$

A_n = the area of the daughter m/z for the native compound
A_l = the area of the daughter m/z for the labeled compound
C_l = the amount of the labeled compound in the calibration standard (pg)
C_n = the amount of the native compound in the calibration standard (pg)

The concentration of a native compound was calculated according to (3):

$$C_n = (A_n C_l)/(A_l RR) \tag{3}$$

Each pair of compound-IS are listed in Table 1. In experiment 4.3.1 and 4.3.2, relative recoveries are presented as percentage figures of the determined amounts of the added compounds (Tables 1 and 2). In experiment 4.3.3, the relative recoveries are presented as the determined concentration in ng/L for the natural location samples using IS-method (Table 3).

4.4.3. ANOVA

One-way ANOVA were calculated for the means of the absolute recovery in experiment 4.3.1 (Table 1, column 7). F-crit 5.14, at $p = 0.05$. Calculations were made in Excel version 16.21.1.

5. Conclusions

The use of an innovative positive pressure sample loading technique with an in-line filter of finely ground sand, enabled large volume sample loading at a high flow rate with good repeatability. All three flow rates produced highly repeatable results, and this allowed flow rate to increase up to 40 mL/min for a 200 mg, 6 mL, SPE cartridge of reversed phase type without compromising the recoveries. There is, at least as far as for the investigated compounds, a generous flow rate window from 10 to 40 mL/min where sample loading can be conducted. The high flow rate at 40 mL/min is more than four times those previously used, and could mean a substantial contribution for reducing the sample preparation bottleneck. The use of this technology together with the finding of a generous flow rate window allows the sample to (1) be transferred to the SPE-column at the sample location and (2) by the person taking the sample. This further means that only the sample cartridge would be needed to be sent to the laboratory, instead of the whole water sample, as of today's technique. This approach avoids the degradation that can occur in the transported frozen samples, as sample that is transferred to the sample cartridge as early as possible is much more stable. It also saves the cumbersome handling of the frozen samples. Altogether, these improved sample handling steps may also contribute to a greener analytical protocol by a much more simplified overall procedure [23]. It should also be noted that the novel technology can be appropriate for detection techniques other than MS/MS.

Supplementary Materials: Figure S1: Overview of the three different natural sampling locations at Pynten (Kristianstad), Degeberga, and St Olof.

Author Contributions: O.S. conceived and designed the experiments; O.S. performed the experiments; O.S. and E.B. analyzed the data; O.S. contributed reagents/materials/analysis tools; O.S. wrote the paper; O.S. and E.B. edited the paper.

Funding: This research received no external funding.

Acknowledgments: Mattias Arvidsson, Scantec, Sweden, for true dedication, technical assistance and rewarding discussions during technical development of the sample loader.

Conflicts of Interest: The authors declare no conflicts of interest.

References

1. Petrovic, M.; Farré, M.; de Alda, M.L.; Pérez, S.; Postigo, C.; Köck, M.; Radjenovic, J.; Gros, M.; Barcelo, D. Recent trends in the liquid chromatography-mass spectrometry analysis of organic contaminants in environmental samples. *J. Chromatogr. A* **2010**, *1217*, 4004–4017. [CrossRef] [PubMed]

2. *QuickStart Guide to SPE*; Biotage: Uppsala, Sweden, 2013.
3. *Guide to Solid Phase Extraction. Bulletin 910*; Supelco: Bellefonte, PA, USA, 1998.
4. *Beginners Guide to SPE*; Waters: Milford, MA, USA, 2015.
5. Lopez-Serna, R.L.; Petrovic, M.; Barcelo, D. Development of a fast instrumental method for the analysis of pharmaceuticals in environmental and wastewaters based on ultra high performance liquid chromatography (UHPLC)-tandem mass spectrometry (MS/MS). *Chemosphere* **2011**, *85*, 1390–1399. [CrossRef] [PubMed]
6. COMMISSION IMPLEMENTING DECISION (EU) 2015/495 of 20 March. Establishing a Watch List of Substances for Union-Wide Monitoring in the Field of Water Policy Pursuant to Directive 2008/105/EC of the European Parliament and of the Council. L 78/40; Official Journal of the European Union, Publications Office of the European Union: Luxembourg, 2015.
7. COMMISSION IMPLEMENTING DECISION (EU) 2018/840 of 5 June 2018. Establishing a Watch List of Substances for Union-Wide Monitoring in the Field of Water Policy Pursuant to DIRECTIVE 2008/105/EC of the European Parliament and of the Council. L 141/61; Official Journal of the European Union, Publications Office of the European Union: Luxembourg, 2018.
8. Petrovic, M. Methodological challenges of multi-residue analysis of pharmaceuticals in environmental samples. *Trends Environ. Anal. Chem.* **2014**, *1*, e25–e33. [CrossRef]
9. Svahn, O.; Björklund, E. Simple, fast and inexpensive large "whole water" volume sample SPE-loading using compressed air and finely ground sand. *Anal. Methods VL IS* **2019**, *1431*, 64–73. [CrossRef]
10. Suri, R.P.S.; Singh, T.S.; Chimchirian, R.F. Effect of process conditions on the analysis of free and conjugated estrogen hormones by solid-phase extraction–gas chromatography/mass spectrometry (SPE–GC/MS). *Environ. Monit. Assess.* **2011**, *184*, 1657–1669. [CrossRef] [PubMed]
11. Wickramasekara, S.; Hernández-Ruiz, S.; Abrell, L.; Arnold, R.; Chorover, J. Natural dissolved organic matter affects electrospray ionization during analysis of emerging contaminants by mass spectrometry. *Anal. Chim. Acta* **2012**, *717*, 77–84. [CrossRef]
12. Petrie, B.; Youdan, J.; Barden, R.; Kasprzyk-Hordern, B. Multi-residue analysis of 90 emerging contaminants in liquid and solid environmental matrices by ultra-high-performance liquid chromatography tandem mass spectrometry. *J. Chromatogr. A* **2016**, *1431*, 1–15. [CrossRef] [PubMed]
13. Gracia-Lor, E.; Martínez, M.; Sancho, J.V.; Peñuela, G.; Hernández, F. Multi-class determination of personal care products and pharmaceuticals in environmental and wastewater samples by ultra-high performance liquid-chromatography-tandem mass spectrometry. *Talanta* **2012**, *99*, 1011–1023. [CrossRef]
14. Hladik, M.L.; Kolpin, D.W.; Kuivila, K.M. Widespread occurrence of neonicotinoid insecticides in streams in a high corn and soybean producing region. USA. *Environ. Pollut.* **2014**, *193*, 189–196. [CrossRef]
15. Baker, D.R.; Kasprzyk-Hordern, B. Critical evaluation of methodology commonly used in sample collection, storage and preparation for the analysis of pharmaceuticals and illicit drugs in surface water and wastewater by solid phase extraction and liquid chromatography-mass spectrometry. *J. Chromatogr. A* **2011**, *1218*, 8036–8059. [CrossRef] [PubMed]
16. Öller SSinger, H.P.; Fässler, P.; Müller, S.R. Simultaneous quantification of neutral and acidic pharmaceuticals and pesticides at the low-ng/l level in surface and waste water. *J. Chromatogr. A* **2001**, *911*, 225–234. [CrossRef]
17. Nowara, A.; Burhenne, J.; Spiteller, M. Binding of Fluoroquinolone Carboxylic Acid Derivatives to ClayMinerals. *J. Agric. Food Chem.* **1997**, 1459–1463. [CrossRef]
18. Pérez-Fernández, V.; Rocca, L.M.; Tomai, P.; Fanali, S.; Gentili, A. Recent advancements and future trends in environmental analysis: Sample preparation, liquid chromatography and mass spectrometry. *Anal. Chim. Acta* **2017**, *983*, 9–41. [CrossRef] [PubMed]
19. Svahn, O.; Björklund, E. Increased electrospray ionization intensities and expanded chromatographic possibilities for emerging contaminants using mobile phases of different pH. *J. Chromatogr. B* **2016**, *1033–1034*, 128–137. [CrossRef] [PubMed]
20. *Method 1694: Pharmaceuticals and Personal Care Products in Water, Soil, Sediment, and Biosolids by HPLC/MS/MS*; EPA: Washington, DC, USA, 2007.
21. Svahn, O. *Tillämpad Miljöanalytisk Kemi för Monitorering Och Åtgärder av Antibiotika- Och Läkemedelsrester i Vattenriket*; Lund University: Lund, Sweden, 2016.

22. DURAN, Group. Available online: http://www.duran-group.com/en/products-solutions/laboratory-glassware/products/laboratory-glass-bottles/ (accessed on 10 March 2018).
23. Płotka-Wasylka, J. A new tool for the evaluation of the analytical procedure: Green Analytical Procedure Index. *Talanta* **2018**, *181*, 204–209. [CrossRef] [PubMed]

© 2019 by the authors. Licensee MDPI, Basel, Switzerland. This article is an open access article distributed under the terms and conditions of the Creative Commons Attribution (CC BY) license (http://creativecommons.org/licenses/by/4.0/).

Article

New Vortex-Synchronized Matrix Solid-Phase Dispersion Method for Simultaneous Determination of Four Anthraquinones in Cassiae Semen

Luyi Jiang [1,†], Jie Wang [2,†], Huan Zhang [1], Caijing Liu [1], Yiping Tang [2,*] and Chu Chu [1,*]

1. College of Pharmaceutical Science, Zhejiang University of Technology, Hangzhou 310014, China; 15700082652@163.com (L.J.); saynoproblem@163.com (H.Z.); 15958027383@163.com (C.L.)
2. College of Materials Science and Engineering, Zhejiang University of Technology, Hangzhou 310014, China; wangjie930724@163.com
* Correspondence: tangyiping@zjut.edu.cn (Y.T.); chuchu@zjut.edu.cn (C.C.); Tel.: +86-571-88320219 (Y.T.); +86-571-88320913 (C.C.)
† The authors contribute equally to this work.

Academic Editor: Nuno Neng
Received: 24 February 2019; Accepted: 29 March 2019; Published: 3 April 2019

Abstract: In this study, a green ionic-liquid based vortex-synchronized matrix solid-phase dispersion (VS-MSPD) combined with high performance liquid chromatography (HPLC) method was developed as a quantitative determination method for four anthraquinones in Cassiae Semen. Two conventional adsorbents, C_{18} and silica gel were investigated. The strategy included two steps: Extraction and determination. Wasted crab shells were used as an alternative adsorbent and ionic liquid was used as an alternative solvent in the first step. Factors affecting extraction efficiency were optimized: A sample/adsorbent ratio of 2:1, a grinding time of 3 min, a vortex time of 3 min, and ionic liquid ([Domim]HSO_4, 250 mM) was used as eluent in the VS-MSPD procedure. As a result, the established method provided satisfactory linearity (R > 0.999), good accuracy and high reproducibility (RSD < 4.60%), and it exhibited the advantages of smaller sample amounts, shorter extraction time, less volume of elution solvent, and was much more environmental-friendly when compared with other conventional methods.

Keywords: vortex-synchronized matrix solid-phase dispersion; crab shells; ionic liquids; anthraquinones; Cassiae Semen

---

1. Introduction

As seafood consumption has increased, shellfish cultivation has emerged as an expanding economic activity in the world, accounting for more than 40% of all marine aquaculture production [1]. Crab, is one of the top ten highest consumed sea food products and continues to compete well against other seafood proteins. However, only a small portion (20–30%) of the weight of crab has been consumed as food on tables, the rest is generally considered to be garbage and is discarded at will [2]. Consequently, a public health problem caused by the wastage of crab shell resources and the deterioration of the environment, which is due to their low biodegradation has arisen [3]. There is an urgent need to develop methods for treating wasted crab shells and making this bio waste valuable.

Crab shell powder (CSP) of marine waste was used for biofungicide production about ten years ago [4,5]. Recently, a good way of recycling wasted crab shells is its application in the removal of pollutants as an adsorbent [1,2,6–13]. It has been reported that crab shells consist of chitin (or chitosan), calcium carbonate along with some proteins. Among them, chitosan, possessing available amino groups on its polymeric chain, shows high metal binding capability and has been used for the adsorption of heavy metals (Cu, Pb, Cd, Ni, Cr, Au, Se, V, Eu, Co, Ce) [6–12], As(V) [14], and other substances such as phosphate [13] and dye (Congo Red) [1,2]. Moreover, the advantages of being used as an adsorbent include the low cost, great mechanical strength, rigid structure, ability to withstand drastic conditions and high biocompatibility [14,15]. So far, however, there have only been a few reports on the extraction of bioactive compounds in a complex matrix.

Cassiae Semen, derived from the ripe seed of *Cassia obtusifolia* L. or *C. tora* L. is an officially edible and medicinal plant in China [16]. It is commonly drunk as a kind of healthy tea for its function in liver protection, bacteriostasis, catharsis and immune adjustment, eyesight improvement, diuresis, antitumor, and antioxidation in many southeast Asian countries [17–19]. It has also been used as a slimming tea to aid metabolic processes by helping detoxify the body [20]. A variety of bioactive anthraquinones, including aurantio-obtusin, chryso-obtusin, obtusin, aloe-emodin, rhein, emodin, chrysophanol, physcion and glucosides, have been reported in Cassiae Semen [19,21]. Among them, chrysophanol and aurantio-obtusin were selected as markers for quality control of Cassiae Semen in the Chinese Pharmacopoeia (2015 ed). However, chrysophanol is non-specific in Cassiae Semen, therefore more bioactive markers are in urgent need to be used for its quality evaluation. The development of efficient and sensitive methodologies to determine those bioactive compounds is also needed.

Several extraction techniques, including heating reflux extraction or ultrasonic extraction, have been reported in the literature for the extraction of anthraquinones from Cassiae Semen [17–20,22–25]. Those methods either need a high volume of toxic solvents, such as methanol or chloroform, or a long extraction time. Additionally, the analytes are exposed to the chemicals in a consecutive clean-up step. Recently, using environmentally-friendly solvents, keeping acceptable accuracy and extraction efficiency, has become the trend of the development of modern methods. Therefore, VS-MSPD technique based on a matrix solid phase dispersion by using vortex agitation instead of the solid phase extraction step, seems to be a good choice. Several studies have shown that it leads to acceptable results with the advantage of high speed, simplicity, and a low cost material in complex matrices [26–32].

MSPD was developed for the treatment of solid, semisolid and highly viscous samples by dispersing the matrix in a solid phase adsorbent [33]. Recent advances in MSPD include the fact that using cheaper, greener solid supports, nanoparticles or environmental-friendly eluents instead of traditional ones [34–37]. In this study, a green vortex-synchronized matrix solid-phase dispersion (VS-MSPD) combined with a HPLC method has been developed to determine four anthraquinones, including aurantio-obtusin, chryso-obtusin, obtusifolin and emodin in Cassiae Semen samples. The experimental procedure consisted of two steps, which were extraction and elution. In the extraction step, solid waste from the sea food industry, CSP, was firstly applied as an alternative adsorbent to extract natural components in the food matrix. Moreover, green ionic liquids (ILs)such as [Bmim]PF$_6$, [Domim]Cl, [Domim]NO$_3$ and [Domim]HSO$_4$ were used instead of a harmful organic solvent for extraction for the first time. The main objective of this study was to develop a method that needed smaller sample amounts, a shorter extraction time, a smaller volume of elution solvent, and was much more environmental-friendly. The developed method has shown great potential on rapid extraction and determination of natural products from complex samples.

2. Results and Discussion

2.1. Characterization of Crab Shell Powder (CSP)

SEM analysis was carried out to observe the morphology of CSP, as show in Figure 1. The size of CSP was less than 80 μm, which agreed with the size of a standard sieve with 200 mesh. Additionally, from the high magnification shown in Figure 1b, the morphology of CSP was very loose and porous indicating the high surface area, which benefits the adsorption process.

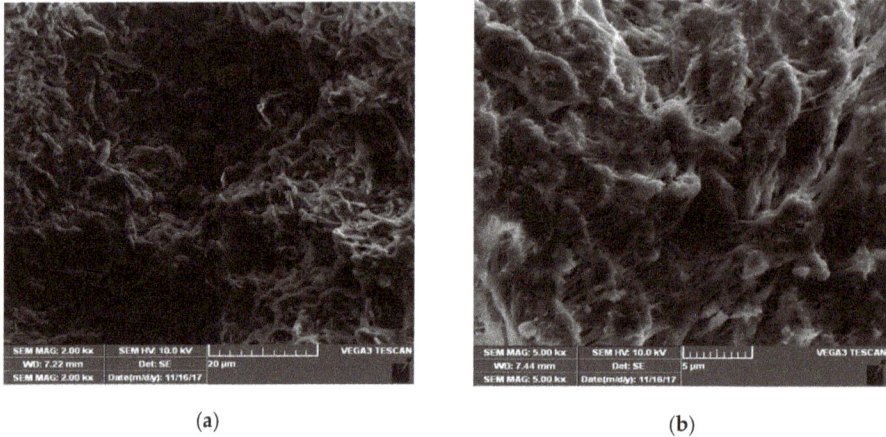

Figure 1. SEM images of CSP at (**a**) low and (**b**) high magnification.

A FT-IR spectrum was also applied to evaluate CSP, as shown in Figure 2. There were six main peaks in the spectra. The sharp peaks located at 1414, 874, 713 cm^{-1} corresponded to CO_3^{2-}, which confirmed the existence of calcite [38]. The strong peak around 3344 cm^{-1} was O-H stretching vibration [39] and N-H stretching vibration of chitin [40]. The peak around 2961 cm^{-1} was ascribed to C-H stretching vibration [41]. The peak around 1070 cm^{-1} was due to C-O stretching vibration of carbohydrate [42]. This implied that chitin was the main composition of CSP.

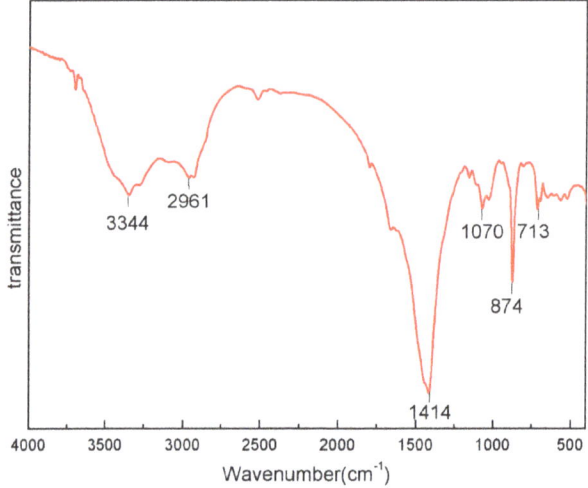

Figure 2. FT-IR spectrum for CSP.

2.2. VS-MSPD Optimization

MSPD was developed to extract compounds from solid, semisolid and highly viscous samples by disrupting and dispersing the matrix in a solid phase adsorbent [33]. The sample components were dissolved by the bonded organic phase and better dispersed on the surface of the support. The surface area of the extracted sample was increased, and the sample was dispersed in the organic phase according to their respective polarities. Furthermore, to obtain the optimal extraction efficiency of the four anthraquinones on VS-MSPD, some factors were investigated, including the type of adsorbent, sample/adsorbent ratio, the grinding time, the vortex time, the concentration and the type of ionic liquid and volume of ionic liquid. All of these parameters were evaluated in detail and each test was repeated in triplicate. Analysis of variance (ANOVA) and Duncan's test were conducted among different groups by SPSS software. The comparisons with $P < 0.05$ were considered significant.

The extraction efficiency of an MSPD procedure depends on the type of adsorbent used [36]. The mechanism of adsorption, the choice of eluent and the interfering compounds remaining on its surface are affected by the different physicochemical properties of adsorbent. In the current experiment, three different adsorbents, including CSP and two conventional adsorbents, C_{18} and silica gel, were investigated. The mixture with the sample/adsorbent ratio of 1:1 was ground for 3 min and eluted with 250 mM [Domim]NO$_3$ aqueous solution by 3 min-vortex. As shown in Figure 3a, C_{18} provided a relatively poor response for all of the anthraquinones, which may be ascribed to the hydrophobic characteristics and the non-polar nature of the reversed-phase C_{18} particles. Silica gel with abundant Si-O-Si and Si-OH groups accelerated the rate of forming hydrogen bonds between sorbent affording higher extraction efficiency than C_{18} to four anthraquinones [37]. CSP showed similar extraction efficiency to silica gel. It might be explained that after base treatment, chitosan was the main residue contained in CSP, which possessed available -NH$_2$ and -OH groups on its polymeric chain and therefore showed high binding capability with anthraquinones [2]. Moreover, the cellulose-like backbone of CSP may also help to adsorb anthraquinones [13]. For the purpose of making the bio waste valuable, CSP was selected as an optimal adsorbent in the following study.

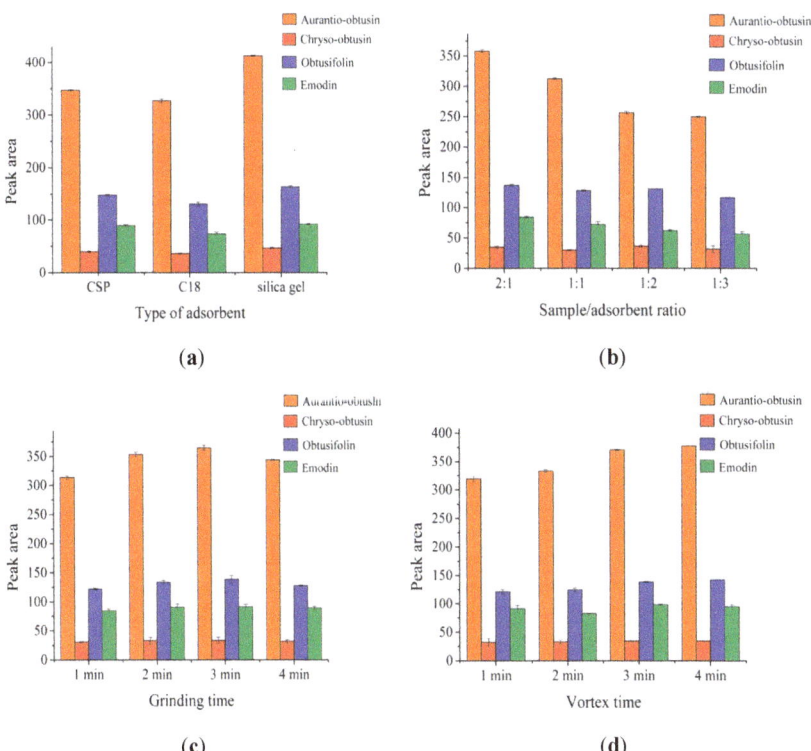

Figure 3. (**a**) Effect of different adsorbents in VS-MSPD process; (**b**) effect of sample/adsorbent ratio in VS-MSPD process; (**c**) effect of grinding time in VS-MSPD process; (**d**) effect of vortex time in the VS-MSPD process.

The sample/adsorbent ratio is a significant factor affecting extraction efficiency since the contact surface/active sites on the adsorbent surface should be sufficient to trap the target compounds [43]. The mixture of CSP and sample was ground for 3 min and eluted with 250 mM [Domim]NO$_3$ aqueous solution by 3 min-vortex. It was seen from Figure 3b that the ratio 2:1 (mg of samples/mg of adsorbent) was found to be optimal for all investigated anthraquinones. Moreover, the extraction efficiency decreased when the ratio changed to 1:1, 1:2 or 1:3, which might be ascribed to the stronger adsorption capacity and the harder elution with the increasing amount of CSP. The decreased extraction efficiency might also be due to increasing hydrogen bonding and electrostatic interaction between the four target compounds and CSP [36]. Thus, the sample/adsorbent ratio of 2:1 was finally chosen in the subsequent experiment.

An entire adsorption requires a sufficient contraction time to get a homogeneous mixing [37]. The mixture of CSP and sample (2:1) was ground at different times and vortexed with a 250 mM [Domim]NO$_3$ aqueous solution over 3 min. Figure 3c vividly depicts a different extraction efficiency with an increased grinding time. The result indicated that the extraction efficiency enhanced when the grinding time increased from 1 min to 3 min, which may be ascribed to the much stronger contraction between four analytes and CSP. However, the response of all the target compounds decreased while the grinding time prolonged to 4 min, which might have been due to a long grinding time leading to an overly strong extraction of the four anthraquinones, which increased the difficulty of elution. Therefore, grinding for 3 min was used for further investigations.

An adequate vortex time can elute four analytes from CSP completely [37]. The mixture of CSP and sample with the ratio of 2:1 was ground for 3 min and vortexed with a 250 mM [Domim]NO$_3$ aqueous solution for 3 min. As shown in Figure 3d, the extraction efficiency raised when the vortex time increased from 1 min to 3 min. However, the extraction efficiency of a 4 min-vortex was no more significantly increased compared to that of a 3 min-vortex, which indicated that 3 min of vortex was enough in this method. Therefore, a vortex time of 3 min was selected in the following study.

The use of solvents for both the environment and human beings associated with sample preparation is one of the goals of green Analytical Chemistry. Ionic liquids possess unique chemico-physical properties, and are a kind of green solvent which has been applied widely in chemical synthesis, catalysis, separation sciences and electrochemistry [36]. The mixture of CSP and sample (2:1) was ground for 3 min and eluted with 250 mM different kinds of ILs aqueous solution by 3 min-vortex. In this work, four types of ILs, including [Bmim]PF$_6$, [Domim]Cl, [Domim]HSO$_4$ and [Domim]NO$_3$, were studied to obtain a satisfactory extraction efficiency. As seen in Figure 4, [Bmim]PF$_6$ had the best elution efficiency of aurantio-obtusin, but the response of chryso-obtusin, obtusifolin and emodin was not very good, which might have been due to the diverse polarity between [Bmim]PF$_6$ and the three target compounds. The other three ILs, [Domim]Cl, [Domim]HSO$_4$ and [Domim]NO$_3$, although consisting of the same long-chain cation, had different elution efficiencies on four target compounds, in which the extraction efficiency of [Domim]HSO$_4$ and [Domim]NO$_3$ was better than [Domim]Cl. That might be ascribed to the weaker electrostatic interaction between the anion Cl$^-$ and analytes. The results showed that a better extraction efficiency could be obtained with [Domim]HSO$_4$ as the elution solvent, which might have accounted for the dominating π–π and hydrogen bonding interactions from the HSO$_4^-$ anion and the stronger electrostatic interaction between the anion HSO$_4^-$ and analytes [37]. Furthermore, interactions with analytes might also be possible due to the ion-dipole and the inclusion complexation, which might affect the extraction efficiency [36]. Finally, [Domim]HSO$_4$ was selected as the eluent in further experiments.

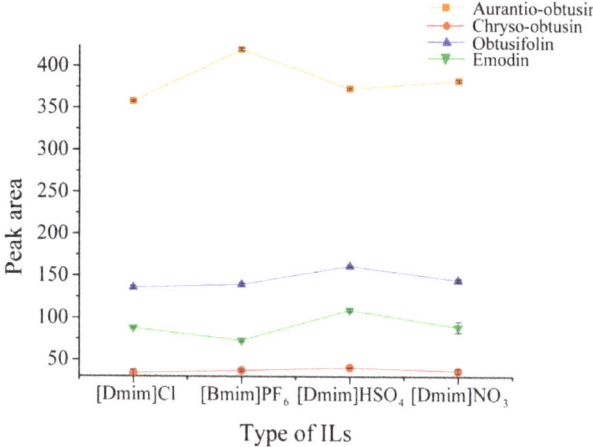

Figure 4. Effect of different ILs in VS-MSPD process.

The concentration of IL was another crucial factor in the elution procedure, because a suitable concentration can result in a satisfactory extraction of the analytes. The mixture of CSP and sample according to the ratio of 2:1, was ground for 3 min and eluted with different concentrations of [Domim]HSO$_4$ aqueous solution by 3 min-vortex. It was found in Figure 5 that the peak areas of four target compounds increased from 150 mM to 250 mM, which might have been due to the solubility and elution capacity of the solvent, which were enhanced with an increased ionic liquid concentration. Upon further increases of [Domim]HSO$_4$ concentration, the extraction efficiency was not much more significantly increased, which might be accountable to the increase in the solution viscosity resulting in the difficulty with the mass transfer of the anthraquinones from the Cassiae Semen matrix into [Domim]HSO$_4$ aqueous solution [36]. On the basis of the above results, the concentration of 250 mM was chosen as the best compromise.

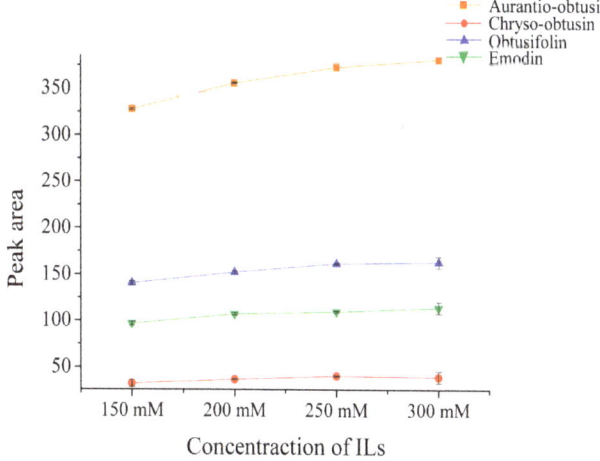

Figure 5. Effect of concentration ILs in VS-MSPD process.

2.3. Method Validation

Method validation was characterized on the basis of linearity, LODs, LOQs, precision, reproducibility and recovery. The results are demonstrated in Tables 1 and 2. The calibration curves of four analytes were built in a form of $y = ax + b$ with peak area (y) and the concentration of the analyte (x). Good correlation coefficients (R > 0.999) were obtained for all analytes. The LOD values were estimated to be 1.40, 2.00, 1.04 and 0.44 µg·mL^{-1} for aurantio-obtusin, chryso-obtusin, obtusifolin and emodin, respectively. The LOQs were based on S/N = 10, which ranged from 2.20 to 13.30 µg·mL^{-1}. Additionally, the precision values of intra-day and inter-day shown in Table 1 were less than 3.59% and 2.90%, respectively. Moreover, the reproducibility values of the method were estimated to be 1.79%, 4.60%, 2.41% and 3.03% for aurantio-obtusin, chryso-obtusin, obtusifolin and emodin, respectively. The stability values were in the 2.00–3.15% range, which indicated the sample solution was stable in 24 h. As listed in Table 2, the average recoveries of four analytes were all in the range of 91.30–106.40% with relative standard deviations (RSDs) of 0.98–3.98%. These results indicated that crab shell-based VS-MSPD using ILs as eluent coupled with HPLC method exhibited satisfactory linearity, good accuracy and high reproducibility.

Table 1. Linear regression data, limit of detection (LOD), limit of quantitation (LOQ), precision, reproducibility and stability of the entire analytical method.

Analytes	Linear Regression Data		Range (µg mL^{-1})	LOD (µg mL^{-1})	LOQ (µg mL^{-1})	Precision (RSD%)		Reproducibility (RSD%)	Stability (RSD%)
	Calibration Curve	R				Intra-Day ($n = 6$)	Inter-Day ($n = 3$)		
Aurantio-obtusin	y = 10.048x + 6.8409	0.9997	3.09–123.60	1.40	4.00	1.26	2.50	1.79	2.47
Chryso-obtusin	y = 2.5094x − 1.4986	0.9997	2.00–40.00	2.00	13.30	3.59	2.90	4.60	2.75
Obtusifolin	y = 13.724x + 0.7768	0.9999	1.04–41.60	1.04	3.50	1.17	2.90	2.41	3.15
Emodin	y = 35.053x + 1.3954	0.9998	1.32–13.20	0.44	2.20	1.91	2.90	3.03	2.00

Table 2. The recoveries of four analytes.

Analytes	Original (µg)	Spiked (µg)	Found (µg)	Recoveries (%)	RSD (%)
		9.11	27.33	94.34	3.78
Aurantio-obtusin	18.22	18.22	36.44	103.68	2.85
		27.33	45.55	99.67	2.67
		4.20	12.60	92.90	3.97
Chryso-obtusin	8.40	8.40	16.80	100.00	3.98
		12.60	21.00	98.22	3.88
		2.93	8.79	91.30	2.54
Obtusifolin	5.86	5.86	11.72	102.72	1.48
		8.79	14.65	106.40	2.93
		0.77	2.31	94.60	3.72
Emodin	1.54	1.54	3.08	105.19	0.98
		2.31	3.85	99.60	3.39

2.4. Application to Real Samples

The developed VS-MSPD method was successfully applied for the analysis of four anthraquinones in Cassiae Semen. As shown in Table 3, the contents of aurantio-obtusin, chryso-obtusin, obtusifolin and emodin in Cassiae Semen were in the range of 0.40–2.95 mg·g^{-1}, 0.1–0.33 mg·g^{-1}, 0.12–0.48 mg·g^{-1} and 0.02–0.07 mg·g^{-1}, respectively. The contents of the four anthraquinones were reported as mean ± standard deviation from five independent batches. Analysis of variance (ANOVA) and Duncan's test were conducted among different batches by SPSS software. With the value of $P < 0.05$, the difference between different batches was considered significant. A typical HPLC chromatogram demonstrating the separation of four analytes is depicted in Figure 6. It was interesting to find the contents of four target compounds in the processed samples were higher than those in the raw materials, which was consistent with the results reported by Guo [24]. These results demonstrated that the VS-MSPD method was feasible for the analysis of complicated semen food samples.

Table 3. Contents (mg·g^{-1}) (mean ± SD) of four compounds in five batches of Cassiae Semen samples by VS-MAPD method.

Sample Number	Aurantio-Obtusin	Chryso-Obtusin	Obtusifolin	Emodin
1 [a]	1.48 ± 0.18	0.11 ± 0.01	0.23 ± 0.03	0.03 ± 0.01
2 [b]	2.93 ± 0.03	0.30 ± 0.04	0.45 ± 0.03	0.06 ± 0.01
3 [a]	0.42 ± 0.02	0.21 ± 0.02	0.12 ± 0.01	0.02 ± 0.00
4 [b]	2.92 ± 0.02	0.23 ± 0.04	0.47 ± 0.02	0.06 ± 0.00
5 [a]	1.69 ± 0.20	0.11 ± 0.02	0.27 ± 0.04	0.02 ± 0.01

[a] Raw Cassiae Semen. [b] Fried Cassiae Semen.

2.5. Comparison with Other Methods

A comparison of the extraction method, analytes, sample amounts, solvents, and extraction time with other reported methods was performed to estimate the analytical performance with the developed method. The results are summarized in Table 4. It was easily found that the new developed VS-MSPD method required smaller sample amounts, a shorter extraction time and less volume of elution solvent. More importantly, the use of recycled wasted crab shells as an adsorbent and the ionic liquid as elution solvent were more environmentally-friendly than other methods. Overall, on the basis of the aforementioned results, this green VS-MSPD method combined with HPLC is rapid, sensitive, simple and promising for the extraction and determination of anthraquinones in complex food matrices.

Table 4. Comparison of VS-MSPD method with other methods in the determination of compounds in Cassiae Semen.

No.	Extraction Method	Analytes	Sample Amounts (g)	Type of Solvent	Solvent Volume (mL)	Extraction Time (min)	Reference
1	Heating reflux extraction	chrysophanol; aurantio-obtusin	0.5	methanol	50	120	[22]
2	Heating reflux extraction	aurantio-obtusin; obtusin; chrysophanol; emodin; chrysophanol; physcion	0.5	90% methanol	50	120	[23]
3	Ultrasonic extraction	emodin; rhein; autrantio-obtusin	0.2	chloroform	60	40	[19]
4	Heating reflux extraction	aurantio-obtusin; obtusifolin; questin; SC-1; rhein; emodin; SC-2	50	75% methanol	500	60	[21]
5	Accelerated solvent extraction (ASE)	aurantio-obtusin; aloe-emodin; rhein; emodin; physcion; chrysophanol	0.2	acetonitrile	11	8	[20]
6	Ultrasonic extraction	emodin; aloe-emodin; rhein	2.0	ethanol-chloroform (1:1)	75	90	[25]
7	Microwave assisted extraction	aloe-emodin; rhein; emodin; chrysophanol; physcion	0.1	10% Genapol X-080	6	3	[17]
8	Heating reflux extraction	2-gluco-aurantioobtusin; casside; rhein; torachrysoneglucosides; physcion; 2-gluco-chrysoobtusin; aurantio-obtusin; chryso-obtusin; 1-desmethylobtusin; obtusin; aloe-emodin; emodin; chrysophanol	0.5	70% ethanol	50	180	[18]
9	VS-MSPD	aurantio-obtusin; chryso-obtusin; obtusifolin; emodin	0.02	250 mM [Domim]HSO$_4$	1	10	This work

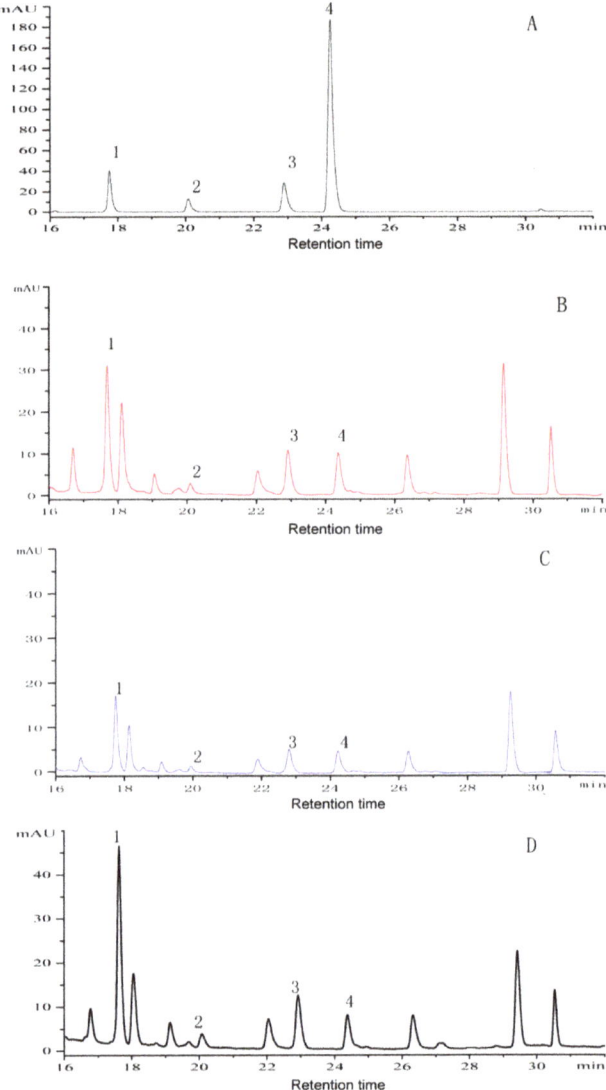

Figure 6. HPLC chromatograms of four anthraquinones in Cassiae Semen (**A**) standard solution of four anthraquinones, (**B**) extracted using heating reflux extraction, (**C**) extracted using normal miniaturized MSPD, (**D**) extracted using VS-MSPD. Peaks: 1 = Aurantio-obtusin, 2 = Chryso-obtusin, 3 = Obtusifolin, 4 = Emodin.

3. Materials and Methods

3.1. Chemicals and Reagents

The ionic liquids used in the study included 1-butyl-3-methylimidazolium hexafluorophosphate ([Bmim]PF_6), 1-dodecyl-3-methyl chloride imidazole ([Domim]Cl), 1-dodecyl-3-methyl-1-*H*-imidazolium hydrogensulfate ([Domim]HSO_4), 1-dodecyl-3-methyl-1- *H*-imidazolium nitrate ([Domim]NO_3), were offered by Chengjie Chemical Co., Ltd. (Shanghai, China). C_{18} and silica gel were purchased from Shanghai Chengya Chemical Co., Ltd. (Shanghai, China). Standard substances including aurantio-obtusin, chryso-obtusin, obtusifolin and emodin were supplied from Chengdu Must Bio-Technology Co., Ltd. (Chengdu, China). Their purities were all higher than 98% and their structures are shown in Figure 7. Purified water was purchased from Wahaha Group Ltd. (Shanghai, China). HPLC grade methanol and acetonitrile were obtained from American Tedia Co., Ltd. (Shanghai, China). Formic acid was of analytical reagent grade from Shanghai Lingfeng Chemical Reagent Co., Ltd. (Shanghai, China). All reagents for HPLC were filtrated through a 0.22 µm filter.

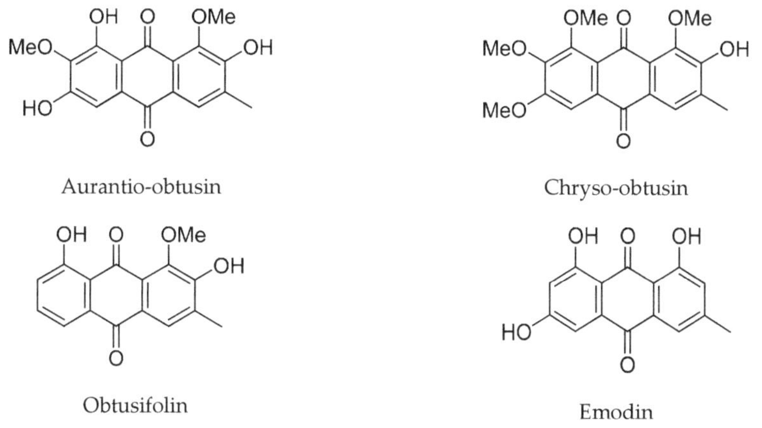

Figure 7. Structures of the four standard substances.

3.2. Preparation and Characterization of CSP

Crab shells were obtained from a local restaurant supplier (Hangzhou, China), and then rinsed with tap water in order to remove other debris and slime. After that, they were washed again with purified water and dried in an electric oven at 60 °C for about 12 h to a constant weight. The dried crab shells were crushed into powder and sieved with 100 meshes. Then, base treatment was conducted using 20% (wt) aqueous NaOH at 100 °C for about 1h in order to remove redundant protein, and the powders were washed with purified water and dried in an electric oven at 60 °C for about 12 h. The treated powders were crushed again, passed over 200 meshes, and stored in dry vacuum packs for ready use as an adsorbent. A scanning electron microscope (TESCAN VEGA 3 SBH) was used to examine the surface morphology of CSP. FT-IR spectra was investigated to analyze CSP by a Nicolet 6700 FT-IR Spectrometer (TA Instruments, Wilmington, DE, USA) in the range of 4000 to 400 cm^{-1}.

3.3. HPLC Analysis

Chromatographic analyses of aurantio-obtusin, chrys-oobtusin, obtusifolin and emodin were performed with Agilent 1260 series (Agilent, Santa Clara, CA, USA) using an ultraviolet (UV) detection set at 440 nm. The mobile phase consisted of 95% acetonitrile-5% water (v/v) (A) and 5% acetonitrile-95% water (v/v) (B), which all contained 0.1% formic acid with a flow rate of 1 mL·min^{-1}. The gradient elution was set as follows: 0–15 min, 5–55% A; 15–18 min, 55–55% A; 18–26 min, 55–80% A; 26–27 min, 80–100% A; 27–31 min, 100–100% A. A Shimadzu Inersustain C$_{18}$ column (4.6 × 250 mm, 5 µm) was used with the column temperature hold set at 25 °C. The injection volume was 20 µL.

3.4. Preparation of Standard Solution

The mixed stock solution was prepared in methanol respectively including 123.6 µg·mL^{-1} of aurantio-obtusin, 40.0 µg·mL^{-1} of chryso-obtusin, 41.6 µg·mL^{-1} of obtusifolin and 13.2 µg·mL^{-1} of emodin. The prepared standard solution was stored in a freezer at 4 °C.

3.5. VS-MSPD Procedure and Normal MSPD Procedure

Dried Cassiae Semen material was crushed and passed through 50 meshes. About 20 mg of sample powder and 20 mg of adsorbent (CSP, C_{18} and silica) were slightly placed into an agate mortar and ground with a pestle for 3 min. The obtained mixtures were transferred into a 10 mL centrifuge polypropylene tube. Then different types of ionic liquids were added, each were 1 mL. The mixture was entirely whirled with vortex for 3 min and filtrated through 0.22 μm filter. The filter eluent was collected in a 1.5 mL centrifuge tube. After centrifuging at 13,000 rpm for 6 min, the supernatant was collected for the HPLC analysis.

Approximately 20 mg of Cassiae Semen powder and 20 mg of CSP were placed into an agate mortar and blended using a pestle until a visually homogeneous mixture was obtained (for 3 min). Following complete dispersal, the mixture was carefully transferred into a 1 mL SPE column with a frit at the bottom. Then a second frit was pressed slightly on the top of the blend with a syringe plunger. The target analytes were eluted afterwards with 1 mL of [Domim]Cl aqueous solution (250 mM) in a vacuum. The eluent was collected in a 1.5 mL centrifuge tube and centrifuged at 13,000 rpm for 6 min. The supernatant was subjected to the HPLC analysis.

3.6. Heating Reflux Extraction (HRE)

HRE was performed based on the method documented in Chinese Pharmacopoeia (2015 ed) with minor modification [22]. Briefly, 0.2 g of Cassiae Semen powder was dissolved in 10 mL of methanol. Then the mixture was extracted by heating reflux for 2 h. After the mixture was extracted, the weight loss of the solution was compensated with methanol. The final solution was passed through a 0.22 μm membrane filter, and analyzed by HPLC.

3.7. Validation of Analytical Procedure

Linearity was evaluated by analyzing a series of concentrations of mixed standard solution (including aurantio-obtusin, chryso-obtusin, obtusifolin and emodin). After analysis, four calibration curves were constructed by plotting peak areas (y) versus concentration (x, mg·mL^{-1}). Moreover, correlation coefficients were obtained by least squares regression. The limits of detection (LODs) and limits of quantification (LOQs) of the proposed method were assessed by the analysis of the four standards at a signal-to-noise ratio of 3 and 10, respectively. Intra-day precision RSDs were determined by six replicate injections of the mixed standard solution analyzed on the same day. Inter-day RSDs were calculated with the detection of the mixed standard solution for three consecutive days. Repeatability of the method was examined with the determination of six replicates of the same sample. Stability was obtained by evaluating a sample solution at 2, 4, 8, 16 and 24 h. Recovery was determined by standard addition method. A certain number of standards including low and high concentration were added into a certain amount of Cassiae Semen powder respectively. The samples were extracted and analyzed with the proposed method.

4. Conclusions

In this study, a green sorbent, recycling wasted crab shell, was applied as the dispersing agent and environmentally-friendly extraction solvent IL ([Domim]HSO$_4$) was used as an eluent of MSPD for the simultaneous extraction of four anthraquinones in Cassiae Semen samples. Compared to the Chinese National Standard (Chinese Pharmacopiea 2015 ed) and other conventional methods, the presented method exhibited advantages of green, smaller sample amounts, a shorter extraction time and less volume of elution solvent. Furthermore, satisfactory linearity, good accuracy and high reproducibility were achieved, indicating that the developed VS-MSPD method which is a quantitative determination method of the anthraquinones can provide an efficient and powerful tool for the extraction of natural products in real samples.Thus, the method is significant in the quality control of edible and medicinal Cassiae Semen.

Author Contributions: All authors designed the study and interpreted the results. L.J. and J.W. did the experiment and drafted the manuscript. H.Z. and C.L. were responsible for the design of the work. Y.T. and C.C. supervised the project, analyzed, and interpreted the data and critically revised the manuscript.

Funding: This research was funded by the National Science Foundation of China, grant number 81503219.

Acknowledgments: The authors gratefully acknowledge the financial support of this research by the National Science Foundation of China (No. 81503219).

Conflicts of Interest: The authors declare no conflicts of interest.

References

1. Dai, L.Q.; Yao, Z.T.; Yang, W.Y.; Xia, M.S.; Ye, Y. Crab shell: A potential high-efficiency and low-cost adsorbent for dye wastewater. *Fresen. Environ. Bull.* **2017**, *26*, 4991–4998.
2. Tamtam, M.R.; Vudata, V.B.R. Biosorption of Congo Red from aqueous solution by crab shell residue: A comprehensive study. *Springerplus* **2016**, *5*, 537. [CrossRef]
3. Shahidi, F.; Synowiecki, J. Isolation and characterization of nutrients and value-added products from snow crab (Chionoecetesopilio) and shrimp (Pandalus borealis) processing discards. *J. Agric. Food Chem.* **1991**, *39*, 1527–1532. [CrossRef]
4. Wang, S.L.; Shih, I.L.; Liang, T.W.; Wang, C.H. Purification and characterization of two antifungal chitinases extracellularly produced by Bacillus amyloliquefaciens V656 in a shrimp and crab shell powder medium. *J. Agric. Food Chem.* **2002**, *50*, 2241–2248. [CrossRef]
5. Wang, S.L.; Hsiao, W.J.; Chang, W.T. Purification and characterization of an antimicrobial chitinase extracellularly produced by Monascuspurpureus CCRC31499 in a shrimp and crab shell powder medium. *J. Agric. Food Chem.* **2002**, *50*, 2249–2255. [CrossRef]
6. Liu, D.; Li, Z.; Zhu, Y.; Kumar, R. Recycled chitosan nanofibril as an effective Cu(II), Pb(II) and Cd(II) ionic chelating agent: Adsorption and desorption *performance. Carbohyd. Polym.* **2014**, *111*, 469–476. [CrossRef]
7. Vijayaraghavan, K.; Palanivelu, K.; Velan, M. Crab shell-based biosorption technology for the treatment of nickel-bearing electroplating industrial effluents. *J. Hazard. Mater.* **2005**, *119*, 251–254. [CrossRef]
8. Zhang, H.; Sun, C.; Zhang, Z.; Zhao, Y. Adsorptive characteristics of chromium (VI) ion from aqueous solutions using the modified crab shell. In Proceedings of the 2015 4th International Conference on Sensors, Measurement and Intelligent Materials, AtlantisPress, Paris, France, January 2016; pp. 732–735.
9. Niu, H.; Volesky, B. Characteristics of anionic metal species biosorption with waste crab shells. *Hydrometallurgy* **2003**, *71*, 209–215. [CrossRef]
10. Cadogan, E.I.; Lee, C.H.; Popuri, S.R.; Lin, H.Y. Efficiencies of chitosan nanoparticles and crab shell particles in europium uptake from aqueous solutions through biosorption: Synthesis and characterization. *Int. Biodeter. Biodegr.* **2014**, *95*, 232–240. [CrossRef]
11. Vijayaraghavan, K.; Palanivelu, K.; Velan, M. Biosorption of copper(II) and cobalt(II) from aqueous solutions by crab shell particles. *Bioresource Technol.* **2006**, *97*, 1411–1419. [CrossRef] [PubMed]
12. Vijayaraghavan, K.; Balasubramanian, R. Single and binary biosorption of cerium and europium onto crab shell particles. *Chem. Eng. J.* **2010**, *163*, 337–343. [CrossRef]
13. Jeon, D.J.; Yeom, S.H. Recycling wasted biomaterial, crab shells, as an adsorbent for the removal of high concentration of phosphate. *Bioresource Technol.* **2009**, *100*, 2646. [CrossRef]
14. Jeon, C. Removal of As(V) from aqueous solutions by waste crab shells. *Korean J. Chem. Eng.* **2011**, *28*, 813–816. [CrossRef]
15. Zhang, Z.; Zhang, X.; Yu, S.; Zhang, Y.; Wang, N.; Sun, C. Kinetic and thermodynamic analysis of adsorption of arsenic (III) with waste crab shells. *J. Water Supply Res. T.* **2014**, *63*, 642–649. [CrossRef]
16. Bu, Z.; Lv, L.; Li, X.; Chu, C.; Tong, S. pH-zone-refining elution-extrusion countercurrent chromatography: Separation of hydroxyanthraquinones from Cassiae semen. *J. Sep. Sci.* **2017**, *40*, 4281–4288. [CrossRef] [PubMed]
17. Shi, Z.H.; Geng, X.M.; Jiang, H.X.; Zhang, H.Y. Microwave assisted micellar extraction-HPLC determination of anthraquinone derivatives in Cassiae Semen. *J. Liq. Chromatogr. R. T.* **2010**, *33*, 1369–1380. [CrossRef]
18. Zhang, W.D.; Wang, Y.; Wang, Q.; Yang, W.J.; Gu, Y.; Wang, R. Quality evaluation of Semen Cassiae (*Cassia obtusifolia* L.) by using ultra-high performance liquid chromatography coupled with mass spectrometry. *J. Sep. Sci.* **2015**, *35*, 2054–2062. [CrossRef] [PubMed]

19. Wang, N.; Wu, Y.; Wu, X.; Liang, S.; Sun, H. A novel nonaqueous capillary electrophoresis method for effective separation and simultaneous determination of aurantio-obtusin, emodin and rhein in semen cassiae and cassia seed tea. *Anal. Methods* **2014**, *6*, 5133–5139. [CrossRef]
20. Wang, N.; Su, M.; Liang, S.; Sun, H. Investigation of six bioactive anthraquinones in slimming tea by accelerated solvent extraction and high performance capillary electrophoresis with diode-array detection. *Food Chem.* **2016**, *199*, 1–7. [CrossRef]
21. Chen, Q.D.; Xu, R.; Xu, Z.N.; Cen, P.L. Progress in studies of active constituents of anthraquinones and their biological activities from Semen Cassiae. *Chin. J. Mod. Appl. Pharm.* **2003**, *20*, 120–124. [CrossRef]
22. National Pharmacopoeia Committee. *Pharmacopoeia of People's Republic of China (1)*; Medical Science and Technology Press: Beijing, Chna, 2015; p. 145.
23. Zhang, J.; Zhang, Z.; Mi, B.; Qu, Y. Determination of nine constituents in Cassiae Semen from different sources by HPLC. *Chin. J. Pharm. Anal.* **2013**, *33*, 1665–1671. [CrossRef]
24. Guo, R.; Wu, H.; Yu, X.; Xu, M.; Zhang, X.; Tang, L. Simultaneous determination of seven anthraquinone aglycones of crude and processed semen cassiae extracts in rat plasma by UPLC-MS/MS and its application to a comparative pharmacokinetic study. *Molecules* **2017**, *22*, 1803. [CrossRef] [PubMed]
25. Zheng, W.; Wang, S.; Chen, X.; Hu, Z. Identification and determination of active anthraquinones in Chinese teas by micellar electrokinetic capillary chromatography. *Biomed. Chromatogr.* **2004**, *18*, 167–172. [CrossRef]
26. Soares, K.L.; Caldas, S.S.; Primel, E.G. Evaluation of alternative environmentallyfriendly matrix solid phase dispersion solid supports for the simultaneous extraction of 15 pesticides of different chemical classes from drinking water treatment sludge. *Chemosphere* **2017**, *182*, 547–554. [CrossRef]
27. Vieira, A.A.; Caldas, S.S.; Escarrone, A.L.V.; Jlo, A.; Primel, E.G. Environmentally friendly procedure based on VA-MSPD for the determination of booster biocides in fish tissue. *Food Chem.* **2018**, *242*, 475–480. [CrossRef]
28. Caldas, S.S.; Bolzan, C.M.; Menezes, E.J.D.; Escarrone, A.L.V.; Martins, C.D.M.G.; Bianchini, A.; Primel, E.G. A vortex-assisted MSPD method for the extraction of pesticide residues from fish liver and crab hepatopancreas with determination by GC-MS. *Talanta* **2013**, *112*, 63–68. [CrossRef]
29. Hertzog, G.I.; Soares, K.L.; Caldas, S.S.; Primel, E.G. Study of vortex-assisted MSPD and LC-MS/MS using alternative solid supports for pharmaceutical extraction from marketed fish. *Anal. Bioanal. Chem.* **2015**, *407*, 4793–4803. [CrossRef]
30. Escarrone, A.L.V.; Caldas, S.S.; Soares, B.M.; Martins, S.E.; Primel, E.G.; Nery, L.E.M. A vortex-assisted MSPD method for triclosan extraction from fish tissues with determination by LC-MS/MS. *Anal. Methods* **2014**, *6*, 8306–8313. [CrossRef]
31. León-González, M.E.; Rosales-Conrado, N. Determination of ibuprofen enantiomers in breast milk using vortex-assisted matrix solid-phase dispersion and direct chiral liquid chromatography. *J. Chromatogr. A* **2017**, *1514*, 88–94. [CrossRef]
32. Caldas, S.S.; Soares, B.M.; Abreu, F.; Castro, Í.B.; Fillmann, G.; Primel, E.G. Antifouling booster biocide extraction from marine sediments: A fast and simple method based on vortex-assisted matrix solid-phase extraction. *Environ. Sci. Pollut. R.* **2018**, *25*, 7553–7565. [CrossRef]
33. Wang, Y.; Sun, Y.; Xu, B.; Li, X.; Wang, X.; Zhang, H.; Song, D. Matrix solid-phase dispersion coupled with magnetic ionic liquid dispersive liquid–liquid microextraction for the determination of triazine herbicides in oilseeds. *Anal. Chim. Acta* **2015**, *888*, 67–74. [CrossRef]
34. He, Z.; Yang, H. Colourimetric detection of swine-specific DNA for halal authentication using gold nanoparticles. *Food Control.* **2018**, *88*, 9–14. [CrossRef]
35. Yu, X.; Yang, H. Pyrethroid residue determination in organic and conventional vegetables using liquid-solid extraction coupled with magnetic solid phase extraction based on polystyrene-coated magnetic nanoparticles. *Food Chem.* **2017**, *217*, 303–310. [CrossRef] [PubMed]
36. Xu, J.J.; Yang, R.; Ye, L.H.; Cao, J.; Cao, W.; Hu, S.S.; Peng, L.Q. Application of ionic liquids for elution of bioactive flavonoid glycosides from lime fruit by miniaturized matrix solid-phase dispersion. *Food Chem.* **2016**, *204*, 167–175. [CrossRef]
37. Du, K.Z.; Li, J.; Bai, Y.; An, M.R.; Gao, X.M.; Chang, Y.X. A green ionic liquid-based vortex-forced MSPD method for the simultaneous determination of 5-HMF and iridoid glycosides from *Fructus Corni* by ultra-high performance liquid chromatography. *Food Chem.* **2018**, *244*, 190–196. [CrossRef] [PubMed]
38. Tang, X.; Li, Z.; Chen, Y. Adsorption behavior of Zn(II) on calcinated Chinese loess. *J. Hazard. Mater.* **2009**, *161*, 824–834. [CrossRef]

39. Njoku, V.O.; Foo, K.Y.; Asif, M.; Hameed, B.H. Preparation of activated carbons from rambutan (Nepheliumlappaceum) peel by microwave-induced KOH activation for acid yellow 17 dye adsorption. *Chem. Eng. J.* **2014**, *250*, 198–204. [CrossRef]
40. Ifuku, S.; Nogi, M.; Abe, K.; Yoshioka, M.; Morimoto, M.; Saimoto, H.; Yano, H. Preparation of chitin nanofibers with a uniform width as alpha-chitin from crab shells. *Biomacromolecules* **2009**, *10*, 1584–1588. [CrossRef]
41. Cárdenas, G.; Cabrera, G.; Taboada, E.; Miranda, S.P. Chitin characterization by SEM, FTIR, XRD, and 13C cross polarization/mass angle spinning NMR. *J. Appl. Polym. Sci.* **2004**, *93*, 1876–1885. [CrossRef]
42. Huang, Y.S.; Yu, S.H.; Sheu, Y.R.; Huang, K.S. Preparation and thermal andanti-UV properties of Chitosan/Mica Copolymer. *J. Nanomater.* **2010**, *5*, 65. [CrossRef]
43. Sowa, I.; Wójciak-Kosior, M.; Strzemski, M.; Sawicki, J.; Staniak, M.; Dresler, S. Silica Modified with polyaniline as a potential sorbent for matrix solid phase dispersion (MSPD) and dispersive solid phase extraction (d-SPE) of plant samples. *Materials* **2018**, *11*, 467. [CrossRef] [PubMed]

Sample Availability: Sample Availability: Not available.

© 2019 by the authors. Licensee MDPI, Basel, Switzerland. This article is an open access article distributed under the terms and conditions of the Creative Commons Attribution (CC BY) license (http://creativecommons.org/licenses/by/4.0/).

Article

Poly (Octadecyl Methacrylate-Co-Trimethylolpropane Trimethacrylate) Monolithic Column for Hydrophobic in-Tube Solid-Phase Microextraction of Chlorophenoxy Acid Herbicides

Wenbang Li [1], Fangling Wu [1], Yongwei Dai [2], Jing Zhang [3], Bichen Ni [1] and Jiabin Wang [1,*]

1. Institute of Biomedical and Pharmaceutical Technology, Fuzhou University, Fuzhou 350002, China; N165720012@fzu.edu.cn (W.L.), flwu16@fudan.edu.cn (F.W.); N185720006@fzu.edu.cn (B.N.)
2. Fuzhou Jiachen Biotechnology Co., Ltd., Fuzhou 350008, China; daiyoungwei@163.com
3. Fujian Inspection and Testing Centre for Agricultural Product Quality and Safety, Fuzhou 350003, China; fjsy0591@sina.com
* Correspondence: jbwang@fzu.edu.cn; Tel.: +86-591-8372-5260

Academic Editor: Nuno Neng
Received: 28 March 2019; Accepted: 25 April 2019; Published: 29 April 2019

Abstract: Chlorophenoxy acid herbicides (CAHs), which are widely used on cereal crops, have become an important pollution source in grains. In this work, a highly hydrophobic poly (octadecyl methacrylate-co-trimethylolpropane trimethacrylate) [poly (OMA-co-TRIM)] monolithic column has been specially prepared for hydrophobic in-tube solid-phase microextraction (SPME) of CAHs in rice grains. Due to the hydrophobicity of CAHs in acid conditions, trace CAHs could be efficiently extracted by the prepared monolith with strong hydrophobic interaction. Several factors for online hydrophobic in-tube SPME, including the length of the monolithic column, ACN and trifluoroacetic acid percentage in the sampling solution, elution volume, and elution flow rate, were investigated with respect to the extraction efficiencies of CAHs. Under the optimized conditions, the limits of detection of the four CAHs fell in the range of 0.9–2.1 µg/kg. The calibration curves provided a wide linear range of 5–600 µg/kg and showed good linearity. The recoveries of this method ranged from 87.3% to 111.6%, with relative standard deviations less than 7.3%. Using this novel, highly hydrophobic poly (OMA-co-TRIM) monolith as sorbent, a simple and sensitive online in-tube SPME-HPLC method was proposed for analysis of CAHs residue in practical samples of rice grains.

Keywords: chlorophenoxy acid herbicides; HPLC; hydrophobic in-tube solid-phase microextraction; poly (OMA-co-TRIM) monolithic column; rice grains

1. Introduction

In order to fulfill the world's growing demand for food, herbicide application to cereal crops is imperative. Because of their low cost, high weeding effectiveness and little effect on cereal crops, chlorophenoxy acid herbicides (CAHs), represented by 2,4-dichlorophenoxyacetic acid (2,4-D), are extensively used on cereal crops for the selective control of most broadleaf weeds [1]. For the high absorption of CAHs by crops, CAHs have become an important pollution source in cereal grains, which cause potential pollution to food with further potential toxicity against humans. In June 2015, the World Health Organization's International Agency for Research on Cancer confirmed its 1987 classification of 2,4-D as a possible carcinogen [2]. Men who work with 2,4-D are at risk for abnormally shaped sperm and thus fertility problems [3]. Therefore, it has become important work to detect and monitor the residual level of CAHs in cereal grains by developing reliable analytical methods that are simple, highly sensitive, and cost-effective.

To date, the main techniques for the quantitative detection of CAHs in different samples include gas chromatography (GC) [4,5], high performance liquid chromatography (HPLC) [6–8], surface-enhanced raman spectroscopy (SERS) [9], capillary electrophoresis (CE) [10,11], and enzyme-linked immunosorbent assay (ELISA) [12]. Among them, GC and HPLC are generally adopted as the determination methods in many standards [13,14]. When detected by GC, the additional derivatization of CAHs is necessary. Thus, HPLC is more simple and suitable for the determination of CAHs. On the other hand, for the low concentrations and matrix interferences, sample preparation is indispensable prior to the chromatographic analysis. Several sample preparation methods, including liquid–liquid extraction [15], soxhlet extraction [16], solid-phase extraction (SPE) [17,18], suspended liquid-phase microextraction [19], molecularly imprinted solid-phase extraction [20,21], dispersive micro-solid phase extraction [22], matrix solid-phase dispersion [23], and microwave-assisted solvent extraction [24,25], have been carried out for the extraction of CAHs from a variety of matrixes. However, considerable amounts of time and solvent are required in these sample preparation processes.

Solid-phase microextraction (SPME), due to its advantages of high sensitivity, solvent-free extraction, simplicity, and easy online coupling to HPLC, has been recognized as a reliable method of sample preparation, especially for analysis in complex sample matrixes (e.g., biological and food samples) [26]. As the core of SPE or SPME, the selection of the appropriate sorbent is crucial. Diverse SPE sorbents, such as highly cross-linked polystyrene-divinylbenzene sorbent [17], molecularly imprinted polymers [20,21], and cationic gemini surfactant-resorcinol-aldehyde resin [22], have been utilized for the extraction of CAHs. Monolithic columns, which could be synthesized in situ in the capillaries or tubes, have attracted increasing interests as an SPME sorbent because of their inherent advantages of facile preparation, diverse surface chemistry, increased enrichment capacity, and high porosity [27]. Recently, a multiple monolithic fiber which was reinforced by graphene has been proposed for extraction of CAHs [28]. In our previous work, a poly (octadecyl methacrylate-co-ethylene dimethacrylate) [poly (OMA-co-EDMA)] monolithic column was fabricated for the extraction of CAHs from water samples [29]. Nevertheless, due to the higher complexity of sample matrixes of cereal grains than water, the unsatisfied extraction efficiency and worse renewability of poly (OMA-co-EDMA) monolith occurred and further showed its inapplicability facing the complex samples of cereal grains. Considering the hydrophobicity of CAHs in acid conditions, the monolithic column with higher hydrophobicity should be favored to enhance extraction efficiency for SPME.

To this end, trimethylolpropane trimethacrylate (TRIM) with higher hydrophobicity was specially selected as an alternative crosslinker to fabricate a novel hydrophobic poly (octadecyl methacrylate-co-trimethylolpropane trimethacrylate) [poly (OMA-co-TRIM)] monolithic column for an online hydrophobic in-tube SPME of CAHs in rice grains. The characterization of the poly (OMA-co-TRIM) monolithic column was carried out in detail. Important SPME parameters, including the length of the monolithic column, the percentage of trifluoroacetic acid and acetonitrile (ACN) in sampling solution, elution volume, and elution flow rate, were optimized to obtain the best extraction efficiency. Under the optimized condition, an online in-tube SPME-HPLC method using the poly (OMA-co-TRIM) monolithic column as the sorbent was developed and applied for the determination of trace CAHs residue in rice grains.

2. Results and Discussion

2.1. Characterization of the Monolithic Column for Hydrophobic in-Tube SPME

2.1.1. Morphology

In this work, the detailed morphology of the poly (OMA-co-TRIM) monolithic column (Figure 1a) was captured by scanning electron microscope (SEM). In this SEM image, there were many mesopores and flow-through channels inlaid in the network skeleton of column bed. When compared with poly (OMA-co-EDMA) monolith in Figure 1b, the flow-through channels in the poly (OMA-co-TRIM) monolith were larger; they could provide large flow paths for sampling solutions and elution solutions to pass through in a fast flow rate, which would be favorable for the further extraction process.

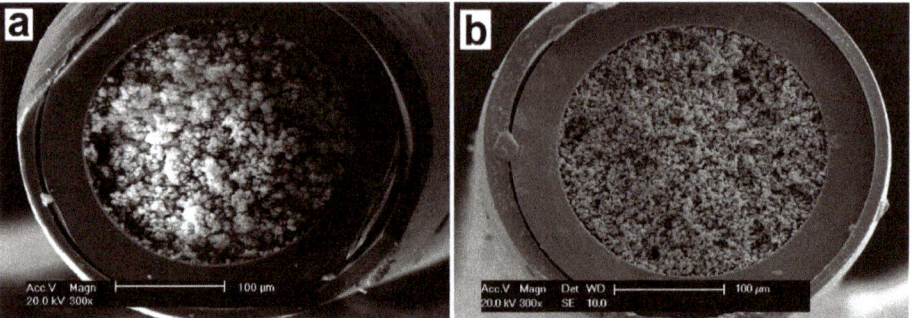

Figure 1. SEM images of (**a**) the poly (OMA-co-TRIM) monolithic column and (**b**) the poly (OMA-co-EDMA) monolithic column.

2.1.2. Permeability and Porosity

The permeability of the obtained monolith was investigated. A 5 cm column cut from the prepared monolithic column was connected to a μHPLC pump and the backpressure was measured by pumping water or ACN. The permeability (K) of the column was calculated with Darcy's equation, K = $F\eta L/(\pi r^2 \Delta P)$ [30], where F is the flow rate of mobile phase, η is the viscosity of mobile phase, L is the effective length of column, r is the inner radius of column, and ΔP is the pressure drop of column. The measured permeability of the poly (OMA-co-TRIM) monoliths were 1.25×10^{-13} m^2 by using ACN as the mobile phase and 2.03×10^{-13} m^2 by using water as the mobile phase. When measured in the same conditions, the permeability of the poly (OMA-co-EDMA) monoliths were 1.90×10^{-13} m^2 and 2.80×10^{-13} m^2, respectively. This result indicated that the permeability of this prepared poly (OMA-co-TRIM) monolith was better than the reported poly (OMA-co-EDMA) monolith and further showed its feasibility for in-tube SPME.

In addition, the pore size distribution of the poly (OMA-co-TRIM) monolithic column was also characterized by the mercury intrusion method (Figure S1). The most probable pore size of the monolith was 0.92 μm. Furthermore, the specific pore volume of the monolith was 1.36 mL/g, and the porosity was 56.34%. These data indicated the high porosity of the obtained poly (OMA-co-TRIM) monolith.

2.1.3. Loading Capacity

In this work, breakthrough curves for 2,4-D using frontal analysis were utilized to evaluate the loading capacity of the obtained monolith. As shown in Figure 2, the breakthrough curves of 2,4-D by using poly (OMA-co-TRIM) monolith and poly (OMA-co-EDMA) monolith both exhibited an obvious rise, demonstrating the typical kinetic adsorption. The void time for calculation was examined by flushing 200 ng/mL thiourea in the same monolithic column and subtracting from the total consumed time for saturating the monolithic column. The saturation times for 2,4-D on the above two monolithic columns were 24.5 and 39.3 min, respectively. The equation (Q = cvt/V [31]) was utilized for the calculation of the loading capacity. In this equation, Q is the loading capacity (μg/cm^3), c is the concentration of the analyte (mg/mL), v is the flow rate (mL/min), t is the saturation time (min), and V is the volume of monolithic column (cm^3). When the flow rate was 10 μL/min, the loading capacity of 2,4-D on the above two monolithic columns were 499.3 and 801.0 μg/cm^3, which illustrated the higher loading capacity of the obtained poly (OMA-co-TRIM) monolith. As a result, to gain a similar extraction efficiency of the analytes, the length of the poly (OMA-co-TRIM) monolith must be shorter than the poly (OMA-co-EDMA) monolith.

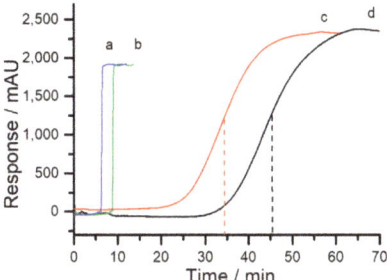

Figure 2. Breakthrough curves on different monoliths: (a) For thiourea (marker for void time) on poly (OMA-co-TRIM) monolith; (b) for thiourea (marker for void time) on poly (OMA-co-EDMA) monolith; (c) for 2,4-D on the poly (OMA-co-EDMA) monolith; (d) for 2,4-D on the poly (OMA-co-TRIM) monolith. Experimental conditions: The monolith, 20 cm × 250 μm i.d.; mobile phase, 98% (v/v) 0.1% trifluoroacetic acid (TFA) solution and 2% (v/v) ACN; pump flow rate, 10 μL/min; the concentration of thiourea, 200 ng/mL; the concentration of 2,4-D, 200 ng/mL; UV detection wave, 223 nm.

2.1.4. Renewability

Apart from the residual CAHs, there are still many unknown things from the sample matrices of rice grains absorbed on the SPME monolith. After extraction, careful cleanup is necessary to make the SPME monolithic column renewable. In this work, ACN was utilized as the cleanup solution. The renewability of the SPME monolith was investigated by the decrease of the peak area of 2,4-D after several operation cycles. As represented in Figure S2, the decrease of the peak area on the poly (OMA-co-EDMA) monolith is bigger than the poly (OMA-co-TRIM) monolith, which denoted the better renewability of the poly (OMA-co-TRIM) monolith. This phenomenon should be attributed to the higher permeability and the higher loading capacity of the poly (OMA-co-TRIM) monolith. Herein, the length of the poly (OMA-co-TRIM) monolith used was shorter than the poly (OMA-co-EDMA) monolith. Due to the higher permeability and the shorter length of the poly (OMA-co-TRIM) monolith, a faster pump flow rate could be adopted for in-tube SPME. In this case, the contact time between the sample solution and the monolithic sorbent was shortened and led to the decrease of the absorbed impurities. On the other hand, owing to the higher hydrophobicity of the poly (OMA-co-TRIM) monolith, its extraction abilities toward CAHs was enhanced and further ensured the extraction efficiency. Therefore, the poly (OMA-co-TRIM) monolith should be a suitable sorbent for in-tube SPME of CAHs from rice grains.

2.2. Optimization of Some Important Parameters for Hydrophobic in-Tube SPME

2.2.1. Length of the Monolithic Column

The amount of monolithic sorbent depended on the length of the monolithic column. In this work, the length of the monolithic column was investigated from 7 to 15 cm. In Figure 3a, while keeping other parameters constant, the extraction efficiencies of the analytes increased with the length of the monolithic column. When the length increased up to 15 cm, the extraction efficiencies only increased a little. The results indicated that the sufficient extraction of CAHs could be obtained with 12 cm of the poly (OMA-co-TRIM) monolithic column. In addition, since the maximum of the pressure was constant in the system, using a short monolithic column meant that a faster flow rate was operated for the extraction, which led to less absorbed impurities, easier cleanup, and a shorter analytical time for the whole process. Thus, the length of 12 cm was chosen for the monolithic column in the following experiments.

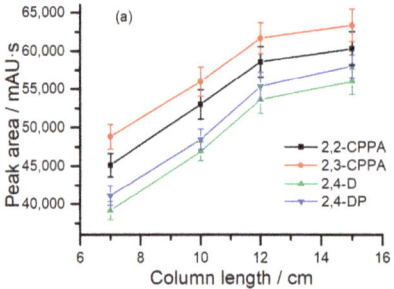 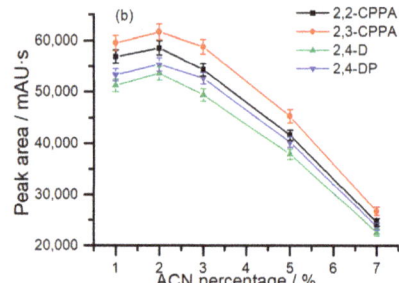

Figure 3. (a) Effect of the length of the monolithic column on the extraction efficiencies of the analytes; (b) effect of the ACN percentage in the sampling solution on the extraction efficiencies of the analytes. Abbreviation: 2-(2-chloro)-phenoxy propionic acid (2,2-CPPA), 2-(3-chloro)-phenoxy propionic acid (2,3-CPPA) and 2-(2,4-dichlorophenoxy)-propionic acid (2,4-DP).

2.2.2. ACN Percentage in Sampling Solution

The ACN percentage in the sampling solution was a vital factor to affect the extraction behaviors of the hydrophobic compounds on the hydrophobic sorbent. The effect of the ACN percentage on the extraction efficiencies was investigated in the range of 1–7% (v/v). As seen in Figure 3b, the extraction efficiencies of the CAHs increased first and then decreased along with the increase in ACN percentage. The adsorption amount of CAHs (n) can be calculated by the equation $n = K\,V_m\,c_0$ [32], in which K is the distribution coefficient, c_0 is the initial concentration of the analytes, and V_m represents the volume of the monolithic stationary phase. Based on this equation, K and c_0 influence the adsorption amount of CAHs. K decreases with the increase of the ACN percentage. When the ACN content was too low, the solubility of the CAHs was reduced, which caused the decrease of c_0 and affected the final amount of the analytes. These two factors led to the result illustrated in Figure 3b. Therefore, the ACN percentage in the sampling solution of 2% was used for higher extraction efficiency.

2.2.3. Trifluoroacetic Acid Percentage in Sampling Solution

Since the ionization state of CAHs can be controlled by the pH value of sampling solution, the percentage of trifluoroacetic acid (TFA) in the sampling solution is one of the most critical factors for their extraction. As shown in Figure 4, the extraction efficiencies of target compounds were significantly improved with the increase of TFA percentage from 0% to 0.1% (pH 2.02). When the TFA percentage was increased to 0.2% (pH 1.39), the extraction efficiencies were almost unchanged. With the increase of the TFA percentage in the sampling solution, the ionization degree of the acidic CAHs were reduced, leading to the enhancement of the hydrophobicities of the analytes and resulting in the increase of hydrophobic extraction efficiencies. Considering that the sampling solution with a high TFA percentage is harmful to the HPLC pump, the TFA percentage in the sampling solution was selected as 0.1% in subsequent experiments.

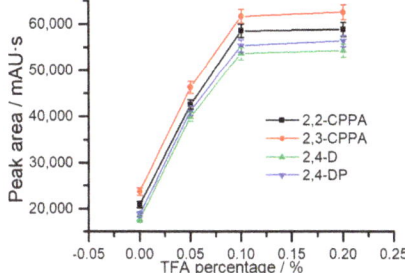

Figure 4. Effect of the trifluoroacetic acid (TFA) percentage in the sampling solution on the extraction efficiencies of the analytes.

2.2.4. Elution Volume

The elution volume is an important parameter affecting extraction efficiency due to its effect on the amount of the analytes injected into the HPLC column. Required by the online coupling system, the mobile phase (ACN: 0.1% TFA solution = 45:55, v/v) was directly applied as the elution solution for CAHs. The elution volume investigated ranged from 200 to 600 µL. As shown in Figure 5a, the extraction efficiency increased rapidly, with the elution volume increasing from 200 to 400 µL. Along with the increase in the elution volume, the extraction efficiency increased slowly. As a result, 400 µL was adopted as the optimal elution volume in this work.

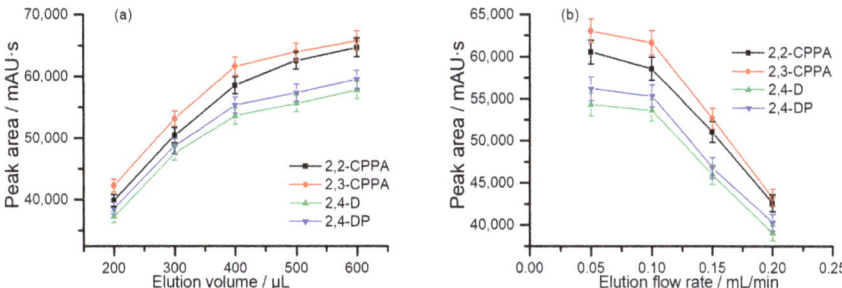

Figure 5. (a) Effect of elution volume on the extraction efficiencies of the analytes; (b) effect of elution flow rate on the extraction efficiencies of the analytes.

2.2.5. Elution Flow Rate

In the present work, the elution flow rate was investigated in the range of 0.05–0.2 mL/min. As can be seen in Figure 5b, the extraction efficiencies became worse with the increasing of the elution flow rate. This result might be attributed to that the faster elution flow rate can shorten the contact time between the poly (OMA-co-TRIM) monolithic column and the elution solution, leading to the decline of extraction efficiencies. Considering that the slow elution flow rate can extend the SPME time, as well as the total analysis time, 0.1 mL/min was utilized as the elution flow rate in the online hydrophobic in-tube SPME process.

2.3. *Method Validation*

Under the optimized conditions, the analytical performance of the developed online hydrophobic in-tube SPME method using the poly (OMA-co-TRIM) monolithic column as the sorbent was further evaluated. As shown in Figure 6a, the spiked CAHs in a blank sample of rice grains could be well extracted, separated, and detected. In comparison with the direct injection mode of HPLC (Figure 6c), a drastic peak height enhancement of target compounds and obvious reduction of matrix interference was achieved, which indicated the strong hydrophobic extraction effect and purification effect of the poly (OMA-co-TRIM) monolithic column. Herein, the enrichment factor (F_E), which is defined as the ratio of the peak area of analytes with in-tube SPME-HPLC to that with the direct injection mode of HPLC, was utilized to evaluate the extraction effect of the poly (OMA-co-TRIM) monolith. As listed in Table 1, the F_E values of the analytes ranged from 83.6 to 98.3, which indicated the excellent hydrophobic extraction effect of the poly (OMA-co-TRIM) monolith. By using the poly (OMA-co-EDMA) monolithic column as the SPME sorbent (Figure 6b), a lower peak height of the analytes and distinct baseline noise from sample matrix occurred. It is worth noting that, in this case, the first two analytes could not be well separated due to the existence of something unknown from the extracted elution. These results confirmed the unsatisfied extraction efficiency and the inapplicability of poly (OMA-co-EDMA) monolith facing the complex samples of rice grains.

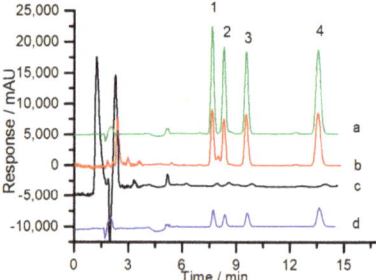

Figure 6. Chromatograms of four chlorophenoxy acid herbicides (CAHs) obtained from a spiked sample of rice grains (spiked at 200 µg/kg) by the proposed online hydrophobic in-tube solid-phase microextraction (SPME)-HPLC method (a) using the poly (OMA-co-TRIM) monolith as the sorbent, (b) using the poly (OMA-co-EDMA) monolith as the sorbent, (c) by direct injection mode of HPLC, and (d) obtained from practical sample of rice grains by the proposed online hydrophobic in-tube SPME-HPLC method using the poly (OMA-co-TRIM) monolith as the sorbent. SPME operating parameters: The length of the monolithic column, 12 cm; sampling solution, 0.1% TFA aqueous solution/ACN = 98/2 (v/v); sampling flow rate, 0.10 mL/min; sampling time, 5 min; elution flow rate, 0.10 mL/min; elution volume: 400 µL. Peak identifications: 1: 2,2-CPPA; 2: 2,3-CPPA; 3: 2,4-D; 4: 2,4-DP.

Table 1. Linearity, linear range, R^2, LOD and F_E of analytes in the samples of rice grains.

Analytes	Linearity [1]	R^2	Linear Range (µg/kg)	LOD (µg/kg)	F_E
2,2-CPPA	Y = 533.0x + 1060.7	0.9957	5–600	1.5	85.7
2,3-CPPA	Y = 507.4x + 1017.7	0.9977	5–600	2.1	83.6
2,4-D	Y = 554.6x + 513.4	0.9973	5–600	1.2	96.7
2,4-DP	Y = 592.2x + 563.4	0.9996	5–600	0.9	98.3

[1] x, Concentration of analyte (µg/kg); y, peak area (mAU·s).

The linearity, linear range, limits of detection (LODs), and correlation coefficients (R^2) were determined for four CAHs in the samples of rice grains and listed in Table 1. The linear range was in the range of 5–600 µg/kg for four CAHs with correlation coefficients ranging from 0.9957 to 0.9996. LODs were detected as the concentration of the analytes at S/N = 3 and ranged from 0.9 to 2.1 µg/kg, which could meet the requirement of the standards [13,33,34] when determining trace amount of CAHs. The recovery studies were carried out with blank samples of rice grains being spiked with the CAHs at three concentrations of 25, 150, and 250 µg/kg, respectively. For each concentration level, three replicate experiments were made. As summarized in Table 2, the recoveries were in the range of 87.3–111.6%, with relative standard deviations (RSDs) less than 7.3%. These results demonstrated the strong applicability of the poly (OMA-co-TRIM) monolith, as well as the excellent sensitivities and recoveries of the proposed method for analysis of CAHs residue in rice grains.

Table 2. Recoveries of four CAHs in rice grains.

Analyte	Spiked/(μg/kg)	Found/(μg/kg)	Recovery [1]/%	RSD/% (n = 3)
2,2-CPPA	25	21.8	87.3	5.2
	150	142.3	94.9	2.7
	250	254.0	101.6	1.2
2,3-CPPA	25	24.6	98.4	4.7
	150	134.0	89.3	2.3
	250	259.0	103.6	0.6
2,4-D	25	27.9	111.6	6.9
	150	151.2	100.8	3.7
	250	232.3	92.9	1.8
2,4-DP	25	26.4	105.6	7.3
	150	145.4	96.9	5.5
	250	244.3	97.7	3.4

[1] Recovery = found/spiked × 100%.

2.4. Reproducibility

The reproducibility of the poly (OMA-co-TRIM) monolith for hydrophobic in-tube SPME was investigated with respect to the peak areas of four CAHs. As listed in Table 3, while using the same one monolithic column for five replicate extractions, the resultant RSDs ranged from 0.6% to 2.5%. The batch-to-batch reproducibility for the poly (OMA-co-TRIM) monoliths was also evaluated in terms of the RSDs of the peak areas of the analytes, with the RSDs being less than 9.3%. The above results indicated that good reproducibility and repeatable extraction performance of the developed monolithic columns.

Table 3. RSDs for the peak areas on different poly (OMA-co-TRIM) monoliths.

Analytes	Run-to-Run Reproducibility (n = 5)	Batch-to-Batch Reproducibility (n = 3)
2,2-CPPA	1.3	7.9
2,3-CPPA	0.6	6.9
2,4-D	1.9	5.1
2,4-DP	2.5	9.3

2.5. Comparison of the Proposed Method with the Standard Method of LC MS

The application of the proposed method was further evaluated by the comparison with the standard method of LC-MS [14]. As listed in Table 4, the LODs of the proposed method were closed to the standard method of LC-MS, while the analytical time was shortened to a third of the standard method, indicating the briefness and sensitivity of the online hydrophobic in-tube SPME-HPLC method using poly (OMA-co-TRIM) monolithic column as the sorbent.

Table 4. Comparison of the proposed online hydrophobic in-tube SPME-HPLC method with the standard method of LC-MS.

Methods	Retention Time (min)				LOD (μg/kg) [1]			
	2,2-CPPA	2,3-CPPA	2,4-D	2,4-DP	2,2-CPPA	2,3-CPPA	2,4-D	2,4-DP
This method	8.2	8.7	10.1	14.3	1.2	0.9	2.1	1.5
LC-MS	24.8	26.1	26.2	27.0	-	-	3.0	1.0

[1] The LOD values of HPLC-MS were found from the agriculture standards of the People's Republic of China (NY/T 1434-2007).

The proposed method was applied to determine the trace CAHs (2,2-CPPA, 2,3-CPPA, 2,4-D and 2,4-DP) in practice rice grains (Figure 6d). Rice grain samples, which were sprayed with these herbicides during planting, were quantified and calculated. The residual contents of 2,2-CPPA, 2,3-CPPA, 2,4-D, and 2,4-DP detected by the proposed online hydrophobic in-tube SPME-HPLC method were 19.8 µg/kg, 21.4 µg/kg, 29.3 µg/kg, and 34.6 µg/kg, respectively, while the values detected by the standard method of LC-MS were 22.3 µg/kg, 20.1 µg/kg, 29.4 µg/kg, and 38.4 µg/kg, respectively. This shows the strong reliability of the proposed method. Furthermore, all the residual contents of the CAHs were lower than the maximum residue limits (MRLs) in the standards of different countries. This means that the CAHs residues meet the standard requirements, while the CAHs were sprayed according to the instructions.

3. Materials and Methods

3.1. Chemicals and Materials

2,4-Dichlorophenoxyacetic acid (2,4-D), 2-(2,4-dichlorophenoxy)-propionic acid (2,4-DP), 2-(2-chloro)-phenoxy propionic acid (2,2-CPPA), and 2-(3-chloro)-phenoxy propionic acid (2,3-CPPA) were all purchased from Sigma (St. Louis, MO, USA). OMA, TRIM, 2,2-azobisisobutyronitrile (AIBN), and 3-(trimethoxysilyl)-propyl methacrylate (γ-MAPS) were bought from Acros (Morris Plains, NJ, USA). Cyclohexanol and 1,4-butanediol were supplied by Tianjin Guangfu Fine Chemical Research Institute (Tianjin, China). Deionized water was obtained by using a Millipore Milli-Q purification system (Milford, MA, USA). Acetonitrile (ACN) and methanol (Chemical Reagent Corporation, Shanghai, China) were of HPLC grade. Trifluoroacetic acid (TFA), hydrochloric acid, sodium hydroxide, ammonium acetate, magnesium sulphate, and ethylene diamine-N-propyl silane (PSA) were purchased from Shanghai Chemical Reagents Corporation (Shanghai, China). Herein, 0.1 mol/L hydrochloric acid solution and 0.1 mol/L sodium hydroxide solution were used for the pre-treatment of the capillary inner wall, according to the literature [35]. The fused-silica capillaries with dimensions of 250 µm i.d. were obtained from Refine Chromatography Ltd. (Yongnian, China). The rice seeds were afforded by the College of Crop Science, Fujian Agriculture, and Forestry University (Fuzhou, China).

3.2. Apparatus

The HPLC-DAD system (Shimadzu, Kyoto, Japan), consisting of a Shimadzu LC-20AD pump, a Shimadzu SPD-M20A photodiode array detector, and a Shimadzu DGU-20A5 degasser, was employed for determining CAHs. Shimadzu LCsolution software was used for system control and data analysis. The HPLC conditions for this work were referred to our previous work [29]. ACN/0.1% TFA solution (45:55, v/v) was introduced as the mobile phase, eluting the analytes at 1 mL/min for HPLC analysis. The detection wavelength was set at 223 nm and the temperature of in-tube SPME and HPLC was set at 30 °C. A Gemini 5u C18 column (250 × 4.6 mm) from Phenomenex Inc. (Torrance, CA, USA) was used for the separation.

The LC-MS analysis was performed with an Agilent 1200/6460 LC-MS system (Palo Alto, CA, USA) with electron spray ionization in the negative ionization mode. A 150 × 2.1 mm Gemini 5u C18 column was used for separation with a flow rate of 0.3 mL/min and column temperature of 40 °C. A 20 µL injection volume was used. A solvent gradient between A = 5 mmol/L ammonium acetate methanol solution and B = 5 mmol/L ammonium acetate aqueous solution was used. The gradient elution program was as follows: From 0 to 30 min, A was increasing from 10% to 80%, while B was decreasing from 90% to 20%. The calculated molecular ions (198, 198, 219, 233 mz^{-1}) of the analytes were collected.

An LC-10AD pump (Shimadzu, Kyoto, Japan) was utilized to construct the online hydrophobic in-tube SPME-HPLC system and rinse the monolithic column. GL-16 centrifuge with highest speed of 16,000 rpm was purchased from Anting Technology (Shanghai, China).

3.3. Preparation of the Poly (OMA-co-TRIM) Monolithic Column

In order to improve the stability of the monolithic columns, the inner wall of a capillary was treated with a bifunctional reagent, γ-MAPS, according to the procedure reported previously [35]. The pre-polymerization mixture consisted of a monomer OMA (360 mg, 18%, *w/w*), a crosslinker TRIM (240 mg, 12%, *w/w*), porogenic solvents cyclohexan (1120 mg, 56%, *w/w*), 1,4-butanediol (280 mg, 14%, *w/w*), and initiator AIBN (1.8 mg, 0.3%, *w/w*, with respect to monomer and crosslinker). After purging with a N_2 stream for 30 min to remove the oxygen, the mixture was allowed to fill in the capillary. The capillary was immediately sealed at both ends with rubbers and the reaction was initiated in a water bath at 60 °C for 24 h. The prepared monolithic capillary was washed with methanol to remove the unreacted components and porogenic solvents.

3.4. Rice Cultivation

Breeding started in late April. In late May, the seedlings were transplanted to the two isolated paddy fields (one for CAHs-sprayed and one for control). According to the farmers' normal tillage requirements, the fertilizers used were urea, phosphate, potash, and ammonium hydrogen carbonate. The CAHs were sprayed between the tillering stage and the jointing stage, with a dosage of 1.0 kg/ha (about twice the normal dosage). The rice grains were collected after rice maturity.

3.5. Sample Preparation

The samples of the rice grains were prepared based on the method from the agriculture standards of the People's Republic of China (NY/T 1434-2007) [14], with some modifications. Approximately 50 g of rice grains were ground at room temperature. A portion of 15 g ground sample was weighed into a 50 mL Teflon centrifuge tube and covered by 30 mL of ACN-TFA (99.9:0.1, *v/v*) and 6 g $MgSO_4$. Next, the mixture was shaken by the vortex mixer for 1 min. The sample extract was centrifuged at 4000 rpm for 1 min. 10 mL of upper layer extract was then transferred to a 15 mL centrifuge tube containing 0.6 g $MgSO_4$ and 0.25 g PSA. The mixture was vortexed for 1 min and centrifuged at 7000 rpm for 1 min. After centrifugation, 3 mL of the supernatant was collected and dried under N_2 stream at 50 °C. Finally, the dried residues were reconstituted with a solution containing 0.1% TFA solution (98%, *v/v*) and ACN (2%, *v/v*) (the solution of MS was ACN) to bring the final sample volume to 3 mL. Finally, the sample solution was filtered by using 0.22 μm needle filters for analysis.

3.6. Online Hydrophobic in-Tube SPME-HPLC System

The online hydrophobic in-tube SPME-HPLC system was constructed on the basis of our previous works [36]. The poly (OMA-co-TRIM) monolithic column (12 cm × 0.25 mm, i.d.) was used as microextraction sorbent, the schematic diagram of the online hydrophobic in-tube SPME-HPLC system used for the study was illustrated in Figure S3 and the procedure used for this online system is described in the supplementary material.

4. Conclusions

In this work, a novel poly (OMA-co-TRIM) monolithic column was specially prepared and utilized as a hydrophobic sorbent for online hydrophobic in-tube SPME. Due to the excellent extraction and purification effects towards hydrophobic analytes, the developed monolithic column showed its applicability for hydrophobic microextraction. After the optimization of the important factors for in-tube SPME, a simple and sensitive online in-tube SPME-HPLC method using this poly (OMA-co-TRIM) monolithic column as sorbent was proposed for the determination of trace CAHs in rice grains. Low detection limits (0.9–2.1 ng/g) for four CAHs were obtained and the sensitivity of the proposed method could reach that of LC-MS. Moreover, this work shows that hydrophobic in-tube SPME based on this poly (OMA-co-TRIM) monolith could be a feasible approach for the determination of hydrophobic target compounds in complex practical samples.

Supplementary Materials: The following are available online: Figure S1: Pore size distribution profile of the poly (OMA-co-TRIM) monolith by the mercury intrusion method; Figure S2: The peak area of 2,4-D after several operation cycles; Figure S3: Construction of the online hydrophobic in-tube SPME-HPLC system.

Author Contributions: Preparation and characterization of the monolithic column, W.L. and F.W.; optimization for the online hydrophobic in-tube SPME, W.L. and B.N.; rice cultivation, Y.D.; sample preparation, J.Z. and B.N.; method validation, W.L. and F.W.; comparative research with LC-MS, J.Z.; writing—original draft preparation, W.L. and F.W.; writing—review and editing, J.W. All authors read and approved the final manuscript.

Funding: This work was funded by the National Natural Science Foundation of China (Grant No. 31771893 and 31401609), and Science and Technology Plan Project of Fuzhou (Grant No. AFZ2019S10008).

Conflicts of Interest: The authors declare no conflict of interest.

References

1. Krizek, B.A. Making bigger plants: Key regulators of final organ size. *Curr. Opin. Plant Biol.* **2009**, *12*, 17–22. [CrossRef] [PubMed]
2. Loomis, D.; Guyton, K.; Grosse, Y.; El Ghissaiu, F.; Bouvard, V.; Benbrahim-Tallaa, L.; Guha, N.; Mattock, H.; Straif, K. Carcinogenicity of lindane, DDT, and 2,4-dichlorophenoxyacetic acid. *Lancet Oncol.* **2015**, *16*, 891–892. [CrossRef]
3. National Institute for Occupational Safety and Health. The effects of workplace hazards on male reproductive health. Available online: https://www.cdc.gov/niosh/docs/96-132/ (accessed on 27 March 2019).
4. Hou, X.; Han, M.; Dai, X.H.; Yang, X.F.; Yi, S.G. A multi-residue method for the determination of 124 pesticides in rice by modified QuEChERS extraction and gas chromatography-tandem mass spectrometry. *Food Chem.* **2013**, *138*, 1198–1205. [CrossRef] [PubMed]
5. Nie, J.; Chen, F.; Song, Z.; Sun, C.; Li, Z.; Liu, W.; Lee, M. Large volume of water samples introduced in dispersive liquid-liquid microextraction for the determination of 15 triazole fungicides by gas chromatography-tandem mass spectrometry. *Anal. Bioanal. Chem.* **2016**, *408*, 7461–7471. [CrossRef]
6. Yeh, M.K.; Lin, S.L.; Leong, M.I. Determination of phenoxyacetic acids and chlorophenols in aqueous samples by dynamic liquid-liquid-liquid microextraction with ion-pair liquid chromatography. *Anal. Sci.* **2011**, *27*, 49–54. [CrossRef]
7. Shamsipur, M.; Fattahi, N.; Pirsaheb, M. Simultaneous preconcentration and determination of 2,4-D, alachlor and atrazine in aqueous samples using dispersive liquid–liquid microextraction followed by high-performance liquid chromatography ultraviolet detection. *J. Sep. Sci.* **2012**, *35*, 2718–2724. [CrossRef]
8. Kittlaus, S.; Schimanke, J.; Kempe, G.; Speer, K. Development and validation of an efficient automated method for the analysis of 300 pesticides in foods using two-dimensional liquid chromatography–tandem mass spectrometry. *J. Chromatogr. A* **2013**, *1283*, 98–109. [CrossRef]
9. Xu, Y.; Kutsanedzie, F.Y.H.; Hassan, M.M.; Li, H.; Chen, Q. Synthesized Au NPs@silica composite as surface-enhanced Raman spectroscopy (SERS) substrate for fast sensing trace contaminant in milk. *Spectrochim. Acta A* **2019**, *206*, 405–412. [CrossRef] [PubMed]
10. Tabani, H.; Fakhari, A.R.; Zand, E. Low-voltage electromembrane extraction combined with cyclodextrin modified capillary electrophoresis for the determination of phenoxy acid herbicides in environmental samples. *Anal. Methods* **2013**, *5*, 1548–1555. [CrossRef]
11. Zhang, Y.; Jiao, B. Dispersive liquid–liquid microextraction combined with online preconcentration MEKC for the determination of some phenoxyacetic acids in drinking water. *J. Sep. Sci.* **2013**, *36*, 3067–3074. [CrossRef]
12. Deng, A.P.; Yang, H. A multichannel electrochemical detector coupled with an ELISA microtiter plate for the immunoassay of 2,4-dichlorophenoxyacetic acid. *Sens. Actuators B* **2007**, *124*, 202–208. [CrossRef]
13. National Standards of People's Republic of China. GB/T 5009 175-2003: Determination of 2,4-D in grains and vegetables. Available online: http://down.foodmate.net/standard/sort/3/2853.html (accessed on 27 March 2019).
14. Agriculture standards of the People's Republic of China. NY/T 1434-2007: Multi-residue determination of 2,4-D and other 12 Herbicides by LC/MS, 2007. Available online: http://down.foodmate.net/standard/sort/5/14681.html (accessed on 27 March 2019).
15. Gure, A.; Megersa, N.; Retta, N. Ion-pair assisted liquid-liquid extraction for selective separation and analysis of multiclass pesticide residues in environmental waters. *Anal. Methods* **2014**, *6*, 4633–4642. [CrossRef]

16. David, M.D.; Campbell, S.; Li, Q.X. Pressurized Fluid Extraction of Nonpolar Pesticides and Polar Herbicides Using In Situ Derivatization. *Anal. Chem.* **2000**, *72*, 3665–3670. [CrossRef]
17. Moret, S.; Sánchez, J.M.; Salvadó, V.; Hidalgo, M. The evaluation of different sorbents for the preconcentration of phenoxyacetic acid herbicides and their metabolites from soils. *J. Chromatogr. A* **2005**, *1099*, 55–63. [CrossRef]
18. Barchanska, H.; Danek, M.; Sajdak, M.; Turek, M. Review of sample preparation techniques for the analysis of selected classes of pesticides in plant matrices. *Crit. Rev. Anal. Chem.* **2018**, *48*, 467–491. [CrossRef]
19. Hassan, J.; Shamsipur, M.; Es'haghi, A.; Fazili, S. Determination of chlorophenoxy acid herbicides in water samples by suspended liquid-phase microextraction–liquid chromatography. *Chromatographia* **2011**, *73*, 999–1003. [CrossRef]
20. Yun, Y.H.; Shon, H.K.; Yoon, S.D. Preparation and characterization of molecularly imprinted polymers for the selective separation of 2,4-dichlorophenoxyacetic acid. *J. Mater. Sci.* **2009**, *44*, 6206–6211. [CrossRef]
21. Zaidi, S.A. Recent developments in molecularly imprinted polymer nanofibers and their applications. *Anal. Methods* **2015**, *7*, 7406–7415. [CrossRef]
22. Zhang, S.; Xu, T.; Liu, Q.; Liu, J.; Lu, F.; Yue, M.; Li, Y.; Sun, Z.; You, J. Cationic gemini surfactant-resorcinol-aldehyde resin and its application in the extraction of endocrine disrupting compounds from food contacting materials. *Food Chem.* **2019**, *277*, 404–413. [CrossRef]
23. Chu, X.G.; Hu, X.Z.; Yao, H.Y. Determination of 266 pesticide residues in apple juice by matrix solid-phase dispersion and gas chromatography–mass selective detection. *J. Chromatogr. A* **2005**, *1063*, 201–210. [CrossRef]
24. Hogendoorn, E.A.; Huls, R.; Dijkman, E. Microwave assisted solvent extraction and coupled-column reversed-phase liquid chromatography with UV detection: Use of an analytical restricted-access-medium column for the efficient multi-residue analysis of acidic pesticides in soils. *J. Chromatogr. A* **2001**, *938*, 23–33. [CrossRef]
25. Fernandez-Alvarez, M.; Llompart, M.; Lamas, J.P.; Lores, M.; Garcia-Jares, C.; Garcia-Chao, M.; Dagnac, T. Simultaneous extraction and cleanup method based on pressurized solvent extraction for multiresidue analysis of pesticides in complex feed samples. *J. Agric. Food Chem.* **2009**, *57*, 3963–3973. [CrossRef]
26. Wang, T.; Chen, Y.; Ma, J.; Qian, Q.; Jin, Z.; Zhang, L.; Zhang, Y. Attapulgite nanoparticles-modified monolithic column for hydrophilic in-tube solid-phase microextraction of cyromazine and melamine. *Anal. Chem.* **2016**, *88*, 1535–1541. [CrossRef]
27. Beloti, L.G.M.; Miranda, L.F.C.; Queiroz, M.E.C. Butyl methacrylate-co-ethylene glycol dimethacrylate monolith for online in-tube SPME-UHPLC-MS/MS to determine chlopromazine, clozapine, quetiapine, olanzapine, and their metabolites in plasma samples. *Molecules* **2019**, *24*, 310. [CrossRef]
28. Pei, M.; Shi, X.; Wu, J.; Huang, X. Graphene reinforced multiple monolithic fiber solid-phase microextraction of phenoxyacetic acid herbicides in complex samples. *Talanta* **2019**, *191*, 257–264. [CrossRef]
29. Wang, J.; Wu, F.; Zhao, Q. Synchronous extraction and determination of phenoxy acid herbicides in water by on-line monolithic solid phase microextraction-high performance liquid chromatography. *Chin. J. Chromatogr.* **2015**, *33*, 849–855. [CrossRef]
30. Gusev, I.; Huang, X.; Horvath, C. Capillary columns with in situ formed porous monolithic packing for micro high-performance liquid chromatography and capillary electrochromatography. *J. Chromatogr. A* **1999**, *855*, 273–290. [CrossRef]
31. Wang, F.; Dong, J.; Jiang, X.; Ye, M.; Zou, H. Capillary trap column with strong cation-exchange monolith for automated shotgun proteome analysis. *Anal. Chem.* **2007**, *79*, 6599–6606. [CrossRef]
32. Zhang, W.; Zhou, W.; Chen, Z. Graphene/polydopamine-modified polytetrafluoroethylene microtube for the sensitive determination of three active components in Fructus Psoraleae by online solid-phase microextraction with high-performance liquid chromatography. *J. Sep. Sci.* **2014**, *37*, 3110–3116. [CrossRef]
33. Commission Regulation (EU) No 1317/2013, Amending Annexes II, III and V to Regulation (EC) No 396/2005 of the European Parliament and of the Council as regards maximum residue levels for 2,4-D, beflubutamid, cyclanilide, diniconazole, florasulam, metolachlor and S-metolachlor, and milbemectin in or on certain products. Available online: http://down.foodmate.net/standard/sort/44/39676.html (accessed on 27 March 2019).
34. National Standards of People's Republic of China. GB 2763-2016: National food safety standard- Maximum residue limits for pesticides in food. Available online: http://down.foodmate.net/standard/sort/3/50617.html (accessed on 27 March 2019).

35. Wu, R.A.; Zou, H.F.; Fu, H.J. Separation of peptides on mixed mode of reversed-phase and ion-exchange capillary electrochromatography with a monolithic column. *Electrophoresis* **2002**, *23*, 1239–1245. [CrossRef]
36. Wu, F.; Wang, J.; Zhao, Q.; Jiang, N.; Lin, X.; Xie, Z.; Li, J.; Zhang, Q. Detection of trans-fatty acids by high performance liquid chromatography coupled with in-tube solid-phase microextraction using hydrophobic polymeric monolith. *J. Chromatogr. B.* **2017**, *1040*, 214–221. [CrossRef]

Sample Availability: Not available.

© 2019 by the authors. Licensee MDPI, Basel, Switzerland. This article is an open access article distributed under the terms and conditions of the Creative Commons Attribution (CC BY) license (http://creativecommons.org/licenses/by/4.0/).

Article

Butyl Methacrylate-Co-Ethylene Glycol Dimethacrylate Monolith for Online in-Tube SPME-UHPLC-MS/MS to Determine Chlopromazine, Clozapine, Quetiapine, Olanzapine, and Their Metabolites in Plasma Samples

Luiz G. M. Beloti, Luis F. C. Miranda and Maria Eugênia C. Queiroz *

Departamento de Química, Faculdade de Filosofia Ciências e Letras de Ribeirão Preto, Universidade de São Paulo, 14040-901 Ribeirão Preto, SP, Brazil; lgmbeloti@usp.br (L.G.M.B.); luisfelipe.c22@usp.br (L.F.C.M.)
* Correspondence: mariaeqn@ffclrp.usp.br; Tel.: +55-16-36034465

Academic Editor: Nuno Neng
Received: 20 December 2018; Accepted: 12 January 2019; Published: 16 January 2019

Abstract: This manuscript describes a sensitive, selective, and online in-tube solid-phase microextraction coupled with an ultrahigh performance liquid chromatography-tandem mass spectrometry (in-tube SPME-UHPLC-MS/MS) method to determine chlopromazine, clozapine, quetiapine, olanzapine, and their metabolites in plasma samples from schizophrenic patients. Organic poly(butyl methacrylate-co-ethylene glycol dimethacrylate) monolith was synthesized on the internal surface of a fused silica capillary (covalent bonds) for in-tube SPME. Analyte extraction and analysis was conducted by connecting the monolithic capillary to an UHPLC-MS/MS system. The monolith was characterized by scanning electron microscopy (SEM) and Fourier transform infrared spectrometry (FTIR). The developed method presented adequate linearity for all the target antipsychotics: R^2 was higher than 0.9975, lack-of-fit ranged from 0.115 to 0.955, precision had variation coefficients lower than 14.2%, and accuracy had relative standard error values ranging from −13.5% to 14.6%, with the exception of the lower limit of quantification (LLOQ). The LLOQ values in plasma samples were 10 ng mL^{-1} for all analytes. The developed method was successfully applied to determine antipsychotics and their metabolites in plasma samples from schizophrenic patients.

Keywords: in-tube SPME; UHPLC-MS/MS; organic-based monoliths; antipsychotics; plasma samples; schizophrenic' patients

1. Introduction

Schizophrenia is a severe and chronic mental disorder characterized by profound disruptions in thinking, which consequently affects language, perception, and the sense of self [1]. This disorder is characterized by positive (psychotic behaviors), negative (disruptions to normal emotions and behaviors), and cognitive symptoms (changes in memory or other aspects of thinking) [1]. Atypical antipsychotics are the mainstay of treatment prescribed to schizophrenic patients. Compared to classic neuroleptics, these drugs induce fewer extrapyramidal syndromes [1,2].

Studies have suggested that the pharmacokinetics of atypical antipsychotics involve large inter- and intra-individual differences among patients (age, gender, lifestyle, genetic and metabolic characteristics, and drug interactions). Therefore, therapeutic drug monitoring (TDM) can be extremely useful to establish an effective individual therapeutic dose that maintains plasma drug concentrations within a targeted therapeutic range, thereby avoiding an overdose [3,4].

Liquid chromatography–tandem mass spectrometry (LC-MS/MS) is a highly sensitive and selective technique to analyze drugs in biological fluids quantitatively. Biological fluids are complex matrixes containing endogenous macromolecules that can irreversibly adsorb on the analytical column stationary phase surface, which reduces chromatographic separation efficiency. Moreover, during MS/MS analysis, nonvolatile solutes can suppress ionization (electrospray ionization), which decreases the LC-MS/MS method sensitivity [5]. These difficulties call for sample preparation to diminish the matrix effect. This step increases not only the sensitivity, but also the selectivity of the LC-MS/MS method.

In-tube solid-phase microextraction (in-tube SPME) is a sample preparation technique that uses a capillary column, as extraction device, directly coupled to a LC system. The in-tube SPME-LC system is fast to operate, easy to automate, and environmentally friendly (organic solvent is only used in the mobile phase). Automated methods always provide better accuracy and precision than manual procedures [5–7]. Capillary columns with different selective stationary phases (coating), including restricted access media (RAM), molecularly imprinted polymers (MIP), immunosorbents, and monolithic materials, have been used for in-tube SPME systems [8–12].

In-tube SPME-LC methods with different organic monolithic capillaries have been applied for analysis of several analytes. For example: poly (methacrylic acid–ethylene glycol dimethacrylate) for basic drugs in human serum [13], and amphetamines in urine [14], NH_2-MIL-53(Al) incorporated poly(styrene-divinylbenzene-methacrylic acid) (poly(St-DVB-MAA)) for estrogens in human urine [15], and (N-isopropylacrylamide-co-ethylene dimethacrylate) for acid, basic, and neutral compounds [16]. Monolith materials with good control of porosity, diverse surface chemistry, and frit-free are easy to prepare by in situ polymerization [15–17]. The monolithic porous structure facilitates convective mass transfer (which is preferable during extraction) with reasonably low pressure. Organic-based monoliths are stable within the entire pH range and biocompatible with biological samples [17,18].

This manuscript describes an in-tube SPME-UHPLC-MS/MS with an organic poly(butyl methacrylate-co-ethylene glycol dimethacrylate) monolithic capillary to determine antipsychotics (chlopromazine, clozapine, quetiapine, and olanzapine) and their metabolites (desmethyl chlorpromazine, 7-hydroxy-chlorpromazine, N-desmethyl clozapine, N-desmethyl olanzapine, and norquetiapine) in plasma samples from schizophrenic patient. Figure 1 illustrates the metabolic and biotransformation pathways of these antipsychotics.

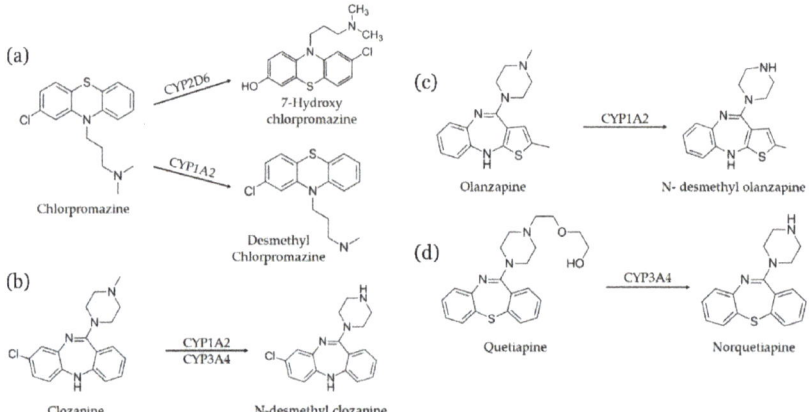

Figure 1. Biotransformation pathways of (**a**) Chlorpromazine, (**b**) Clozapine, (**c**) Olanzapine, and (**d**) Quetiapine. CYP = cytochrome P450 complex. CYP1A2, CYP2D6, and CYP3A4 are isoenzymes of cytochrome P450 complex.

2. Results and Discussion

2.1. Organic Poly(Butyl Methacrylate-Co-Ethylene Glycol Dimethacrylate) Monolith Capillary Preparation and Characterization

Although the proposed monolith synthesis was based on classical radicalar procedures [19,20], the innovation of this work is related to direct coupling of the poly (butyl methacrylate-co-ethylene glycol dimethacrylate monolith capillary (in-tube SPME) to the LC-MS/MS system.

The cross-linking monomer (type and crosslink density), the porogenic solvent (type and amount), and the ratio between the functional and the cross-linking monomers substantially influence the polymer surface area, pore volume, pore size, and porosity [17]. Pore size distribution must be adjusted during monolith preparation, so that the monolith fits the desired application [18].

Table 1 illustrates optimization of the conditions of the organic poly(butyl methacrylate-co-ethylene glycol dimethacrylate) monolith capillary synthesis procedure. The porogenic solvent controls the organic monolith porosity [21,22]. The porogenic solvents 1-propanol and 1,4-butenodiol employed here produced a homogenous pre-polymer solution containing the monomers. A slight increase in the amount of porogenic solvent generated larger pores, which improved monolithic capillary permeability and favored sample percolation through the capillary under the in-tube SPME system low pressure.

Table 1. Optimization of the organic poly(butyl methacrylate-co-ethylene glycol dimethacrylate) monolith capillary synthesis procedure.

Monolith	Monomer/Porogen (%m/m)	EGDMA:BMA (%m/m)	Porogens BUT:PRO (%m/m)	AIBN	Permeability
M1	40:60	70:30	25:65	1%	Poor
M2	40:60	55:45	30:60	1%	Poor
M3	35:65	40:60	57:43	1%	Good
M4	35:65	50:50	57:43	1%	Good
M5	35:65	30:70	57:43	1%	Good

BUT = 1,4-butanediol; PRO = 1-propanol; AIBN = 2,2-azobisisobutylnitrile, BMA = butyl methacrylate, EGDMA = ethyleneglicol dimetacrylate.

Alteration in the cross-linking monomer percentage in relation to the functional monomer modified monolith porosity. High cross-linking monomer concentration decreased both pore size and permeability.

On the basis of analyte extraction efficiency (Figure 2), the functional and cross-linking monomer percentages were optimized (Table 1). The M3 monolithic capillary (Figure 2) was selected for the in-tube SPME-LC analysis because it gave the highest extraction efficiency and adequate permeability.

Figure 3 shows the cross-sectional SEM images of poly(butyl methacrylate-co-ethylene glycol dimethacrylate) monolith capillary at 200× and 10,000× magnification. The monolith exhibited continuous (greater homogeneity) coating with interconnected macropores, which allowed the sample to be percolated through the capillary at low pressure. The monolith was clearly tightly attached to the capillary inner-wall.

The monolith was fixed to the inner capillary surface by covalent bonds, which dismissed the need for frits [22]. As a result of this chemical attachment and its structure, the monolith exhibited mechanical and chemical stability, as well as biocompatibility with biological samples [22,23]. The proposed monolithic capillary was reused more than one hundred times without observing significant changes in sorption capacity.

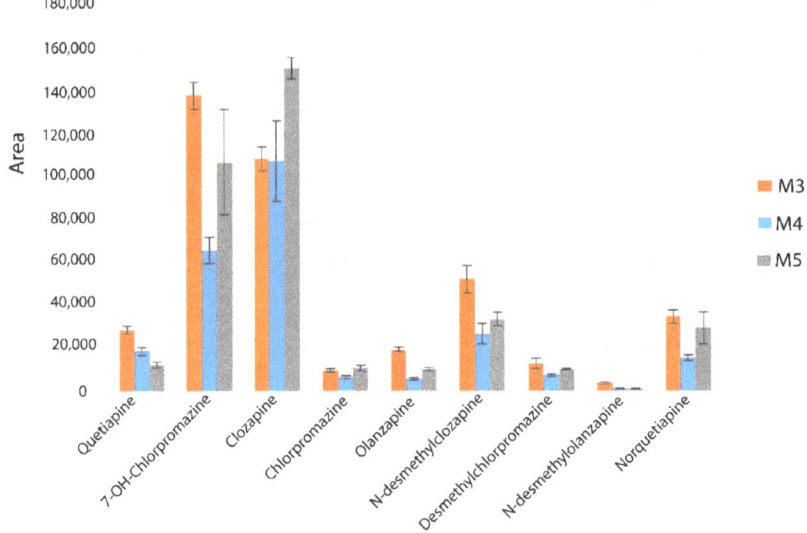

Figure 2. The effect of different synthesis procedures on the performance of in-tube SPME-MS/MS method (Table 1 describes the synthesis conditions).

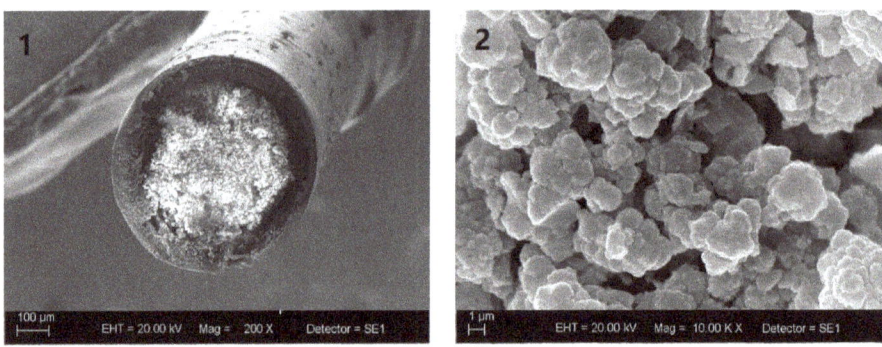

Figure 3. Scanning electron microscope images of the cross-section of poly(butyl methacrylate-co-ethyleneglicol dimethacrylate) monolith capillary at (**1**) 200× magnification and (**2**) 10,000× magnification.

Figure 4 depicts the monolithic phase FTIR spectra. The bands at 2960.7 and 1454.4 cm^{-1} indicated bond between carbon sp^3 and hydrogen atoms. The band at 1728.6 cm^{-1} corresponded to carbonyl bond. The bands at 1388.9, 1254, and 1157.1 cm^{-1} were ascribed to symmetric and asymmetric ester C-O bond vibrations. The spectra also displayed bands at 1637.6 cm^{-1}, due to residual vinyl C=C bond stretching; at 814.36 and 751.24 cm^{-1}, attributed to cis-substituted vinyl group vibration; and at 959.12 cm^{-1}, assigned to C-H bond out-of-plane vibration. The band at 3443.8 cm^{-1} referred to adsorbed water hydroxyl groups, a consequence of the monolith phase high porosity. The FTIR spectra confirmed incorporation of both monomers, butyl methacrylate and ethylene glycol dimethacrylate, in the monolithic capillary structure [24,25].

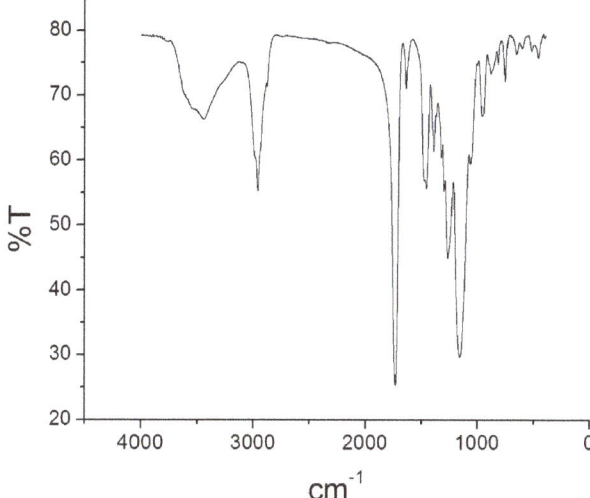

Figure 4. Fourier transform infrared spectrometry spectra of the poly(butyl methacrylate-co-ethyleneglicol dimethacrylate) monolith capillary.

2.2. Plasma Sample Pre-Treatment

After protein precipitation, the dried extract was reconstituted with 100 µL of 5 mM ammonium acetate solution at different pH values. At pH 10, analytes were in non-ionized or the partially ionized form, which improved hydrophobic interactions between the analytes and the monolithic capillary.

As reported in the literature, the poly(butyl methacrylate-co-ethylene glycol dimethacrylate) monolith is selective for hydrophobic analytes, such as polycyclic aromatic hydrocarbons in smoked meat products [19] and cyclophosphamide and busulfan in whole blood samples [20].

The pre-treatment step boosted extraction efficiency because it attenuated the matrix effect.

2.3. Analytical Validation

Analytes were detected by ESI–MS/MS in the SRM and in the positive ionization modes. The protonated molecules of the analytes [M + H]$^+$ were fragmented by collision-induced dissociation (CID). Two product ions of each analyte were selected as transitions in the SRM detection mode: one for quantitative and the other for qualitative purposes (Table 2).

Table 2. Ions transitions, instruments settings, and retention times for each antipsychotic and their metabolites.

Analyte	Precursor Ion (m/z)	Quantifier Ion (m/z)	Ce (eV)	DP (v)	Qualifier Ion (m/z)	Retention Time
Chlorpromazine	319.0	85.9	38	18	57.9	9.27
Chlorpromazine-d3	324.0	60.9	34	42	89.0	9.25
Clozapine	327.0	270.0	44	30	191.9	8.62
Olanzapine	313.0	256.0	22	20	84.0	5.95
Quetiapine	384.0	253.0	36	18	221.0	8.66
Quetiapine-d8	392.2	225.9	38	48	257.8	8.64
Desmethyl chlorpromazine	304.9	72.0	30	14	43.9	9.19
7-hydroxy-chlorpromazine	335.0	85.9	30	34	57.8	8.71
N-desmethyl clozapine	313.0	191.9	28	52	69.9	8.52
N-desmethyl olanzapine	299.0	197.9	26	28	212.9	5.46
Norquetiapine	296.0	209.9	54	26	138.9	8.61

Method selectivity was assessed by comparing a blank plasma sample chromatogram with the chromatogram of a blank plasma sample spiked with the target analytes at concentrations corresponding to the lower limit of quantification (LLOQ) (Figure 5).

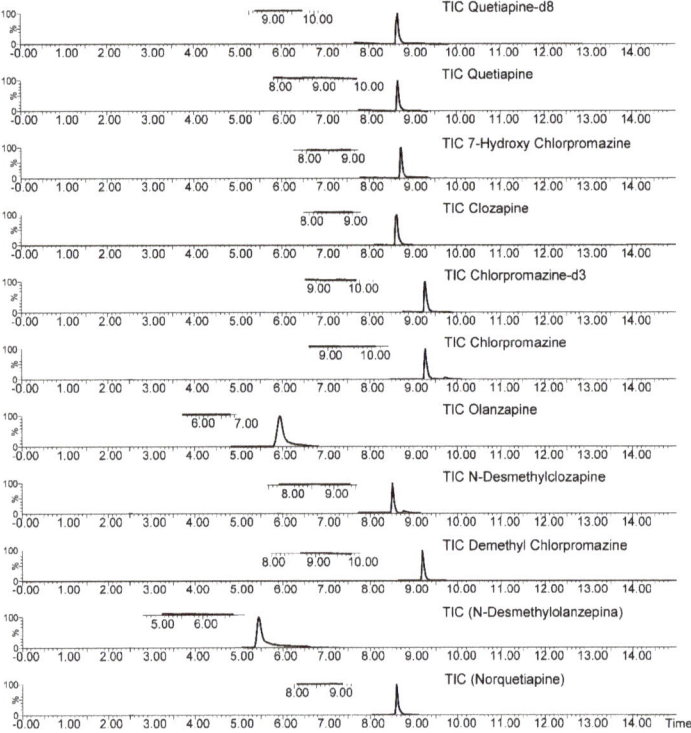

Figure 5. LC-MS/MS TIC (Total Ions Current) chromatograms of a drug-free plasma sample (subscript on the left) and drug-free plasma sample spiked with target drugs at the LLOQ concentrations.

The developed method was linear from the LLOQ (10 ng mL^{-1}) to the upper limit of quantification (ULOQ) (200 ng mL^{-1} to 700 ng mL^{-1}); the coefficient of determination was higher than 0.9975 (Table 3). The lack-of-fit test confirmed method linearity. Calibration standards ($n = 5$) presented coefficient of variation (CV%) lower than 15%. This linear range included therapeutic intervals.

Table 3. Linearity of the SPME-UHPLC-MS/MS method.

Analyte	Linearity		
	R^2	Internal Standart	Lack of Fit Test
Chlorpromazine	0.9986	chlorpomazine-d3	0.848
Clozapine	0.9997	quetiapine-d8	0.226
Olanzapine	0.9989	quetiapine-d8	0.146
Quetiapine	0.9981	quetiapine-d8	0.888
Desmethyl chlorpromazine	0.9975	chlorpomazine-d3	0.420
7-hydroxy-chlorpromazine	0.9992	chlorpomazine-d3	0.166
N-desmethyl clozapine	0.9989	quetiapine-d8	0.196
N-desmethyl olanzapine	0.9997	quetiapine-d8	0.955
Norquetiapine	0.9985	quetiapine-d8	0.115

* p-value at a significance level of 0.05.

The developed method presented accuracy, with RSE values ranging from −19.4 to 19.9% (intra-assay) and from −18.9 to 19.3% (inter-assay), as well as precision, with CV values ranging from 0.7 to 14.2 (intra- and inter-assay) (Table 4).

Table 4. Accuracy, precision, and matrix effects of the SPME-UHPLC-MS/MS method.

Analyte	Concentration (ng mL^{-1})	Accuracy		Precision		Matrix Effects (%CV)
		Intra-Assay (%RSE) $n = 5$	Inter-Assay	Intra-assay (%CV) $n = 5$	Inter-Assay	
Chlorpromazine	10	−19.4	−18.9	1.7	0.9	
	30	3.5	1.9	2.5	0.7	0.6
	200	−3.1	−10.9	4.8	7.8	
	300	−0.1	0.4	7.7	0.7	12.1
	400	−2.5	0.2	10.0	7.0	
Clozapine	10	14.7	14,7	14.2	3.3	
	30	14.6	6.7	9.1	12.1	3.6
	350	−1.0	−5.9	3.5	1.1	
	500	−2.5	−3.2	2.4	1.0	1.9
	700	1.0	−0.7	0.7	3.8	
Olanzapine	10	18.4	19.3	9.0	12.1	
	30	13.3	9.7	3.2	3.7	6.7
	100	−1.9	−4.2	1.2	8.3	
	150	5.7	6.2	4.8	3.3	7.9
	200	2.5	1.8	3.5	1.0	
Quetiapine	10	−18.8	−14.5	2.6	3.2	
	30	13.6	−10.5	7.7	1.6	1.5
	250	−9.5	−8.4	2.5	5.2	
	500	4.7	0.7	4.1	10.2	9.1
	600	−1.4	0.3	2.0	8.1	
Desmethyl chlorpromazine	10	−6.3	−9.5	1.7	4.5	
	30	−13.4	−13.5	1.6	2.7	9.9
	100	−2.2	−2.1	3.1	3.6	
	150	4.6	−1.8	2.4	6.8	10.6
	200	−7.9	−12.8	9.1	2.3	
7-hydroxy-chlorpromazine	10	−5.0	8.8	11.9	3.6	
	30	0,9	−4.4	1.5	4.3	13.5
	100	2.0	0.4	8.3	2.7	
	150	−3.9	5.5	3.0	6.6	14.3
	200	2.9	5.0	3.6	3.2	
N-desmethyl clozapine	10	7.5	10.4	0.8	2.3	
	30	−8.8	−0.1	9.1	7.3	2.1
	200	−4.3	−10.3	2.2	13.1	
	300	1.0	12.6	6.3	14.2	2.2
	500	−1.7	−2.5	1.3	10.0	
N-desmethyl olanzapine	10	11.0	15.6	0.7	6.0	
	25	−1.3	4.3	13.5	5.1	5.3
	100	−2.4	−3.6	9.1	3.0	
	150	0.8	12.8	10.1	3.6	6.1
	200	1.0	4.3	3.5	8.6	
Norquetiapine	10	19.9	18.4	4.1	3.4	
	30	7.5	7.5	9.5	9.2	3.1
	100	−4.6	−5.4	4.4	1.5	
	150	0.04	−6.1	6.7	11.7	5.3
	200	1.5	−1.4	1.0	7.2	

The method matrix effect was evaluated by using CV values of the IS-normalized matrix effect that were lower than 15% (Table 4). Residual carryover in blank plasma samples following ULOQ analysis exhibited values lower than 3% of the analyte LLOQ signal, or 0.05% of the IS LLOQ signal.

Comparing the in-tube SPME-UHPLC method with literature methods (Table 5), the proposed method presented lower LLOQ values than the values obtained with the MEPS-UHPLC [26] method. The proposed method presented the lowest chromatographic separation time (Table 5). Moreover, the proposed method used reduced plasma sample volume (300 µL) and provided simultaneous determination of different antipsychotics and their metabolites.

Table 5. Comparison of the in-tube SPME-UHPLC-MS/MS method with other counterparts to determine antipsychotics and their metabolites in biological samples.

Analytes	Matrix	Sample Amount (μL)	Chromatographic Separation (min)	Analytical Technique	LLOQ (ng mL^{-1})	References
Aripiprazole Olanzapine Paliperidone Ziprasidone	Plasma	200	8	Protein precipitation	10.0 (olanzapine)	Park et al. 2018 [27]
chlorpromazine, haloperidol, levomepromazine, olanzapine, risperidone, and sulpiride	Plasma	500	7	SPE (Oasis HLB)	13.2 (chlorpromazine) 2.9 (olanzapine)	Khelfi et al. 2018 [28]
Haloperidol, olanzapine, chlorpromazine, quetiapine, clozapine	Plasma	200	10	column switching LC-MS/MS (hybrid monolith with cyano groups)	0.075–0.188	Domingues et al. 2015 [8]
Clozapine, risperidone, and metabolites	Urine	500	10	MEPS (C18) UHPLC-PDA	100.0	Gonçalves et al. 2015 [26]
chlopromazine, clozapine, quetiapine, olanzapine and metabolites	Plasma	300	4.5	In-tube SPME-UHPLC-MS/MS	10.0	This work

2.4. Determination of Antipsychotics and Their Metabolites in Plasma Samples from Schizophrenic Patients

The proposed method was successfully applied to determine the target antipsychotics in plasma samples from three schizophrenic patients undergoing therapy with atypical antipsychotics (Table 6). Consequently, this method could be used for therapeutic drug monitoring.

Table 6. Concentrations of the target antipsychotic and their metabolites in plasma samples from Schizophrenic patients.

Drugs	Therapeutic Drug Monitoring Interval (ng mL^{-1})	Plasma Concentrations			
		Patient 1	Patient 2	Patient 3	Patient 4
Chlorpromazine	30–300	-	-	-	329
Clozapine	350–600	-	-	528	-
Olanzapine	20–80	85	-	-	-
Quetiapine	100–500	-	400	-	-
Desmethyl chlorpromazine	-	-	-	-	12
7-hydroxy-chlorpromazine	-	-	-	-	40
N-desmethyl clozapine	-	-	-	328	-
N-desmethyl olanzapine	-	19	-	-	-
Norquetiapine	-	-	61	-	-

3. Materials and Methods

3.1. Standards and Reagents

Chlorpromazine, clozapine, olanzapine, and quetiapine were purchased from Cerilliant (Round Rock, TX, USA). Desmethylchlorpromazine and 7-hydroxy-chlorpromazine were acquired from TRC Canada (Toronto, ON, Canada). N-desmethylclozapine was obtained from Sigma-Aldrich (St. Louis, MO, USA). N-desmethylolanzapine was purchased from Santa Cruz Biotechnology (Dallas, TX, USA). Norquetiapine was acquired from Biovision (Milpitas, CA, USA). The internal standards chlorpromazine-d3 and quetiapine-d8 were obtained from Cerilliant (Round Rock, TX, USA). Butyl methacrylate (BMA) (99%), ethyleneglicol dimetacrylate (EGDMA, 98%), 1-propanol (HPLC grade), vinyltrimethoxysilane (VTMS) 1,4-butendiol (99%), and 2,2-azobisisobutylnitrile (AIBN) were purchased from Sigma–Aldrich (St. Louis, MO, USA). Fused silica capillary (530 µm i.d. × 10 cm) was acquired from NST (São Paulo, Brazil). Acetonitrile, methanol (HPLC grade), ammonium acetate, and formic acid were supplied by JTBaker (Phillisburg, NJ, USA). Hydrochloric acid and sodium hydroxide were purchased from Sigma–Aldrich (St. Louis, MO, USA). Aqueous solutions were prepared with ultrapure water from a Milli-Q, Millipore system (18.2 MΩ cm) (São Paulo, SP, Brazil).

3.2. Organic Poly(Butyl Methacrylate-Co-Ethylene Glycol Dimethacrylate) Monolith Capillary Preparation

The monolith was synthesized based on published literature [19,20], with modifications. The fused silica capillary was pretreated to activate surface silanol groups. The capillary was initially rinsed with 0.2 mol L^{-1} HCl for 30 min, which was followed by water until the outlet solution achieved pH 7.0. Subsequently, the capillary was flushed with 1 mol L^{-1} NaOH for 2 h, which was followed by water and methanol for 30 min. Finally, the capillary was purged with nitrogen at 160 °C for 3 h prior to use. To achieve covalent binding between the polymer materials and the capillary inner wall, the capillary was modified with vinyltrimethoxysilane solution. The activated capillary was then silanized as previously reported by Ho, T.D. et al. [29]. The capillary was filled with VTMS, sealed with silicon rubbers, and reacted at 85 °C for 2 h. The silylated capillary was rinsed with MeOH and purged with nitrogen at 60 °C in a GC oven for 3 h, to give the vinyl-functionalized capillary.

To perform polymerization, different BMA (functional monomer), EGDMA (cross-linking monomer), AIBN (radicalar initiator), and 1-propanol and 1,4-butenodiol (porogenic solvents) proportions (Table 1) were mixed (vortex for 1 min), sonicated for 10 min, and purged with a nitrogen stream for 10 min. The activated capillary was filled with this mixture, and both capillary ends were sealed with silicon rubbers. The polymerization reaction was kept at 60 °C for 20 h. Finally, the capillary was rinsed with methanol for 2 h to remove unreacted monomers, porogens, and any other soluble compounds from the pores.

3.3. Organic Poly(Butyl Methacrylate-Co-Ethylene Glycol Dimethacrylate) Monolith Capillary Characterization

Scanning Electron Microscopy was employed to evaluate organic monolith surface morphology. Samples were submitted to carbon evaporation and were coated with gold for 180 s in a Bal-Tec SCD050 Sputter (Fürstentum, Liechtenstein). The samples were then analyzed in a Carl Zeiss EVO5O scanning electron microscope (Cambridge, UK). The chemical groups present on the monolith were identified by Fourier Transform Infrared Spectroscopy (FTIR) on the Shimadzu-IRPrestige-21 (ABB Bomem series MB100) spectrometer; KBr pellets were used.

3.4. Plasma Samples

The developed in-tube SPME-UHPLC-MS/MS method was optimized and validated with drug-free plasma (blank samples) from volunteers that had not been exposed to any drug for at least 72 h. These blank plasma samples and plasma samples from patients undergoing therapy with antipsychotics were kindly supplied by the Psychiatric Nursing staff of the University Hospital of Ribeirão Preto Medical School, University of São Paulo, Brazil. The plasma samples were collected in agreement with the criteria established by the Ethics Committee of Ribeirão Preto Medical School, University of São Paulo, Brazil. The plasma samples from schizophrenic patients were collected and stored at −80 °C for six months.

3.5. Plasma Sample Pre-Treatment

Initially, plasma sample (300 µL) was spiked with internal standard solutions (Chlorpromazine-d3 and Quetiapine-d8 at 50 ng mL^{-1}). Plasma proteins were precipitated with acetonitrile (600 µL), and then after centrifugation for 20 min, 800 µL of the supernatant was collected and dried in the vacuum concentrator (Eppendorf, Brazil). The dried extract was reconstituted with ammonium acetate/ammonium hydroxide solution (5 mmol L^{-1}). Considering the monolith sorption capacity, different pH values (4.0, 7.0, and 10.0) of this solution were evaluated. Then, 10 µL of this solution was injected into the in-tube SPME-UHPLC-MS/MS system.

3.6. LC-MS/MS Conditions

LC–MS/MS analyses were performed on a Waters ACQUITY UPLC H-Class system coupled to the Xevo® TQ-D tandem quadrupole mass spectrometrer (Waters Corporation, Milford, MA, USA); a Z-spray source operating in the positive mode was used. The optimum parameters were: capillary voltage of 3.20 kV, source temperature of 150 °C, desolvation temperature of 400 °C, desolvation gas flow of 700 L h^{-1} (N_2 99.9% purity), and cone gas flow of 150 L h^{-1} (N_2 99,9% purity). Analytes were analyzed in the selected reaction monitoring (SRM) mode. Argon (99.9999% purity) was employed as collision gas, and the dwell time for each transition was set to 0.049 seconds. The analytes were separated on an ACQUITY UPLC CSH C18 (1.7 µm, 2.1 × 100mm) column at 40 °C, and data were acquired by using the MassLynx V4.1 Software (Waters Corporation, Milford, MA, USA).

3.7. In-Tube SPME-UHPLC-MS/MS Configuration

Two columns were connected by means of a six-port valve, as shown in Figure 6. The monolithic column (first dimension) was connected to the quaternary pump (QSM), and the analytical column (second dimension) was connected to the binary pump (BSM). In the first step, the six-port valve was set in position 1, which allowed the columns to be conditioned with the initial mobile phase composition (Table 7). Then, 10 µL of the sample solution was injected into the system. Water was used as mobile phase to percolate the sample solution through the monolithic capillary at a flow rate of 100 µL min^{-1}. In this step, analytes were pre-concentrated, and macromolecules from plasma samples were eluted for waste. After 2.0 min, the six-port valve was set to position 2. The target drugs from the monolithic capillary were eluted to the analytical column using the mobile phase (BSM pump), consisted of (A) 10 mmol L^{-1} ammonium acetate (with 0.1% formic acid) and (B) acetonitrile (80:20 v/v), at a flow rate of 100 µL min^{-1}. At 6.50 min, the six-port valve was set to position 1. Using the same mobile phase, from 6.51 to 13.00 min, the chromatographic separation occurred at 300 µL min^{-1}. From 7.0 to 16.5 min, acetonitrile was percolated through the monolithic capillary for clean-up. After this time, both columns were conditioned with the initial mobile phase composition for the following sample injection (Table 7). Different mobile phases were evaluated to establish the highest analyte sorption and the best analyte resolution within the minimum analysis time.

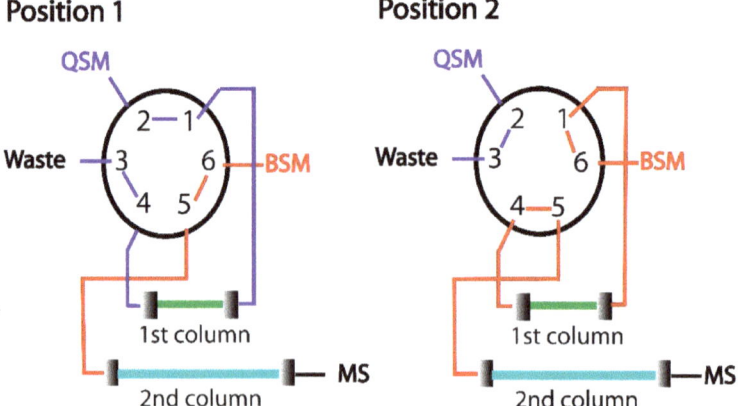

Figure 6. Scheme of in-tube SPME-UHPLC-MS/MS system. (**1**) Sample clean-up and sorption of the analytes, and (**2**) elution of the analytes from the 1st to 2nd column. QSM: quaternary pump, BSM: binary pump, MS: mass spectrometry, 1st column: monolithic capillary, 2nd column: analytical column.

Table 7. Chromatography conditions for In-tube SPME-UHPLC-MS/MS analysis.

QSM A = Water, B = Acetonitrile					BSM A = 10 mmol L^{-1} Ammonium acetate (0.1% Formic Acid) B = Acetonitrile
T (min)	Pump	Flow Rate (µL min^{-1})	%A	Valve Position	Comments
0.0	QSM	100	100	1	Sample cleanup and drug pre-concentration into monolithic capillary
0.0	BSM	100	15	1	Analytical column conditioning
2.0	BSM	100	80	2	Analyte elution from the monolith column to the chromatographic column
5.50	BSM	300	80	1	Beginning of chromatography separation on the analytical column.
5.50	QSM	100	0	1	Cleanup of monolithic capillary
13.00	BSM	100	15	1	End of analytical separation and start of column conditioning for the next sample injection
13.0	QSM	300	100	1	Start of monolithic column conditioning for the next sample injection

QSM: quaternary pump, and BSM: binary pump.

3.8. Analytical Validation

Based on current international guidelines of the European Medicines Agency (EMA) and Food and Drug Administration (FDA), the in-tube SPME-LC-MS/MS method was validated.

Using linear regression of the ratio between the peak area of the target drugs and the IS (Y-axis) peak area versus the nominal drug concentrations (X-axis, ng mL^{-1}), the calibration curves were generated.

The lower limit of quantitation (LLOQ) is the lowest concentration in the calibration curve that can be quantitatively determined with suitable precision and accuracy.

Accuracy was evaluated from relative squared error (RSE) values. Precision was estimated from the CV values of the analyses of the blank plasma samples spiked with drugs at five different concentrations ($n = 5$), namely LLOQ, low quality control (QC), medium QC, high QC, and upper limit of quantitation (ULOQ).

Matrix effects were investigated by using eight lots of blank plasma obtained from different sources spiked with drugs at low QC and high QC concentrations. The matrix factor (MF) was evaluated for each matrix lot, by comparing the drugs responses in the presence of matrix with those in the absence of matrix. The IS normalized MF was also calculated by dividing the drug MF by the IS MF. The CV of the IS-normalized MF calculated from the eight matrix sources should not exceed 15%.

The carry-over should be evaluated by analyzing a blank sample following the highest concentration calibration standard. The response in the blank sample obtained after measurement of the highest concentration standard should not be greater than 20% of the analyte response at the LLOQ and 5% of the response of internal standard.

4. Conclusions

The organic poly(butyl methacrylate-co-ethylene glycol dimethacrylate) monolith developed herein exhibited low backpressure, high permeability, and adequate sorption to determine antipsychotics and their metabolites in plasma samples at sub-therapeutic levels.

Compared to other microextracion techniques, the automate in-tube SPME system allows direct coupling of the microextraction step to chromatographic systems, which not only increases the accuracy and precision, but also reduces the organic solvent consumption and analysis time.

The in-tube SPME-UHPLC-MS/MS method exhibited good selectivity and sensitivity due to analyte pre-concentration in monolithic capillary. This method was successfully applied to determine antipsychotics and their metabolites in plasma samples from schizophrenic patients.

Author Contributions: L.G.M.B.: formal analysis, investigation, methodology, writing, validation, and original draft preparation. L.F.C.M.: investigation, methodology and writing. M.E.C.Q.: supervision, methodology, project administration, funding acquisition, review and editing.

Funding: This research was funded by FAPESP (Fundação de Amparo à Pesquisa do Estado de São Paulo, 2017/02147-0), INCT-TM (465458/2014-9) (Instituto Nacional de Ciência e Tecnologia Translacional em Medicina) and CAPES (Coordenação de Aperfeiçoamento de Pessoal de Nível Superior).

Conflicts of Interest: The authors declare no conflict of interest.

References

1. Lieberman, J.A.; Perkins, D.; Belger, A.; Chakos, M.; Jarskog, F.; Boteva, K.; Gilmore, J. The early stages of schizophrenia: Speculations on pathogenesis, pathophysiology, and therapeutic approaches. *Biol. Psychiatry* **2001**, *50*, 884–897. [CrossRef]
2. Seeman, P.; Kapur, S. Schizophrenia: More dopamine, more D2 receptors. *Proc. Natl. Acad. Sci. USA* **2000**, *97*, 7673–7675. [CrossRef] [PubMed]
3. Patteet, L.; Morrens, M.; Maudens, K.E.; Niemegeers, P.; Sabbe, B.; Neels, H. Therapeutic drug monitoring of common antipsychotics. *Ther. Drug Monit.* **2012**, *34*, 629–651. [CrossRef] [PubMed]
4. Baumann, P.; Hiemke, C.; Ulrich, S.; Eckermann, G.; Gaertner, I.; Gerlach, M.; Kuss, H.-J.; Laux, G.; Müller-Oerlinghausen, B.; Rao, M. The AGNP-TDM expert group consensus guidelines: Therapeutic drug monitoring in psychiatry. *Pharmacopsychiatry* **2004**, *37*, 243–265. [CrossRef] [PubMed]
5. Queiroz, M.; Melo, L. Selective capillary coating materials for in-tube solid-phase microextraction coupled to liquid chromatography to determine drugs and biomarkers in biological samples: A review. *Anal. Chim. Acta* **2014**, *826*, 1–11. [CrossRef] [PubMed]
6. Kataoka, H.; Narimatsu, S.; Lord, H.L.; Pawliszyn, J. Automated in-tube solid-phase microextraction coupled with liquid chromatography/electrospray ionization mass spectrometry for the determination of β-blockers and metabolites in urine and serum samples. *Anal. Chem.* **1999**, *71*, 4237–4244. [CrossRef]
7. Kataoka, H. Automated sample preparation using in-tube solid-phase microextraction and its application–A review. *Anal. Bioanal. Chem.* **2002**, *373*, 31–45. [CrossRef]
8. Domingues, D.S.; de Souza, I.D.; Queiroz, M.E.C. Analysis of drugs in plasma samples from schizophrenic patients by column-switching liquid chromatography-tandem mass spectrometry with organic–inorganic hybrid cyanopropyl monolithic column. *J. Chromatogr. B* **2015**, *993*, 26–35. [CrossRef]
9. Marchioni, C.; de Souza, I.D.; Grecco, C.F.; Crippa, J.A.; Tumas, V.; Queiroz, M.E.C. A column switching ultrahigh-performance liquid chromatography-tandem mass spectrometry method to determine anandamide and 2-arachidonoylglycerol in plasma samples. *Anal. Bioanal. Chem.* **2017**, *409*, 3587–3596. [CrossRef]
10. Chaves, A.R.; Queiroz, M.E.C. Immunoaffinity in-tube solid phase microextraction coupled with liquid chromatography with fluorescence detection for determination of interferon α in plasma samples. *J. Chromatogr. B* **2013**, *928*, 37–43. [CrossRef]
11. Souza, I.D.; Melo, L.P.; Jardim, I.C.; Monteiro, J.C.; Nakano, A.M.S.; Queiroz, M.E.C. Selective molecularly imprinted polymer combined with restricted access material for in-tube SPME/UHPLC-MS/MS of parabens in breast milk samples. *Anal. Chim. Acta* **2016**, *932*, 49–59. [CrossRef] [PubMed]
12. Xu, W.; Su, S.; Jiang, P.; Wang, H.; Dong, X.; Zhang, M. Determination of sulfonamides in bovine milk with column-switching high performance liquid chromatography using surface imprinted silica with hydrophilic external layer as restricted access and selective extraction material. *J. Chromatogr. A* **2010**, *1217*, 7198–7207. [CrossRef] [PubMed]
13. Fan, Y.; Feng, Y.-Q.; Da, S.-L.; Shi, Z.-G. Poly (methacrylic acid–ethylene glycol dimethacrylate) monolithic capillary for in-tube solid phase microextraction coupled to high performance liquid chromatography and its application to determination of basic drugs in human serum. *Anal. Chim. Acta* **2004**, *523*, 251–258. [CrossRef]
14. Fan, Y.; Feng, Y.-Q.; Zhang, J.-T.; Da, S.-L.; Zhang, M. Poly (methacrylic acid-ethylene glycol dimethacrylate) monolith in-tube solid phase microextraction coupled to high performance liquid chromatography and analysis of amphetamines in urine samples. *J. Chromatogr. A* **2005**, *1074*, 9–16. [CrossRef] [PubMed]
15. Luo, X.; Li, G.; Hu, Y. In-tube solid-phase microextraction based on NH_2-MIL-53 (Al)-polymer monolithic column for online coupling with high-performance liquid chromatography for directly sensitive analysis of estrogens in human urine. *Talanta* **2017**, *165*, 377–383. [CrossRef]
16. Ma, Q.; Chen, M.; Shi, Z.G.; Feng, Y.Q. Preparation of a poly (N-isopropylacrylamide-co-ethylene dimethacrylate) monolithic capillary and its application for in-tube solid-phase microextrac-tion coupled to high-performance liquid chromatography. *J. Sep. Sci.* **2009**, *32*, 2592–2600. [CrossRef] [PubMed]

17. Liu, K.; Aggarwal, P.; Lawson, J.S.; Tolley, H.D.; Lee, M.L. Organic monoliths for high-performance reversed-phase liquid chromatography. *J. Sep. Sci.* **2013**, *36*, 2767–2781. [CrossRef]
18. Svec, F. Preparation and HPLC applications of rigid macroporous organic polymer monoliths. *J. Sep. Sci.* **2004**, *27*, 747–766. [CrossRef]
19. Liu, W.; Qi, J.; Yan, L.; Jia, Q.; Yu, C. Application of poly (butyl methacrylate-co-ethylene glycol dimethacrylate) monolith microextraction coupled with high performance liquid chromatography to the determination of polycyclic aromatic hydrocarbons in smoked meat products. *J. Chromatogr. B* **2011**, *879*, 3012–3016. [CrossRef]
20. Skoglund, C.; Bassyouni, F.; Abdel-Rehim, M. Monolithic packed 96-tips set for high-throughput sample preparation: Determination of cyclophosphamide and busulfan in whole blood samples by monolithic packed 96-tips and LC-MS. *Biomed. Chromatogr.* **2013**, *27*, 714–719. [CrossRef]
21. Viklund, C.; Svec, F.; Fréchet, J.M.; Irgum, K. Monolithic,"molded", porous materials with high flow characteristics for separations, catalysis, or solid-phase chemistry: Control of porous properties during polymerization. *Chem. Mater.* **1996**, *8*, 744–750. [CrossRef]
22. Coufal, P.; Čihák, M.; Suchankova, J.; Tesařová, E.; Bosakova, Z.; Štulík, K. Methacrylate monolithic columns of 320 μm ID for capillary liquid chromatography. *J. Chromatogr. A* **2002**, *946*, 99–106. [CrossRef]
23. Zhang, M.; Wei, F.; Zhang, Y.-F.; Nie, J.; Feng, Y.-Q. Novel polymer monolith microextraction using a poly (methacrylic acid-ethylene glycol dimethacrylate) monolith and its application to simultaneous analysis of several angiotensin II receptor antagonists in human urine by capillary zone electrophoresis. *J. Chromatogr. A* **2006**, *1102*, 294–301. [CrossRef] [PubMed]
24. Smith, B.C. *Infrared Spectral Interpretation: A Systematic Approach*; CRC Press: Boca Raton, FL, USA, 1998.
25. Coates, J. Interpretation of infrared spectra, a practical approach. *Encycl. Anal. Chem.* **2000**, *12*, 10815–10837.
26. Gonçalves, J.L.; Alves, V.L.; Conceição, C.J.; Teixeira, H.M.; Câmara, J.S. Development of MEPS–UHPLC/PDA methodology for the quantification of clozapine, risperidone and their major active metabolites in human urine. *Microchem. J.* **2015**, *123*, 90–98. [CrossRef]
27. Park, D.; Choi, H.; Jang, M.; Chang, H.; Woo, S.; Yang, W. Simultaneous determination of 18 psychoactive agents and 6 metabolites in plasma using LC–MS/MS and application to actual plasma samples from conscription candidates. *Forensic Sci. Int.* **2018**, *288*, 283–290. [CrossRef] [PubMed]
28. Khelfi, A.; Azzouz, M.; Abtroun, R.; Reggabi, M.; Alamir, B. Determination of Chlorpromazine, Haloperidol, Levomepromazine, Olanzapine, Risperidone, and Sulpiride in Human Plasma by Liquid Chromatography/Tandem Mass Spectrometry (LC-MS/MS). *Int. J. Anal. Chem.* **2018**, in press. [CrossRef] [PubMed]
29. Ho, T.D.; Toledo, B.R.; Hantao, L.W.; Anderson, J.L. Chemical immobilization of crosslinked polymeric ionic liquids on nitinol wires produces highly robust sorbent coatings for solid-phase microextraction. *Anal. Chim. Acta* **2014**, *843*, 18–26. [CrossRef] [PubMed]

Sample Availability: not available.

© 2019 by the authors. Licensee MDPI, Basel, Switzerland. This article is an open access article distributed under the terms and conditions of the Creative Commons Attribution (CC BY) license (http://creativecommons.org/licenses/by/4.0/).

Article

A Sample Preparation Technique Using Biocompatible Composites for Biomedical Applications

Huifang Liu [1,2,†], Geun Su Noh [1,2,†], Yange Luan [1,2], Zhen Qiao [1,2], Bonhan Koo [1,2], Yoon Ok Jang [1,2] and Yong Shin [1,2,*]

[1] Department of Convergence Medicine, Asan Medical Institute of Convergence Science and Technology (AMIST), University of Ulsan College of Medicine, 88 Olympicro-43gil, Songpa-gu, Seoul 05505, Korea; liuhuifang.1229@gmail.com (H.L.); ngs90@hanmail.net (G.S.N.); luanyange@gmail.com (Y.L.); qiaozhen90@hotmail.com (Z.Q.); qhsgksdlek@naver.com (B.K.); jangyo17@daum.net (Y.O.J.)

[2] Biomedical Engineering Research Center, Asan Institute of Life Sciences, Asan Medical Center, 88 Olympicro-43gil, Songpa-gu, Seoul 05505, Korea

* Correspondence: shinyongno1@amc.seoul.kr; Tel.: +82-2-3010-4193

† These authors (H.L & G.N) contributed equally to this work.

Academic Editor: Nuno Neng
Received: 25 March 2019; Accepted: 3 April 2019; Published: 3 April 2019

Abstract: Infectious diseases, especially pathogenic infections, are a growing threat to public health worldwide. Since pathogenic bacteria usually exist in complex matrices at very low concentrations, the development of technology for rapid, convenient, and biocompatible sample enrichment is essential for sensitive diagnostics. In this study, a cucurbit[6]uril (CB) supermolecular decorated amine-functionalized diatom (DA) composite was fabricated to support efficient sample enrichment and in situ nucleic acid preparation from enriched pathogens and cells. CB was introduced to enhance the rate and effectiveness of pathogen absorption using the CB-DA composite. This novel CB–DA composite achieved a capture efficiency of approximately 90% at an *Escherichia coli* concentration of 10^6 CFU/mL within 3 min. Real-time PCR analyses of DNA samples recovered using the CB–DA enrichment system showed a four-fold increase in the early amplification signal strength, and this effective method for capturing nucleic acid might be useful for preparing samples for diagnostic systems.

Keywords: sample preparation; nanocomposite; pathogenic; enrichment; nucleic acid isolation

1. Introduction

Pathogenic infections result in diseases caused by toxins released by pathogenic organisms, and such infections are a growing threat to human health and public health worldwide [1–3]. Currently, pathogen identification and therapeutic approaches play important roles in controlling infections. However, the traditional gold standard diagnostic method, i.e., culturing and colony counting, is limited by long waiting times (and thus wasted time), as culturing of most clinical bacterial pathogens requires 1–2 days (much longer times are required for several bacterial species), and low efficiency due to contamination and significant experimental error [4–6]. Rapid and effective detection technologies are especially and urgently needed for an early-stage diagnosis, at which time there are low concentrations of the target pathogen. Therefore, new technologies based on the use of novel materials for sample preparation and biosensors for highly sensitive detection are emerging.

Among the emerging technologies, 'enrichment technology' is playing an increasingly important role in both sample preparation and sensor diagnosis amplification [7–9]. Nanomaterials that can be used as nanosorbents and activators for sample preparation have received considerable attention in various applications. A key advantage of such nanomaterials is their usefulness as sorbents. Their large surface area combined with a potential to modify their surfaces with various special reactive groups can increase their chemical affinity to target compounds [10–12]. A supermolecular modified diatomaceous earth (DE) composite platform has been used for molecular encapsulation in water treatment and in a broad range of biological systems that require rapid results and high stability [6,11]. In previous studies of functionalized DE composites, the extraordinary three-dimensional porous structure of DE has been shown to supply a massive surface area; furthermore, the well-known process of amino functionalization activates DE, while the addition of the pumpkin-shaped molecule cucurbituril (CB) maximizes its encapsulation performance by improving its dispersion [3,10,13–15].

Here, an efficient method for fabricating a cucurbituril-decorated, amine-modified diatom composite (CB–DA) is proposed, and the usefulness of this CB–DA composite for efficient sample enrichment and in situ nucleic acid preparation from pathogens and cells is demonstrated. The well-characterized process of amino functionalization of the surface of DE renders it a universal tool for targeting molecules, and cucurbit[6]uril (CB) has been introduced to enhance the ability of CB–DA composites to rapidly and efficiently adsorb molecules [3,16,17]. We have also verified that modification with 3-aminopropyl-methyl-diethoxysilane (APDMS) is more effective than traditional modification with 3-aminopropyl-triethoxysilane (APTES). The characterization of the surface charge (as represented by the zeta potential) indicates that the charge conferred by APDMS modification is about twice that conferred by APTES modification. Furthermore, the efficiency of the CB–DA composite to enrich for pathogens and cells was examined using three approaches. First, we tested the absorbance of supernatants collected after CB–DA enrichment. The results show that the enrichment efficiency was as high as 90% within 3 min, even at an *Escherichia coli* concentration of 10^6 CFU/mL. Second, the morphology of the composites precipitated by the CB–DA system showed that numerous eukaryotic cells adhered to the surface of the CB–DA. Third, we also compared the performance of real-time PCR using amplified DNA, samples collected from the CB–DA enrichment system, and DNA extracted via a commercial column system as templates. The results showed approximately a four-fold increase in the early amplification signal. In summary, we have confirmed that this CB–DA composite system can provide improved performance for biosample preparation for early diagnosis.

2. Results and Discussion

2.1. Design and Principle of the CB–DA Biocompatible Composite

As we reported in our previous study, cucurbituril-based diatom composites (CB–DA) exhibit a strong host–guest interaction that supports molecular encapsulation [11]. An efficient method for fabricating CB–DA to diagnose pathogen is proposed in this study. By examining the DE using scanning electron microscopy (SEM) (JEOL JSM-7500F, Tokyo, Japan) and dynamic light scattering (DLS) (DynaPro NanoStar, Wyatt), we confirmed that the size of the DE particles was well distributed between 10 and 20 μm—an appropriate size range for this project (Figure 1A,B). APTES (3-aminopropyl-triethoxysilane)-modified DE (also known as DA) has an enhanced pathogen enrichment property [11]. Here, we focus on DE surface medication with the similar chemical of diethyl amino polydimethylsiloxane (3-aminopropyl-methyl-diethoxysilane (APDMS)) [12]. The structural formula of APDMS is shown in Figure 1C, and the chain length of the organic amino compound can be seen. A diagram of cucurbit[6]uril [18,19] with a portal diameter of 3.9 Å, a cavity diameter of 5.8 Å, and a height of 9.1 Å is shown in Figure 1D.

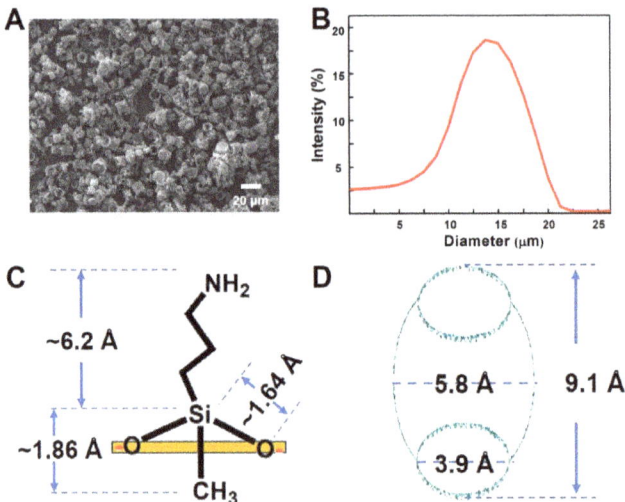

Figure 1. Characterization of the studied materials. (**A**) Scanning electron microscopy (SEM) image of diatomaceous earth (DE). (**B**) Dynamic light scattering (DLS) analysis of the size distribution of the DE. (**C**) Short chain length of amino organic compound (3-aminopropyl-methyl-diethoxysilane). (**D**) Diagram of cucurbit[6]uril, portal diameter (3.9 Å), cavity diameter (5.8 Å), and height (9.1 Å).

2.2. Preparation and Characterization of DA

A schematic of the process flow for amino functionalization of diatomaceous earth (DA) with different chemicals is shown in Figure 2A. The diethyls of APDMS are different from the triethyls of APTES for the surface modification of DA. Two points of the bonding site from APDMS is expected to be more efficient at bonding than APTES with three points (Figure 2A). The surface charges of the composites (reported as the zeta potential) were measured to estimate the efficiency of the amino modification (Figure 2B,C). Pure DE exhibited a negative surface charge. Amine groups surrounding the inner and outer surfaces of the DE skeleton can enhance its chemical stability, allow it to be used for extended periods of time, and lead to a robust coating with saline via covalent bond formation. Serial doses of APDMS or APTES, ranging from 50 µL to 400 µL, were added to 500 µL (100 mg/mL) of DE. The zeta potentials shown in Figure 2B,C show that a modification ratio of 1:2 (250 µL into 500 µL of DE) led to a good modification efficiency for both APDMS and APTES. Overall, APDMS modification was more efficient than APTES modification under the same conditions, perhaps due to differences in the contact surface area of the molecules. Modification with APDMS (which contains diethyl groups) requires two sites, while modification with APTES (which contains triethyl groups) requires three sites. To optimize the APDMS modification time (Figure 2D), the modification reaction was monitored from 30 to 180 min. Here, the modification ratio of 1:2 (250 µL into 500 µL DE) was used to measure the optimization of the APDMS modification time. Notably, according to the APTES modification procedure [11], a 120 min incubation time for the modification has been fixed in dosage studies. A modification time of 90 min was chosen for use in the fabrication process. Taken together, these results show that APDMS modification resulted in the presence of two free chains on the DA surface that could be involved in additional linkages.

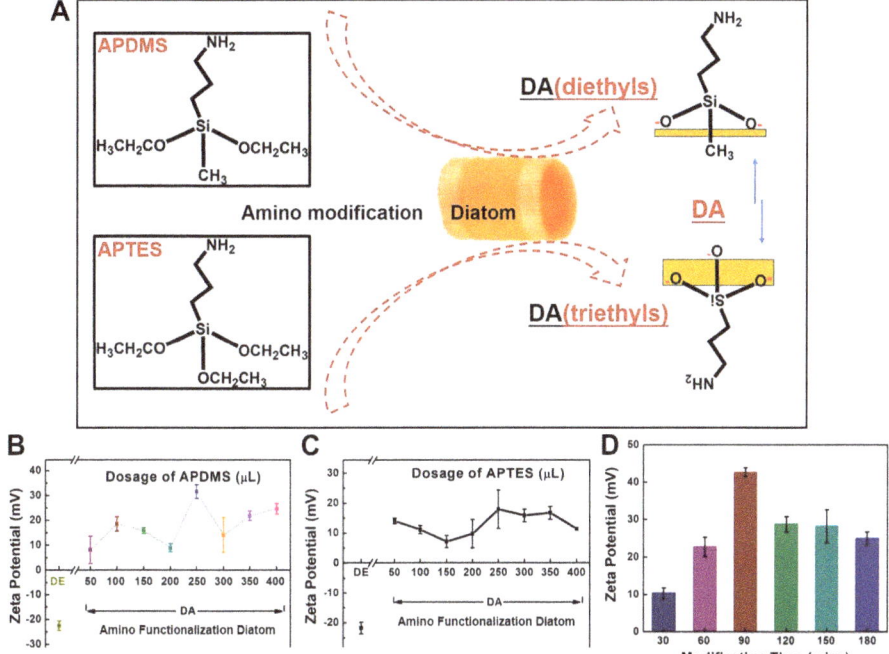

Figure 2. (**A**) Schematic of the process flow for amino functionalization of diatomaceous earth (DA). The diatom substrate was modified with either 3-aminopropyl-methyl-diethoxysilane (APDMS) or 3-aminopropyl-triethoxysilane (APTES). (**B,C**) Optimized conditions for amino functionalization of DE via APDMS (diethyls) and APTES (triethyls). Doses (μL) of APDMS and APTES reacted with 500 μL, 50 mg·L^{-1} DE in 1.5 mL tubes for 120 min. (The volume ratios ranged from 1/11 to 8/18.) APDMS and APTES (50 μL to 400 μL) were tested serially. (**D**) Optimization of the APDMS modification time. The zeta potentials of DA products with different APDMS modification times ranging from 30 to 180 min.

2.3. Preparation and Characterization of CB–DA Biocompatible Composites

A schematic representation of the cucurbituril coating of APDMS-modified DE (CB–DA) is shown in Figure 3A. The two free chains of the APDMS on the substrate are encapsulated in the cavity of the cucurbit[6]uril (CB), which is a key property of this second DE surface modification. Chemical-optical spectrum analysis of the composites was performed to assess the modification status. SEM images were used to assess the morphology of the DE and CB–DA, as shown in Figure 3B. The pores on the DE surface are open, and the pore diameter is less than 100 nm (Figure 3B, top). However, the CB–DA is rougher with blocked pores on the surface (Figure 3B, bottom). To assess the electrostatic properties of the CB–DA in solution, the zeta potentials of the CB–DA were measured (Figure 3C). Due to the uniform size of the DE particles, we ignored the size effect on the surface charge. Notably, the zeta potential of the CB–DA composite was higher than that of the DA, likely reflecting the diverse anchor bindings between the DA and CB, which may include the reported possible anchor linking/ion-dipole interaction between the carbonyl groups of the CB portals. Furthermore, the positively charged amine groups in the CB–DA composite could enhance the absorbency efficiency of the CB–DA conjugate during its interaction with other molecules via enhanced covalent bonding, physical adsorption, electrostatic interaction, and heterogeneous surface binding [11,19–21]. In the Fourier-transform infrared (FTIR) spectrum analysis of the composite (Figure 3D), the absorption peak at 1450 cm^{-1} was attributed to asymmetric stretching vibrations in the Si–O–Si bonds, and the peak at 1410 cm^{-1}

resulted from the Si–CH$_2$ bond (black curve). In addition, the absorption peaks at 3295 and 1180 cm^{-1} can be attributed to Si–OH and C–N bonding, respectively, on the surface of the pure DE. After the surface modification to form DA, the well-defined absorption bands at 1100, 2250, and 2720 cm^{-1} represent C–C–C bonding, C–H bonding, and O–H bonding, respectively. The stretching vibrations at 2850–3000 cm^{-1} from the CH, CH$_2$, and CH$_3$ groups and that at 2720 cm^{-1} from the aldehyde (C–H) in the CB–DA group (blue curve) verified the presence of a CB–DA supermolecule.

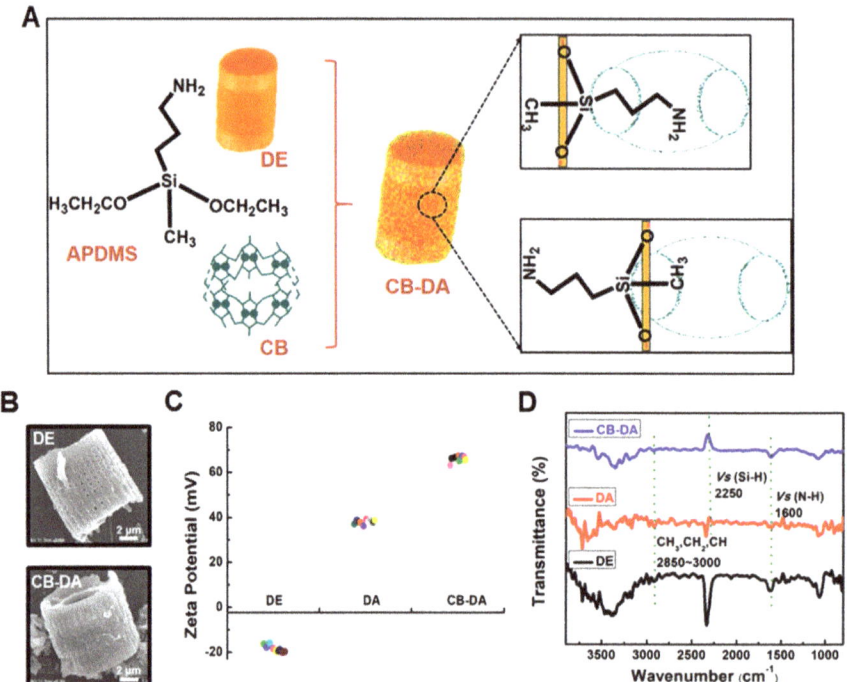

Figure 3. (**A**) Schematic representation of the cucurbituril modification of the amino-functionalized diatomaceous earth (CB-DA). The two free chains of APDMS on the substrate are encapsulated in the cavity of the cucurbit[6]uril (CB). (**B**) Scanning electron microscopy (SEM) images of DE and CB-DA. (**C**) Zeta potentials of the prepared materials: pure DE, amino-functionalized diatomaceous earth (DA), and cucurbituril-modified amino-functionalized diatomaceous earth (CB-DA). (**D**) Fourier-transform infrared (FTIR) spectrum analysis of the materials with dye. Pure DE (DE, black line), amine-modified DE (DA, red line), cucurbituril-coated amine-modified DE (CB-DA, blue line).

2.4. Cell and Pathogen Enrichment Using the Biocompatible Composite

A schematic of the pathogen enrichment process is shown in Figure 4. Electrostatic interactions between the positive surface of the CB–DA and the negative charge of the cell membrane form bridges that facilitate absorption (Figure 4A). The pathogen–composite complex precipitates easily. To assess the enrichment capacity, a UV spectrophotometer (Libra 22 UV) was used to measure the absorbances of the supernatants from the tested pathogen samples containing *E. coli* (10^6 CFU, 2 mL) after treatment with DA alone or after the CB–DA enrichment process. As shown in Figure 4B, the lower absorbance of the supernatant following CB–DA enrichment indicates that the CB–DA composite achieved a 90% capture efficiency within 3 min at an *E. coli* concentration of 10^6 CFU/mL. Furthermore, SEM images of HCT-116 cells bound to the CB–DA surface are shown in Figure 4C. These experiments confirm that CB–DA is useful for the biocompatible enrichment of pathogens and cells.

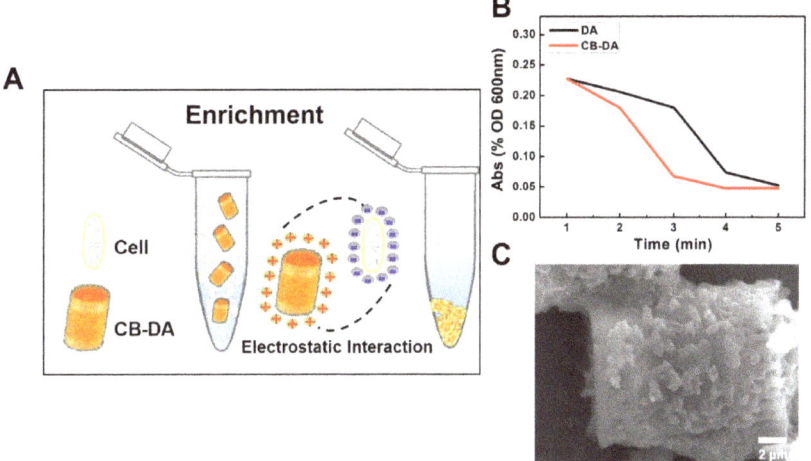

Figure 4. Pathogen enrichment schematic and demonstration. (**A**) Enrichment schematic; the electrostatic interaction between the positive surface of the CB–DA and the negative charge from the cell membrane. (**B**) The supernatant absorbances of the tested pathogen samples after DA and CB–DA *E. coli* enrichment (CFU 10^6, 2 mL). (**C**) Cell enrichment demonstration. SEM images of the HCT-116 cells adhered to the surface of CB-DA.

2.5. Nucleic Acid Isolation Using the Biocompatible Composite

To further confirm that the CB–DA composite rapidly and effectively adsorbed the bacteria, the fluorescence signals from real-time PCR analyses of amplified DNA extracted from the supernatant and precipitate (Figure 5A) of the *E. coli* (CFU 10^4, 1 mL) enrichment using a Qiagen kit (100 µL of tested sample) are shown in Figure 5B. The inset figure shows the melting-curve plots, which represent the amplification products from the systems (black line: 10^4, 100 µL as a positive control; red line: supernatant from the CB-DA-treated sample, 100 µL; blue line: enrichment with CB-DA, 100 µL of precipitate; and green line: distilled water (DW) as a negative control). As shown in Figure 5B, the cycle threshold (Ct) value was approximately two cycles earlier for pathogen enrichment by the CB–DA enrichment system than that of using the kit. According to RT-qPCR amplification theory, two cycles earlier corresponds to a template concentration that is four-fold higher [22]. In addition, we simply measured the amount of DNA (low molecular weight from salmon sperm, 31149-10G-F) captured by DA or CB–DA in 5 min. DNA (100 µL; 0.1 mg/mL) was added to 1 mL of DA (50 mg/mL) or CB–DA (50 mg/mL), and the surface charges of the DA–DNA and CB–DA–DNA mixtures after washing out the free DNA are shown in Figure 5C. The reduction in the surface charge of the CB–DA–DNA indicates that the CB–DA composite is more effective at capturing the nucleic acid, and this effect may be due to the previously reported covalent bonding, physical adsorption, electrostatic interactions, and heterogeneous surface binding intrinsic to supermolecular family members [23,24].

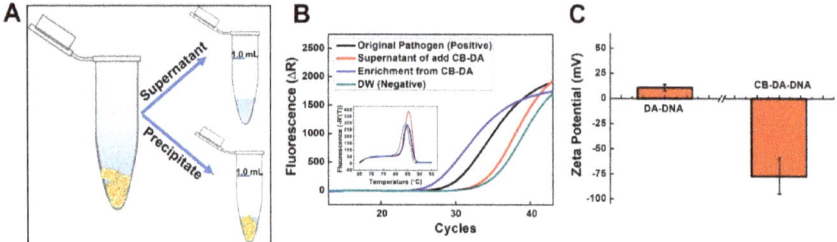

Figure 5. (**A**) A diagram of the supernatant and precipitate from the enrichment system. (**B**) Fluorescence signals from real-time PCR analyses of amplified DNAs extracted from the supernatant and precipitate following *E. coli* enrichment (CFU 10^4, 1 mL) using a Qiagen kit (100 μL of the tested sample). The inset figure shows the melting-curve plots representing the amplification products from the systems. Black line—10^4 CFU in 100 μL as a positive control; red line—supernatant with CB–DA addition, 100 μL; blue line—enrichment from CB–DA (100 μL of precipitate); and green line—distilled water (DW) as a negative control. (**C**) Zeta potential-based comparison of the nucleic acid capture efficiencies of the composites, i.e., DA-DNA and CB-DA-DNA.

3. Experimental Section

3.1. Chemicals and Reagents

All of the reagents were of an analytical grade and were used without further purification. Ammonium hydroxide solution (28% NH_3 in H_2O, 99.99% trace metals basis), 3-aminopropyltriethoxysilane (APTES), and 3-aminopropyl(diethoxy)methylsilane 97% (APDMS) were purchased from Sigma-Aldrich (St. Louis, MO, USA). Cucurbit[6]uril hydrate ($C_{36}H_{36}N_{24}O_{12}$, 94544-1G-F) was also obtained from Sigma-Aldrich. Proteinase K solution (Mat. No. 1014023, Qiagen, Germany) is commonly used to digest proteins and remove contaminants in nucleic acid preparations. The QIAamp DNA buffer system was from Qiagen (Hilden, Germany). The biocompatible DE (powder) and sodium bicarbonate ($NaHCO_3$) used in the nucleic acid extraction were purchased from Sigma-Aldrich. Milli-Q water, ethanol (95–100%), and phosphate-buffered saline (PBS) (10×, pH 7.4) (Thermo Fisher Scientific, Waltham, MA, USA) were used in all experiments.

3.2. Biological Samples

The eukaryotic cells (HCT-116 colorectal cancer cells) were maintained in plastic culture dishes in high-glucose Dulbecco's Modified Eagle's Medium (DMEM) (Life Technologies, Carlsbad, CA, USA) supplemented with 10% fetal calf serum in a 37 °C humidified incubator with 5% ambient CO_2. After culturing, genomic DNA was extracted from the cells using a spin column-based kit (Qia-kit) and a nanocomposite method. Prokaryotic *E. coli* (ATCC 25922) cells were inoculated into either nutrient broth medium or Luria-Bertani medium and then incubated overnight at 37 °C with shaking. The primers for the downstream analyses of the eukaryotic and prokaryotic cell are listed in Table 1.

Table 1. Primers used in this study.

	Primer	Sequences (5′→3′)	Annealing Temp. (°C)
E coli	Ecoli-rodA-195F Ecoli-rodA-195R	GCA AAC CAC CTT TGG TCG CTG TGG GTG TGG ATT GAC AT	58

3.3. Preparation of Biocompatible Composites

The gravity-powered washing method was used to remove fragments from the commercial DE to prepare uniform DE. For the production of the amine-modified diatomaceous earth (DA), the modification efficiencies of two types of amino polydimethylsiloxanes were compared.

The electrokinetic potentials of the composite surfaces were assessed by measuring the zeta potentials. Briefly, serial aliquots of APTES or APDMS were added dropwise into 1 mL aliquots of 95% ethanol solution, followed by manual shaking for 3 min at room temperature (RT). Subsequently, 50 mg of DE was added to the amino solution with stirring. The optimal reaction ratio and modification time were determined. The amino-functionalized DA was washed, collected by centrifugation, and dried in a vacuum overnight at RT. The DA powder was stored in a reagent bottle. Preparation of the CB–DA was performed via the microwave method. Briefly, 50 mg of DA was dissolved in 1 mL of DI water to form a 50 mg/mL DA solution. CB (25 mg) [6] was added to 1 mL of DI water, and this solution was then sonicated for 1 min using an ultrasonic instrument. Subsequently, 20 µL of the 25 mg/mL CB solution was added dropwise into 2 mL of prepared DA solution, followed by heating in a microwave oven for 1 min. The double-functionalized CB–DA was washed and collected by centrifugation and then dried in a vacuum overnight at RT. The CB–DA powder was stored in a reagent bottle.

3.4. Characterization of the Biocompatible Composites

The morphologies of the DE, DA, and CB–DA were characterized using field-emission scanning electron microscopy (FE-SEM) (JEOL JSM-7500F) to confirm a uniform size distribution and decoration of the DE with CB. The zeta potentials of the materials were acquired via dynamic light scattering (DLS) (DynaPro NanoStar, Wyatt, GA, USA). Fourier transform infrared spectroscopic analysis (FTIR) (JASCO 6300, JASCO, Easton, MD, USA) was performed on unmodified DE, DA, and CB–DA to obtain information about the chemical modification.

3.5. Cell and Pathogen Enrichment and Nucleic Acid Capturing

A Libra 22 UV/visible spectrophotometer was used to estimate the effectiveness of the composites for pathogen enrichment by measuring the absorbance of the cell and pathogen solutions at 600 nm. A vortex mixer (T5AL, 60 Hz, 30 W, 250 V) was used for mixing the sample wells in the enrichment system. A CF-5 centrifuge (100–240 Vas, 50/60 Hz, 8 W, 5500 rpm) was used for sample collection. QIAamp DNA/RNA Mini Kits (DNA Mini Kit: Cat No. 51304 and RNA Mini Kit: Cat No. 74104, Hilden, Germany) were used as spin column-based methods for isolating nucleic acids (following the manufacturer's instructions) from both the supernatant and precipitate from the CB–DA enrichment system. An AriaMx real-time PCR system (Agilent Technologies, Santa Clara, CA, USA) was used to confirm and estimate the enrichment efficiency.

4. Conclusions

In this study, an efficient fabrication method for a supermolecular modified diatom composite (CB–DA) is presented, and this CB–DA composite was then used for efficient sample enrichment and in situ nucleic acid preparation from enriched prokaryotic and eukaryotic cells. The well-known amino functionalization technology was improved, as verified by surface charge characterization. APDMS modification, which involves diethyl groups, requires two sites for modification, while APTES modification, which involves triethyl groups, requires three sites for modification. Our results show that the latter approach leads to a higher density of amino modification. Our results also show that the supermolecule formation was maximized in the CB–DA composite, and that the encapsulation performance of the composite was improved via enhanced dispersion. We also showed that the novel CB–DA composite achieved a 90% capture efficiency within 3 min even at an *E. coli* concentration of only 10^6 CFU/mL. We also observed that real-time PCR analysis of amplified DNA isolated using the CB–DA enrichment system showed a four-fold enhancement of the early signal, which might be useful for sample preparation to support early diagnosis. This study provides new insight into amino functionalization and lays a foundation for the further development of sample preparation techniques for human disease diagnostics and molecular encapsulation in drug delivery systems.

Author Contributions: H.L. and G.S.N. conceived the research. Y.S. supervised the whole project. H.L. and G.S.N. designed the experiments. Y.S., G.S.N., H.L., Y.L., and Z.Q. performed the analysis and made interpretations of the data. B.K. and Y.O.J. provided the chemicals and biological samples. Y.S., G.S.N., and H.L. wrote and edited the manuscript. All authors read and approved the final manuscript.

Funding: This study was supported by the Ministry of Science, ICT, and Future Planning (MSIP) through the National Research Foundation of Korea (NRF) (2017R1A2B4005288), Republic of Korea.

Conflicts of Interest: The authors declare no conflict of interest.

References

1. Koo, B.; Hong, K.H.; Jin, C.E.; Kim, J.Y.; Kim, S.-H.; Shin, Y. Arch-shaped multiple-target sensing for rapid diagnosis and identification of emerging infectious pathogens. *Biosens. Bioelectron.* **2018**, *119*, 79–85. [CrossRef]
2. Hahm, B.-K.; Kim, H.; Singh, A.K.; Bhunia, A.K. Pathogen enrichment device (PED) enables one-step growth, enrichment and separation of pathogen from food matrices for detection using bioanalytical platforms. *J. Microbiol. Methods* **2015**, *117*, 64–73. [CrossRef]
3. Zhao, F.; Koo, B.; Liu, H.; Jin, C.E.; Shin, Y. A single-tube approach for in vitro diagnostics using diatomaceous earth and optical sensor. *Biosens. Bioelectron.* **2018**, *99*, 443–449. [CrossRef] [PubMed]
4. Jin, C.E.; Koo, B.; Lee, T.Y.; Han, K.; Lim, S.B.; Park, I.J.; Shin, Y. Simple and low-cost sampling of cell-free nucleic acids from blood plasma for rapid and sensitive detection of circulating tumor DNA. *Adv. Sci.* **2018**, *5*, 1800614. [CrossRef] [PubMed]
5. Sun, Y.; Yang, S.; Koh, Y.-S.; Kim, Y.; Li, W. Isolation and Identification of Benzochroman and Acylglycerols from Massa Medicata Fermentata and Their Inhibitory Effects on LPS-Stimulated Cytokine Production in Bone Marrow-Derived Dendritic Cells. *Molecules* **2018**, *23*, 2400. [CrossRef]
6. Liu, H.; Zhao, F.; Jin, C.E.; Koo, B.; Lee, E.Y.; Zhong, L.; Yun, K.; Shin, Y. Large Instrument-and Detergent-Free Assay for Ultrasensitive Nucleic Acids Isolation via Binary Nanomaterial. *Anal. Chem.* **2018**, *90*, 5108–5115. [CrossRef] [PubMed]
7. Li, L.; Shen, C.; Huang, Y.-X.; Li, Y.-N.; Liu, X.-F.; Liu, X.-M.; Liu, J.-H. A New Strategy for Rapidly Screening Natural Inhibitors Targeting the PCSK9/LDLR Interaction In Vitro. *Molecules* **2018**, *23*, 2397. [CrossRef]
8. Jin, C.E.; Koo, B.; Lee, E.Y.; Kim, J.E.; Kim, S.H.; Shin, Y. Simple and label-free pathogen enrichment via homobifunctional imidoesters using a microfluidic (SLIM) system for ultrasensitive pathogen detection in various clinical specimens. *Biosens. Bioelectron.* **2018**, *111*, 66–73. [CrossRef] [PubMed]
9. Murray, J.; Sim, J.; Oh, K.; Sung, G.; Lee, A.; Shrinidhi, A.; Thirunarayanan, A.; Shetty, D.; Kim, K. Enrichment of specifically labeled proteins by an immobilized host molecule. *Angew. Chem. Int. Ed.* **2017**, *56*, 2395–2398. [CrossRef] [PubMed]
10. Barooah, N.; Kunwar, A.; Khurana, R.; Bhasikuttan, A.C.; Mohanty, J. Stimuli-Responsive Cucurbit [7] uril-Mediated BSA Nanoassembly for Uptake and Release of Doxorubicin. *Chem. Asian J.* **2017**, *12*, 122–129. [CrossRef] [PubMed]
11. Liu, H.; Luan, Y.; Koo, B.; Lee, E.Y.; Joo, J.; Dao, T.N.T.; Zhao, F.; Zhong, L.; Yun, K.; Shin, Y. Cucurbituril-based Reusable Nanocomposites for Efficient Molecular Encapsulation. *ACS Sustain. Chem. Eng.* **2019**, *7*, 5440–5448. [CrossRef]
12. Lu, H.-T. Synthesis and characterization of amino-functionalized silica nanoparticles. *Colloid J.* **2013**, *75*, 311–318. [CrossRef]
13. Kim, K.; Selvapalam, N.; Ko, Y.H.; Park, K.M.; Kim, D.; Kim, J. Functionalized cucurbiturils and their applications. *Chem. Soci. Rev.* **2007**, *36*, 267–279. [CrossRef] [PubMed]
14. Huang, Y.; Wang, J.; Xue, S.-F.; Tao, Z.; Zhu, Q.-J.; Tang, Q. Determination of thiabendazole in aqueous solutions using a cucurbituril-enhanced fluorescence method. *J. Incl. Phenom. Macrocyclic Chem.* **2012**, *72*, 397–404. [CrossRef]
15. Huang, Z.; Zhang, H.; Bai, H.; Bai, Y.; Wang, S.; Zhang, X. Polypseudorotaxane constructed from cationic polymer with cucurbit [7] uril for controlled antibacterial activity. *ACS Macro Lett.* **2016**, *5*, 1109–1113. [CrossRef]

16. Chao, C.-L.; Huang, H.-W.; Huang, H.-C.; Chao, H.-F.; Yu, S.-W.; Su, M.-H.; Wang, C.-J.; Lin, H.-C. Inhibition of Amyloid Beta Aggregation and Deposition of Cistanche tubulosa Aqueous Extract. *Molecules* **2019**, *24*, 687. [CrossRef]
17. Li, H.; Li, W.; Liu, F.; Wang, Z.; Li, G.; Karamanos, Y. Detection of tumor invasive biomarker using a peptamer of signal conversion and signal amplification. *Anal. Chem.* **2016**, *88*, 3662–3668. [CrossRef]
18. Lee, J.W.; Samal, S.; Selvapalam, N.; Kim, H.-J.; Kim, K. Cucurbituril homologues and derivatives: New opportunities in supramolecular chemistry. *Acc. Chem. Res.* **2003**, *36*, 621–630. [CrossRef]
19. Xiao, X.; Li, W.; Jiang, J. Porphyrin-cucurbituril organic molecular porous material: Structure and iodine adsorption properties. *Inorg. Chem. Commun.* **2013**, *35*, 156–159. [CrossRef]
20. Pennakalathil, J.; Jahja, E.; Özdemir, E.S.; Konu, O.; Tuncel, D. Red emitting, cucurbituril-capped, pH-responsive conjugated oligomer-based nanoparticles for drug delivery and cellular imaging. *Biomacromolecules* **2014**, *15*, 3366–3374. [CrossRef]
21. Zhou, X.; Su, X.; Pathak, P.; Vik, R.; Vinciguerra, B.; Isaacs, L.; Jayawickramarajah, J. Host–Guest Tethered DNA Transducer: ATP Fueled Release of a Protein Inhibitor from Cucurbit [7] uril. *J. Am. Chem. Soc.* **2017**, *139*, 13916–13921. [CrossRef] [PubMed]
22. Zhao, F.; Lee, E.Y.; Shin, Y. Improved Reversible Cross-Linking-Based Solid-Phase RNA Extraction for Pathogen Diagnostics. *Anal. Chem.* **2018**, *90*, 1725–1733. [CrossRef] [PubMed]
23. El-Sheshtawy, H.S.; Chatterjee, S.; Assaf, K.I.; Shinde, M.N.; Nau, W.M.; Mohanty, J. A Supramolecular Approach for Enhanced Antibacterial Activity and Extended Shelf-life of Fluoroquinolone Drugs with Cucurbit [7] uril. *Sci. Rep.* **2018**, *8*, 13925. [CrossRef] [PubMed]
24. Hou, C.; Zeng, X.; Gao, Y.; Qiao, S.; Zhang, X.; Xu, J.; Liu, J. Cucurbituril as a Versatile Tool to Tune the Functions of Proteins. *Isr. J. Chem.* **2018**, *58*, 286–295. [CrossRef]

Sample Availability: Samples of the compounds are available from the authors.

© 2019 by the authors. Licensee MDPI, Basel, Switzerland. This article is an open access article distributed under the terms and conditions of the Creative Commons Attribution (CC BY) license (http://creativecommons.org/licenses/by/4.0/).

Article

Evaluation of Polyvinyl Alcohol/Pectin-Based Hydrogel Disks as Extraction Phase for Determination of Steroidal Hormones in Aqueous Samples by GC-MS/MS

Naiara M. F. M. Sampaio, Natara D. B. Castilhos, Bruno C. da Silva, Izabel C. Riegel-Vidotti and Bruno J. G. Silva *

Department of Chemistry, Federal University of Paraná, Curitiba/PR 81530-900, Brazil; naiaramfms@gmail.com (N.M.F.M.S.); nataraduane@gmail.com (N.D.B.C.); brunosvik@gmail.com (B.C.d.S.); iriegel@gmail.com (I.C.R.-V.)
* Correspondence: bruno@quimica.ufpr.br; Tel.: +55-41-3361-3299

Academic Editor: Nuno Neng
Received: 22 November 2018; Accepted: 21 December 2018; Published: 22 December 2018

Abstract: A new extraction phase based on hydrogel disks of polyvinyl alcohol (PVOH) and pectin was proposed, characterized and evaluated for the extraction of six steroidal hormones (estriol, estrone, 17β-estradiol, 17α-ethinylestradiol, progesterone, and testosterone) in aqueous samples with subsequent determination by gas chromatography-tandem mass spectrometry (GC-MS/MS) after the derivatization procedure. The developed extraction procedure was based on the solid phase extraction (SPE) technique, but employed hydrogel as the sorbent phase. The effects of several parameters, including the amount and composition of the sorbent phase, pH, sample volume, flow rate, and gel swelling over the extraction efficiency, were evaluated. Gels with lower swelling indexes and larger amounts of sorbent ensured higher extraction yields of analytes. The main benefits of using the PVOH/pectin-based hydrogel as the extraction phase are the ease of synthesis, low-cost preparation, and the possibility of reusing the extraction disks. Limits of quantification of 0.5 μg L^{-1} for estrone and 17β-estradiol, and 1 μg L^{-1} for testosterone, 17α-ethinylestradiol, progesterone, and estriol were obtained. Accuracy values ranged from 80% to 110%, while the inter-assay precision ranged from 0.23% to 22.2% and the intra-assay from 0.55% to 12.3%. Since the sorbent phase has an amphiphilic character, the use of hydrogels is promising for the extraction of medium-to-high polarity compounds.

Keywords: gas chromatography; hydrogel; hormones; pectin; polyvinyl alcohol; sample preparation

1. Introduction

Solid phase extraction (SPE) is the most popular sample preparation technique for analyte concentration and removal of matrix interferents, with several devices (cartridges and disks) and sorbent phases developed [1,2]. However, commercial phases, such as octadecylsilane (C18), have several disadvantages, including poor selectivity that leads to co-extraction of interferents, and difficulty in extracting polar compounds from aqueous matrices, since they present hydrophobic character.

To overcome these problems, there are commercially available extraction phases capable of acting in a wider range of polarity, such as Oasis HLB® and Strata-X®. In these cases, the retention mechanisms of the analytes occur through hydrophobic interactions, π-π interactions, and hydrogen bonding. Even so, SPE cartridges are expensive, especially those with sorbent phases with a hydrophilic–hydrophobic balance. Moreover, these devices are disposable and usually used for a single application [2].

Significant efforts have recently been made to develop new materials with high sorption capacities, selectiveness, stability (longer lifetimes) and low costs [3]. For the past 50 years, hydrogels have been applied in the medical and pharmaceutical fields for the development of artificial organs [4], tissue engineering [5], and controlled drug delivery systems [6]. Nevertheless, the use of hydrogels as sorbent phases has recently drawn attention, mainly because of their structures formed by highly hydrophilic polymeric networks that are capable of absorbing and retaining large amounts of water without dissolving [7]. The possibility of modulating the chemical and physical-chemical properties of these materials enables the production and use of more selective phases.

Despite these attractive characteristics, the use of hydrogels in developing new extraction phases only started within the past decade and is still scarcely explored. In the literature, few works can be found that employ hydrogels for extracting organochlorine pesticides [8] or organic contaminants [9] from water samples using a polymer-coated hollow fiber microextraction technique. Hydrogels have also been associated with others materials, such as chitosan for recognition and separation of albumin bovine serum [10], zirconia nanoparticle-decorated calcium alginate hydrogel fibers [11], and acrylated composite hydrogels [12] for the extraction of organophosphorus pesticides and methyl blue, respectively. In these applications, the interaction between the analytes and the extraction phase occurs on the material surface (hollow fiber and dispersive extractions). Until now, no work has been carried out to explore the interactions that occur inside the hydrogel, promoted by the permeation of the sample through the gel phase, as occurs in SPE-packed phases (cartridges or disks).

In the present work, a new extraction phase is proposed based on disks of hydrogel, which is an innovative material for extraction purposes and presents some advantages over the solid sorbent phases employed in SPE, since gels have rheological properties that are intermediate between that of solid and liquid materials. This approach consists of the following three steps: (1) gel hydration, (2) sample percolation, and (3) elution of the analytes. In this case, the extraction process does not require conditioning of the extraction phase with an organic solvent and allows for the use of high flow rates, thus reducing the time for sample preparation.

Highlighting the potential of hydrogel-based materials as sorbent phases, hydrogels of polyvinyl alcohol (PVOH) and the biopolymer pectin were developed. PVOH is a semi-crystalline hydrophilic synthetic polymer produced by the hydrolysis of polyvinyl acetate (PVA). The PVOH structure contains several –OH groups and is, therefore, a highly hydrophilic and water-soluble polymer that exhibits excellent mechanical properties, chemical stability and the ability to form films [13]. Pectin is a polysaccharide extracted from plants and is mainly composed of α(1-4)-linked galacturonic acid (GalA) units containing varying amounts of methyl ester substituents (methoxylation degree), depending on the pectin's origin and purification method [6,14]. Moreover, the presence of hydrophilic sites, such as –OH and/or –COOH in the PVOH and pectin, is expected to promote the extraction of compounds of high polarity.

The proposed PVOH/pectin-based hydrogel disk as an extraction device was characterized and applied for the extraction of medium-to-high polarity organic compounds from aqueous samples. For this purpose, six steroidal hormones were selected as the analytes (estriol (E3), estrone (E1), 17β-estradiol (E2), 17α-ethinylestradiol (EE2), progesterone (PRO), and testosterone (TES)) and the determination was carried out using gas chromatography-tandem mass spectrometry (GC-MS/MS).

2. Results and Discussion

2.1. Physical-Chemical Characterization of the Hydrogel

The hydrogel disk performance was characterized in relation to several parameters, including mass, composition, morphological aspects, and swelling index (SI) of the hydrogel. Hydrogel disks prepared only with PVOH (>99% hydrolyzed) did not allow any water permeation through the gel.

As highly hydrolyzed PVOH was employed, a gel with a high degree of reticular closure was obtained, thus hindering the permeation of water through the hydrogel.

The addition of pectin into the hydrogel increased the amphiphilic character of the sorbent phase, due to presence of methyl ester groups in the pectin structure. This assured high extraction efficiencies for all hormones, especially for progesterone, which has the highest octanol/water partition coefficient (log Ko/w 3.87 [15]) among the analytes. The addition of this natural polymer to PVOH proved to be advantageous since it allowed for permeation of the samples and extraction of the analytes and did not affect the excellent mechanical properties of PVOH.

The swelling degree of hydrogels is dependent on several factors, such as the chemical structure, molar mass, composition and, degree of crosslinking of the polymer matrix [16]. Figure 1a shows the swelling behavior of P5PC1, P10PC2, P15PC3, P20PC4, P15PC2, and pure PVOH gels crosslinked with citric acid. For the differentiation of each gel composition studied, the acronym PxPCy was adopted, where x corresponds to the initial concentration (%, w/v) of PVOH dispersion and y the initial concentration (%, w/v) of pectin dispersion. In Figure 1a, it can be observed that for the hydrogels containing pectin, the equilibrium swelling was reached within the first 30 min, whereas for pure PVOH, 90 min was required. In addition, the presence of pectin resulted in higher water uptake, while for PVOH the SI was 104.3%. Gels containing pectin presented values ranging from 277.2% (P5PC1) to 135.2% (P20PC4). P5PC1 and P10PC2 disks absorbed the largest amounts of water and obtained the lowest extraction efficiencies in comparison to the others. This fact indicates that there is an optimal SI that favors the interactions between the analytes and the sorbent phase.

The swelling process of gels is governed by the mobility of the polymeric network, which defines the distance between the chains and, consequently, the volume available for the permeation of the solvent and the transport of solutes [17]. Therefore, for hydrogels that absorb larger volumes of water, the distance between the chains of the polymeric network is greater, which impairs the retention of analytes by the sorbent phase.

The water loss process was studied at room temperature (25 °C) and under refrigeration (4 °C). However, since both temperatures had practically the same behavior, except that the mass equilibrium was reached faster at 25 °C, only the results for room temperature are presented as a graph (Figure 1b). The hydrogels presented similar behavior regarding the water loss process, with the highest rate being ~80% for P5PC1 and the lower rate being ~50% for PVOH. For the gels prepared with a lower polymer mass, the diffusion process of the solvent in and out of the polymer network was more pronounced. Higher diffusion rates did not favor the extraction of the hormones, since these compounds have high affinity with the aqueous matrix, and a shorter contact time of the sample with the sorbent phase may have contributed to the low extraction efficiencies of the P5PC1 and P10PC2 gels.

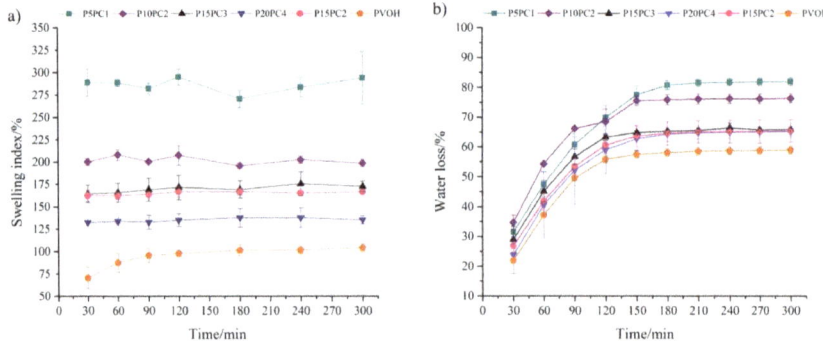

Figure 1. (a) Swelling index (SI) (wt.%) and (b) water loss (wt.%) of the hydrogels (n = 3) at 25 °C.

The morphological aspects of pure PVOH, pure pectin and PVOH/pectin phases are shown with photographs in Figure 2a–f and scanning electron microscopy (SEM) images in Figure 2g–i. It can be noted in Figure 2i that the presence of spherical particles corresponds to the pectin phase. The spherical shape of the pectin particles indicates that the two polymeric phases are immiscible. The addition of pectin into the hydrogels enhanced the PVOH surface area, leaving PVOH functional groups available to interact with the analytes and also assigning a more amphiphilic character to the sorbent phase. The average thickness of the hydrogel disks was between 150 and 200 µm.

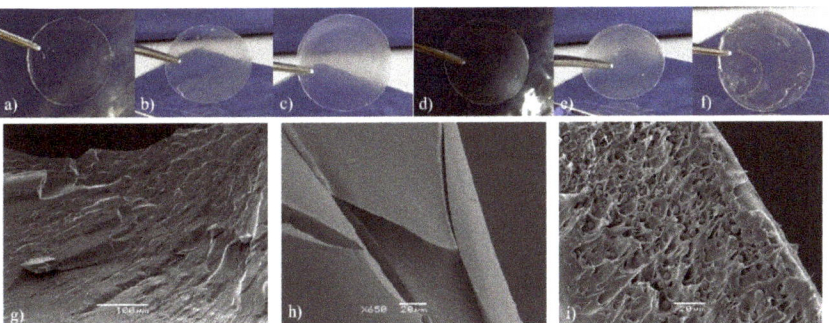

Figure 2. Hydrogel disk pictures of (**a**) pure polyvinyl alcohol (PVOH); (**b**) P5PC1; (**c**) P10PC2; (**d**) P15PC2; (**e**) P15PC3 and (**f**) P20PC4. Scanning electron microscope (SEM) images of (**g**) pure PVOH 250×; (**h**) pure pectin 650x and (**i**) PVOH/pectin extraction disk (P15PC2) 750×.

The hydrogel-SPE efficiency (Figure 3a) was evaluated as a function of the total amount of sorbent phase. PVOH/pectin hydrogel disks were prepared, maintaining a ratio of 5:1, but increasing the polymer mass and resulting in extraction phases with different mass: P5PC1 (19.03 mg), P10PC2 (44.62 mg), P15PC3 (52.67 mg) and P20PC4 (63.86 mg). Different ratios of the initial concentrations of the dispersions were also used as follows: 3.3:1 (P10PC3), 7.5:1 (P15PC2), and 10:1 (P10PC1). This variation in the ratio of PVOH to pectin was studied since the addition of pectin can alter the polarity of the extraction phase, being able to interfere in the retention mechanisms of the analytes. Another reason relates to the swelling index that depends on the composition of the gel and has an influence on the extraction efficiency.

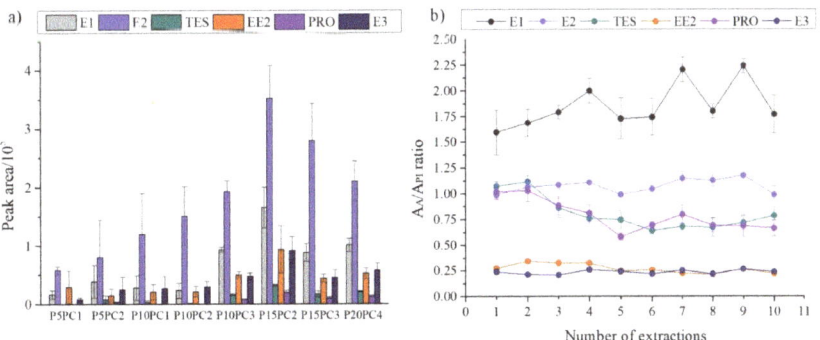

Figure 3. (**a**) Peak areas (n = 3) of the hormones (1 µg mL^{-1}) extracted from ultrapure water according to the hydrogel disk used, where P = PVOH and PC = pectin. (**b**) AA/AIS ratios of 10 consecutive extractions with the same hydrogel disk (n = 3) for each hormone. AA: analyte peak area; AIS: internal standard peak area.

It was found that higher amounts of sorbent led to an increase in the extraction efficiency up to P15PC2. This fact may be linked to an increase in the amount of pores, which makes analyte desorption more difficult. In addition, the P20PC4 gel (Figure 2f) was less uniform than the others due to the high viscosity of the initial mixture (PVOH at 20% and pectin at 4% w/v). Another important aspect was that the use of low amounts of polymers (P5PC1 and P10PC2, Figure 2b,c, respectively) produced very moldable gels that folded inside the holder, which could cause the sample to percolate by preferential flow paths. The highest extraction efficiency was achieved for the P15PC2 hydrogel disk (Figure 2d), which was adopted as the sorbent phase for the hydrogel-SPE and employed in the subsequent studies.

2.2. Optimization of Hydrogel-Solid Phase Extraction (SPE) Conditions

A multivariate study was carried out through a 2^3 full factorial design using the P15PC2 disk as the extraction phase. Higher extraction efficiency was obtained when the pH, sample volume and flow rate conditions were 7.5, 100 mL and 4 mL min^{-1}, respectively. Pareto charts (Figure S1—Supplementary Material) shows that only E1, E2, EE2 and E3 presented significant main and interaction effects.

2.2.1. Effect of pH

The pH was studied regarding the stability of the gel and the extraction process. In the pH range from 3.0 to 9.0, all the sorbent disks were stable. Therefore, for the extraction process, pH values of 4.5, 6.0 and 7.5 were employed. The pH effect was not relevant for any hormone; however, second and third-order effects could be observed (Figure S1—Supplementary Material). In this pH range, all the hormones are in their non-ionized form (pKa between 10.2 and 15.1 [18]), which guarantees a higher affinity of the analytes with the sorbent phase.

Therefore, the pH will also influence the degree of opening or closing of the polymeric network, since the degree of swelling of the hydrogel phase may be altered due to ionization or dissociation of functional groups. For example, when raising the pH, the OH$^-$ ions hydrolyze the remaining acetates (pKa = 4.76) of PVOH. Then, charges present in the polymer chain, due to the ionized groups, repel each other and the degree of swelling increases [19]. By opening the network, due to the water absorption, it is assumed that there will be an increase in the rate of solvent diffusion into the network. It was observed that higher diffusion rates had a negative effect on the extraction efficiency (in terms of peak area), because of the insufficient contact time of the hormones with the hydrogel, which is a preferable closer network; therefore, higher pH values led to an improvement in the extraction rates, which was due to an increase in the ionic strength, leading to neutralization of the charges

2.2.2. Effect of Sample Volume and Flow Rate on the Extraction Efficiency

Increasing the sample volume from 100 to 200 mL did not promote an increase in the response. Instead for most of the analytes, the AA/AIS ratio was lower when the sample volume was 200 mL. Higher sample volumes can lead to a higher enrichment of analytes; however, they can also cause desorption of compounds from the sorbent phase, even in the extraction step, especially when these compounds are polar and have great affinity with water.

The flow rate effect was significant for the four hormones (E1, E2, EE2, and E3) and had the greatest influence over the extraction efficiency of the PVOH/pectin hydrogel. As the compounds studied have medium to high polarity, a longer contact time of the sample with the hydrogel, due to the use of lower flow rates (2 mL min^{-1}), was not beneficial to the extraction process because it promoted the elution of hormones by the sample. Thus, higher flow rate (4 mL min^{-1}) was selected.

2.2.3. Optimization of Desorption Volume

The desorption step was studied in order to use the lowest volume of organic solvent possible, as well as evaluating the possibility of a memory effect in consecutive extractions using the same extraction disk. For this purpose, the use of six elution steps (1.0 mL of methanol for each one) was evaluated in triplicate. The number of elutions was determined by establishing that the accumulated peak area should be higher than 90%, which was reached with a total of five elutions.

The proposed method required 5 mL of organic solvent for the elution step; meanwhile, the literature usually reports the use of 8 to 15 mL [20–23] of organic solvents when employing cartridges in SPE; for commercial disks, this amount is even higher and can reach 20 mL [23,24]. Therefore, hydrogel-SPE is an interesting approach because it does not require conditioning of the extraction phase with organic solvents and uses smaller amounts of elution solvent, which makes the method environmentally friendly and reduces time spent in concentrating the eluate.

2.3. Quantitative Parameters and Analytical Comparison with other Methods

From the beginning of this study, one of the main characteristics for the hydrogel extraction phase was that it could be reusable, which would guarantee not only commercial appeal to this new material, but also greater ease and acceptance of the new phases based on hydrogel for the SPE technique. Moreover, the reuse of the hydrogel disk also guarantees robustness of the proposed extraction phase. For all hormones, there was a difference of up to 20% in the standard deviation between the results of one extraction and another. E1, E2, and E3 had a deviation of only 10% (Figure 3b). These results show that it is possible to reuse the hydrogel extraction disks, which is an advantage in relation to commercial sorbent phases employed in SPE, where reuse is normally not feasible.

Chromatograms of the standard solutions (pink line) and the application of P15PC2 hydrogel as the extraction phase for SPE (black line) for steroidal hormones can be seen in Figure 4. Enrichment factors of 500-fold were obtained based on the sample volume (100 mL) and the final volume of the extract (200 µL). Limits of quantification (LOQ) were estimated using the lowest point of the analytical curve with a relative standard deviation (RSD) less than 20%. The proposed method presented a linear relationship from the LOQ to 100 µg L^{-1} with all correlation coefficients higher than 0.99. For E1 and E2, the LOQ concentration was 0.5 µg L^{-1}, while for TES, EE2, PRO, and E3, the LOQ value was 1.0 µg L^{-1}. The RSD ($n = 3$) values for the LOQ were lower than 12% and the accuracy was between 85.1% and 107.3% for all analytes. The limit of detection (LOD) ranged from 0.15 (E1, E2) to 0.30 µg mL^{-1} (TES, EE2, PRO, and E3), which was based on a signal-to-noise ratio 3.

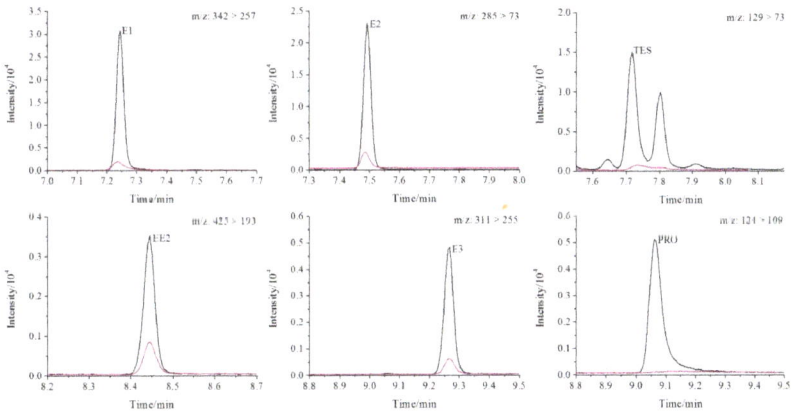

Figure 4. Chromatograms (multiple reaction monitoring, MRM) for a standard solution of hormones (500 µg L^{-1}) (pink) and ultrapure water spiked with hormones (500 µg L^{-1}) after extraction with polyvinyl alcohol (PVOH)/pectin disk (P15PC2) at optimized conditions (black).

In our study, the LOQ values were higher than those described in the literature using commercial SPE phases for analysis of hormones in environmental samples [21,23,25,26]. However, take into consideration the low thickness of the developed hydrogel disks (around 150 µm), we believe that the results are promising, especially in terms of precision, accuracy, low cost, and robustness. Since the thickness of the commercial SPE disk is about 500 µm, some modifications to the proposed device can be performed to achieve a lower LOQ. In future studies, for example, two strategies will be evaluated: i) increasing the diameter or thickness of the hydrogel-based disks, leading to a larger mass, and obtaining higher enrichment factors due to the increase in sample volume and ii) using the sandwich configuration, where more than one disk is placed inside the device in order to improve the detectability of the technique. In addition, the possibility of modulating the hydrogel disk by varying its chemical constitution can be evaluated for a group of target compounds, which may guarantee greater detectability of the analytical method.

Two concentration levels were evaluated for the intra (40.0 and 80.0 µg L^{-1}) and inter-assay (50.0 and 100.0 µg L^{-1}) precision experiments (performed in triplicate). The RSD values were lower than 13%, except for the PRO inter-assay at 50 µg L^{-1} (22%). The hydrogel-SPE technique proved to be reproducible for the extraction of hormones from aqueous samples since the RSD value was less than 13%, using different extraction disks for each precision assay. In addition, accuracy results of 80% to 110% demonstrated that it was possible to use the proposed hydrogel as the extraction phase to quantify organic compounds of medium to high polarity present in aqueous matrices. Table 1 summarizes these results.

In this work, precision and accuracy were similar to those obtained in previous studies. For example, Migowska et al. [27] extracted E1, E2, EE2, and E3 from surface water with an Oasis HLB cartridge with determination by GC-MS, where the RSD values for intra- and inter-day assays were lower than 14.2%, while the trueness values were up to 119.2%. In addition, the device developed in this work is inexpensive when compared to commercial devices. In fact, the average cost of a PVOH/pectin hydrogel disk is about $0.15 per disk, taking in to account only the cost of the polymers, since the polycarbonate holder is reusable.

Table 1. Precision and accuracy of the hydrogel-solid phase extraction (SPE) procedure.

Analytes	Concentration Evaluated/μg L^{-1}	Intra-Assay Precision (n = 3)			Concentration Evaluated/μg L^{-1}	Inter-Assay Precision (n = 3)		
		Measured Conc./μg L^{-1}	RSD/%	Accuracy/%		Measured Conc./μg L^{-1}	RSD/%	Accuracy/%
E1	40.0	38.6 ± 0.21	0.5	96.7	50.0	50.3 ± 2.9	5.7	100.5
	80.0	71.8 ± 2.9	4.1	89.7	100.0	102.6 ± 1.9	1.9	102.6
E2	40.0	33.1 ± 1.5	1.9	82.6	50.0	50.5 ± 3.1	6.2	100.9
	80.0	63.9 ± 1.5	2.4	80.0	100.0	99.8 ± 8.29	8.3	99.8
TES	40.0	38.1 ± 4.0	10.4	95.4	50.0	50.3 ± 3.0	6.0	100.6
	80.0	72.8 ± 5.2	7.2	91.0	100.0	98.7 ± 3.8	3.8	98.7
EE2	40.0	37.6 ± 2.0	5.4	94.0	50.0	46.9 ± 2.8	6.0	93.7
	80.0	85.1 ± 7.6	8.9	106.3	100.0	100.8 ± 4.8	4.7	100.8
PRO	40.0	35.6 ± 4.1	11.5	88.9	50.0	49.1 ± 10.9	22.2	98.1
	80.0	83.7 ± 10.3	12.3	104.7	100.0	102.4 ± 6.8	6.6	102.4
E3	40.0	40.4 ± 3.0	7.3	101.0	50.0	49.9 ± 1.7	3.4	99.7
	80.0	85.0 ± 6.5	7.6	106.2	100.0	108.2 ± 8.6	7.9	108.2

2.4. Analysis of Real Samples

Finally, the hydrogel-SPE device was applied for the determination of hormones in surface waters. The Belém River passes through the central region of Curitiba, which has a high population density and receives both domestic sewage and solid waste produced in this region. Estrone, 17β-estradiol and 17α-ethynylestradiol were quantified at concentrations of 1.11 to 1.58 µg L^{-1}, while testosterone reached 3.58 to 5.84 µg L^{-1}, which were the highest concentrations among all the studied hormones. Progesterone could also be identified in all samples, but at concentrations below its LOQ. Previous studies have shown a similar concentration for hormones in the same region. For example, Ide et al. [28] detected E2 and EE2 at concentrations of 1.42 and 1.48 µg L^{-1}, respectively. These results prove that the developed SPE device based on hydrogel disks can be successfully used to extract steroidal hormones from water samples.

3. Materials and Methods

3.1. Chemicals and Supplies

The hormone standards (>98% purity) used were 17α-ethynylestradiol (EE2), estriol (E3), estrone (E1), progesterone (PRO), testosterone (TES) (Fluka), and 17β-estradiol (E2) (Sigma-Aldrich, Saint Louis, MI, USA). Bisphenol Ad16 (BPAd$_{16}$) (99.9%, Supelco, Bellefonte, PA, USA) was used as an internal standard. Ethyl acetate and methanol (J. T. Baker, HPLC grade) were used for the extraction procedure and chromatographic analysis. Ultrapure water was obtained from a Milli-Q system controlled at 18.2 MΩ cm (Millipore, São Paulo, SP, Brazil). Single stock solutions for each hormone and bisphenol Ad16 with concentrations of 1000 and 400 mg L^{-1}, respectively, were prepared in methanol. Aqueous samples containing all hormones and the internal standard were prepared daily by spiking the stock solution with ultrapure water.

The reagents used for the derivatization step, N-methyl-trimethylsilyltrifluoroacetamide (MSTFA) and trimethylsilylimidazole (TMSI), were purchased from Sigma-Aldrich. The polymers used for the hydrogels were PVOH (MW 89,000–98,000, >99% hydrolyzed), purchased from Sigma-Andrich, and a commercial citrus pectin (PC) obtained from the Municipal Market of Curitiba (Curitiba, Brazil). The PC was purified by dialyzing against distilled water through a 12–14 kDa MW cut-off (Spectra/Por Cellulose Ester) for 72 h, followed by lyophilization. The dialyzed pectin is a high methoxyl pectin, as reported elsewhere [14]. Citric acid (Qhemis) was used as the crosslinking agent.

3.2. Preparation of the Hydrogel Disks

First, aqueous dispersions of PVOH (5%, 10%, 15% and 20%, w/v) and dialyzed PC (1%, 2%, 3% and 4%, w/v) were prepared using ultrapure water. Then, the PVOH and pectin dispersions were mixed in a 1:1 (wt.) proportion. Table S1 (Supplementary Material) shows the initial concentrations of the dispersions employed in the preparation of the hydrogel disks studied. The resultant dispersion was stirred at room temperature until complete mixing of the polymers. Next, citric acid, used as a crosslinking agent, was added under agitation into the dispersion. In this step, a final concentration of 10% with respect to the total weight of the polymers was reached. Finally, 1 g of the dispersion was cast in a 10 mL beaker, used as a template, and left in an oven at 60 °C until completely dry. Then, the dried hydrogel was cut into disks with a stainless-steel blade. Hydrogels employing only PVOH or only pectin were prepared according to the same above mentioned procedure.

3.3. Physical-Chemical Characterization of the Hydrogel

The swelling index (SI) of the hydrogels was determined according to Equation (1) as follows:

$$SI(\%) = \frac{W_f - W_i}{W_i} 100 \qquad (2)$$

The dried gels were weighed (W_i) and immersed in 30 mL of ultrapure water. Then, the mass of the swollen gels (W_f) was monitored for 300 min, when a constant mass was reached. Before weighing, in order to remove excess fluid, the hydrogels were slightly pressed against an absorbing paper.

The hydrogel water loss percentage (syneresis) was measured at room temperature (25 °C) and in a refrigerator (4 °C). First, the gels were immersed in ultrapure water for 90 min and slightly pressed against an absorbing paper. The gels were weighed at their equilibrium swelling (W_{eq}) and after every 30 min interval (W_t) for 300 min. The water loss percentage was calculated by Equation (2) as follows:

$$\text{Water loss}(\%) = \frac{W_{eq} - W_t}{W_{eq}} \cdot 100 \qquad (2)$$

3.4. Scanning Electron Microscopy (SEM) Analysis

The dry hydrogels were frozen in liquid nitrogen and fractured. Images of the fractured cross-section were obtained after metallization with gold (Balzers Union sputter-coater, model SCD 030). The scanning electron microscope was a JOEL' JSM-6360LV model, operated at 10 kV.

3.5. Hydrogel-SPE Procedure

The hydrogel-SPE assembly consisted of a polycarbonate reusable syringe filter holder, which is where the hydrogel disk was placed. A syringe was attached to the holder without a plunger through which the sample was discharged (Figure 5).

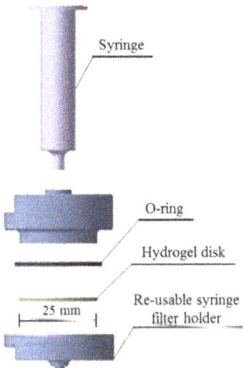

Figure 5. Schematic of the hydrogel-SPE device.

Before the extraction, each dried hydrogel disk was positioned inside a reusable polycarbonate syringe filter holder (Sartorius) with a 25 mm diameter and hydrated with 10 mL of ultrapure water. Aqueous samples were passed through the hydrogel disk using a vacuum manifold (model Visiprep, Supelco, Bellefonte, PA, USA). Then, the disk was air-dried under a vacuum, and the elution of the analytes was carried out at a flow rate of 1 mL min^{-1}. Finally, the extracts were completely dried and reconstituted with 200 µL of methanol and carried on to the derivatization procedure.

To evaluate the better extraction conditions, a two-level full factorial design with three factors (2^3) was performed. The optimized parameters were the following: pH (4.5, 6.0, and 7.5), sample volume (100, 150, and 200 mL), and flow rate (2, 3, and 4 mL min^{-1}). The peak area values were used has response. Factorial design calculations and Pareto charts were performed by Statistica software (Mathworks, Natick, MA, USA, version 7.0.1).

Moreover, the optimization process of the elution step was carried out taking into consideration the number of elutions (1, 2, 3, 4, 5 and 6 elutions) of 1.0 mL of methanol. The optimization process was accomplished in triplicate.

3.6. Derivatization Step and Gas Chromatography-Tandem Mass Spectrometry (GC-MS/MS) Analysis

For the derivatization step, 100 µL of the methanolic extracts were completely dried, 30 µL of MSTFA:TMSI (99:1, v/v) were added to the samples, and the silylation reaction was carried out in the same oven at 70 °C for 30 min. Then, 70 µL of ethyl acetate were added to the samples and 1 µL was injected into the GC-MS/MS system.

The GC-MS/MS analysis was carried out on a Shimadzu system consisting of a GC2010 Plus gas chromatograph hyphenated to a TQ8040 triple quadrupole mass spectrometer (Shimadzu, Kyoto, Japan). The GC was equipped with an AOC5000 auto-injector and a split-splitless injector. The chromatographic separation occurred in a GC column SH-RTX-5 ms (30 m × 0.25 mm × 0.25 µm) with helium 5.0 as the carrier gas at a flow rate of 1.0 mL min^{-1}. A sample volume of 1 µL was injected in a split ratio of 1:10. Both the injector and ion source temperatures were set at 250 °C. The interface temperature was 300 °C. The GC oven temperature was kept at 220 °C for 2 min, followed by 20 °C min^{-1} to 280 °C (1 min), 2 °C min^{-1} to 290 °C and 20 °C min^{-1} to 300 °C (1.5 min). The total analysis time was 13 min. The triple quadrupole mass spectrometer was operated in multiple reaction monitoring (MRM) mode with electron ionization at 70 eV. For each analyte, one quantification and two confirmation transitions were determined (Table S2—Supplementary Material).

Although the goal of this paper was not to describe a validated analytical method, the feasibility of the application of the developed device for quantitative analysis was evaluated. For this purpose, the steroidal hormones were chosen as target analytes, and some merit parameters were studied, such as limit of quantification (LOQ), RSD, accuracy, linear range, and correlation coefficient (r). Moreover, to evaluate the reuse, and consequently, the robustness of the hydrogel disks, ten subsequent extractions with the same disk were performed under the same extraction and analysis conditions described previously. Between the extractions, the material was cleaned with methanol and ultrapure water, avoiding the memory effect.

3.7. Surface Water Samples

Surface water samples were collected at three locations along the Belém River, in the city of Curitiba, Brazil. The sampling was carried out using 4 L amber glass bottles previously cleaned and rinsed with the sample. All water samples were filtered through a 0.6 µm fiberglass membrane (Macherey-Nagel, Düren, Germany), stored in amber glass bottles at 4 °C for a maximum period of 24 h and analyzed according to the developed method. Before the extraction, the surface water samples were spiked with the IS (BPA-d$_{16}$), resulting in a concentration of 25.0 ng mL^{-1}.

4. Conclusions

In this work, an amphiphilic, reproducible, free of memory effects, and cost-effective sorbent phase was obtained. The use of hydrogels as the sorbent phase showed potential for the extraction of media and high-polarity compounds from aqueous matrices, which has been a challenge for extraction techniques in general. The evaluation of the physical-chemical properties of the polyvinyl alcohol and pectin (a natural polymer) hydrogel proved that the water content within the hydrogel matrix and its functional groups are very important to ensure satisfactory polar interactions with the analytes.

Compared with previous studies on SPE/GC-MS for the analysis of organic contaminants in water, similar precisions and accuracies were achieved with the device developed in this work, but the PVOH/pectin extraction phase was more stable than commercial SPE phases, allowing for reutilization of the extraction device. However, further studies are necessary to reach LOQ values as low as those obtained in SPE when using commercial extraction phases. On the other hand, the new hydrogel-based material developed was successfully used to determine steroidal hormones in surface waters. Therefore, using an easily prepared hydrogel (in disk form) can be an alternative for extraction techniques, especially due to its characteristics of modulating and containing several polar groups in its structure, as well as the simplicity of its preparation and its low cost.

Supplementary Materials: Figure S1: Pareto charts of the effects of pH, sample volume (V), and flow rate (F) on the extraction efficiency. Table S1: Hydrogel disk composition. Table S2: Optimized parameters for tandem MS.

Author Contributions: N.S., N.C., B.S., I.V., and B.d.S. conceived and designed the experiments; N.S. performed the experiments; N.S., I.V., and B.S. analyzed the data; and N.S., I.V., and B.S. wrote the paper.

Funding: This work was supported by the Brazilian funding agencies CAPES and CNPq (Grants 308635/2011-6 and CNPq: 442541/2014-7).

Acknowledgments: The authors thank the Electron Microscopy Center (CME) of the Federal University of Paraná for the SEM images.

Conflicts of Interest: The authors declare no conflict of interest.

References

1. Ghani, M.; Palomino Cabello, C.; Saraji, M.; Manuel Estela, J.; Cerdà, V.; Turnes Palomino, G.; Maya, F. Automated solid-phase extraction of phenolic acids using layered double hydroxide–alumina–polymer disks. *J. Sep. Sci.* **2018**, *41*, 2012–2019. [CrossRef] [PubMed]
2. Smith, R.M. Before the injection—modern methods of sample preparation for separation techniques. *J. Chromatogr. A* **1000**, 3–27. [CrossRef]
3. Augusto, F.; Carasek, E.; Silva, R.G.; Rivellino, S.R.; Batista, A.D.; Martendal, E. New sorbents for extraction and microextraction techniques. *J. Chromatogr. A* **2010**, *1217*, 2533–2542. [CrossRef] [PubMed]
4. Lee, C.-T.; Kung, P.-H.; Lee, Y.-D. Preparation of poly(vinyl alcohol)-chondroitin sulfate hydrogel as matrices in tissue engineering. *Carbohydr. Polym.* **2005**, *61*, 348–354. [CrossRef]
5. Childs, A.; Li, H.; Lewittes, D.M.; Dong, B.; Liu, W.; Shu, X.; Sun, C.; Zhang, H.F. Fabricating customized hydrogel contact lens. *Sci. Rep.* **2016**, *6*, 34905. [CrossRef] [PubMed]
6. Moreira, H.R.; Munarin, F.; Gentilini, R.; Visai, L.; Granja, P.L.; Tanzi, M.C.; Petrini, P. Injectable pectin hydrogels produced by internal gelation: PH dependence of gelling and rheological properties. *Carbohydr. Polym.* **2014**, *103*, 339–347. [CrossRef] [PubMed]
7. Hoffman, A.S. Hydrogels for biomedical applications. *Adv. Drug Deliv. Rev.* **2002**, *64*, 18–23. [CrossRef]
8. Basheer, C.; Suresh, V.; Renu, R.; Lee, H.K. Development and application of polymer-coated hollow fiber membrane microextraction to the determination of organochlorine pesticides in water. *J. Chromatogr. A* **2004**, *1033*, 213–220. [CrossRef]
9. Castilhos, N.D.B.; Sampaio, N.M.F.M.; da Silva, B.C.; Riegel-Vidotti, I.C.; Grassi, M.T.; Silva, B.J.G. Physical-chemical characteristics and potential use of a novel alginate/zein hydrogel as the sorption phase for polar organic compounds. *Carbohydr. Polym.* **2017**, *174*, 507–516. [CrossRef]
10. Basheer, C.; Vetrichelvan, M.; Valiyaveettil, S.; Lee, H.K. On-site polymer-coated hollow fiber membrane microextraction and gas chromatography-mass spectrometry of polychlorinated biphenyls and polybrominated diphenyl ethers. *J. Chromatogr. A* **2007**, *1139*, 157–164. [CrossRef]
11. Zare, M.; Ramezani, Z.; Rahbar, N. Development of zirconia nanoparticles-decorated calcium alginate hydrogel fibers for extraction of organophosphorous pesticides from water and juice samples: Facile synthesis and application with elimination of matrix effects. *J. Chromatogr. A* **1473**, 28–37. [CrossRef]
12. Bardajee, G.R.; Azimi, S.; Sharifi, M.B.A.S. Application of central composite design for methyl red dispersive solid phase extraction based on silver nanocomposite hydrogel: Microwave assisted synthesis. *Microchem. J.* **2017**, *133*, 358–369. [CrossRef]
13. Park, J.W.; PARK, J.S.; Ruckenstein, E.L.I. On the Viscoelastic Properties of Poly (vinyl alcohol) and Chemically Crosslinked Poly (vinyl alcohol). *Appl. Ploym.* **2001**, *82*, 1816–1823. [CrossRef]
14. Lopes, L.C.; Simas-Tosin, F.F.; Cipriani, T.R.; Marchesi, L.F.; Vidotti, M.; Riegel-Vidotti, I.C. Effect of low and high methoxyl citrus pectin on the properties of polypyrrole based electroactive hydrogels. *Carbohydr. Polym.* **2017**, *155*, 11–18. [CrossRef] [PubMed]
15. Albero, B.; Sánchez-Brunete, C.; Miguel, E.; Pérez, R.A.; Tadeo, J.L. Analysis of natural-occurring and synthetic sexual hormones in sludge-amended soils by matrix solid-phase dispersion and isotope dilution gas chromatography-tandem mass spectrometry. *J. Chromatogr. A* **2013**, *1283*, 39–45. [CrossRef] [PubMed]
16. Peppas, N.A.; Bures, P.; Leobandung, W.; Ichikawa, H. Hydrogels in pharmaceutical formulations. *Eur. J. Pharm. Biopharm.* **2000**, *50*, 27–46. [CrossRef]

17. Gerald, G.; Margarita, G.; Joerg, S.; Gunnar, S.; Karl-Friedrich, A.; Andreas, R. Chemical and pH sensors based on the swelling behavior of hydrogels. *Sens. Actuators B Chem.* **2005**, *111–112*, 555–561.
18. Bizkarguenaga, E.; Ros, O.; Iparraguirre, A.; Navarro, P.; Vallejo, A.; Usobiaga, A.; Zuloaga, O. Solid-phase extraction combined with large volume injection-programmable temperature vaporization-gas chromatography-mass spectrometry for the multiresidue determination of priority and emerging organic pollutants in wastewater. *J. Chromatogr. A* **2012**, *1247*, 104–117. [CrossRef]
19. Jeannine, E.E.; Mara, M.; Jun, N.; Christopher, N.B. Structure and swelling of poly(acrylic acid) hydrogels: Effect of pH, ionic strength, and dilution on the crosslinked polymer structure. *Polymer* **2004**, *45*, 1503–1510.
20. Trinh, T.; Harden, N.B.; Coleman, H.M.; Khan, S.J. Simultaneous determination of estrogenic and androgenic hormones in water by isotope dilution gas chromatography-tandem mass spectrometry. *J. Chromatogr. A* **2011**, *1218*, 1668–1676. [CrossRef]
21. Rafika, B.S.; Sopheak, N.; Ibtissem, G.-A.; Salma, B.; Maïwen, L.C.; Dalila, B.H.-C.; Malika, T.-A.; Michele, T.; Baghdad, O. Simultaneous Detection of 13 Endocrine Disrupting Chemicals in Water by a Combination of SPE-BSTFA Derivatization and GC-MS in Transboundary Rivers (France-Belgium). *Water Air Soil Pollut.* **2017**, *228*, 2. [CrossRef]
22. Zhang, K.; Fent, K. Determination of two progestin metabolites (17α-hydroxypregnanolone and pregnanediol) and different classes of steroids (androgens, estrogens, corticosteroids, progestins) in rivers and wastewaters by high-performance liquid chromatography-tandem mass spe. *Sci. Total Environ.* **2018**, *610–611*, 1164–1172. [CrossRef] [PubMed]
23. Caban, M.; Lis, E.; Kumirska, J.; Stepnowski, P. Determination of pharmaceutical residues in drinking water in Poland using a new SPE-GC-MS(SIM) method based on Speedisk extraction disks and DIMETRIS derivatization. *Sci. Total Environ.* **2015**, *538*, 402–411. [CrossRef]
24. Noppe, H.; Verheyden, K.; Gillis, W.; Courtheyn, D.; Vanthemsche, P.; De Brabander, H.F. Multi-analyte approach for the determination of ng L-1levels of steroid hormones in unidentified aqueous samples. *Anal. Chim. Acta* **2007**, *586*, 22–29. [CrossRef] [PubMed]
25. Fernando, F.S.; Igor, C.P.; Cassiana, C.M.; Wilson, F.J. Assessing selected estrogens and xenoestrogens in Brazilian surface waters by liquid chromatography-tandem mass spectrometry. *Microchem. J.* **2010**, *96*, 92–98.
26. Zorica, D.; Svetlana, D.; Ivana, V.B.; Mila, D.L. Determination of sterols and steroid hormones in surface water and wastewater using liquid chromatography-atmospheric pressure chemical ionization-mass spectrometry. *Microchem. J.* **2017**, *135*, 39–47.
27. Migowska, N.; Caban, M.; Stepnowski, P.; Kumirska, J. Simultaneous analysis of non-steroidal anti-inflammatory drugs and estrogenic hormones in water and wastewater samples using gas chromatography-mass spectrometry and gas chromatography with electron capture detection. *Sci. Total Environ.* **2012**, *441*, 77–88. [CrossRef] [PubMed]
28. Alessandra, H.I.; Rodrigo, A.O.; Luana, O.M.; Jorge, D.C.P.; Júlio, C.R.D.A. Occurrence of Pharmaceutical Products, Female Sex Hormones and Caffeine in a Subtropical Region in Brazil. *Clean (Weinh)* **2017**, *45*, 334. [CrossRef]

Sample Availability: Samples of the compounds are not available from the authors.

© 2018 by the authors. Licensee MDPI, Basel, Switzerland. This article is an open access article distributed under the terms and conditions of the Creative Commons Attribution (CC BY) license (http://creativecommons.org/licenses/by/4.0/).

Article

Multi-Spheres Adsorptive Microextraction (MSAµE)—Application of a Novel Analytical Approach for Monitoring Chemical Anthropogenic Markers in Environmental Water Matrices

Ana R. M. Silva, Nuno R. Neng * and José M. F. Nogueira *

Centro de Química e Bioquímica, Centro de Química Estrutural, Faculdade de Ciências, Universidade de Lisboa, Campo Grande, Ed. C8, 1749-016 Lisbon, Portugal; rita.mendao@ipleiria.pt
* Correspondence: ndneng@fc.ul.pt (N.R.N.); nogueira@fc.ul.pt (J.M.F.N.);
  Tel.: +351-21-7500-000 (N.R.N.); +351-21-7500-088 (J.M.F.N)

Academic Editor: Simone Morais
Received: 28 January 2019; Accepted: 28 February 2019; Published: 7 March 2019

Abstract: Multi-spheres adsorptive microextraction using powdered activated carbons (ACs) was studied as a novel enrichment approach, followed by liquid desorption and high-performance liquid chromatography with diode array detection (MSAµE(AC)-LD/HPLC-DAD) to monitor caffeine (CAF) and acetaminophen (ACF) traces in environmental matrices. In this study, commercial activated carbons (N, N_{OX}, and R) were tested, with the latter showing a much better performance for the analysis of both anthropogenic drugs. The main parameters affecting the efficiency of the proposed methodology are fully discussed using commercial AC(R). Textural and surface chemistry properties of the ACs sample were correlated with the analytical results. Assays performed on 30 mL of water samples spiked at 10 µg L^{-1} under optimized experimental conditions, yielding recoveries of 75.3% for ACF and 82.6% for CAF. The methodology also showed excellent linear dynamic ranges for both drugs with determination coefficients higher than 0.9976, limits of detection and quantification of 0.8–1.2 µg L^{-1} and 2.8–4.0 µg L^{-1}, respectively, and suitable precision (RSD < 13.8%). By using the standard addition method, the application of the present method to environmental matrices, including superficial, sea, and wastewater samples, allowed very good performance at the trace level. The proposed methodology proved to be a feasible alternative for polar compound analysis, showing to be easy to implement, reliable, and sensitive, with the possibility to reuse and store the analytical devices loaded with the target compounds for later analysis.

Keywords: sorbent-based techniques; multi-spheres adsorptive micro-extraction (MSAµE); floating sampling technology; caffeine and acetaminophen tracers; environmental water matrices

1. Introduction

Pharmaceuticals and personal care products (PPCPs) have generated one of the most important emerging environmental issues, particularly in aquatic systems [1]. Wastewaters from municipal treatment plants, livestock farms, veterinary drug use, and residues from hospitals and pharmaceutical manufactures are the best known sources of pollutants discharged into the water environment [2], affecting aquatic wildlife forms and producing changes that threaten the sustainability of the ecosphere [3]. Continuous releases and chronic exposure to these chemicals not only can result in subtle effects on aquatic species but could also pose a risk to human health associated with consuming contaminated drinking water over a lifetime [4]. For instance, analgesic, anti-inflammatory, antibiotic, antiepileptic, antimicrobial, and antacid drugs are good examples of some of the most used PPCP types worldwide [5]. Nevertheless, due to the huge number of PPCPs as well as the very low concentrations in which they usually occur in the environment, it is frequent to monitor particular anthropogenic chemical markers during pollution control programs. For instance, caffeine (CAF) and acetaminophen (ACF) are good examples of PPCPs tracers, since they are widely used anthropogenic drugs and the levels usually discharged from wastewaters into the environment are also very significant [3].

Owing to the great environmental impact, it is mandatory to develop cheap, simple, rapid, and sensitive analytical strategies for monitoring trace levels of these anthropogenic chemical markers in the environment. State-of-the-art analytical methodologies for monitoring trace levels of PPCPs are usually based on sample enrichment techniques prior to chromatographic or hyphenated systems. In recent years, sample preparation methods are characterized by simplification and high throughput to enhance selectivity and sensitivity [6]. Therefore, for the determination of trace levels of PPCPs in environmental water matrices, solid-phase extraction (SPE) [3] and solid phase microextraction (SPME) [7] are well-established techniques that have showed great performance. Recently, adsorptive microextraction (AµE) techniques were introduced as a novel analytical approach that uses nanostructured sorbents and operates under the floating sampling technology [8,9]. For the implementation of these innovative static microextraction techniques, two configurations were originally designed, namely, analytical devices having bar or multi-spheres geometry. The present contribution proposes the application of a new analytical approach, namely multi-spheres adsorptive microextraction coated with several commercial activated carbons (MSAµE(AC), Figure 1) followed by high performance liquid chromatography with diode array detection (HPLC-DAD), for monitoring trace levels of two PPCPs markers (ACF and CAF, Figure 2) in environmental water matrices. According to previous exploratory data, MSAµE devices present much better stability compared to bar geometry devices when they are subjected to a more aggressive sample matrix because, in this case, thermal supporting promotes much higher robustness from the fixation point of view.

Figure 1. Image of device for multi-spheres adsorptive microextraction coated with several commercial activated carbons (MSAµE(AC)) used in the present work.

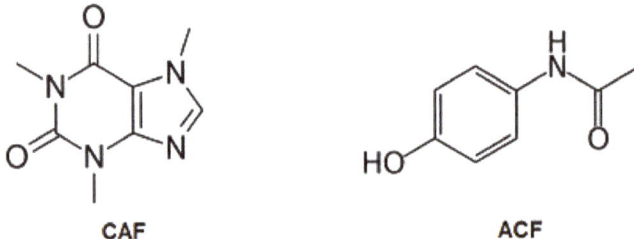

Figure 2. Chemical structures of the two markers of pharmaceuticals and personal care products PPCPs studied: caffeine (CAF) and acetaminophen (ACF).

2. Results and Discussion

2.1. Characterization of the ACs

The commercial ACs used in this work were characterized by texture and surface chemistry (Table S1). From the pH_{PZC} values, AC(N) has basic surface and AC(R and N_{OX}) have a slightly acidic surface. From the data depicted in Table S1, approximately 60% of the total volume of the ACs are constituted by microporous and 40% by mesoporous, which was in the same order of magnitude found in the literature [10].

2.2. HPLC-DAD Optimization

In this study, ACF and CAF were selected as anthropogenic tracer models. We evaluated the HPLC-DAD conditions, such as UV/vis spectra and chromatographic parameters. A wavelength (λ_{max}) of 205 nm was selected, since it maximizes the DAD response for the target compounds. Using the chromatographic conditions described in Section 3.5, a good response was attained for the two tracers with suitable resolution and analytical time < 7 min. Instrumental sensitivity was also checked through the limits of detection (LODs) and quantification (LOQs) for both target analytes and calculated with a signal-to-noise ratio (S/N) of 3 and 10, respectively, resulting in LODs and LOQs of 120 and 400 µg L^{-1} for ACF, and 89 and 297 µg L^{-1} for CAF, respectively. Determination coefficients of 0.9999 were achieved with instrumental calibration having concentrations ranging from 0.6 to 10.0 mg L^{-1} (n = 6). Furthermore, instrumental precision was also assessed through repeated injections resulting in relative standard deviations (RSDs) lower than 8.2% and no carry-over was observed (<LODs).

2.3. Optimization of MSAµE(AC)-LD Assays

Several parameters affecting the microextraction and back-extraction stages efficiency of the MSAµE(AC)-LD approach, such as equilibrium time, agitation speed, matrix characteristics, and liquid desorption (LD) conditions, were tested and optimized [6].

In a first approach, we selected the most appropriate sorbent phase to microextract the target compounds. Since the pH of the extraction process plays an important role in the recovery yield, three ACs (R, N, and N$_{OX}$) were assayed at five different pH values (2.0, 4.0, 6.0, 8.0, and 10.0), using standard conditions (microextraction: 17 h (1000 rpm) and back-extraction: formic acid (30 min)). This assessment is related with two factors, i.e., the analyte ionization and the charges on the ACs surface at a given pH. Therefore, the variation of pH affects the dissociation of the ACF and CAF molecules, as well as the surface chemistry of the sorbent involved due to the dissociation of the surface functional groups. The AC surface may be either positively or negatively charged depending on the nature of the AC. At a given pH, the AC surface and the analyte species may coexist in a complex system, in which similar or opposite charges may be present. However, as the ACs tested have distinct pH$_{PZC}$ values (Table S1), the influence of the pH will be different. Figure 3 shows how the pH influences the recovery yield for each AC tested. For ACF (Figure 3a), the results show that to optimize the adsorption-desorption process, we must work in a pH between 4 and 8; once in this range, the difference observed was negligible, assuring that the ACF has not charged. The different performance of the three ACs was more pronounced with the surface chemistry than with the porous structure. In fact, N had the highest total, meso, and microporous volume (Table S1) despite its lower recovery yields. The ACs with more acidic characteristics (R and N$_{OX}$) showed similar performance. Concerning CAF, AC(N$_{OX}$) presented lower recovery and AC(N) had the maximum recovery value at pH 6.0. The lower recovery was reached at pH 10.0 and this decrease was more pronounced for ACF. This can be explained through the pK_a values, where ACF (pK_a 9.38) should present a negative charge at pH 10.0, while for CAF the pK_a is 10.40, which is higher than 10.0. AC(N) presented a different behavior with CAF and the recovery did not decrease at pH 10.0. In short, we can state that the AC(R) is the sorbent that allows a higher performance to simultaneously microextract both probe molecules.

The LD conditions that ensure complete back-extraction for the two tracers from the devices were also optimized. Formic acid and MeOH were tested to assess the LD performance. The results obtained showed that formic acid have better desorption capacity. Subsequently, the back-extraction time was also evaluated for both tracers by testing 15, 30, and 45 min of sonification time, from which 30 min seemed long enough without observing any carry-over effect. The evaporation step is essential for solvent switch, being necessary to carefully check for possible analyte losses during this process. For such purpose, three concentration levels were evaluated, where no significant losses were observed.

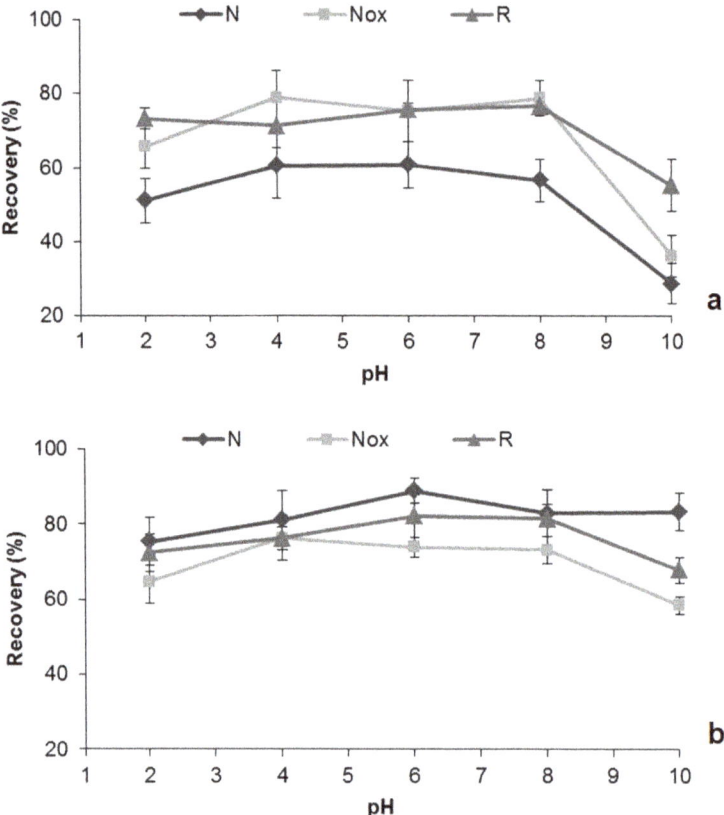

Figure 3. Effect of matrix pH on the recovery of ACF (**a**) and CAF (**b**) by MSAµE(AC) followed by liquid desorption and high-performance liquid chromatography with diode array detection (MSAµE(AC)-LD/HPLC-DAD), using different activated carbons (ACs) phases (R, N, and N_{OX}).

For MSAµE(AC(R)) the number of multi-spheres can be chosen according to the expected content level for each target analyte. Consequently, five AC(R) masses (1.1, 1.7, 2.1, 2.5, and 3.2 mg) were tested. The data showed that 2.1 mg is enough to extract the target compounds at trace level (Figure 4a). According to the sorptive-based theory [11], extraction time and agitation speed are important parameters to better control the recovery conditions; during the extraction process, the multi-sphere devices are floating under the vortex created by the Teflon stir bar. We started with stirring rates of 750, 1000, and 1250 rpm, and the results obtained were negligible (data not shown). Subsequently, the equilibrium time was optimized (1000 rpm) by performing assays at 0.5, 1, 2, 4, 6, and 17 h, at room temperature. Figure 4b depicts the data obtained, where it can be observed that the recovery yields increase up to 17 h of extraction. Although slow kinetics are noticed, we decided to fix this parameter for further experiments, since this methodology can be performed overnight without any special requirements.

In accordance with previous works [10,12], the characteristics of the aqueous medium, i.e., ionic strength and polarity characteristics, are also important parameters with substantial effects on extraction efficiency. The ionic strength and polarity were modified through the addition of NaCl and MeOH (5, 10, and 15%) onto matrix media. Furthermore, the addition of MeOH reduced the recovery yield, with higher performance for ACF (Figure S1a). On the other hand, the ionic strength increment produced a very slight decrease on the recovery (Figure S1b).

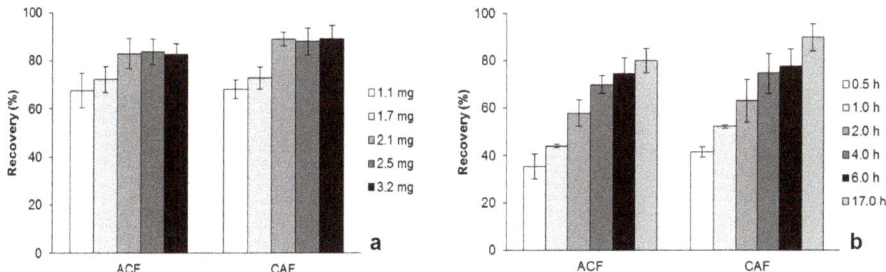

Figure 4. Effect of AC(R) mass (**a**) and equilibrium time (**b**) on the recovery of ACF and CAF by MSAμE(AC(R)-LD/HPLC-DAD.

2.4. Validation of the MSAμE(AC(R))-LD/HPLC-DAD Method

To obtain the recovery yields under optimized experimental conditions (microextraction stage: 2.1 mg of AC(R), 17 h (1000 rpm, pH 7.0); LD: formic acid (30 min)), assays were performed on ultra-pure water samples spiked at 10 μg L^{-1}. The recovery yields achieved were higher than 75.0% (Table 1), indicating that both tracers have a high affinity to the selected AC.

Table 1. Comparison of average recovery yields of both tracers achieved by the current method and other microextraction techniques.

Compounds	Microextraction Techniques	Recovery (%)	Refs.
CAF	MSAμE	75.3	This work
	SPE	99.3	13
	SPE	88.0	14
	SPME	98.5	15
ACF	MSAμE	82.6	This work
	SPE	11.3	16
	SPE	47.0	17

The lifetime of the microextraction device is also a very important parameter for practical utilization. The coating can be damaged by acidic and alkaline solutions as well as organic solvents. Therefore, the extraction capacity was evaluated after immersing the microextraction device in MeOH, n-hexane, 0.1 M HCl, and 0.1 M NaOH. From the data obtained (Figure S2), CAF-extraction ability showed no obvious decline after the microextraction devices were tested in different solvents for 48 h. For ACF, the recovery slightly decreased for the microextraction devices dipped in n-hexane. After each application, the microextraction devices can be reused and the number of possible uses dependents on the matrix type. After 10 applications of ACF and CAF using ultra-pure water matrix, the data indicate no obvious difference in efficiency (Figure 5a). It was also demonstrated that the microextraction devices can be stored with the analytes at 4 °C for 8 days without solute degradation (Figure 5b). This characteristic opens interesting features for on-site microextraction sampling, allowing the shipment of loaded devices to the laboratory after enrichment, instead of samples.

To validate the proposed methodology, parameters such as linearity, LOD, LOQ, and precision were evaluated. For linearity, the determination coefficients (r^2) were higher than 0.9976 with concentration levels ranging from 4.0 to 40.0 μg L^{-1} ($n = 5$). For LODs and LOQs, values of 0.8–1.2 μg L^{-1} and 2.8–4.0 μg L^{-1} were achieved, respectively. Precision was evaluated using within and between-day repeatability with three concentrations levels (4.0, 20.0, and 40.0 μg L^{-1}, $n = 6$) and expressed as relative standard deviation (RSD = 13.8%). According to the requirements of Directive 98/83/EC [13], a value under 25% for organic compounds in water matrices may be considered acceptable.

Finally, we compared the average recovery yields of the proposed methodology with those of other microextraction techniques already established (Table 1). As can be observed, the proposed methodology revealed recovery levels for ACF almost two-fold better and of the same order of magnitude for CAF when compared with other well-established microextraction techniques. It must also be emphasized that the proposed methodology uses small amounts of sample during the extraction stage and less organic solvent during the back-extraction stage than many other microextraction approaches reported in the literature (e.g., SPE) [14–18].

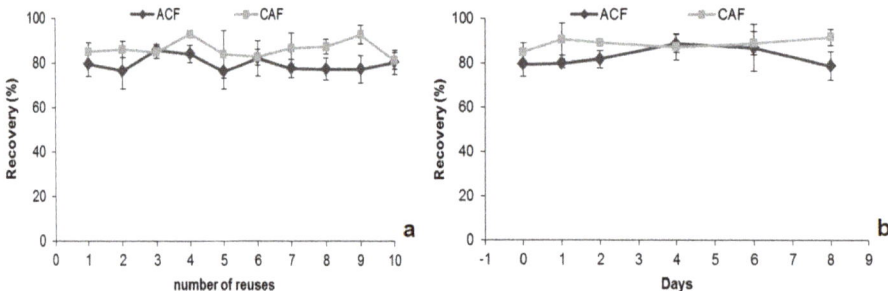

Figure 5. Comparison of recoveries of ACF and CAF after several reuses (**a**) and influence of storage time (**b**).

2.5. Application to Environmental Water Matrices

The efficiency of sample preparation can be affected by the complexity of the matrix involved. For instance, substantial levels of dissolved or suspended inorganic or organic matter contained in water matrices may interfere with the extraction process of the target compounds; therefore, the yields might drastically change from sample to sample [19]. In this work, the standard addition method (SAM) was used to minimize the occurrence of matrix effects. Table 2 summarizes the data obtained by SAM assays on the different matrices studied under optimized experimental conditions. When comparing the responses obtained when AC(R) is used in ultra-pure, surface, sea, and treated wastewater matrices, a very slow decrease in the signal for target compounds is noticed. The matrix effects were evident in the treated wastewater samples, thus justifying the quantification performed by the SAM in this kind of sample for this type of target compounds. Figure 6 exemplifies chromatograms obtained from assays performed on spiked (10 µg L^{-1}) ultra-pure (a), surface (b), sea (c), and urban wastewater (d) samples by MSAµE(AC(R))-LD/HPLC-DAD under optimized experimental conditions, where good sensitivity and remarkable selectivity are noticed.

In short, it must be emphasized that the present methodology is easy to work-up and environmentally friendly, and requires low sample and solvent volumes. Therefore, this new methodology presents advantages in relation to other well established microextraction approaches [20].

Table 2. Regression parameters obtained with the standard addition method (SAM) and average recovery yields for ACF and CAF in ultra-pure, surface, sea, and wastewater matrices obtained by MSAµE(AC(R))-LD/HPLC-DAD under optimized experimental conditions.

Compounds	Recovery (%); (r^2)			
	Ultra-Pure Water	Surface Water	Sea Water	Wastewater
ACF	75.3 ± 9.5 (0.9983)	64.8 ± 8.3 (0.9960)	73.1 ± 2.9 (0.9970)	50.1 ± 2.2 (0.9953)
CAF	82.6 ± 2.9 (0.9976)	83.5 ± 1.1 (0.9935)	74.6 ± 2.7 (0.9975)	54.7 ± 5.6 (0.9989)

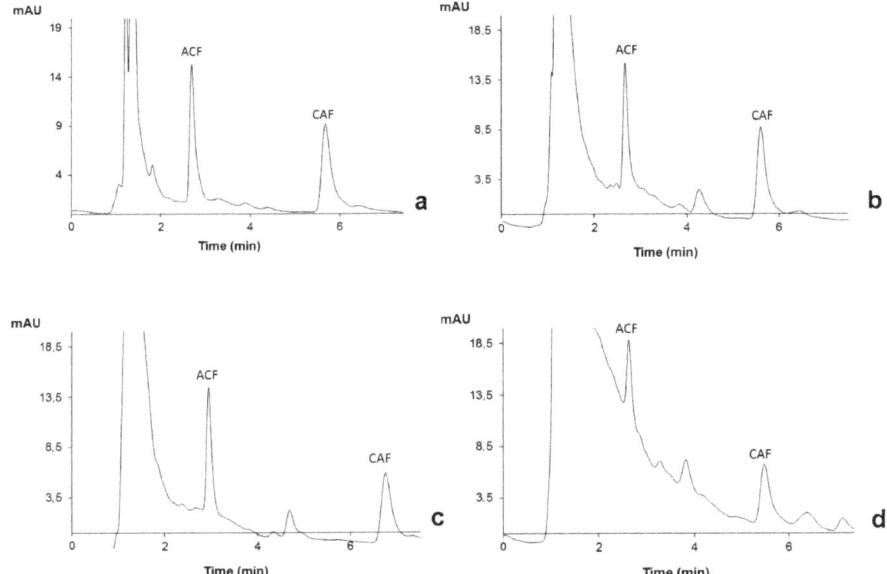

Figure 6. Chromatograms obtained from assays performed on ultra-pure water (**a**), superficial water (**b**), sea water (**c**), and urban wastewater (**d**) by MSAµE(AC(R))-LD/HPLC-DAD under optimized experimental conditions.

3. Materials and Methods

3.1. Samples and Reagents

HPLC-grade methanol (MeOH, 99.9%), formic acid (99%) and *n*-hexane (96%) were purchased from Merck (Darmstadt, Germany). Ultra-pure water was obtained from Milli-Q water purification systems (Millipore, Burlington, MA, USA). Sodium chloride (NaCl, 99.9%) was supplied by José M.G. dos Santos (Lisbon, Portugal). Hydrochloric acid (HCl, 37%) and phosphoric acid (85%) were obtained from Riedel-de Haën (Seelze, Germany). Sodium hydroxide (NaOH, 98.0%) was obtained from AnalaR BDH Chemicals (London, UK). ACF (98.0%) and CAF (99.0%) were supplied from Sigma Aldrich (Darmstadt, Germany). A stock solution of each compound (200 mg L^{-1}) was prepared in MeOH and stored refrigerated at −20 °C. An intermediate standard solution containing both compounds at a concentration of 10 mg L^{-1} was obtained after mixing individual stock solutions, diluting with MeOH/water (25/75%) and stored at 4 °C. The commercial AC(N) was supplied by Salmon & Cia. (Lisbon, Portugal) and the AC(R) from Riedel-de Haën (Seelze, Germany). The AC(N$_{OX}$) was obtained by oxidized AC(N) with nitric acid (HNO$_3$, 20%). All environmental samples were collected in Portugal previously filtered and stored refrigerated at 4 °C until their analysis. Discrete wastewater samples were collected from the effluent of an urban treatment plant located at Cartaxo after all wastewater treatment; surface water samples were collected from Tagus river bank at Lisbon (Oriente); sea water samples were collected in Algarve seaside.

3.2. ACs Characterization

The procedure of textural and surface chemistry characterization of all ACs (N, N$_{OX}$, and R) was performed as described by Mestre et al. [21].

3.3. MSAμE Devices Preparation

The microextraction devices were lab-made (Figure 1) as previously reported [9]. Before use, the microextraction devices were cleaned through a sonification treatment using MeOH and ultra-pure water.

3.4. Recovery Assays and Method Validation

Typical assays were performed with 30 mL of ultra-pure water spiked with a standard mixture at the desired concentration, followed by the introduction of the microextraction device and a conventional Teflon magnetic bar.

Several parameters of MSAμE(AC), such as coating phase (N, N_{OX} or R), sorbent amount (1.1, 1.7, 2.1, 2.5, and 3.2 mg), equilibrium time (0.5, 1, 2, 4, 6, and 17 h), agitation speed (750, 1000, and 1250 rpm), pH (2, 4, 6, 8, and 10), organic modifier (MeOH; 5, 10, and 15%, v/v), ionic strength (NaCl; 5, 10, and 15%, w/v), desorption solvent (MeOH and formic acid), and desorption time (15, 30, and 45 min) were systematically studied and optimized ($n = 3$). After back-extraction, the devices were removed, and the stripping solvent was evaporated to dryness under purified nitrogen (>99.5%). The residues were redissolved in 200 μL of MeOH/water (25/75%) and afterwards, the vials were closed with a seal and placed on the auto-sampler for HPLC-DAD analysis. Blank assays were also carried out without spiking. More detail can be found in our previous reports [6,10,12]. For testing chemical and mechanical stability, the microextraction devices were immersed in vials loaded with MeOH, n-hexane, HCl (0.1 M), and NaOH (0.1 M) solutions at room temperature. After 48 h, the microextraction devices were taken out for the extraction assay. The analyte recoveries (10 μg L^{-1}) before and after dipping in different solutions were compared. The SAM was used for quantification on real matrices, and 30 mL of wastewater, sea, and river water samples were used. To guarantee maximum control of the analytical methodology, the samples were first and foremost fortified with four working standards to produce the corresponding spiking levels for the target compounds (4–40 μg L^{-1}) and blank assays ("zero-point") were also performed without spiking. All assays were done in triplicate.

3.5. HPLC-DAD Settings

An Agilent 1100 Series LC system (Agilent Technologies, Waldbronn, Germany) equipped with a C18 column (Tracer excel, 150 mm × 4.0 mm, 5 μm particle size, Teknokroma, Barcelona, Spain) and LC3D ChemStation software (version Rev.A.10.02[1757], Waldbronn, Germany) was used for the HPLC-DAD analysis. The mobile phase consisted in a mixture of 25% (v/v) MeOH solution and 0.1% phosphoric acid aqueous solution (flow: 1 mL min^{-1}). The column temperature was maintained at 25 °C, the injection volume was 20 μL with a draw speed of 200 μL min^{-1} and the detector was set at 205 nm.

4. Conclusions

The MSAμE(AC(R))-LD/HPLC-DAD methodology proposed in this work offers a new practical alternative method for analysis of PPCPs tracers such as CAF and ACF in environmental water samples. Under optimized experimental conditions, good analytical performance was attained as well as suitable detection limits and linear dynamic ranges. The textural and surface properties of the ACs were correlated with the analytical results for a better understanding of the overall process. By using the SAM, the established methodology showed a good response for the analysis of ACF and CAF in water samples such as surface, sea, and wastewater matrices. Furthermore, MSAμE(AC) is cost-effective and easy to work-up, demonstrating to be a remarkable analytical tool for trace analysis of priority and emerging pollutants in environmental water matrices. The novel methodology proposed herein is compliant with green analytical chemistry principles and a good alternative to overcome the limitations of other technologies for trace analysis of polar compounds.

Supplementary Materials: The following are available online. Table S1. Nanotextural and chemical characteristics of the commercial ACs. Figure S1. Effect of matrix polarity (a) and ionic strength (b) on the recovery of ACF and CAF by MSAµE(AC(R)-LD/HPLC-DAD. Figure S2. Evaluation of extraction capacity after immersing the devices in different solvents.

Author Contributions: A.R.M.S. performed the design of experiments; N.R.N. performed the original draft preparation, review, and editing; J.M.F.N. performed the review, editing, and supervision.

Funding: This research was funded by Fundação para a Ciência e a Tecnologia (Portugal) (UID/MULTI/00612/2013), for postdoctoral (SFRH/BPD/86071/2012) and PhD (SFRH/BD/40926/2007) grants.

Acknowledgments: The authors also thank Ana Carvalho (Faculdade de Ciências, Universidade de Lisboa) for the characterization of the ACs.

Conflicts of Interest: The authors declare no conflict of interest.

References

1. Kot-wasik, A.; De, J.; Namies, J. Analytical Techniques in Studies of the Environmental Fate of Pharmaceuticals and Personal-Care Products. *TrAC-Trend Anal. Chem.* **2007**, *26*, 557–568. [CrossRef]
2. Sim, W.; Lee, J.; Lee, E.; Shin, S.; Hwang, S.; Oh, J. Occurrence and distribution of pharmaceuticals in wastewater from households, livestock farms, hospitals and pharmaceutical manufactures. *Chemosphere* **2011**, *82*, 179–186. [CrossRef] [PubMed]
3. Sui, Q.; Cao, X.; Lu, S.; Zhao, W.; Qiu, Z.; Yu, G. Occurrence, Sources and Fate of Pharmaceuticals and Personal Care Products in the Groundwater: A Review. *Emerg. Contam.* **2015**, *1*, 14–24. [CrossRef]
4. Al-odaini, N.A.; Pauzi, M.; Ismail, M.; Surif, S. Multi-Residue Analytical Method for Human Pharmaceuticals and Synthetic Hormones in River Water and Sewage Effluents by Solid-Phase Extraction and Liquid Chromatography-Tandem Mass Spectrometry. *J. Chromatogr. A* **2010**, *1217*, 6791–6806. [CrossRef] [PubMed]
5. Wang, J.; Wang, S. Removal of Pharmaceuticals and Personal Care Products (PPCPs) from Wastewater: A Review. *J. Environ. Manag.* **2016**, *182*, 620–640. [CrossRef] [PubMed]
6. Neng, N.R.; Nogueira, J.M.F. Development of a Bar Adsorptive Micro-Extraction-Large-Volume Injection-Gas Chromatography-Mass Spectrometric Method for Pharmaceuticals and Personal Care Products in Environmental Water Matrices. *Anal. Bioanal. Chem.* **2012**, *402*, 1355–1364. [CrossRef] [PubMed]
7. Wang, S.; Oakes, K.D.; Bragg, L.M.; Pawliszyn, J.; Dixon, G.; Servos, M.R. Chemosphere Validation and Use of in Vivo Solid Phase Micro-Extraction (SPME) for the Detection of Emerging Contaminants in Fish. *Chemosphere* **2011**, *85*, 1472–1480. [CrossRef] [PubMed]
8. Nogueira, J.M.F. Microextração Adsortiva Em Barra (BAµE): Um Conceito Analítico Inovador Para Microextração Estática. *Sci. Chromatogr.* **2014**, *5*, 275–283. [CrossRef]
9. Neng, N.R.; Silva, A.R.M.; Nogueira, J.M.F. Adsorptive Micro-Extraction Techniques-Novel Analytical Tools for Trace Levels of Polar Solutes in Aqueous Media. *J. Chromatogr. A* **2010**, *1217*, 7303–7310 [CrossRef] [PubMed]
10. Neng, N.R.; Mestre, A.S.; Carvalho, A.P.; Nogueira, J.M.F. Powdered Activated Carbons as Effective Phases for Bar Adsorptive Micro-Extraction (BAµE) to Monitor Levels of Triazinic Herbicides in Environmental Water Matrices. *Talanta* **2011**, *83*, 1643–1649. [CrossRef] [PubMed]
11. David, F.; Sandra, P. Stir Bar Sorptive Extraction for Trace Analysis. *J. Chromatogr. A* **2007**, *1152*, 54–69. [CrossRef] [PubMed]
12. Neng, N.R.; Nogueira, J.M.F. Determination of Short-Chain Carbonyl Compounds in Drinking Water Matrices by Bar Adsorptive Micro-Extraction (BAµE) with in Situ Derivatization. *Anal. Bioanal. Chem.* **2010**, *398*, 3155–3163. [CrossRef] [PubMed]
13. European Commission. Council Directive 98/83/EC. *Off. J. Eur. Commun.* **1998**, *L330*, 32–54.
14. Mokh, S.; El Khatib, M.; Koubar, M.; Daher, Z.; Al Iskandarani, M. Innovative SPE-LC-MS/MS Technique for the Assessment of 63 Pharmaceuticals and the Detection of Antibiotic-Resistant-Bacteria: A Case Study Natural Water Sources in Lebanon. *Sci. Total Environ.* **2017**, *609*, 830–841. [CrossRef] [PubMed]
15. De Kesel, P.M.M.; Lambert, W.E.; Stove, C.P. An Optimized and Validated SPE-LC-MS/MS Method for the Determination of Caffeine and Paraxanthine in Hair. *Talanta* **2015**, *144*, 62–70. [CrossRef] [PubMed]

16. Paíga, P.; Lolić, A.; Hellebuyck, F.; Santos, L.H.M.L.M.; Correia, M.; Delerue-Matos, C. Development of a SPE-UHPLC-MS/MS Methodology for the Determination of Non-Steroidal Anti-Inflammatory and Analgesic Pharmaceuticals in Seawater. *J. Pharm. Biomed. Anal.* **2015**, *106*, 61–70. [CrossRef] [PubMed]
17. Maldaner, L.; Jardim, I.C.S.F. Determination of Some Organic Contaminants in Water Samples by Solid-Phase Extraction and Liquid Chromatography-Tandem Mass Spectrometry. *Talanta* **2012**, *100*, 38–44. [CrossRef] [PubMed]
18. Müller, C.; Vetter, F.; Richter, E.; Bracher, F. Determination of Caffeine, Myosmine, and Nicotine in Chocolate by Headspace Solid-Phase Microextraction Coupled with Gas Chromatography-Tandem Mass Spectrometry. *J. Food Sci.* **2014**, *79*, T251–T255. [CrossRef] [PubMed]
19. Rodil, R.; Muniategui-lorenzo, S. Multiresidue Analysis of Acidic and Polar Organic Contaminants in Water Samples by Stir-Bar Sorptive Extraction–Liquid Desorption–Gas Chromatography–Mass Spectrometry. *J. Chromatogr. A* **2007**, *1174*, 27–39.
20. Togola, A. Multi-Residue Analysis of Pharmaceutical Compounds in Aqueous Samples. *J. Chromatogr. A* **2008**, *1177*, 150–158. [CrossRef] [PubMed]
21. Mestre, A.S.; Pires, J.; Nogueira, J.M.F.; Carvalho, A.P. Activated Carbons for the Adsorption of Ibuprofen. *Carbon* **2007**, *45*, 1979–1988. [CrossRef]

Sample Availability: Not available.

© 2019 by the authors. Licensee MDPI, Basel, Switzerland. This article is an open access article distributed under the terms and conditions of the Creative Commons Attribution (CC BY) license (http://creativecommons.org/licenses/by/4.0/).

Article

Solvent Front Position Extraction with Semi-Automatic Device as a Powerful Sample Preparation Procedure Prior to Quantitative Instrumental Analysis

Anna Klimek-Turek *, Kamila Jaglińska, Magdalena Imbierowicz and Tadeusz Henryk Dzido

Physical Chemistry Department, Medical University of Lublin, Chodźki 4A, 20-093 Lublin, Poland; kamila.skop@umlub.pl (K.J.); magdalenaimbierowiczumlub@gmail.com (M.I.); tadeusz.dzido@umlub.pl (T.H.D.)
* Correspondence: anna.klimek@umlub.pl; Tel.: +48-81448-7206; Fax: +48-81448-7208

Academic Editor: Nuno Neng
Received: 14 March 2019; Accepted: 4 April 2019; Published: 6 April 2019

Abstract: The new prototype device is applied to the Solvent Front Position Extraction (SFPE) sample preparation procedure. The mobile phase is deposited onto the chromatographic plate adsorbent layer by the pipette, which is moved, according to programmed movement path, by a 3D printer mechanism. The application of the prototype device to SFPE procedure leads to the increased repeatability of the results and significant reduction of the analysis time in comparison to the classical procedure of chromatogram development. Additionally, the new equipment allows use procedures that are not possible to run using the classic chromatogram development. In this paper, the results of manual and semi-automatic sample preparation with SFPE are compared and the possible application of this prototype device is discussed.

Keywords: sample preparation with TLC/HPTLC; solvent front position extraction; solvent delivery with a moving pipette; automation; LC–MS/MS

---

1. Introduction

Sample preparation is an integral part of any analytical method, influences all subsequent steps of the analysis, and has a relevant impact on the precision and accuracy of the results [1]. Unfortunately, it is still the most labor-intensive step of the analytical procedure, consuming two-thirds of the total analysis time [2,3]. This is in strong contrast to modern, fast chromatographic methods, mass spectrometric detection, and makes sample preparation a limiting step of the analysis. In the light of these considerations, it is not surprising that there is a growing interest in the automation of this process [4]. According to the result of a survey carried out among analysts, the percentage of respondents using automated sample preparation jumped significantly, from 29% in 2013 to 39% in 2015 [4]. This is perfectly understandable if one takes into account that the application of automation sample preparation techniques reduces sample manipulation, and provides high recovery, throughput, and speed of analysis [4].

Many automated systems operate in off-line mode, when the sample preparation procedure is independent of further analysis [5]. In recent years, there has been a significant increase of interest in on-line processing techniques, where both sample preparation and instrumental analysis operate as a single system [6]. On-line mode allows the determination of target compounds automatically in a fast and efficient way. The online coupling of sample preparation techniques with LC began from the connection of solid phase extraction (SPE) to liquid chromatography (LC) [7]. Recently, a variety of online sample preparation techniques have been emerging, such as solid phase microextraction (SPME) [8–10], microextraction in packed syringe (MEPS) [11,12], stir bar sorptive extraction (SBSE) [13], liquid phase microextraction (LPME) [14], accelerated solvent extraction, (ASE) [15], microdialysis (MD) [16], microwave assisted extraction (MAE) [17], supercritical fluids extraction (SFE) [18], and other modern technologies which can overcome weaknesess and shortcomings of the traditional techniques of sample preparation.

Undoubtedly, among the online liquid sample preparation techniques, the column-based SPE is the most commonly used one [19]. In this technique, the one mobile phase is used to introduce the sample into the first column called the 'extraction column' and then the components are washed out to the second column, the so-called analytical one, using a different mobile phase. The direction of the mobile phase flow through the extraction and analytical column is the same [20]. The other technique uses a back-flush mode. Then the sample solution is introduced in the extraction column and the analytes are retained in the head of it. In the subsequent stage, the analytes are washed out to the analytical column by a mobile phase flowing in the opposite direction with respect to the flow during sample introduction into the extraction column [21,22].

The major advantage of SPME technique is its relatively easy automation. The technique integrates sampling, extraction, preconcentration, and injection into one analytical system, which is simple, efficient, and solvent-free [23].

Micro-extraction packed sorbent is also the kind of miniaturization of conventional SPE from milliliter to microliter bed volumes [24]. The manipulation of the small volumes is realized with a precision gas-tight syringe. The automated online MEPS-LC can be achieved by connecting the syringe to an autosampler [25].

A quite different approach to the sample preparation procedure is used in Dried Blood Spot (DBS) analysis [26,27]. This method has recently gained attention in bioanalytical laboratories.

In this method, after application to special cards, the samples are dried, collected, transported, and stored in a dry form. In order to enable analysis, each card is extracted for the isolation of target analytes, and the obtained solution typically undergoes LC–MS/MS analysis. Traditional DBS techniques are highly labor-intensive, require manual card punching, extracting the sample from the blot paper, centrifugation, and then transfer of the supernatant to a vial for further analysis. Fortunately, in recent years automated DBS card extraction and analysis systems have been developed in order to meet or exceed the performance of manual methods for sensitivity, accuracy, reproducibility, and precision in the analysis [28].

In our previous papers, we presented a solvent front position extraction (SFPE) technique, which is based on planar solid phase extraction clean-up concept for pesticide residue analysis developed by Oellig and Schwack [29,30]. In the papers thin-layer chromatography was used to sample preparation for the determination of solutes followed by HPLC (High Performance Liquid Chromatography) [31] or MS (Mass Spectrometry) analysis [32]. SFPE procedure was used to separate substances of interest (acebutolol, aminophenazone, acetaminophen, caffeine, theophylline, tramadol, ciprofloxacin, acetylsalicylic acid) and their internal standard (acetanilide) from other matrix components (bovine serum) and to form a single spot/zone containing them at the solvent front position on a chromatographic plate. We obtained very promising results. Some disadvantages of the presented technique (e.g., time-consuming, relatively high values of percent relative standard deviation (%RSD)) were mostly due to a large number of manual operations. In this work, we present the next stage of our research which makes feasible the semi-automation of the proposed technique. We propose a new prototype device which is applied to SFPE procedure. A part of manual operations are replaced by using prototype device; additionally, new equipment allow the use of procedures that are not possible to run using the classic chromatogram development. We compare the results of manual and semi-automatic sample preparation with SFPE and discuss the prospects for the approach mentioned.

2. Results and Discussion

In our previous articles [31,32], we proposed a new approach to biological sample preparation for instrumental analysis. The results obtained using the presented SFPE procedure were very promising.

However, the number of manual operations needed to perform the SFPE procedure seems to be the main limitation. The necessity of cutting the plates into fragments corresponding to the sample track followed by chromatogram development from two opposite sides along the short edge of the plate towards its middle considerably prolongs the time of the procedure and makes it quite tedious. Unfortunately, in manual execution of SFPE procedure, this stage has to be performed because it eliminates the influence of the radial chromatography effect (occurring during the sample application) on spreading of substances of interest in their spots in the start line and in the solvent front position. Therefore, in order to make this stage more user-friendly, a prototype semiautomatic device for delivering the eluent to the chromatographic plate has been designed. Details of this prototype device construction are presented in the Experimental section.

The new prototype allows delivering the mobile phase at a controlled velocity to the chromatographic plate by the pipette moved slightly above the adsorbent layer. The similar type of device based on the 3D printer's mechanism has been reported by our group to test separation efficiency of substances on chromatographic plate with adsorbent layer face down [33]. The equipment presented in the paper is somewhat different in construction and is especially adapted to sample preparation with SFPE. First of all, the mobile phase is delivered to the adsorbent layer face up. This makes it so that any horizontal TLC chamber can be easily adapted to perform SFPE, i.e., by production a slit in glass cover plate of the chamber. In the mentioned paper, the horizontal chromatographic chamber was equipped with a slit (a gap for pipette movement) produced in its bottom. In the case of the proposed prototype, the chromatographic plate may be placed in a standard horizontal chromatographic chamber. Additionally, the pipette is to move along three axes at various speeds, which allows delivery of the mobile phase to any position onto the chromatographic plate and development of chromatograms in any desired direction. This property of the device is very essential for SFPE, especially at the stage of substance zone narrowing, which is documented and discussed in the paper.

2.1. Preliminary Research

As was described in the previous paper [32], the accuracy and precision of the SFPE technique are based on the assumption that both sample components and the internal standard after chromatograms developments were evenly focused (preconcentrated) in the zone of the solvent front. In order to fulfil that requirement, the substance (or substances) and the internal standard should show the value of the coefficient ^nR_F (relative distance migration of a solute after the n-th development) equal to at least 0.99. Therefore, to fulfil this requirement, the number of chromatogram developments has to be determined before application of SFPE to sample preparation.

The relative migration distance of all substances increases according to number of developments when the mobile phase is delivered below the start line. This is valid if R_F (retardation factor) of substance after the first development is greater than 0. The dependence of ^nR_F vs. the number of developments using prototype device is presented in Figure 1A. The ^nR_F refers to the experiments, when each development of the chromatogram was performed by delivering the mobile phase onto the adsorbent layer of chromatographic plate by pipette moved along the path located between the lower edge of the chromatographic plate and the start line—path position Y1 (Figure 8). One can see that the ^nR_F increases with each chromatogram development for all substances except ciprofloxacin (that stays in the start line in this chromatographic system applied). Similar results were obtained by developing the chromatograms in a conventional horizontal DS chamber [32]. As was mentioned above, the prototype device gives the important advantage of developing the chromatogram from any chosen solvent entry position on the adsorbent layer. This feature was used during the experiments. The first chromatogram development was carried out as described above, however, the subsequent developments were caried out by delivering the mobile phase onto the adsorbent layer of the chromatographic plate by the pipette moved along the path parallel to the X-axis at path position Y2 (Figure 8). For such chromatogram development the relative migration distance have been denoted as $^nR_F'$. In this case $^nR_F'$ of the substances increases with each consecutive chromatogram development too. However, it is applied to the substances tramadol, acetylsalicylic acid, aminophenazone, caffeine, theophylline, acetaminophen, acetanilide, which showed zones located above the path position Y2. On the other side $^nR_F'$ of acebutol decreases with each subsequent development because its zone was below the path position Y2 of the moving pipette.

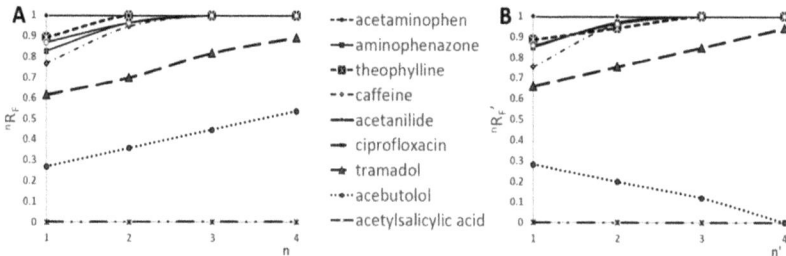

Figure 1. (**A**) ^nR_F of solutes vs. the number of developments. (**B**) $^nR_F'$ of solutes vs. the number of developments; biological matrix sample; chromatographic plate: HPTLC Silica gel 60 F254, mobile phase: methanol.

The delivering of the mobile phase to any position on the chromatographic plate is very important for SFPE technique especially in case of samples with reach matrix because it prevents elution of the interfering compounds to the solvent front zone and facilitates preconcentration of components of interest in this zone.

It also can significantly shorten the time of analysis. It is not necessary to multiply develop chromatogram from the bottom edge of the plate, but from the path position of the moving pipette located above the zones of unwanted matrix substances.

Based on the results presented in Figure 1B, the time necessary for four-fold chromatogram development, when the path position of the the mobile phase delivery onto the adsorbent layer of the chromatographic plate was Y2 (above the starting spots), had been reduced by 40% compared to the classical four-fold chromatogram development, while keeping the obtained results on similar level of accuracy.

The main goal of our research was to determine if the prototype device would have been streamlined the process of SFPE sample preparation. Therefore, we did not focus on the determination of all substances in the sample, but rather on presenting this possibility by the prototype device in this regard. Based on the results presented in Figure 1, it can be concluded that the minimum number of developments, that guarantees the values ^nR_F (or $^nR_F'$) higher or equal to 0.99 for at least six of the nine analytes, is four. Consequently, in the following experiments, all chromatograms were developed four times. It means that six components (acetylsalicylic acid, aminophenazone, caffeine, theophylline, acetaminophen, acetanilide) of nine in the test mixture are used for presentation of quantitation with semiautomatic procedure. In the next paper, we will show an optimization procedure for quantitation of all mixture components investigated in the paper.

2.2. Quantitation Result with SFPE Procedure

2.2.1. Methanolic Sample

During the SFPE procedure the samples were applied as a drops on the chromatographic plate using automatic pipette. This method of sample application is convenient because it does not require advanced equipment and can also be carried out outside the laboratory. It is also an ideal choice for samples which cannot be applied to the chromatographic plate by an automatic applicator because of their viscosity or the amount of contaminants. Unfortunately, such a method of sample application results in radial chromatography, especially when the substances of interest differ in retention. In the previous study [32], we showed that in order to obtain appropriate quantification results the effect of radial chromatography occurring at the sample application stage (Figure 2A) should be eliminated. The procedure of this effect elimination using the prototype device is presented in the Experimental section as the narrowing of the starting spots. The start spots after the narrowing procedure are presented in Figure 2B. One can see that the zones of the starting spot are considerably narrowed (except the zone of ciprofloxacin, because these substances do not migrate in the applied chromatographic system). The use of the prototype device allowed narrowing of the substance zones of many samples applied on the plate, with minimal involvement of manual operations. In the previous paper, each sample track had to be cut into a single chromatographic plate strip to manually perform the narrowing procedure.

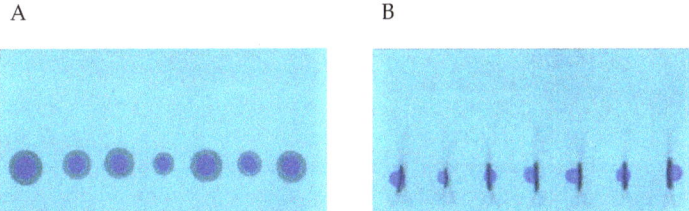

Figure 2. The mixture (methanol sample) chromatograms: (**A**) after sample application; (**B**) after spot narrowing procedure. HPTLC Silica gel 60 F254 (Merck), mobile phase: methanol.

After the narrowing procedure (S), chromatograms were developed using SPK3 mode and then the solutes were extracted from solvent front position with methanol using the CAMAG TLC–MS interface. The results of the substance determination are presented in Table 1. %RSD and the percentage difference of the results obtained by the dLCMS method and by the LCMS combined with the SFPE technique exceeds 5% for tramadol and aminophenazone only. The use of the device improved the results for the theophylline, caffeine, and acetylsalicylic acid [32]. Unfortunately, as in the previous research, there was a problem with tramadol and aminophenazone determination. Less satisfactory results obtained for aminophenazone are mostly due to the specific substance properties—probably the analyte decomposes during the analysis. This substance is "air and light sensitive" [34] what can explain the problem. The similar problem with tramadol also occurred in our previous work [32] and could be solved by changing the pH of the mobile phase [35].

Table 1. The values of relative standard deviation (%RSD) of average values ($n = 7$) of substance/acetanilide peak area ratios, the relative percentage difference between the results obtained by direct LCMS analysis and by LCMS analysis followed substances and internal standard extraction from final solvent front position by methanol with TLC–MS Interface (SFPE procedure). dLCMS and Mx are the average values of the substance/acetanilide peak area ratio obtained by direct LCMS and by LCMS analysis followed the extraction from final solvent front position by methanol with TLC–MS Interface (SFPE procedure), respectively. Chromatographic plate HPTLC Silica gel 60 F254.

	Methanolic Sample		Serum Sample	
	%RSD	100(Mx-dLCMS)/dLCMS	%RSD	100(Mx-dLCMS)/dLCMS
Tramadol	44.35	−74.9	24.65	−27.2
Acetylsalicylic acid	3.12	3.01	4.76	3.21
Aminophenazone	53.98	−51.24	24.92	−32.09
Caffeine	3.17	4.47	3.29	0.78
Theophylline	4.96	2.93	4.98	−3.15
Acetaminophen	4.52	4.03	4.98	−2.85

It is particularly noteworthy that the time of implementing the SFPE technique, especially the stage of narrowing the start zones, is considerably shortened. The narrowing time is about 10 s per sample, while in the case of manual execution of this procedure it is about 40 s (omitting the time needed to dry the plates between the two consecutive developments and cutting the plates).

The 'manual' method of the narrowing starting zone used in the our previous research prolonged the SFPE procedure not only at this but at developing chromatograms stage as well. In such a case, chromatograms of each applied sample had to be developed separately. For seven sample analyses, the development time was extended seven times in comparison with the situation when all seven samples are developed simultaneously on a single plate.

Of course, it is possible to put the cut pieces of the chromatographic plate in several chambers and develop chromatograms at the same time, but this requires careful keeping track of several solvent fronts at the same time, and thus it requires a lot of laboratory personnel involvement.

Last but not least—using prototype devices for narrowing starting spots and chromatograms development will can make the SFPE procedure fully-automatic.

2.2.2. Serum Samples

The next step was the application of prototype device to the SFPE procedure for substance quantitation in a sample with a biological matrix such as bovine serum. Figure 3A shows the spots formed after the application of serum on the chromatography plate (applied volume was 10 µL per spot). At this stage, contrary to the methanol samples, the effect of radial chromatography is less visible, what is probably suppressed by matrix components. This effect, however, occurs, which is more clearly seen at the wavelength 366 nm (Figure 3B). This allowed us to suppose that quantitation results obtained by any conventional chromatogram development, without narrowing the start zone, would not have been accurate.

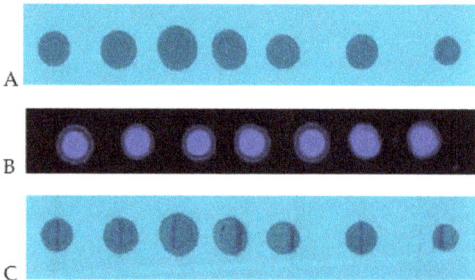

Figure 3. Mixture (serum sample) starting spots: (**A**) after sample application, λ = 254 nm; (**B**) after sample application, λ = 366 nm; (**C**) after starting spot narrowing, λ = 254 nm. HPTLC Silica gel 60 F254 (Merck), mobile phase: methanol.

In order to eliminate the effect of radial chromatography, the prototype device was used for narrowing the starting spot (as in case of the methanol sample discussed above). The spots after the narrowing procedure are presented in Figure 3C. One can see that the substances are gathered in a narrow zone. Consequently, after the chromatogram development, the substances are concentrated in a small area (Figure 4D), which could be almost completely covered by the head of TLC–MS interface.

The chromatograms were developed using the SPK3 mode. The obtained chromatograms are presented in Figure 5. The percentage difference of the results obtained by the dLCMS method and by the LCMS combined with the SFPE procedure are lower than 5%. However, this value is considerably exceeded for tramadol and aminophenazone only (Table 1). The explanation for this phenomena is discussed above.

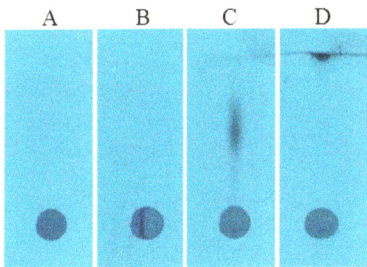

Figure 4. The mixture (serum sample) chromatograms: (**A**) after sample application; (**B**) after starting spot narrowing; (**C**) after first development; (**D**) after fourth development. Stationary phase: HPTLC Silica gel 60 F254. (Merck), mobile phase: methanol.

It should be underlined that the determination results are far better in comparison to our previous work [32]. The %RSD decreased below 5% for caffeine, theophylline and acetylsalicylic acid. It is worth to stress that the number of chromatogram developments in this work is four, while in the previous work it was 6. Using the prototype machine we managed to get a %RSD value below 5 for four substances, while in the previous work only for one (acetaminophen).

The advantages of the prototype device are analogous to those discussed in the case of methanol samples. Additionally, the possibility of delivering mobile phase to any place on the chromatographic plate allows chromatogram development above the application zone. It is very important in the case of samples with reach matrix, when the repeated wetting of matrix zones leads to the elution of the interfering compounds to the solvent front zone.

Moreover, it should be mentioned that the 'edge' effect (encountered in the classical chromatogram developments) did not occur when the prototype device was used at controlled velocity of the mobile phase. This is a very desirable feature, especially if the SFPE technique will be fully automated.

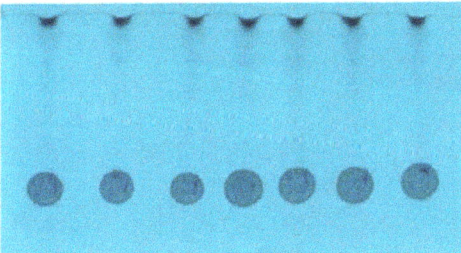

Figure 5. The serum sample chromatograms after development using SPK3 procedure; stationary phase. HPTLC Silica gel 60 F254 (Merck), mobile phase: methanol.

3. Materials and Method

3.1. Materials and Reagents

Chromatographic plates, HPTLC Silica gel 60 F254, 10 × 20 cm, were supplied by Merck (Darmstadt, Germany). Methanol, water and formic acid LC–MS grade were purchased from Merck (Darmstadt, Germany). Acetaminophen, acetanilide, aminophenazone, caffeine, acetylsalicylic acid, and theophylline were purchased from Sigma–Aldrich (St. Louis, MO, USA). Acetobutolol was purchased from Biomedicals, USA (Santa Ana, CA, USA), ciprofloxacin from Sreepathi Pharmaceutical Ldt. India (Telangana, India), Tramadol from Inogent Laboratories, India (Telengana). Bovine serum was purchased from Biomed (Lublin, Poland).

3.2. Preparation of Internal Standard and Analyte Solutions

The stock solutions were prepared by dissolving proper amounts of each substance in methanol and stored in a refrigerator at 4 °C. Solutions of the substances mentioned were prepared by adding 10 µL of each stock solution to 1 mL of serum and methanol and finally were in the range of substances concentrations typically found in real blood/serum samples (Table 2).

Table 2. Final concentration of substances in sample.

Substance	mg/L
Acebutolol	20.6
Ciprofloxacin	8.0
Tramadol	15.8
Acetylsalicylic acid	141.0
Aminophenazone	15.2
Caffeine	20.2
Theophylline	20.3
Acetaminophen	15.1
Acetanilide	15.2

3.3. HPTLC Plate Preparation

HPTLC Silica gel 60 F25410 × 20 cm plates were cut into desired sizes using TLC plate cutter (CAMAG, Wilmington, NC, USA). Before chromatogram development, the plates were washed by immersion in methanol for 1 min. Afterward the plates were dried in the air and activated in an oven at 105–110 °C for 15 min.

3.4. Application of the Samples on the HPTLC Plate

The samples were applied as small single drops on the chromatographic plate using an automatic pipette (PZ HTL S.A., Warszawa, Poland). The volume of drop was about 10 µL.

3.5. Prototype Device for Semiautomatic Sample Preparation Using SFPE Procedure

The experiments were performed using a new prototype device presented in Figure 6. The chromatographic plate (1) was placed in the horizontal Teflon chamber (2) with the adsorbent layer face-up (3). Then, the mobile phase was provided onto the adsorbent layer of the chromatographic plate by the pipette (4), which was moved, according to a programmed movement path, by 3D printer mechanism (6) controlled by computer (8). The tip of pipette was in close contiguity to the adsorbent layer (0.15 mm), not touching it. The pipette was equipped with the capillary (5) and combined with syringe pumps (7) (TYD02-01 Laboratory Syringe Pump, Leadfluid, Baoding, Heibei, China).

The pipette is able to move in three axes at a variety of speeds, which can deliver the mobile phase to any place on the chromatographic plate and develop chromatograms in any desired direction. When the chromatograms were developed, the chromatographic plate was covered with a glass plate (3).

3.6. Narrowing Procedure of the Sample Spots Applied on the Start Line

Single drops of sample solutions were spotted on the start line of the chromatographic plate. The plate was left in the air for 10 min to dry the sample spots. Then the starting sample spots were subjected to narrowing procedure, i.e., elimination of radial chromatography effect generated during sample application. During this procedure, the mobile phase solution was delivered to the chromatographic plate with the pipette moved by the 3D machine along eight paths each perpendicular to the X-axis and located between starting spots (see Figure 7). The length of these paths was determined by Y1 = 5 mm and Y2 = 25 mm, measured from the lower edge of the chromatographic plate, which were the turning points of the pipette's movement.

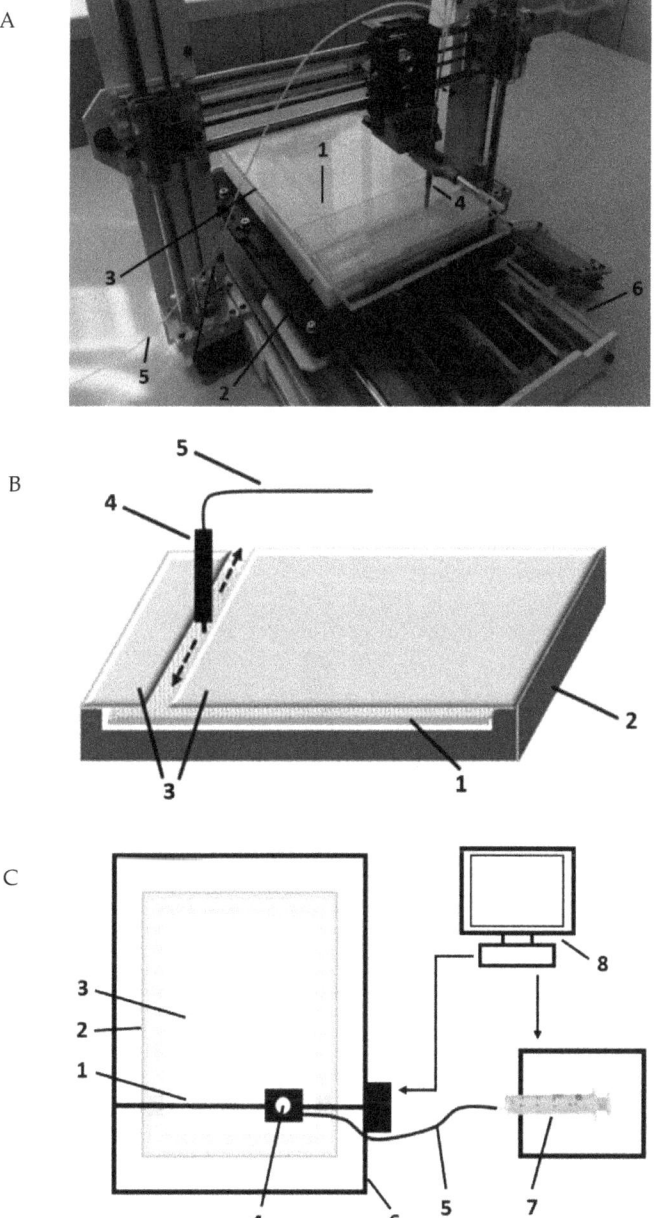

Figure 6. (**A**) The picture of the prototype device. (**B**) Axonometric view of the horizontal chamber with moving pipette. (**C**) Conceptual view of the whole device, 1—chromatographic plate, 2—Teflon horizontal TLC chamber, 3—cover glass plate, 4—moving pipette, 5—capillary, 6—3D machine, 7—syringe pump, 8—computer.

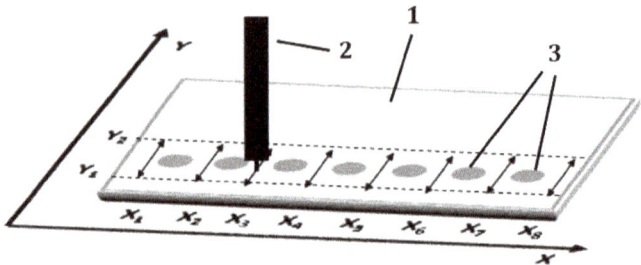

Figure 7. Scheme of the moving pipette paths represented by double-sided arrows for the narrowing procedure of the sample spots applied on the start line. 1—chromatography plate; 2—moving pipette; 3—sample spots applied on the chromatographic plate.

In each of the eight paths ($X_1 \ldots X_8$), the pipette moved eight cycles of "Y1-Y2-Y1". The speed of the pipette movement was 2000 mm/min and velocity of the mobile phase delivery to the adsorbent layer was 5 mL/h. When the procedure above for the all paths was completed the plate was dried in the air for 10 min.

3.7. Planar Chromatogram Development

In each development variant, the pipette tip was moved over the entire width of the chromatography plate (back and forth, parallel to the X-axis and perpendicular to the Y-axis) at constant speed of 2000 mm/min (Figure 8). The distance of the pipette tip from the surface of the adsorbent layer was 0.15mm. The speed of the mobile phase solution delivery to the adsorbent layer was equal to 5 mL/h.

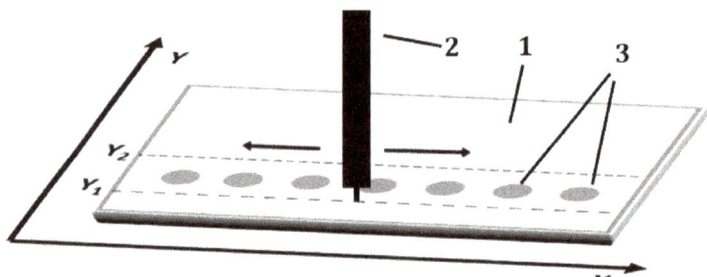

Figure 8. Scheme of the paths represented by the dashed lines for delivering the mobile phase solution to the adsorbent layer with moving pipette driven by 3D machine. 1—chromatography plate; 2—moving pipette; 3—sample spots applied on the chromatographic plate.

Variants of Sample Preparation for Quantitation

- SPK3 variant

Single drops of sample solutions were spotted 15 mm apart from each other on the start line of the chromatographic plates. The start line was located at the distance 15 mm from the chromatographic plate edge. After drying of the spots obtained (10 min) narrowing procedure described above was performed. Then the chromatograms were developed one time in horizontal Teflon chamber using pipette parallel moved to the X-axis at Y = 10 mm from the lower edge of the chromatographic plate. Distance of chromatogram development, measured from start line, was equal to 35mm. Next the chromatograms were developed three times in the chamber applying pipette parallel moved to the X-axis at Y = 25 mm from the edge of the chromatographic plate. In this case, the distance of movement of the mobile phase front was equal to 20 mm. After each development the plates were dried in the air for 10 min. Then the plate was subjected to extraction of the components of interest from the final position of the mobile phase front. The extraction was performed by using the TLC–MS interface from CAMAG, extrahent: methanol. The obtained sample solutions were injected (10 µL) into HPLC column of Agilent Zorbax Eclipse Plus C18 (Santa Clara, CA, USA) for their quantitation.

- dLCMS variant

Substances were determined by LC–MS/MS (Liquid Chromatography Tandem-Mass Spectrometry, Agilent, Santa Clara, CA, USA) technique without SFPE preparation procedure—i.e., the standard solution was directly injected to HPLC column of the Agilent 1290 chromatograph.

3.8. Instrumentation

Thin-layer chromatography experiments were performed using a prototype of 3D machine with controlled eluent flow (Lublin, Boland, Infinum 3D) equipped with the horizontal Teflon chamber for TLC. For plate image documentation the TLC Visualizer, CAMAG (Muttenz, Switzerland), was used. The CAMAG TLC–MS interface, connected with Agilent 1260 Infinity isocratic pump, was used for extraction of the substances from the chromatographic plates. The Agilent1290 Infinity LC System (Santa Clara, CA, USA) connected with Agilent 6460 Triple Quadrupole was used for the LC–MS experiments. The chromatography was performed with the Zorbax Eclipse Plus-C18 column (4.6 × 100 mm, 3.5 µm, Agilent, Santa Clara, CA, USA). Mobile phase: A: 0.1% formic acid in water B: 0.1% formic acid in methanol. The gradient program was as follows: 0 min: 95% A, 5% B; 2 min: 75% A, 25% B; 9 min: 5% A, 95% B; 10min: 95% A, 5% B; 11 min: 95% A, 5% B. Data were acquired in the positive- and negative-ion modes (multiple-reaction monitoring mode) with electrospray probe voltages of 3500 V. The nebulizer gas setting was 15 psi. The ion source was operated at a temperature of 350 °C and a drying gas setting of 7 L/min.

4. Conclusions

In this paper, it has been shown that the application of the prototype device to SFPE procedure leads to increased repeatability of results and significant reduction of analysis time in comparison to the classical procedure of chromatogram development.

The pipette of the prototype device is able to move along three axes at different speeds, which allows delivery of the mobile phase to any position on the chromatographic plate and development of the chromatograms in any direction. This advantage can narrow the substance start zones for many samples on one chromatographic plate without the necessity of cutting the plate into single track strips.

This device enables the mobile phase delivery to any position on the chromatographic plate. This feature is very advantageous for samples with a reach matrix, because it prevents interfering compounds from reaching the solvent front zone during SFPE.

The problem with determination of some substances in the paper could be solved by changing the pH of the mobile phase. This is preliminarily confirmed by our nonpublished investigations. The investigations are under development by our group and we hope to submit such results for publication in the near feature.

The presented results undoubtedly constitute a step in the process of full automation of the sample preparation by SFPE technique.

Author Contributions: Conceptualization, A.K.-T. and T.H.D.; methodology, A.K.-T. and T.H.D.; validation, A.K.-T.; formal analysis, A.K.-T.; investigation, A.K.-T., K.J. and M.I.; data curation, A.K.-T.; writing—original draft preparation, A.K.-T.; writing—review and editing, T.H.D.; visualization, A.K.-T.; supervision, A.K.-T. All authors have read and approved the final version.

Funding: The article was developed using the equipment purchased within the agreement no. POPW.01.03.00-06-010/09-00 Operational Program Development of Eastern Poland 2007–2013, Priority Axis I, Modern Economy, Operations 1.3, Innovations Promotion.

Conflicts of Interest: The authors declare no conflict of interest. The funders had no role in the design of the study; in the collection, analyses, or interpretation of data; in the writing of the manuscript, or in the decision to publish the results.

References

1. Agilent Technologies, Sample Preparation Fundamentals for Chromatography. Available online: https://www.agilent.com/cs/library/primers/public/5991-3326EN_SPHB.pdf (accessed on 13 November 2013).
2. Nováková, L. Advances in Sample Preparation for Biological Fluids. *LCGC* **2016**, *29*, 9–15.
3. LCGC Editors. Overview of Sample Preparation. *LCGC* **2015**, *33*, 46–52.
4. Raynie, D.E. Trends in Sample Preparation. *LCGC N. Am.* **2016**, *34*, 174–188.
5. Brinkman, U.A.; Durig, J.R.; Van Espen, P. *TRAC: Trends in Analytical Chemistry*; Elsevier: Amsterdam, The Netherlands, 1989.
6. Peng, J.; Tang, F.; Zhou, R.; Xie, X.; Li, S.; Xie, F.; Yu, P.; Mu, L. New techniques of on-line biological sample processing and their application in the field of biopharmaceutical analysis. *Acta Pharm. Sin. B* **2016**, *6*, 540–551. [CrossRef]
7. Chen, L.; Wang, H.; Zeng, Q.; Xu, Y.; Sun, L.; Xu, H.; Ding, L. On-line Coupling of Solid-Phase Extraction to Liquid Chromatography—A Review. *J. Chromatogr. Sci.* **2009**, *47*, 614–623. [CrossRef]
8. Vas, G.; Vekey, K. Solid-phase microextraction: A powerful sample preparation tool prior to mass spectrometric analysis. *J. Mass Spectrom.* **2004**, *39*, 233–254. [CrossRef] [PubMed]
9. Hinshaw, J. Solid-Phase Microextraction. *LCGC N. Am.* **2015**, *30*, 904–910.
10. Souza-Silva, É.A.; Gionfriddo, E.; Pawliszyn, J. A critical review of the state of the art of solid-phase microextraction of complex matrices II. Food analysis. *TrAC Trends Anal. Chem.* **2015**, *71*, 236–248. [CrossRef]
11. Abdel-Rehim, M. New trend in sample preparation: On-line microextraction in packed syringe for liquid and gas chromatography applications: I. Determination of local anaesthetics in human plasma samples using gas chromatography–mass spectrometry. *J. Chromatogr. B Anal. Technol. Biomed. Life Sci.* **2004**, *801*, 317–321. [CrossRef]
12. Yanga, L.; Saidb, R.; Abdel-Rehim, M. Sorbent, device, matrix and application in microextraction by packed sorbent (MEPS): A review. *J Chromatogr. B* **2017**, *1043*, 33–43. [CrossRef] [PubMed]
13. Bicchi, C.; Liberto, E.; Cordero, C.; Sgorbini, B.; Rubiolo, P. Stir Bar Sorptive Extraction (SBSE) and Headspace Sorptive Extraction (HSSE): An Overview. *LCGC N. Am.* **2009**, *27*, 376–390.
14. Sarafraz-Yazdi, A.; Amiri, A. Liquid-phase microextraction. *Trends Anal. Chem.* **2010**, *29*, 1–14. [CrossRef]
15. Jones, B.A.; Ezzell, J.L. Accelerated solvent extraction: A technique for sample preparation. *Anal. Chem.* **1996**, *68*, 1033–1039.
16. Nandi, P.; Lunte, S.M. Microdialysis Sampling as a Sample Preparation Method. In *Handbook of Sample Preparation*; Pawliszyn, J., Lord, H.L., Eds.; John Wiley & Sons, Inc: Hoboken, NJ, USA, 2010.
17. Destandau, E.; Michel, T.; Elfakir, C. *Microwave-Assisted Extraction*; The Royal Society of Chemistry Cambridge: Cambridge, UK, 2013; pp. 113–115.
18. Wrona, O.; Rafińska, K.; Możeński, C.; Buszewski, B. Supercritical Fluid Extraction of Bioactive Compounds from Plant Materials. *J AOAC Int.* **2017**, *100*, 1624–1635. [CrossRef] [PubMed]
19. Płotka-Wasylka, J.; Szczepanska, N.; de la Guardia, M.; Namiesnik, J. Modern trends in solid phase extraction: New sorbent media. *TrAC Trends Anal. Chem.* **2016**, *77*, 23–43. [CrossRef]
20. Marinova, M.; Artusi, C.; Brugnoloa, L.; Antonelli, G.; Zaninottoa, M.; Plebani, M. Immunosuppressant therapeutic drug monitoring by LC-MS/MS: Workflow optimization through automated processing of whole blood samples. *Clin. Biochem.* **2013**, *46*, 1723–1727. [CrossRef]
21. Motoyama, A.; Suzuki, A.; Shirota, O.; Namba, R. Direct determination of bisphenol A and nonylphenol in river water by column-switching semi-microcolumn liquid chromatography/electrospray mass spectrometry. *Rapid Commun. Mass Spectrom.* **1999**, *13*, 2204–2208. [CrossRef]
22. Ligang, C.; Hui, W.; Qinglei, Z.; Yang, X.; Lei, S.; Haoyan, X.; Lan, D. On-line Coupling of Solid Phase extraction to liquid chromatography. *J. Chrom. Sci.* **2009**, *47*, 614–623.
23. Zhang, Z.; Yang, M.J.; Pawliszyn, J. Solid-Phase Microextraction. A Solvent-Free Alternative for Sample Preparation. *Anal. Chem.* **2008**, *66*, 844A–853A. [CrossRef]

24. Moein, M.M.; Abdel-Rehim, A.; Abdel-Rehim, M. Microextraction by packed sorbent (MEPS). *TrAC Trends Anal. Chem.* **2015**, *67*, 34–44. [CrossRef]
25. Abdel-Rehim, M.; Altun, Z.; Blomberg, L. Microextraction in packed syringe (MEPS) for liquid and gas chromatographic applications. Part II-Determination of ropivacaine and its metabolites in human plasma samples using MEPS with liquid chromatography/tandem mass spectrometry. *J. Mass Spectrom.* **2004**, *39*, 1488–1493. [CrossRef]
26. Guthrie, R.; Suzi, A. A Simple phenylalanine metod for detecting phenylketonuria In large populations of newborn infants. *Pediatrics* **1963**, *32*, 338–343.
27. Koal, T.; Burhenne, H.; Romling, R. Quantification of antiretroviral drugs In dried blood samples by means of liquaid chromatography/tandem mass spectrometry. *Rapid Commun Mass Spectrom.* **2005**, *19*, 2995–3001. [CrossRef]
28. CAMAG DBS-MS 500-Fully Automated Online Extraction System for LC-MS, MS or Sample Collector Coupling. Available online: https://www.labmate-online.com/news/mass-spectrometry-and-spectroscopy/41/camag-ag/camag-dbs-ms-500-nbspfully-automated-online-extraction-system-for-lc-ms-ms-or-sample-collector-coupling/39570 (accessed on 11 July 2016).
29. Oellig, C.; Schwack, W. Planar solid phase extraction—A new clean-up concept in multi-residue analysis of pesticides by liquid chromatography-mass spectrometry. *J. Chromatogr. A* **2011**, *1218*, 6540–6547. [CrossRef] [PubMed]
30. Oellig, C.; Schwack, W. Planar solid phase extraction clean-up for pesticide residue analysis in tea by liquid chromatography-mass spectrometry. *J. Chromatogr. A* **2012**, *1260*, 42–53. [CrossRef] [PubMed]
31. Klimek-Turek, A.; Sikora, M.; Rybicki, M.; Dzido, T.H. Frontally eluted components procedure with thin layer chromatography as a mode of sample preparation for high performance liquid chromatography quantitation of acetaminophen in biological matrix. *J. Chromatogr. A* **2016**, *1436*, 19–27. [CrossRef] [PubMed]
32. Klimek-Turek, A.; Sikora, E.; Dzido, T.H. Solvent front position extraction procedure with thin-layer chromatography as a mode of multicomponent sample preparation for quantitative analysis by instrumental technique. *J. Chromatogr. A* **2017**, *1530*, 204–210. [CrossRef] [PubMed]
33. Hałka-Grysińska, A.; Skop, K.; Klimek-Turek, A.; Gorzkowska, M.; Dzido, T.H. Thin-layer chromatogram development with a moving pipette delivering the mobile phase onto the surface of the adsorbent layer. *J. Chromatogr. A* **2018**, *1575*, 91–99. [CrossRef] [PubMed]
34. Chemical Book. Available online: https://www.chemicalbook.com/ProductChemicalPropertiesCB1476086_EN.htm (accessed on 11 March 2019).
35. Chwalczuk, A. Application of gradient elution to substance determination using combination of High Performance Liquid Chromatography and Thin Layer Chromatography techniques (original title: Zastosowanie elucji gradientowej do oznaczania substancji z wykorzystaniem łączenia technik wysokosprawnej chromatografii cieczowej i chromatografii planarnej). Master's Thesis, Medical University of Lublin, Lublin, Poland, June 2018.

Sample Availability: Not available.

© 2019 by the authors. Licensee MDPI, Basel, Switzerland. This article is an open access article distributed under the terms and conditions of the Creative Commons Attribution (CC BY) license (http://creativecommons.org/licenses/by/4.0/).

Article

High-Frequency Heating Extraction Method for Sensitive Drug Analysis in Human Nails

Fumiki Takahashi [1,*], Masaru Kobayashi [2], Atsushi Kobayashi [2], Kanya Kobayashi [3,*] and Hideki Asamura [3]

1. Department of Chemistry, Faculty of Science, Shinshu University, 3-1-1 Asahi, Nagano 390-8621, Japan
2. Research Institute of Scientific Criminal Investigation, Nagano Prefectural Police Headquarters, 3916 Nishijo, Matsushiro, Nagano 381-1232, Japan; fumi.fumi.fufumi.0001@gmail.com (M.K.); fumi.fumi.fufumi.0002@gmail.com (A.K.)
3. Department of Legal Medicine, Shinshu University School of Medicine, 3-1-1 Asahi, Matsumoto, Nagano 390-8621, Japan; asamura@shinshu-u.ac.jp
* Correspondence: takahashi@shinshu-u.ac.jp (F.T.); kanya_k@shinshu-u.ac.jp (K.K.); Tel.: +81-263-37-2474 (F.T.)

Academic Editor: Nuno Neng
Received: 27 November 2018; Accepted: 6 December 2018; Published: 7 December 2018

Abstract: *Background:* A simple, sensitive, and rapid extraction method based on high-frequency (H-F) heating was developed for drug analysis in human nails. *Methods:* A human nail was placed in a glass tube with an extraction solvent (methanol and 0.1% formic acid; 7:3, v/v), and a ferromagnetic alloy (pyrofoil) was wrapped in a spiral around the glass tube. Then, the glass tube was placed in a Curie point pyrolyzer, and a H-F alternating voltage (600 kHz) was applied. The sample and extraction solvent were heated at the Curie temperature for 3 min. Different Curie temperatures were applied by changing the pyrofoil (160 °C, 170 °C, 220 °C, and 255 °C). *Results:* The caffeine in the nail was effectively and rapidly extracted into the extraction solvent with the pyrofoil at 220 °C. The peak area obtained for the caffeine using liquid chromatography mass spectrometry (LC-MS/MS) was five times that of what was obtained after conventional ultrasonic irradiation extraction. Because the extraction uses high-pressure and high-temperature conditions in a test tube, the drugs that were strongly incorporated in nails could be extracted into the solvent. The amount of caffeine extracted was independent of the size of the pieces in the sample. *Conclusions:* Therefore, the sensitive determination of target drugs in nails is possible with rapid (20 min, including H-F extraction for 3 min) and simple sample preparation. The developed method was applied to a nail from a patient with hypertension.

Keywords: nail; curie temperature; high-frequency heating; liquid chromatography–tandem mass spectrometry; caffeine; amlodipine

1. Introduction

Blood, urine, and gastric contents are commonly used for drug analysis in forensic and clinical toxicology, and allow for a relatively easy estimation of toxicological risk [1–4]. However, when these samples are not available, such as after the putrefaction and decomposition of the body, hair and nails can be used as alternative samples for forensic and biomedical analysis [5,6]. These alternative specimens provide a long surveillance window for exposure, because drugs incorporated into the hair and nails have a high stability at room temperature [7,8]. Therefore, many applications of hair samples for drug detection have been reported for drugs of abuse, such as methamphetamine [9–11], opiates [12], and cannabis [13], as well as therapeutic drugs, such as benzodiazepines and barbiturates [7,14]. However, on the other hand, there has been little investigation of nail samples for drug analysis. Because nails are harder than hair, the effective extraction of drugs from nails is very difficult. Some reports on drug analysis in nails have used alkaline or enzymatic hydrolysis extraction, but these methods have long extraction times and result in sample destruction [15,16]. Moreover, an effective drug extraction method is required in order to determine the concentrations of drugs in nails at trace levels [6,17,18].

High-frequency (H-F) induction heating is a pyrolysis technique that uses a ferromagnetic alloy and a H-F induction coil [19–21]. The ferromagnetic alloy is placed in an electromagnetic field with a H-F generated using a pyrolyzer, and the alloy is heated to its Curie temperature. The Curie temperature can be reached in as little as a few hundredths of a second. Hence, the combination of H-F heating pyrolysis with GC-MS has been accepted as a powerful tool for the analysis of rubber, plastic, and resin [20,22]. Recently, Kurihara et al. reported the analysis of additives in polymers using H-F heating extraction [23]. Polymers were placed in a glass tube with glass wool, and the tube was wrapped with a ferromagnetic alloy. Subsequently, a H-F alternating voltage was applied to the glass tube, and the additives in the polymer were rapidly extracted onto the glass wool. The sensitive detection of the extracted additives was performed by gas chromatography (GC)-MS. Thermal extraction with H-F heating was applied to the fatty acids [24], antioxidants [25], and polymer resins [26,27].

In our previous work, H-F extraction was applied to drug analysis in nails [28]. Human nails were placed in a glass tube with a solvent, and were heated as described above. A highly effective solvent extraction from the nails was achieved, and this was combined with liquid chromatography mass spectrometry (LC-MS/MS). This approach is simple, rapid, and does not need a sample decomposition such as alkalization. The proposed method has high sensitivity compared with the conventional non-destructive method of ultrasonic irradiation extraction, and the caffeine, amlodipine, probucol, cilostazol, and minoxidil in nails were extracted to the solvent effectively [28]. However, the extraction processes have not been examined quantitatively, including the morphological approach. Because information regarding the morphology of the nail undergoing the H-F heating process is one of the essential solid–liquid solvent extraction process, this knowledge provides the optimization for drug analysis in nails. In addition, the incorporation of drugs to nails under metabolic processes, especially the effect of sample site (finger- and toe-nail), has not been characterized.

In this study, the H-F heating extraction method was optimized using caffeine as a representative drug. Caffeine is an ingredient in stimulant preparations and is often used as an adulterant in drugs of abuse [6]. The effect of the H-F heating process was evaluated using morphological and metabolomic information, the proposed extraction technique had suitability for drug analysis in human nails, because the incorporated property of the drugs is highly bound to the nail. The optimized method was applied to the sensitive determination and characterization of amlodipine in toenails and fingernails from a patient with hypertension.

2. Results and Discussion

2.1. Features of H-F Heating Extraction

In the initial experiments, we applied the H-F heating extraction method to caffeine. A selected ion monitoring (SIM) chromatogram (Figure 1) was obtained of a sample extracted into 200 µL of solvent from 10 mg of nails.

The H-F heating extraction time was fixed at 3 min, and the Curie point temperature was 220 °C (Figure 1A). For the validation of the proposed method, a conventional ultrasonic extraction with indirect irradiation was carried out for 30 min (Figure 1B). The peak at 3.5 min in the SIM chromatogram was assigned to caffeine. The H-F heating method showed a much better sensitivity for caffeine than the ultrasonic extraction method, and the peak area of the caffeine in the SIM chromatogram was more than five times that obtained using the conventional extraction. The incorporated property of the drugs is highly bound to nails, and the effective extraction of drugs was suggested from nails.

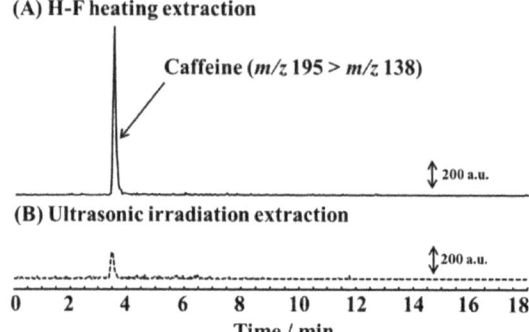

Figure 1. Selected reaction monitoring (SRM) chromatogram for caffeine in nails with (**A**) high-frequency (H-F) heating extraction (3 min; 220 °C) and (**B**) ultrasonic irradiation extraction (30 min). The sample mass was 10 mg and the extraction solvent volume was 200 µL.

To evaluate the H-F heating process, the surface profiles of the nails under extraction were evaluated using SEM.

The nail before extraction was constructed of fine plates that were arranged in layers on the nail surface (Figure 2A). This structure was maintained even after ultrasonic irradiation extraction for 30 min (Figure 2B). This illustrates that ultrasonic irradiation has almost no effect on the nail surface structure, which could explain why it is difficult to effectively extract drugs from nails using the conventional extraction method. However, after H-F heating extraction (220 °C, 3 min), holes appeared on the nail surface (Figure 2C). This shows that the nail structure was affected by the high-temperature and high-pressure conditions generated by the H-F heating in the glass tube. This suggests that the effective extraction of caffeine using H-F heating is related to the increased solid–liquid surface area between the nails and the extraction solvent. However, the fundamental shape of the nail was retained, and non-destructive extraction with a high sensitivity could be achieved using the H-F heating method.

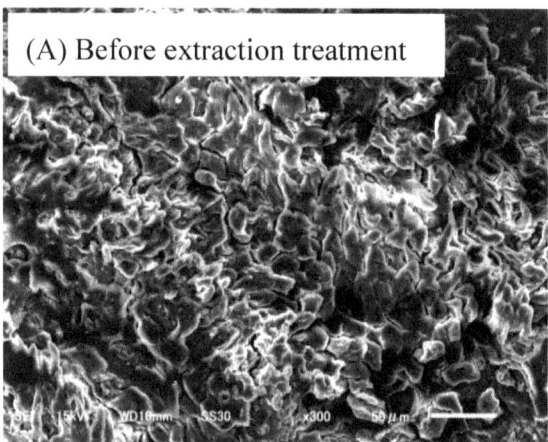

Figure 2. *Cont.*

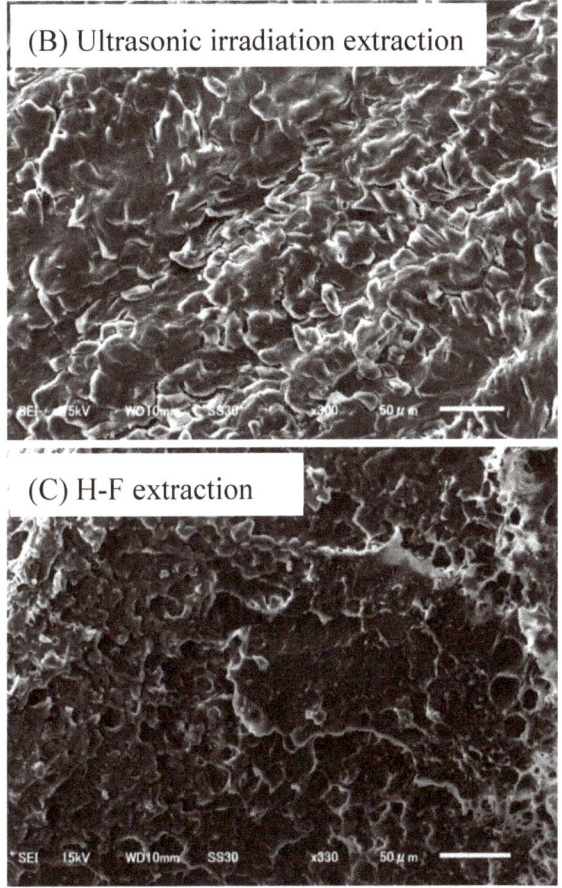

Figure 2. SEM images of the nail surface (**A**) before the extraction treatment, (**B**) after ultrasonic irradiation extraction (30 min), and (**C**) after H-F heating extraction (3 min).

2.2. Optimization of H-F Heating Extraction for Nail Analysis

To optimize the H-F heating extraction, the dependence of the peak area of the caffeine on the extraction conditions was evaluated. The nail mass was 10 mg and the extraction solvent volume was 200 µL.

When we plotted the caffeine peak area as a function of the extraction time for the H-F heating method (Figure 3), we found the caffeine peak area increased with the increase in the extraction time at 170 °C, but the signal was relatively low. By contrast, at 220 °C, the caffeine contained in the nails was rapidly extracted into the solvent within a short time. At this temperature, the largest peak area was obtained with a 3-min extraction time. The amount of caffeine extracted decreased slightly as the extraction time was increased to over 5 min. If the extraction period is too short, the drugs in the nail will not be completely extracted into the solvent. The optimum time for the H-F heating extraction was 3 min.

It is well known that nails contain keratinized proteins in fine plates [6,29], but the chemical and physical properties of nails have not been completely characterized. Fujii et al. fabricated novel keratin fibers and films from actual human hair and nails, and evaluated their chemical stability under thermal treatment [16,30,31]. The chemical structure of the nail was denatured at temperatures of over

170 °C [30]. In the present study, morphological changes occurred and small holes were generated at the nail surface, which enhanced the extraction of the drug into the solvent at 220 °C under H-F heating. On the other hand, temperatures that were too high (255 °C) decreased the caffeine signal (Figure S1), because the thermal decomposition of cthe affeine was reported at 237 °C using differential scanning calorimetric (DSC) analysis [32]. In addition, the fundamental nail structure was destroyed and the solvent became cloudy as the nail dissolved at 255 °C. Caffeine may be incorporated into small particles of the nail rather than extracted into the solvent, and the caffeine signal will decrease.

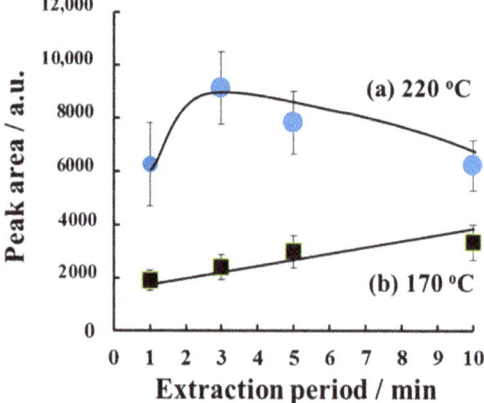

Figure 3. Dependence of the caffeine peak area on the extraction time for extraction at 220 °C ((a) closed circles) and 170 °C ((b) closed squares), respectively. The sample mass was 10 mg and the extraction solvent volume was 200 μL. Error bars represent the standard deviation ($n = 3$).

Next, the dependence of the caffeine peak area on the ratio of methanol (MeOH) in the extraction solvent was evaluated (Figure 4).

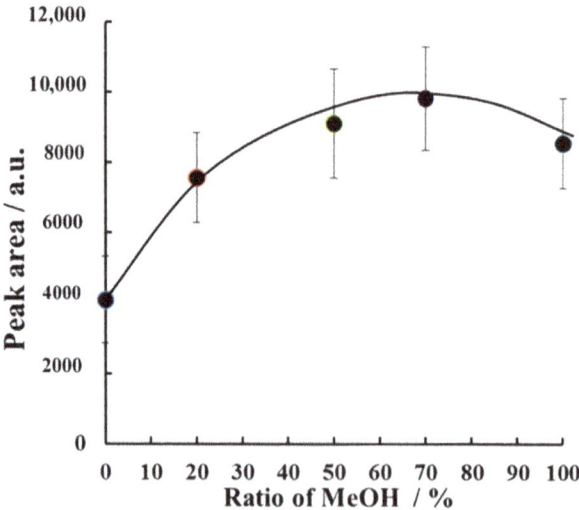

Figure 4. Dependence of the caffeine peak area on the ratio of MeOH in the extraction solvent for H-F heating extraction. The sample mass was 10 mg and the H-F heating extraction was carried out for 3 min at 220 °C. Error bars represent the standard deviation ($n = 3$).

When we used only 0.1% formic acid with no MeOH, a relatively small amount of caffeine was extracted into the solvent. The caffeine peak area increased as the proportion of MeOH increased to 70%. The highest amount of caffeine was extracted with 70% MeOH. When 100% MeOH was used as the extraction solvent, the extraction procedure was relatively difficult, because of the risk of rupture of the glass tube under H-F heating. Therefore, methanol-0.1% formic acid (7:3, v/v) was selected as the extraction solvent.

The dependence of the caffeine peak area on the size of the pieces in the nail sample was investigated by cutting (ca. 0.5 mm) and shredding (ca. 0.1 mm) the samples (Figure 5).

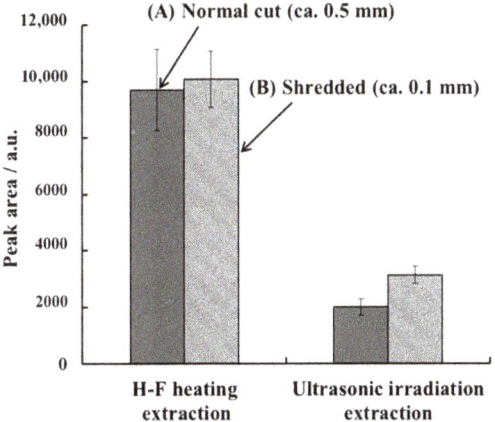

Figure 5. Effect of the size of pieces in the sample (((A) fill pattern = dots) cut to 0.5 mm, ((B) fill pattern = diagonal lines) shredded to 0.1 mm) on the caffeine peak area after H-F heating extraction and ultrasonic irradiation extraction. The sample mass was 10 mg and the extraction solution volume were 200 µL. The other parameters were the same as in Figure 4. Error bars represent the standard deviation (n = 3).

With the ultrasonic irradiation extraction method, the amount of caffeine extracted from the shredded sample was 50% more than that obtained from the cut sample. The surface area of the sample increased after shredding, which could improve the efficiency of the solid–liquid extraction of caffeine into the solvent. A sample pretreatment to increase the surface area is required for the sensitive determination of caffeine using the ultrasonic extraction method [33]. By contrast, in the H-F heating extraction method, no significant changes were observed between the cut and shredded samples for the amount of caffeine extracted. The caffeine incorporated in the nail was effectively extracted into the solvent using H-F heating under high-temperature and high-pressure conditions for both types of samples. Therefore, sample shredding may not be essential for H-F heating extraction. These results show that the H-F heating extraction method can be performed using a simple sample preparation method without any decrease in the sensitivity.

2.3. Application of H-F Heating Extraction to Nails from a Hypertension Patient

The H-F heating extraction was applied for the detection of amlodipine in nails from a hypertension patient. Caffeine and amlodipine were extracted using H-F heating and ultrasonic irradiation (Figure 6).

In the H-F heating extraction results, the peak areas for both caffeine and amlodipine were higher than those obtained with the ultrasonic irradiation method. This result shows that multiple drugs can be effectively extracted into the solvent using H-F heating.

The peak areas of the caffeine and amlodipine from the toenails and fingernails were determined after H-F heating extraction (Figure 7).

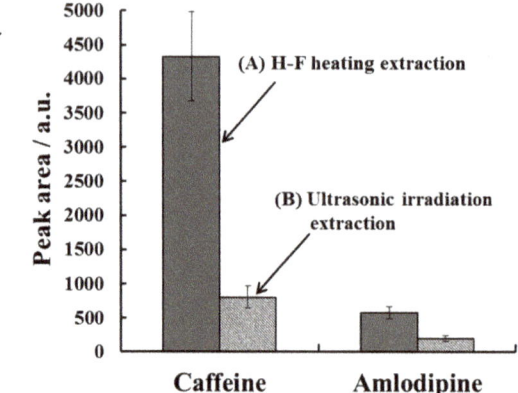

Figure 6. The extraction profiles of caffeine and amlodipine for nails from a hypertension patient. (A) H-F heating extraction (3 min; 220 °C) and (B) ultrasonic irradiation extraction (30 min). The other parameters were the same as in Figure 5. Error bars represent the standard deviation ($n = 3$).

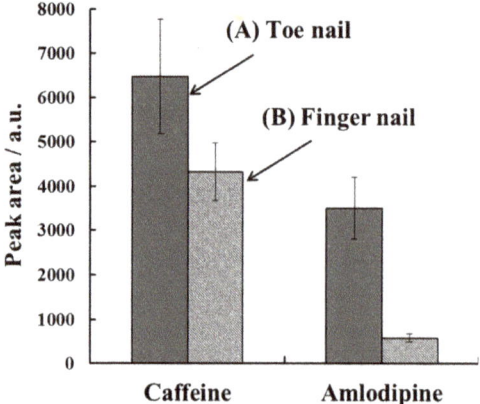

Figure 7. Comparison of the concentrations of caffeine and amlodipine in (A) toenails and (B) fingernails from a hypertension patient. The H-F heating extraction conditions were the same as in Figure 6A. Error bars represent the standard deviation ($n = 3$).

Both the caffeine and amlodipine were more concentrated in the toenails than in the fingernails. Drugs are deposited in the nails from blood or sweat during the germinal matrix generation of nails [29,33]. The growth rates of the fingernails and toenails are approximately 3–5 mm/month and approximately 1 mm/month, respectively [6,29]. Consequently, for samples of the same mass, the toenails will have been exposed to any drugs in the system for longer than fingernails. This correlates with the larger peak signals measured for the toenail samples. However, the concentration ratio for caffeine between the toenails and fingernails was higher than that for amlodipine. It is likely that the external contamination of the nails with caffeine powder increased the caffeine signal. These results show that the toenail analysis may be advantageous over the fingernail analysis for the detection of drugs in cases of continuous drug use. However, the incorporation mechanisms of the drugs in the nails are unclear, and are being examined in our laboratory currently.

The total time required for the extraction of one sample was less than 20 min, with 3 min for the H-F heating extraction. Compared with the conventional ultrasonic irradiation, this time is a shorter and should improve the analytical reliability without the need for sample shredding to increase the surface area. The H-F heating extraction is also advantageous compared with the conventional method, because all of the extraction procedure is carried out in a glass tube without the need for an enzyme or strong reagent. Therefore, the extraction of drugs from nail samples using H-F heating extraction is simpler and should have a lower risk of sample contamination. The correct determination of drugs in nails is practically difficult, because the extraction efficiency influences the quantitative results. However, the proposed method provides an improvement on the extraction efficiency from nails, and the results are useful when the sample decomposition cannot be available. This method will expand the potential applications of forensic and clinical toxicology analysis.

3. Materials and Methods

3.1. Chemicals

Analytical grade caffeine and amlodipine were purchased from Wako Pure Chemical Industries Ltd. (Osaka, Japan), and were used as received. HPLC grade 0.1% (v/v) formic acid and 0.1% (v/v) formic acid-acetonitrile were purchased from Kanto Chemical Ltd. (Tokyo, Japan). All of the other reagents were of the highest grade available, and were purchased from Tokyo Chemical Industry Co. Ltd. (Tokyo, Japan). Stock solutions (200 µg/mL) of caffeine and amlodipine were prepared with methanol, and stored at −20 °C when not in use. Working standard solutions were prepared by the precise dilution of the stock solutions using methanol just before they were required. Water was purified using a Milli-Q Integral purification system (Millipore, Billerica, MA, USA).

3.2. Apparatus

The H-F heating extraction was carried out using a ferromagnetic alloy and a Curie point pyrolyzer (JHP-3S, 600 kHz, Japan Analytical Industry Co. Ltd., Tokyo, Japan). Different ferromagnetic alloys (160 °C, 170 °C, 220 °C, and 250 °C) were purchased from Japan Analytical Industry (Tokyo, Japan). Glass tubes (70 mm × 5 mm i.d.) with one sealed end were purchased from Shimadzu GLC (Tokyo, Japan). A BRASONIC ultrasonic cleaner (80 W; 47 kHz; BRANSON, Danbury, CT, USA) was used for the conventional ultrasonic irradiation extraction [26].

The LC-MS/MS analyses were performed using a Prominence UFLC liquid chromatograph (Shimadzu Co., Kyoto, Japan) combined with a LXQ mass spectrometer (Thermo Fisher Scientific, MA, Waltham, MA, USA). An Inersil® ODS-4 column (75 mm × 2.1 mm i.d.; 3 µm particle size; GL Science Inc., Tokyo, Japan) was used for separation, and kept at 40 °C during the analysis. The mobile phase was a mixture of 0.1% (v/v) formic acid (solvent A) and 0.1% (v/v) formic acid-methanol (solvent B). Both components of the mobile phase were filtered and degassed before use. The elution used a linear gradient of solvent B from 20% to 80% (20 min), with a constant mobile phase flow rate of 0.2 mL min^{-1}. The mass spectra were collected in MS/MS mode. The precursor ion m/z 195 was selected for caffeine and m/z 409 for amlodipine. The product ions were analyzed in the selected reaction monitoring (SRM) mode. The ions at m/z 138 and m/z 238 were selected for the detection of caffeine and amlodipine, respectively.

Scanning electron microscope (SEM) images were obtained by a JSM-6610LV electron microscope (JEOL, Tokyo, Japan), and used to observe the surface profiles of the nails.

3.3. Sample Preparation

Nails were collected from two adult volunteers with their agreement under informed consent. One volunteer (thirties; male) constantly consumed caffeine during daily life. The other volunteer (forties; male) consumed caffeine and took amlodipine as an antihypertensive (10 mg/day). Both their finger- and toe-nail samples were collected using commercial nail clippers during daily hygiene routines. The collected nail samples were cut into 1-mm segments, and mixed until homogenous. Then, the nail samples were washed by ultrasonication with 1% sodium dodecyl sulfate solution for 1 min, with water five times, and then with methanol. After washing, the nails were air dried at room temperature. Figure 8 shows a schematic diagram of the H-F heating extraction method.

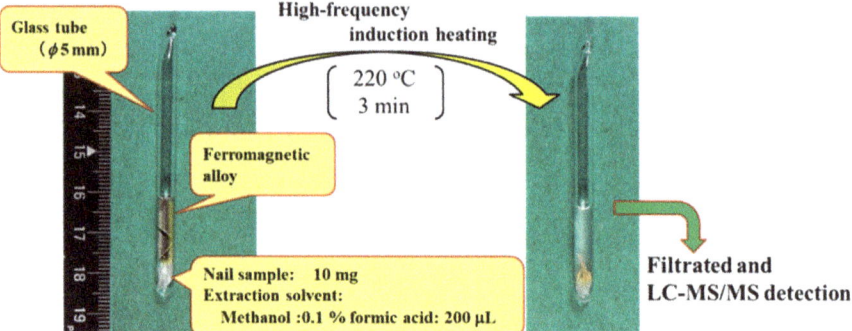

Figure 8. Schematic diagram of the high-frequency (H-F) heating extraction method.

The nails (10 mg) were placed in glass tube with 200 µL of the extraction solvent and the glass tube was sealed over a flame. Next, a ferromagnetic alloy (pyrofoil; 9 mm × 20 mm, Shimadzu GLC Co. Ltd., Tokyo, Japan) was wrapped around the glass tube in a spiral, and the glass tube was placed in the pyrolyzer. A H-F voltage (600 kHz) was applied to the pyrofoil for a few minutes, and any drugs contained in the nails were extracted into the solvent. The extraction temperature was the Curie temperature of the pyrofoil. After extraction, the glass tube was cooled to room temperature. Then, the glass tube was cut through the middle, and the extraction solvent was collected using a microsyringe, and was filtered through a Mini-Uniprep syringeless filter device (pore size 0.45 µm, GE Healthcare UK Ltd., Buckinghamshire, UK). Five microliters of the extract were injected into the LC-MS/MS system.

Part of this work was carried under ethics certification (No. 25M3) from the Japanese Association of Forensic Science and Technology.

4. Conclusions

In this study, a simple and efficient extraction method for the analysis of drugs in nails was developed using H-F heating combined with LC-MS. A significant peak for caffeine was obtained by the proposed method using MeOH:0.1% formic acid (7:3, v/v) as the extraction solvent. Because of the high-pressure and high-temperature conditions in the glass tube under H-F heating, the drugs that are strongly incorporated in the nails will be effectively extracted into the solvent within 3 min. Under the optimized extraction conditions, the extraction of the drugs is not dependent on the sample preparation process (e.g., cutting or shredding). Therefore, H-F heating will be useful for simplifying extraction in forensic and clinical analysis, and this will reduce the overall extraction time and minimize the destruction of the sample. This method was successfully applied to the detection of drugs in nails from a hypertension patient.

Supplementary Materials: The following are available online. Figure S1: Dependence of the caffeine peak area on the extraction temperature for extraction of caffeine from nails.

Author Contributions: Conceptualization, F.T. and K.K.; methodology, F.T. and K.K.; experimental work, F.T., M.K., A.K. and K.K.; analysis of the results, F.T. and K.K.; writing—original draft preparation, F.T.; writing—review and editing, K.K. and H.A.

Funding: This work was supported by Japan Society for the Promotion of Sciences (JSPS), Grant-in-Aid for Young Scientists (B) [grant number JP17K17776].

Acknowledgments: This article was received a support by Shinshu University Gender Equality Promotion Center, Researcher Assistance System. We thank Gabrielle David, from Edanz Group (www.edanzediting.com/ac) for editing a draft of this manuscript.

Conflicts of Interest: The authors declare no conflict of interest.

References

1. Ribeiro, C.; Santos, C.; Gonçalves, V.; Ramos, A.; Afonso, C.; Tiritan, E.M. Chiral Drug Analysis in Forensic Chemistry: An Overview. *Molecules* **2018**, *23*, 262. [CrossRef]
2. Wang, Z.; Lin, H.; Zhu, H.; Yang, N.; Zhou, B.; Wang, C.; Li, P.; Liu, J. Pharmacokinetic and Metabolism Studies of 12-Riboside-Pseudoginsengenin DQ by UPLC-MS/MS and UPLC-QTOF-MSE. *Molecules* **2018**, *23*, 2499. [CrossRef] [PubMed]
3. René, B.; Patrick, B.; Jean-Francois, B.; Pierre, D.; Gilles, P.; Éric, G.; Normand, F. New approach for the determination of ortho-phenylphenol exposure by measurement of sulfate and glucuronide conjugates in urine using liquid chromatography-tandem mass spectrometry. *Anal. Bioanal. Chem.* **2018**, *410*, 7275–7284.
4. Takahashi, F.; Nitta, S.; Shimizu, R.; Jin, J. Electrochemiluminescence and voltammetry of tris(2,2′-bipyridine)ruthenium (II) with amphetamine-type stimulants as coreactants: An application to the discrimination of methamphetamine. *Forensic Toxicol.* **2018**, *36*, 185–191. [CrossRef]
5. Tina, MB.; Franziska, G.; Clarissa, D.V.; Mathias, H.; Markus, R.B.; Thomas, K. Systematic investigations of endogenous cortisol and cortisone in nails by LC-MS/MS and correlation to hair. *Anal. Bioanal. Chem.* **2018**, *410*, 4895–4903.
6. Kintz, P.; Spiehler, V.; Negrusz, A. Alternative specimens. In *Clarke's Analytical Forensic Toxicology*; Jickells, S., Negrusz, A., Eds.; Pharmaceutical Press: London, UK, 2008; Volume 6, pp. 153–190.
7. Suzuki, O.; Hattori, H.; Asano, M. Detection of methamphetamine and amphetamine in a single human hair by gas chromatography/chemical ionization mass spectrometry. *J. Forensic Sci.* **1984**, *29*, 611–617. [CrossRef]
8. Nakahara, Y.; Takahashi, K.; Takeda, Y.; Konuma, K.; Fukui, S.; Tokui, T. Hair analysis for drug-abuse. 2. Hair analysis for monitoring of methamphetamine abuse by isotope-dilution gas-chromatography mass-spectrometry. *Forensic Sci. Int.* **1990**, *46*, 243–254. [CrossRef]
9. Zhu, K.Y.; Leung, K.W.; Ting, A.K.; Wong, Z.C.; Ng, W.Y.; Choi, R.C.; Dong, T.T.; Wang, T.; Lau, D.T.; Tsim, K.W. Microfluidic chip based nano liquid chromatography coupled to tandem mass spectrometry for the determination of abused drugs and metabolites in human hair. *Anal. Bioanal. Chem.* **2012**, *402*, 2805–2815. [CrossRef]
10. Phinney, K.W.; Sander, L.C. Liquid chromatographic method for the determination of enantiomeric composition of amphetamine and methamphetamine in hair samples. *Anal. Bioanal. Chem.* **2004**, *378*, 144–149. [CrossRef]
11. Wang, H.; Wang, Y. Matrix-assisted laser desorption/ionization mass spectrometric imaging for the rapid segmental analysis of methamphetamine in a single hair using umbelliferone as a matrix. *Anal. Chim. Acta.* **2017**, *975*, 42–51. [CrossRef]
12. Miguez-Framil, M.; Moreda-Pineiro, A.; Bermejo-Barrera, P.; Cocho, J.A.; Tabernero, M.J.; Bermejo, A M. Electrospray ionization tandem mass spectrometry for the simultaneous determination of opiates and cocaine in human hair. *Anal. Chim. Acta.* **2011**, *704*, 123–132. [CrossRef] [PubMed]
13. Emidio, E.S.; Prata, V.D.; Dorea, H.S. Validation of an analytical method for analysis of cannabinoids in hair by headspace solid-phase microextraction and gas chromatography-ion trap tandem mass spectrometry. *Anal. Chim. Acta.* **2010**, *670*, 63–71. [CrossRef] [PubMed]
14. Moriya, F. Alternative Specimens. In *Handbook of Practical Analysis of Drugs and Poisons in Human Specimens—Chromatographic Methods*; Jiho: Tokyo, Japan, 2002; Volume 1–2, pp. 8–14.
15. Cappelle, D.; Doncker, D.M.; Gys, C.; Krysiak, K.; Keukeleire, D.S.; Maho, W.; Crunelle, C.L.; Dom, G.; Covaci, A.; Van-Nuijs, A.; et al. A straightforward, validated liquid chromatography coupled to tandem mass spectrometry method for the simultaneous detection of nine drugs of abuse and their metabolites in hair and nails. *Anal. Chim. Acta.* **2017**, *960*, 101–109. [CrossRef] [PubMed]
16. Busardo, F.P.; Gottardi, M.; Tini, A.; Mortali, C.; Giorgetti, R.; Pichini, R. Ultra-High-Performance Liquid Chromatography Tandem Mass Spectrometry Assay for Determination of Endogenous GHB and GHB-Glucuronide in Nails. *Molecules* **2018**, *23*, 2686. [CrossRef] [PubMed]
17. Solimini, R.; Minutillo, A.; Kyriakou, C.; Pichini, S.; Pacifici, R.; Busardo, F.P. Nails in Forensic Toxicology: An Update. *Curr. Pharm. Des.* **2017**, *23*, 5468–5479. [CrossRef] [PubMed]

18. Palmeri, A.; Pichini, S.; Pacifici, R.; Zuccaro, P.; Lopez, A. Drugs in nails: Physiology, pharmacokinetics and forensic toxicology. *Clin. Pharmacokinet.* **2000**, *38*, 95–110. [CrossRef] [PubMed]
19. Simon, W.; Kriemler, P.; Voelimin, J.A.; Steiner, H. Elucidation of the structure of organic compound by thermal fragmentation. *J. Gas Chromatogr.* **1967**, *5*, 53–57. [CrossRef]
20. Tsuge, S.; Ohtani, H.; Watanabe, C. *Pyrolysis-GC/MS Data Book of Synthetic Polymers: Pyrograms, Thermograms and MS of Pyrolyzates*; Elsevier: Amsterdam, the Netherlands, 2011.
21. Michael, N.L. *Pyrolysis and GC in Polymer Analysis*; Marcel Dekker: New York, NY, USA, 1985.
22. Ludovic, H.; Charlotte, H.; Béatrice, B.; Maria, K.; Rachid, A.; Anne-Laure, C.; Michel, L.; Ika, P.P.; Philippe, S.; Alexandre, D.; et al. Optimization, performance, and application of a pyrolysis-GC/MS method for the identification of microplastics. *Anal. Bioanal. Chem.* **2018**, *410*, 6663–6676.
23. Kurihara, K.; Tsuchiya, F.; Takada, K.; Shoji, T. Pretreatment of analytical samples with high-frequency heating. *DIC Tech. Rev.* **2001**, *7*, 21–28.
24. Kurihara, K.; Tanoue, F. Identification of fatty acid composing metal salt of it in resins by alcohol added thermal extraction with high-frequency heating. *Bunseki Kagaku* **2000**, *49*, 265–267. [CrossRef]
25. Kurihara, K.; Tanoue, F. Qualitative analysis of a hindered phenol-type antioxidant by alcohol added thermal extraction with high-frequency heater, analyzing with GC/MS. *Bunseki Kagaku* **2000**, *49*, 205–208. [CrossRef]
26. Kurihara, K.; Tanoue, F. Identification of polyester resin components by alcohol added thermal extraction with high-frequency heating. *Bunseki Kagaku* **2000**, *49*, 269–271. [CrossRef]
27. Nakamura, M.; Tsuchiya, F.; Kurihara, K.; Takahashi, M. Quick molecular weight determination of polyester-polyurethane soft blocks with phenylisocyanate using high-frequency heating technique. *Bunseki Kagaku* **2007**, *56*, 237–240. [CrossRef]
28. Takahashi, F.; Masaru, K.; Atsushi, K.; Kanya, K. Development of high-frequency heating extraction for drug analysis of human nails. *Jpn. J. Forensic Sci. Tech.* **2015**, *20*, 103–112. [CrossRef]
29. Krumbiegel, F.; Hastedt, M.; Tsokos, M. Nails are a potential alternative matrix to hair for drug analysis in general unknown screenings by liquid-chromatography quadrupole time-of-flight mass spectrometry. *Forensic Sci. Med. Path.* **2014**, *10*, 496–503. [CrossRef] [PubMed]
30. Fujii, T.; Takashima, Y.; Takayama, S.; Ito, Y.; Kawasoe, T. Effect of heat treatment of human hair keratin film. *Jpn. J. Cosmetic Sci. Soc.* **2013**, *37*, 165–170.
31. Nakamura, A.; Arimoto, M.; Takeuchi, K.; Fujii, T. A rapid extraction procedure of human hair proteins and identification of phosphorylated species. *Biol. Pharm. Bull.* **2002**, *25*, 569–572. [CrossRef]
32. Wesolowski, M.; Szynkaruk, P. Thermal decomposition of methylxanthines interpretation of the results by PCA. *J. Therm. Anal. Cal.* **2016**, *948*, 40–47.
33. Kuwayama, K.; Miyaguchi, H.; Iwata, Y.; Kanamori, T.; Tsujikawa, K.; Yamamuro, T.; Segawa, H.; Inoue, H. Three-step drug extraction from a single sub-millimeter segment of hair and nail to determine the exact day of drug intake. *Anal. Chim. Acta* **2016**, *948*, 40–47. [CrossRef]

Sample Availability: Samples of the compounds are no more available from the authors. They have been used for analysis.

© 2018 by the authors. Licensee MDPI, Basel, Switzerland. This article is an open access article distributed under the terms and conditions of the Creative Commons Attribution (CC BY) license (http://creativecommons.org/licenses/by/4.0/).

Review

A Review on the Recent Progress in Matrix Solid Phase Dispersion

Xijuan Tu [1,2] and Wenbin Chen [1,2,*]

1. College of Bee Science, Fujian Agriculture and Forestry University, Fuzhou 350002, China; xjtu@fafu.edu.cn
2. MOE Engineering Research Center of Bee Products Processing and Application, Fujian Agriculture and Forestry University, Fuzhou 350002, China
* Correspondence: wbchen@fafu.edu.cn; Tel.: +86-591-83789482

Academic Editor: Nuno Neng
Received: 28 September 2018; Accepted: 24 October 2018; Published: 25 October 2018

Abstract: Matrix solid phase dispersion (MSPD) has proven to be an efficient sample preparation method for solid, semi-solid, and viscous samples. Applications of MSPD have covered biological, food, and environmental samples, including both organic and inorganic analytes. This review presents an update on the development of MSPD in the period 2015~June 2018. In the first part of this review, we focus on the latest development in MSPD sorbent, including molecularly imprinted polymers, and carbon-based nanomaterials etc. The second part presents the miniaturization of MSPD, discussing the progress in both micro-MSPD and mini-MSPD. The on-line/in-line techniques for improving the automation and sample throughput are also discussed. The final part summarizes the success in the modification of original MSPD procedures.

Keywords: sample preparation; matrix solid phase dispersion; sorbent; miniaturization; on-line

1. Introduction

Sample preparation is the key step in analytical workflow [1]. For solid, semi-solid, and viscous samples, procedures of sample preparation generally start with extracting analytes from matrix into homogeneous liquid solvents. Then a consequent clean-up step may be performed to reduce interference compounds in the extract. Finally, an additional enrichment or concentration step may also be required to meet the sensitivity of the analytical technique. Limitations in the classical method are the use of large volumes of solvent, labor-intensive, and time-consuming through the manipulation. Matrix solid phase dispersion (MSPD), first introduced by Barker et al. [2], provides an alternative approach to reduce solvent use and analysis time for preparing solid, semi-solid, and viscous samples [3].

In a typical MSPD procedure, samples are blended with sorbent to obtain homogeneous mixture. The resulting mixture is transferred and packed into an extraction column. Then solvent is passed through the column to carry out washing and elution step for the extraction and isolation of analytes from the matrix. In some case, an additional co-sorbent could be loaded at the bottom of the column to further clean-up the eluent. Generally, the final extract can be analyzed by chromatography based analytical techniques. Compared with classical solvent extraction method, MSPD eliminates steps of repeated centrifugation and/or filtration, and procedures of re-extraction. Different with solid phase extraction (SPE), in which separated solvent extraction procedure is required to make solid samples suitable for loading into a SPE column, MSPD eliminates the solvent extraction step. These would dramatically reduce the consumption of solvent and the required manipulation time for the preparation. There have been extensive reviews regarding the trends and developments of MSPD [3–10]. In this review, we focus attention on the latest developments in MSPD sorbent, miniaturization of MSPD, on-line/in-line techniques, and the modification of original MSPD procedure. Literatures during 2015 and June 2018 are reviewed to avoid the overlap with recent excellent reviews [9,10].

2. Latest Developments in MSPD Sorbent

Molecularly imprinted polymers (MIPs), the synthetic sorbents which exhibit selective binding of target molecular, have been widely used for the extraction of specific compounds [11,12]. In MSPD, sorbent requires to be blended with sample to obtain homogeneous mixture. To improve mechanical strength of MIPs materials, imprinted polymers can be synthesized using other sorbents as carrier. For example, MIPs were prepared on carbon nanotubes (CNTs) for the MSPD preparation of malachite green in aquatic products [13]. Silica gel [14], silica nanoparticles [15], and mesoporous silica [16] also have been reported as the carrier of MIPs to improve the selectivity of MSPD sorbents. Additionally, Wang et al. reported the synthesis of mixed-template MIPs for the extraction of multi-class veterinary drugs [17]. This novel MIPs sorbent was used for the simultaneous MSPD extraction of 20 drugs in meat, including 8 fluoroquinolones, 8 sulfonamides and 4 tetracyclines.

Graphene is one of the carbon-based nanomaterials which shows great promise in sample preparation [18,19]. Graphene provides large surface area and nanosheets morphology for improving adsorption capacity. In addition, the delocalized π electron system in graphene could make it form strong π-stacking interaction with compounds containing aromatic rings. These properties make graphene a good candidate for the adsorption of benzenoid compounds. Sun et al. reported a graphene-encapsulated silica sorbent for the analysis of flavonoids in the leaves of *Murraya panaculata* (L.) Jack [20]. Immobilized on the of surface of silica gel avoided the aggregation and maintained the large surface area and π-electron rich structure graphene during the mechanical blending. Compared with five sorbents (graphene, silica gel, C18, diatomaceous earth, and neutral alumina), graphene-encapsulated silica showed the better extraction efficiency for the target flavonoid compounds.

Phenyltrichlorosilane-functionalized magnesium oxide microspheres were designed by Tan et al. for the extraction of polycyclic aromatic hydrocarbons (PAHs) in soils [21]. This material takes advantage of the high affinity between magnesium oxide and PAHs to enhance the retention of target molecules. Grafting the microspheres with phenyltrichlorosilane reduced the competitive adsorption of chlorine-contained interferences which are widely exist in soil samples. Using hexane and DCM as rinsing and eluting solvent, respectively, seven PAHs were successfully determined in HPLC-FLD with limits of detection (LODs) of 0.02–0.12 µg/kg.

The use of polyethyleneimine (PEI)-modified attapulgite material as MSPD sorbent was reported by Wang et al. for the determination of cadmium in seafood products [22]. Introducing of PEI, which is a cationic polymer with high affinity to cadmium ion, resulted in the high recovery of target ion in complex matrices. High concentration of HNO_3 (50%, v/v) was required to release the cadmium. Determined by atomic absorption spectrometry (AAS), the LOD of cadmium in fish sample was found to be 2.5 µg/kg.

Additionally, sorbents such as mussel shell [23,24], molecular sieve [25,26], microcrystalline cellulose [27], and metal-organic framework materials [28,29] also have been reported. These emerging sorbents are summarized in Table 1.

Table 1. Selected representative studies involving developments in MSPD sorbent.

Sorbent	Analytes	Matrix	MSPD Parameters					Detection	LOD (μg/kg)	LOQ (μg/kg)	Ref.
			Sample Amounts (g)	Sorbent Amounts (g)	Blend Time (min)	Co-Sorbent	Washing Solvent	Elution Solvent				
MIPs	Veterinary drugs	Meat	0.2	0.15	3	0.05 g MIPs	3 mL MeOH/H$_2$O (2:8, v/v)	4 mL MeOH/acetic acid (9:1, v/v)	UPLC-DAD	0.5–3	1.5–6	[12]
CNTs-MIPs	Malachite green	Aquatic products	0.3	0.2	15	None	4 mL 50% aqueous MeOH	3 mL MeOH-acetic acid (98:2, v/v)	HPLC-UV	0.7	n.r.	[13]
CNTs-MIPs	Camptothecin	Herb (*Camptotheca acuminate*)	0.1	0.1	5	None	5 mL 10% aqueous MeOH	4 mL MeOH-acetic acid (95:5, v/v)	HPLC-UV	130 μg/L	n.r.	[30]
Silica gel -MIPs	Degradation products of penicillin	Milk	0.3 mL	0.2	n.r.	None	2 mL DCM	3 mL MeOH-10% acetic acid (9:1, v/v)	HPLC-UV	40/50	130/170	[14]
SiO$_2$-MIP	Acrylamide	Biscuit and bread	0.1	0.15	n.r.	None	1 mL hexane	2.5 mL ACN-MeOH (50:50, v/v)	HPLC-UV	14.5/16.1	40.5/40.1	[15]
Mesoporous silica-MIPs	Ketoprofen	Powder milk	0.05	0.025	n.r.	None	None	1 mL ACN	HPLC -MS/MS	n.r.	n.r.	[16]
Graphene-encapsulated silica	Flavonoids	Herb (*Murraya paniculata* (L.) Jack)	0.025	0.05	3	None	None	5 mL MeOH	UPLC-UV	4–12 μg/L	10–40 μg/L	[20]
PTS-MgO	PAHs	Soils	0.1	0.1	n.r.	0.05 g PTS-MgO	None	4 mL DCM	HPLC-FLD	0.02–0.12	0.07–0.40	[21]
PEI-attapulgite	Cadmium	Seafood	0.21	0.13	n.r.	None	6 mL H$_2$O	8 mL 50%HNO$_3$/H$_2$O (v/v)	AAS	2.5	8.3	[22]
Golden mussel shell	Pesticides and PPCPs	Mussel tissue	2.5	0.5	5	None	None	5 mL ethyl acetate	LC-MS/MS	3–30	10–100	[23]
Mussel shell	Booster biocides	Fish tissue	0.5	0.5	5	None	None	5 mL EtOH	LC-MS/MS	1.5/15	5/50	[24]
Molecular sieves	Flavonoids	Fruit peels	0.025	0.025	2.5	None	None	0.5 mL MeOH	UPLC-UV	20–30 μg/L	70–90 μg/L	[25]
Molecular sieve	Sesquiterpenes	Herb (*Curcuma wenyujin*)	0.2	0.2	2.5	None	None	1 mL MeOH	MEEKC	5–34 μg/mL	16–78 μg/mL	[26]
Microcrystalline cellulose	Triterpenoid acids	Herb (loquat leaves)	0.024	0.024	1	None	None	0.2 × 3 mL EtOH	UHPLC-Q-TOF	19.6–51.6	65.3–171.8	[27]
MOFs	Pesticides	Coconut palm	0.25	1	3	None	None	20 mL ACN	HPLC-DAD	10–50	50–100	[28]
MOFs	Pesticides	Peppers (*Capsicum annuum* L.)	0.5	0.35	n.r.	1 g Na$_2$SO$_4$ + 0.5 g silica	None	10 mL DCM	GC-MS	16.0–67.0	50.3–200.0	[29]

DCM, dichloromethane; CNTs, carbon nanotubes; MIPs, molecularly imprinted polymers; PPCPs, pharmaceutical and personal care products; MOFs, metal-organic frameworks; PTS, phenyltrichlorosilane; PAHs, polycyclic aromatic hydrocarbons; PEI, polyethyleneimine; ACN, acetonitrile. n.r., not reported.

3. Miniaturization of MSPD

In classical MSPD protocol, the sample amount is typically 0.5 g [3]. The miniaturization of MSPD (micro/mini-MSPD) can significantly reduce the sample amount, and consequently the consumption of sorbent, solvent, and preparation time. Developed micro/mini-MSPD methods are summarized in Table 2. For instance, Guerra et al. developed a method based on micro-MSPD combined with LC-MS/MS for the simple and rapid determination of dyes in cosmetic products [31]. The proposed micro-MSPD was carried by grounding 0.1 g cosmetic sample with 0.3 g anhydrous Na_2SO_4 (drying agent) and 0.4 g of Florisil. After transferring the mixture into a glass Pasteur pipette, 2 mL of methanol was eluted to extract nine water-soluble dyes. By using micro-MSPD method, time and solvent consumption in the sample preparation could be reduced.

Taking advantage of high sensitive detection methods, sample amount in recently published mini-MSPD could be reduced to the scale of milligram. Chen et al. reported a sensitive quantification of mercury distribution in fish organ based on the mini-MSPD [32]. The sample amount in this research was as low as 1 mg of organ sample. Multiwall carbon nanotubes (MWCNTs) were used as the sorbent, with amount of 0.5 mg. Mercury species were eluted by 100 µL eluent containing HCOOH and L-cysteine. When combined with a sensitive mercury determination method named single-drop solution electrode glow discharge-induced cold vapor generation combined with atomic fluorescence spectrometry, LOD of 0.01 µg/L was achieved. The consumption of sample, adsorbent, and solvent were all dramatically decreased in this mini-MSPD.

Another example of mini-MSPD was reported by Deng et al. [33], in which only 0.30–0.80 mg of plant samples were ground with 2 mg C18 sorbent in liquid nitrogen to obtain the homogenous mixture. Based on this mini-MSPD and the precolumn derivatization coupled with UPLC-MS/MS determination, phytohormone gibberellins were detected with the limits of quantification (LOQs) of 0.54–4.37 pg/mL. As only sub-milligram sample was required for the determination, a spatial distribution of gibberellins in a single *Arabidopsis thaliana* leaf with resolution of 2×2 mm^2 was profiled.

Table 2. Selected representative studies using miniaturized MSPD.

Analytes	Matrix	MSPD Parameters					Detection	LOD (μg/kg)	LOQ (μg/kg)	Ref.
		Sample Amounts (g)	Sorbent Amounts	Blend Time (min)	Co-Sorbent	Washing Solvent	Elution Solvent			
Dyes	Cosmetic products	0.1	0.4 g Florisil + 0.3 g Na$_2$SO$_4$	n.r.	0.1 g Florisil	None	2 mL MeOH	LC-MS/MS	0.01–11	n.r.	[31]
Photoproducts of cosmetic preservatives	Personal care products	0.1	0.4 g Florisil + 0.4 g Na$_2$SO$_4$	5	0.2 g Florisil	None	1 mL hexane-acetone (1:1, v/v)	GC-MS/MS	31–170	n.r.	[34]
Flavonoids	Lime fruit	0.05	0.15 g Florisil	1	None	None	0.4 mL [Bmin]BF$_4$ aqueous solution (250 mM)	UPLC-UV	4.08/5.04 μg/g	14.01/14.56 μg/g	[35]
Phenolic isomers	Honeysuckle	0.025	0.075 g β-cyclodextrin	2	None	None	0.5 mL MeOH-H$_2$O (80:20, v/v)	UPLC-UV-Q-TOF	1.62–3.33 ng/mL	5.52–11.40 ng/mL	[36]
Inorganic iodine and iodinated amino acids	Seaweed	0.05	0.05 g molecular sieve SBA-15	0.5	None	None	0.4 mL [C12min] Br (200 mM)	UHPLC-UV	3.7–16.7 ng/mL	12.4–55.8 ng/mL	[37]
Phenols	Olive fruits	0.05	0.025 g chitosan	1	None	None	0.5 mL × 3 MeOH-H$_2$O (6:4, v/v)	UHPLC-Q-TOF	69.6–358.4	232–1240.8	[38]
Mercury species	Fish organs	1 mg	0.5 mg MWCNTs	5	0.15 g C18	None	0.1 mL × 2 0.5% L-cysteine and 4% HCOOH	AFS	0.01	n.r.	[22]
Gibberellins	Plant	0.3–0.8 mg	2 mg C18	n.r.	None	None	0.2 mL ACN	UPLC-MS/MS	0.16–1.31 pg/mL	0.53–4.37 pg/mL	[33]
Synthetic dyes	Cosmetics and foodstuffs	0.1	0.4 g C18 + 0.3 g Na$_2$SO$_4$	n.r.	0.1 g C18	None	2 mL MeOH	LC-MS/MS	14.2–95.2	n.r.	[39]
Phenolic acids	Plant preparation (Danshen tablets)	0.024	0.024 3 graphene nanoplatelets	1	None	None	0.2 mL H$_2$O	UHPLC-ECD	1.19–4.62 ng/mL	3.91–15.23 ng/mL	[40]
Lignans	Herbs (Schisandrae Chinensis Fructus)	0.025	0.05 g molecular sieve TS-1	2.5	None	None	0.5 mL MeOH	MEEKC	n.r.	2.77 μg/mL	[41]

MWCNTs, multiwall carbon nanotubes; AFS, atomic fluorescence spectrometry. n.r., not reported.

4. On-Line/In-Line MSPD

On-line/in-line sample preparation techniques that couple sample preparation step and chromatography separation are regarded as a promising technique with advantages of automatable high sample throughput, reducing sample manipulation and contamination, improving precision, and lower regent consumption [42]. On-line/in-line MSPD provides a potential automated way for the sample preparation of solid, semi-solid, and viscous samples.

Rajabi et al. reported an in-line micro-MSPD method for the determination of Sudan dyes in spices [43]. In this in-line MSPD, the filled MSPD column was placed in the mobile phase pathway before the analytical column. Then the mobile phase passed through the MSPD column to elute analytes and subsequently separated in a reverse-phased HPLC. Since the in-line method integrated extraction and separation into one step, this proposed approach was much faster than other reported methods for the determination of Sudan dyes.

Gutiérrez-Valencia et al. developed an on-line MSPD-SPE sample preparation method combined with HPLC-FLD for the analysis of PAHs in bovine tissues [44]. The bovine liver sample (50 mg) was dispersed on C18 sorbent (200 mg). Then the obtained homogenous mixture was packed into a stainless steel cartridge which was connected to a MSPD-SPE-HPLC-FLD system. The SPE column was used to trap and pre-concentrate the target compounds eluted from the MSPD cartridge. Acetonitrile (ACN)-water mixture and pure ACN solution were applied to wash and elute the MSPD cartridge, respectively. However, ACN extract exhibited poor retention of analytes in C18 SPE column. Thus a dynamic mixing chamber was required to dilute the ACN extract with water before pre-concentration to quantitatively transfer PAHs from MPSD cartridge to the SPE column. Finally, the analytes pre-concentrated on the SPE column were eluted through the guard-column and the analytical column with mobile phase and detected by FLD. Compared with off-line MSPD, the on-line MSPD method showed advantages of lower consumption of sample amount and saving of analysis time.

Additionally, an on-line MSPD-HPLC-ICP-MS method for the determination of mercury speciation in fish was reported by Deng et al. [45]. In this on-line MSPD performance, 1 mg fish sample was blended with 2 mg of MWCNTs, then the mixture was transferred into a stainless steel column which was prior loaded with 0.20 g of C18. The eluent solution containing HCl (2%, v/v) and L-cysteine (1.5%, m/v) was loaded by a 100 µL loop through the six-port valve. Then mobile phase flushed the eluent to pass through the MSPD column for the extraction of analytes, which were further separated and detected by HPLC-ICP-MS. It is interesting to notice that the on-line MSPD system consisting of two sequential valves and six stainless steel MSPD columns to improve sample throughput. This on-line system shows the potential of automatable high sample throughput in MSPD method.

5. Modification of Original MSPD

The original MSPD can be modified or combined with other extraction methodologies to improve the extraction yields or simplify the MSPD procedures. The schematic procedure of the original and representative modification of MSPD is shown in Figure 1. For instance, ultrasonic-assisted MSPD (UA-MSPD) was first reported by Ramos et al. to improve the extraction yields by putting MSPD column into ultrasonic bath or sonoreactor after the extraction solvent was loaded into the MSPD column [46]. As summarized in Table 3, UA-MSPD has been introduced for the analysis of multi-class organic contaminants. For example, Albero et al. developed an UA-MSPD method for the analysis of 17 emerging contaminants in vegetables [47]. In this modified method, vegetable samples (2 g) was blended with Florisil (4 g) and magnesium sulfate anhydrous (1 g), then the homogenous mixture was transferred into a 20 mL glass column. Extraction solution of 8 mL EtAc:MeOH (9:1, v/v) containing 3% of NH_4OH were added to the column. After that, column was sonicated for 15 min in an ultrasonic water bath at room temperature for the extraction. Finally, extract was collected under vacuum manifold. Results indicated that better recoveries were obtained with the assistance of sonication.

Table 3. Selected representative studies using ultrasonic assisted MSPD and vortex assisted MSPD.

| Modification | Analytes | Matrix | MSPD Parameters ||||| Detection | LOD (µg/kg) | LOQ (µg/kg) | Ref. |
			Sample Amount (g)	Sorbent Amount	Grind Time (min)	Extraction Time (min)	Elution Solvent				
Ultrasonic assisted	Emerging organic contaminants	Poultry manure	0.5	2 g Florisil + 1 g MgSO$_4$	n.r.	15	8 mL ACN with 3% NH$_4$OH + 10 mL ACN with 4% formic acid	GC-MS/MS	0.9–2.2	2.8–5.5	[48]
Ultrasonic assisted	Emerging contaminants	Vegetables	2	4 g Florisil + 1 g MgSO$_4$	n.r.	15	8 mL EtAc:MeOH (9:1, v/v) containing 3% NH$_4$OH	GC-MS/MS	0.1–0.4	n.r.	[47]
Ultrasonic assisted	Emerging contaminants	Aquatic plants	1	4 g Florisil + 2 g MgSO$_4$	5	15	8 mL EtAc with 3% NH$_4$OH	GC-MS	0.3–2.2	1.0–6.7	[49]
Ultrasonic assisted	Aflatoxins	Rice	1	1 g C18	5	11	4 mL ACN	HPLC-FLD	0.04–0.14 ng/g	0.12–0.56 ng/g	[50]
Vortex assisted	5-HMF and iridoid glycosides	Herb (Fructus Cornii)	0.02	0.04 g silica	3	3	6 mL [Domin]HSO$_4$	UHPLC-UV	0.02–0.08 µg/mL	0.07–0.24 µg/mL	[51]
Vortex assisted	Booster biocides	Marine sediments	2	0.25 g C18	n.r.	1	10 mL MeOH	LC-MS/MS	n.r.	0.5–5	[52]
Vortex assisted	Phenol	Herb (Forsythiae Fructus)	0.02	0.02 g Florisil	3	2	2 mL 10% (v/v) Triton X-114	UHPLC-UV	0.03–0.08 µg/mL	0.08–0.25 µg/mL	[53]
Vortex assisted	Ibuprofen enantiomers	Milk	0.5	0.30 g diatomaceous earth + 0.30 g Na$_2$SO$_4$ + 0.26 g PSA + 0.021 g β-cyclodextrin	5	1	2 mL MeOH	HPLC-UV	0.042/0.045 µg/g	0.14/0.15 µg/g	[54]
Vortex assisted	Pesticides	Drinking water treatment sludge	1.5	0.5 g Chitin	5	1	5 mL ethyl acetate	GC-MS	n.r.	5–500	[55]
Vortex assisted	Pharmaceuticals	Fish tissue	0.5	0.5 g diatomaceous earth + 0.5 g Na$_2$SO$_4$	5	1	5 mL MeOH	LC-MS/MS	1.5–300	5–1000	[56]

5-HMF, 5-hydroxymethyl furfurol. n.r, nor reported.

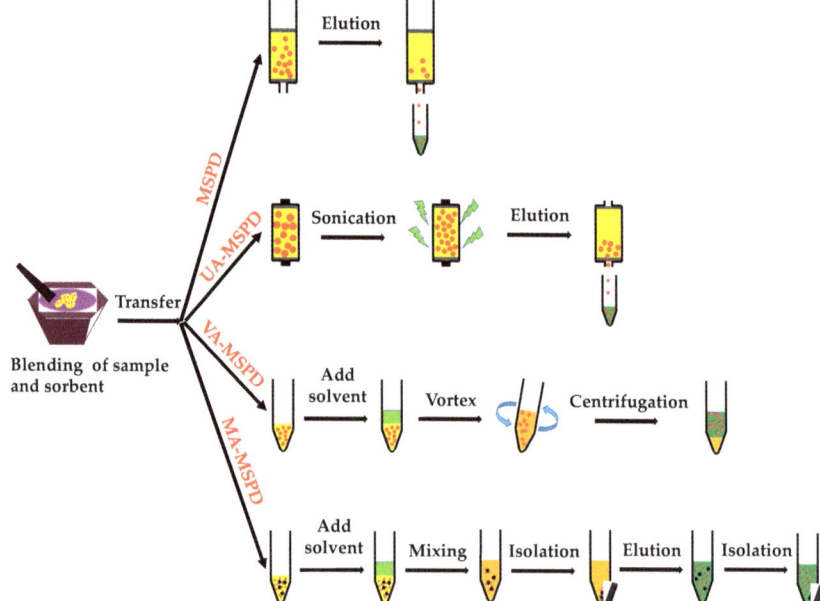

Figure 1. Schematic procedure of original matrix solid phase dispersion (MSPD), ultrasonic-assisted MSPD (UA-MSPD), vortex-assisted MSPD (VA-MSPD), and magnetically-assisted MSPD (MA-MSPD).

Vortex-assisted MSPD (VA-MSPD), in which the step of column elution is replaced by vortex, has been developed to reduce the solvent consumption and analysis time. This simplified MSPD procedure has been found applications in the analysis of phytochemical compounds and organic contaminants (Table 3). For instance, Caldas et al. reported the analysis of antifouling booster biocides in marine sediments by employing VA-MSPD [52]. In the sample preparation procedure, the homogenized mixture of sample and sorbent was added into a centrifuge tube. Then the extraction solvent was added, and the sample was vortexed for 1 min. Finally, the mixture was centrifuged, and the supernatant was collected for the LC-MS/MS analysis. Compared with other extraction methods including ultrasonic extraction, SPE, and microwave extraction, this VA-MSPD exhibited the advantages of shorter extraction time and less solvent consumption.

Another recent progress of the modification is the magnetically-assisted MSPD (MA-MSPD) developed by Fotouhi et al. for the extraction of parabens from breast milks [57]. Modified magnetic nanoparticles were used as the sorbent in the MA-MSPD. Milk sample (200 µL) was blended with poly(indole-thiophene) coated magnetic graphene oxide (MGO@PIT, 50 mg) and drying salt Na_2SO_4 (550 mg). After blending, the homogenous mixture was transferred into water solution and mechanically stirred for the adsorption of parabens. Then the MGO@PIT with target compounds were isolated from the solution by magnet. Subsequently, analytes were desorbed from the sorbent with methanol. Compared with the magnetic liquid-solid extraction (MLSE) [58,59], a hot topic of nanomaterials in sample preparation, the major difference between MA-MSPD and MLSE is the manipulation of sample. For MLSE, analytes in solid sample are extracted into the liquid solution prior to the introduction of magnetic sorbent. While in MA-MSPD, the solid sample is blended with magnetic nanoparticles to obtain the homogenous mixture. The similarity of these two methods is the replacement of column packing and elution with simple magnetic isolation. This would simplify the preparation step and reduce the extraction time. More importantly, magnetic nanoparticles have been demonstrated to be reusable in the liquid-solid extraction [58]. Thus MA-MSPD may provide a solution for the reusability of sorbent in MSPD.

Recently, we reported the combination of Soxhlet extraction and MSPD to develop a Soxhlet assisted MSPD (SA-MSPD) method [60,61]. In this modification method, sample was blended with silica gel following the original MSPD protocol and loaded into a column of constant pressure funnel. Then elution solvent was heated and continuous refluxed and passed through the column for the extraction and isolation of flavonoids. By comparing with conventional solvent extraction and Soxhlet extraction method, SA-MSPD showed the higher extraction yield with shorter extraction time and less consumption of solvent. Moreover, the introduction of sorbent into the Soxhlet enabled this classical method to be with clean-up ability. More recently, this SA-MSPD method was further combined with acid-hydrolysis for the quantification of flavonoid aglycones in bee pollen [61]. The acid hydrolysis SA-MSPD procedure accomplished the extraction and hydrolysis of flavonoid glycosides into one step, and provided a more efficient sample preparation method for the quantification of flavonoid aglycones.

6. Conclusion Remarks

Application fields of MSPD have been extended from the first reported drug residues in biological tissues to the food and environmental analysis, both for organic and inorganic analytes. Development of new sorbent materials for improving the capacity or selectivity is still the exciting research area in MSPD. One of the drawbacks of MSPD is the reusability of the extraction column. Among the emerging MSPD sorbents, modified magnetic nanoparticles are expected to provide the possibility of reusability. Combining high efficient sorbents with ultra-sensitive analytical technologies, miniaturization of MSPD might be found great interests in the analysis of limited or small size samples. Especially, the mini-MSPD may provide more information on the evolution or the spatial distribution of analytes in the sample matrices. In addition, on-line MSPD has shown the possibility of high-throughput analysis in MSPD. This would also be the trend of automation in MSPD. The modification of the original MSPD appears to be simplified the MSPD procedure and could be help for improving the reproducibility of the manipulation.

Author Contributions: X.T. organized the literatures and wrote the manuscript; W.C. conceived and designed the review, and wrote the manuscript. Both authors approved the submitted version.

Funding: This research was funded by Natural Science Foundation of China, grant number 31201861 and 51202030.

Conflicts of Interest: The authors declare no conflict of interest.

References

1. Chen, Y.; Guo, Z.; Wang, X.; Qiu, C. Sample preparation. *J. Chromatogr. A* **2008**, *1184*, 191–219. [CrossRef] [PubMed]
2. Barker, S.A.; Long, A.R.; Short, C.R. Isolation of drug residues from tissues by solid phase dispersion. *J. Chromatogr.* **1989**, *475*, 353–361. [CrossRef]
3. Barker, S.A. Matrix solid phase dispersion (MSPD). *J. Biochem. Biophys. Methods* **2007**, *70*, 151–162. [CrossRef] [PubMed]
4. Barker, S.A. Applications of matrix solid-phase dispersion in food analysis. *J. Chromatogr. A* **2000**, *880*, 63–68. [CrossRef]
5. Barker, S.A. Matrix solid-phase dispersion. *J. Chromatogr. A* **2000**, *885*, 115–127. [CrossRef]
6. Bogialli, S.; Di Corcia, A. Matrix solid-phase dispersion as a valuable tool for extracting contaminants from foodstuffs. *J. Biochem. Biophys. Methods* **2007**, *70*, 163–179. [CrossRef] [PubMed]
7. Moreda-Pineiro, J.; Alonso-Rodriguez, E.; Lopez-Mahia, P.; Muniategui-Lorenzo, S.; Prada-Rodriguez, D.; Romaris-Hortas, V.; Miguez-Framil, M.; Moreda-Pineiro, A.; Bermejo-Barrera, P. Matrix solid-phase dispersion of organic compounds and its feasibility for extracting inorganic and organometallic compounds. *Trends Anal. Chem.* **2009**, *28*, 110–116. [CrossRef]
8. Capriotti, A.L.; Cavaliere, C.; Giansanti, P.; Gubbiotti, R.; Samperi, R.; Lagana, A. Recent developments in matrix solid-phase dispersion extraction. *J. Chromatogr. A* **2010**, *1217*, 2521–2532. [CrossRef] [PubMed]
9. Capriotti, A.L.; Cavaliere, C.; Lagana, A.; Piovesana, S.; Samperi, R. Recent trends in matrix solid-phase dispersion. *Trends Anal. Chem.* **2013**, *43*, 53–66. [CrossRef]

10. Capriotti, A.L.; Cavaliere, C.; Foglia, P.; Samperi, R.; Stampachiacchiere, S.; Ventura, S.; Lagana, A. Recent advances and developments in matrix solid-phase dispersion. *Trends Anal. Chem.* **2015**, *71*, 186–193. [CrossRef]
11. Speltini, A.; Scalabrini, A.; Maraschi, F.; Sturini, M.; Profumo, A. Newest applications of molecularly imprinted polymers for extraction of contaminants from environmental and food matrices: A review. *Anal. Chim. Acta* **2017**, *974*, 1–26. [CrossRef] [PubMed]
12. Ashley, J.; Shahbazi, M.-A.; Kant, K.; Chidambara, V.A.; Wolff, A.; Bang, D.D.; Sun, Y. Molecularly imprinted polymers for sample preparation and biosensing in food analysis: Progress and perspectives. *Biosens. Bioelectron.* **2017**, *91*, 606–615. [CrossRef] [PubMed]
13. Wang, Y.; Chen, L. Analysis of malachite green in aquatic products by carbon nanotube-based molecularly imprinted—Matrix solid phase dispersion. *J. Chromatogr. B* **2015**, *1002*, 98–106. [CrossRef] [PubMed]
14. Luo, Z.; Du, W.; Zheng, P.; Guo, P.; Wu, N.; Tang, W.; Zeng, A.; Chang, C.; Fu, Q. Molecularly imprinted polymer cartridges coupled to liquid chromatography for simple and selective analysis of penicilloic acid and penilloic acid in milk by matrix solid-phase dispersion. *Food Chem. Toxicol.* **2015**, *83*, 164–173. [CrossRef] [PubMed]
15. Arabi, M.; Ghaedi, M.; Ostovan, A. Development of dummy molecularly imprinted based on functionalized silica nanoparticles for determination of acrylamide in processed food by matrix solid phase dispersion. *Food Chem.* **2016**, *210*, 78–84. [CrossRef] [PubMed]
16. Ganan, J.; Morante-Zarcero, S.; Perez-Quintanilla, D.; Sierra, I. Evaluation of mesoporous imprinted silicas as MSPD selective sorbents of ketoprofen in powder milk. *Mater. Lett.* **2017**, *197*, 5–7. [CrossRef]
17. Wang, G.N.; Zhang, L.; Song, Y.P.; Liu, J.X.; Wang, J.P. Application of molecularly imprinted polymer based matrix solid phase dispersion for determination of fluoroquinolones, tetracyclines and sulfonamides in meat. *J. Chromatogr. B* **2017**, *1065*, 104–111. [CrossRef] [PubMed]
18. De Toffoli, A.L.; Soares Maciel, E.V.; Fumes, B.H.; Lancas, F.M. The role of graphene-based sorbents in modern sample preparation techniques. *J. Sep. Sci.* **2018**, *41*, 288–302. [CrossRef] [PubMed]
19. Zhang, B.T.; Zheng, X.; Li, H.F.; Lin, J.M. Application of carbon-based nanomaterials in sample preparation: A review. *Anal. Chim. Acta* **2013**, *784*, 1–17. [CrossRef] [PubMed]
20. Sun, T.; Li, X.; Yang, J.; Li, L.; Jin, Y.; Shi, X. Graphene-encapsulated silica as matrix solid-phase dispersion extraction sorbents for the analysis of poly-methoxylated flavonoids in the leaves of *Murraya panaculata* (L.) Jack. *J. Sep. Sci.* **2015**, *38*, 2132–2139. [CrossRef] [PubMed]
21. Tan, D.; Jin, J.; Li, F.; Sun, X.; Dhanjai; Ni, Y.; Chen, J. Phenyltrichlorosilane-functionalized magnesium oxide microspheres: Preparation, characterization and application for the selective extraction of dioxin-like polycyclic aromatic hydrocarbons in soils with matrix solid-phase dispersion. *Anal. Chim. Acta* **2017**, *956*, 14–23. [CrossRef] [PubMed]
22. Wang, T.; Chen, Y.; Ma, J.; Jin, Z.; Chai, M.; Xiao, X.; Zhang, L.; Zhang, Y. A polyethyleneimine-modified attapulgite as a novel solid support in matrix solid-phase dispersion for the extraction of cadmium traces in seafood products. *Talanta* **2018**, *180*, 254–259. [CrossRef] [PubMed]
23. Rombaldi, C.; de Oliveira Arias, J.L.; Hertzog, G.I.; Caldas, S.S.; Vieira, J.P.; Primel, E.G. New environmentally friendly MSPD solid support based on golden mussel shell: Characterization and application for extraction of organic contaminants from mussel tissue. *Anal. Bioanal. Chem.* **2015**, *407*, 4805–4814. [CrossRef] [PubMed]
24. Vieira, A.A.; Caldas, S.S.; Venquiaruti Escarrone, A.L.; de Oliveira Arias, J.L.; Primel, E.G. Environmentally friendly procedure based on VA-MSPD for the determination of booster biocides in fish tissue. *Food Chem.* **2018**, *242*, 475–480. [CrossRef] [PubMed]
25. Cao, W.; Hu, S.S.; Ye, L.H.; Cao, J.; Pang, X.Q.; Xu, J.J. Trace matrix solid phase dispersion using a molecular sieve as the sorbent for the determination of flavonoids in fruit peels by ultra-performance liquid chromatography. *Food Chem.* **2016**, *190*, 474–480. [CrossRef] [PubMed]
26. Wei, M.; Chu, C.; Wang, S.; Yan, J. Quantitative analysis of sesquiterpenes and comparison of three Curcuma wenyujin herbal medicines by micro matrix solid phase dispersion coupled with MEEKC. *Electrophoresis* **2018**, *39*, 1119–1128. [CrossRef] [PubMed]
27. Cao, J.; Peng, L.Q.; Xu, J.J. Microcrystalline cellulose based matrix solid phase dispersion microextration for isomeric triterpenoid acids in loquat leaves by ultrahigh-performance liquid chromatography and quadrupole time-of-flight mass spectrometry. *J. Chromatogr. A* **2016**, *1472*, 16–26. [CrossRef] [PubMed]

28. De Jesus, J.R.; Wanderley, K.A.; Alves Junior, S.; Navickiene, S. Evaluation of a novel metal-organic framework as an adsorbent for the extraction of multiclass pesticides from coconut palm (*Cocos nucifera* L.): An analytical approach using matrix solid-phase dispersion and liquid chromatography. *J. Sep. Sci.* **2017**, *40*, 3327–3334. [CrossRef] [PubMed]
29. Barreto, A.S.; da Silva Andrade, P.D.C.; Farias, J.M.; Menezes Filho, A.; de Sa, G.F.; Alves Junior, S. Characterization and application of a lanthanide-based metal-organic framework in the development and validation of a matrix solid-phase dispersion procedure for pesticide extraction on peppers (*Capsicum annuum* L.) with gas chromatography-mass spectrometry. *J. Sep. Sci.* **2018**, *41*, 1593–1599. [CrossRef] [PubMed]
30. Liu, H.; Hong, Y.; Chen, L. Molecularly imprinted polymers coated on carbon nanotubes for matrix solid phase dispersion extraction of camptothecin from *Camptotheca acuminate*. *Anal. Methods* **2015**, *7*, 8100–8108. [CrossRef]
31. Guerra, E.; Celeiro, M.; Pablo Lamas, J.; Llompart, M.; Garcia-Jares, C. Determination of dyes in cosmetic products by micro-matrix solid phase dispersion and liquid chromatography coupled to tandem mass spectrometry. *J. Chromatogr. A* **2015**, *1415*, 27–37. [CrossRef] [PubMed]
32. Chen, Q.; Lin, Y.; Tian, Y.; Wu, L.; Yang, L.; Hou, X.; Zheng, C. Single-drop solution electrode discharge-induced cold vapor generation coupling to matrix solid-phase dispersion: A robust approach for sensitive quantification of total mercury distribution in fish. *Anal. Chem.* **2017**, *89*, 2093–2100. [CrossRef] [PubMed]
33. Deng, T.; Wu, D.; Duan, C.; Yan, X.; Du, Y.; Zou, J.; Guan, Y. Spatial profiling of gibberellins in a single leaf based on microscale matrix solid-phase dispersion and precolumn derivatization coupled with ultraperformance liquid chromatography-tandem mass spectrometry. *Anal. Chem.* **2017**, *89*, 9537–9543. [CrossRef] [PubMed]
34. Alvarez-Rivera, G.; Llompart, M.; Garcia-Jares, C.; Lores, M. Identification of unwanted photoproducts of cosmetic preservatives in personal care products under ultraviolet-light using solid-phase microextraction and micro-matrix solid-phase dispersion. *J. Chromatogr. A* **2015**, *1390*, 1–12. [CrossRef] [PubMed]
35. Xu, J.J.; Yang, R.; Ye, L.H.; Cao, J.; Cao, W.; Hu, S.S.; Peng, L.Q. Application of ionic liquids for elution of bioactive flavonoid glycosides from lime fruit by miniaturized matrix solid-phase dispersion. *Food Chem.* **2016**, *204*, 167–175. [CrossRef] [PubMed]
36. Xu, J.J.; Cao, J.; Peng, L.Q.; Cao, W.; Zhu, Q.Y.; Zhang, Q.Y. Characterization and determination of isomers in plants using trace matrix solid phase dispersion via ultrahigh performance liquid chromatography coupled with an ultraviolet detector and quadrupole time-of-flight tandem mass spectrometry. *J. Chromatogr. A* **2016**, *1436*, 64–72. [CrossRef] [PubMed]
37. Cao, J.; Peng, L.Q.; Xu, J.J.; Du, L.J.; Zhang, Q.D. Simultaneous microextraction of inorganic iodine and iodinated amino acids by miniaturized matrix solid-phase dispersion with molecular sieves and ionic liquids. *J. Chromatogr. A* **2016**, *1477*, 1–10. [CrossRef] [PubMed]
38. Peng, L.Q.; Li, Q.; Chang, Y.X.; An, M.; Yang, R.; Tan, Z.; Hao, J.; Cao, J.; Xu, J.J.; Hu, S.S. Determination of natural phenols in olive fruits by chitosan assisted matrix solid-phase dispersion microextraction and ultrahigh performance liquid chromatography with quadrupole time-of-flight tandem mass spectrometry. *J. Chromatogr. A* **2016**, *1456*, 68–76. [CrossRef] [PubMed]
39. Guerra, E.; Llompart, M.; Garcia-Jares, C. Miniaturized matrix solid-phase dispersion followed by liquid chromatography-tandem mass spectrometry for the quantification of synthetic dyes in cosmetics and foodstuffs used or consumed by children. *J. Chromatogr. A* **2017**, *1529*, 29–38. [CrossRef] [PubMed]
40. Peng, L.Q.; Yi, L.; Yang, Q.C.; Cao, J.; Du, L.J.; Zhang, Q.D. Graphene nanoplatelets based matrix solid-phase dispersion microextraction for phenolic acids by ultrahigh performance liquid chromatography with electrochemical detection. *Sci. Rep.* **2017**, *7*, 7496. [CrossRef] [PubMed]
41. Chu, C.; Wei, M.; Wang, S.; Zheng, L.; He, Z.; Cao, J.; Yan, J. Micro-matrix solid-phase dispersion coupled with MEEKC for quantitative analysis of lignans in Schisandrae Chinensis Fructus using molecular sieve TS-1 as a sorbent. *J. Chromatogr. B* **2017**, *1063*, 174–179. [CrossRef] [PubMed]
42. Fumes, B.H.; Andrade, M.A.; Franco, M.S.; Lancas, F.M. On-line approaches for the determination of residues and contaminants in complex samples. *J. Sep. Sci.* **2017**, *40*, 183–202. [CrossRef] [PubMed]

43. Rajabi, M.; Sabzalian, S.; Barfi, B.; Arghavani-Beydokhti, S.; Asghari, A. In-line micro-matrix solid-phase dispersion extraction for simultaneous separation and extraction of Sudan dyes in different spices. *J. Chromatogr. A* **2015**, *1425*, 42–50. [CrossRef] [PubMed]
44. Gutierrez-Valencia, T.M.; Garcia de Llasera, M.P. On-line MSPD-SPE-HPLC/FLD analysis of polycyclic aromatic hydrocarbons in bovine tissues. *Food Chem.* **2017**, *223*, 82–88. [CrossRef] [PubMed]
45. Deng, D.; Zhang, S.; Chen, H.; Yang, L.; Yin, H.; Hou, X.; Zheng, C. Online solid sampling platform using multi-wall carbon nanotube assisted matrix solid phase dispersion for mercury speciation in fish by HPLC-ICP-MS. *J. Anal. At. Spectrom.* **2015**, *30*, 882–887. [CrossRef]
46. Ramos, J.J.; Rial-Otero, R.; Ramos, L.; Capelo, J.L. Ultrasonic-assisted matrix solid-phase dispersion as an improved methodology for the determination of pesticides in fruits. *J. Chromatogr. A* **2008**, *1212*, 145–149. [CrossRef] [PubMed]
47. Albero, B.; Sanchez-Brunete, C.; Miguel, E.; Tadeo, J.L. Application of matrix solid-phase dispersion followed by GC-MS/MS to the analysis of emerging contaminants in vegetables. *Food Chem.* **2017**, *217*, 660–667. [CrossRef] [PubMed]
48. Aznar, R.; Albero, B.; Ana Perez, R.; Sanchez-Brunete, C.; Miguel, E.; Tadeo, J.L. Analysis of emerging organic contaminants in poultry manure by gas chromatography-tandem mass spectrometry. *J. Sep. Sci.* **2018**, *41*, 940–947. [CrossRef] [PubMed]
49. Aznar, R.; Albero, B.; Sanchez-Brunete, C.; Miguel, E.; Martin-Girela, I.; Tadeo, J.L. Simultaneous determination of multiclass emerging contaminants in aquatic plants by ultrasound-assisted matrix solid-phase dispersion and GC-MS. *Environ. Sci. Pollut. Res.* **2017**, *24*, 7911–7920. [CrossRef] [PubMed]
50. Manoochehri, M.; Asgharinezhad, A.A.; Safaei, M. Determination of aflatoxins in rice samples by ultrasound-assisted matrix solid-phase dispersion. *J. Chromatogr. Sci.* **2015**, *53*, 189–195. [CrossRef] [PubMed]
51. Du, K.; Li, J.; Bai, Y.; An, M.; Gao, X. m.; Chang, Y.X. A green ionic liquid-based vortex-forced MSPD method for the simultaneous determination of 5-HMF and iridoid glycosides from Fructus Corni by ultra-high performance liquid chromatography. *Food Chem.* **2018**, *244*, 190–196. [CrossRef] [PubMed]
52. Caldas, S.S.; Soares, B.M.; Abreu, F.; Castro, I.B.; Fillmann, G.; Primel, E.G. Antifouling booster biocide extraction from marine sediments: A fast and simple method based on vortex-assisted matrix solid-phase extraction. *Environ. Sci. Pollut. Res.* **2018**, *25*, 7553–7565. [CrossRef] [PubMed]
53. Du, K.; Li, J.; Tian, F.; Chang, Y.X. Non-ionic detergent Triton X-114 Based vortex- synchronized matrix solid-phase dispersion method for the simultaneous determination of six compounds with various polarities from Forsythiae Fructus by ultra high-performance liquid chromatography. *J. Pharm. Biomed. Anal.* **2018**, *150*, 59–66. [CrossRef] [PubMed]
54. Leon-Gonzalez, M.E.; Rosales-Conrado, N. Determination of ibuprofen enantiomers in breast milk using vortex-assisted matrix solid-phase dispersion and direct chiral liquid chromatography. *J. Chromatogr. A* **2017**, *1514*, 88–94. [CrossRef] [PubMed]
55. Soares, K.L.; Rodrigues Cerqueira, M.B.; Caldas, S.S.; Primel, E.G. Evaluation of alternative environmentally friendly matrix solid phase dispersion solid supports for the simultaneous extraction of 15 pesticides of different chemical classes from drinking water treatment sludge. *Chemosphere* **2017**, *182*, 547–554. [CrossRef] [PubMed]
56. Hertzog, G.I.; Soares, K.L.; Caldas, S.S.; Primel, E.G. Study of vortex-assisted MSPD and LC-MS/MS using alternative solid supports for pharmaceutical extraction from marketed fish. *Anal. Bioanal. Chem.* **2015**, *407*, 4793–4803. [CrossRef] [PubMed]
57. Fotouhi, M.; Seidi, S.; Shanehsaz, M.; Naseri, M.T. Magnetically assisted matrix solid phase dispersion for extraction of parabens from breast milks. *J. Chromatogr. A* **2017**, *1504*, 17–26. [CrossRef] [PubMed]
58. Yu, X.; Yang, H.S. Pyrethroid residue determination in organic and conventional vegetables using liquid-solid extraction coupled with magnetic solid phase extraction based on polystyrene-coated magnetic nanoparticles. *Food Chem.* **2017**, *217*, 303–310. [CrossRef] [PubMed]
59. Yu, X.; Li, Y.X.; Ng, M.; Yang, H.S.; Wang, S.F. Comparative study of pyrethroids residue in fruit peels and fleshes using polystyrene-coated magnetic nanoparticles based clean-up techniques. *Food Control* **2018**, *85*, 300–307. [CrossRef]

60. Ma, S.Q.; Tu, X.J.; Dong, J.T.; Long, P.; Yang, W.C.; Miao, X.Q.; Chen, W.B.; Wu, Z.H. Soxhlet-assisted matrix solid phase dispersion to extract flavonoids from rape (*Brassica campestris*) bee pollen. *J. Chromatogr. B* **2015**, *1005*, 17–22. [CrossRef] [PubMed]
61. Tu, X.J.; Ma, S.Q.; Gao, Z.S.; Wang, J.; Huang, S.K.; Chen, W.B. One-step extraction and hydrolysis of flavonoid glycosides in rape bee pollen based on Soxhlet-assisted matrix solid phase dispersion. *Phytochem. Anal.* **2017**, *28*, 505–511. [CrossRef] [PubMed]

© 2018 by the authors. Licensee MDPI, Basel, Switzerland. This article is an open access article distributed under the terms and conditions of the Creative Commons Attribution (CC BY) license (http://creativecommons.org/licenses/by/4.0/).

Review

Sample Digestion and Combined Preconcentration Methods for the Determination of Ultra-Low Gold Levels in Rocks

Yan-hong Liu, Bo Wan * and Ding-shuai Xue

State Key Laboratory of Lithospheric Evolution, Institute of Geology and Geophysics, Institutions of Earth Science, Chinese Academy of Sciences, Beijing 100029, China; liuyanhong@mail.iggcas.ac.cn (Y.-h.L.); xuedingshuai@mail.iggcas.ac.cn (D.-s.X.)
* Correspondence: wanbo@mail.iggcas.ac.cn; Tel.: +86-010-82998154

Academic Editor: Nuno Neng
Received: 27 March 2019; Accepted: 2 May 2019; Published: 8 May 2019

Abstract: The gold abundance in basic rocks, which normally varies between 0.5 and 5 ppb, has served as a very important indicator in many geoscience studies, including those focused on the planetary differentiation, redistribution of elements during the crustal process, and ore genesis. However, because gold is a monoisotopic element that exhibits a nugget effect, it is very difficult to quantify its ultra-low levels in rocks, which significantly limits our understanding of the origin of gold and its circulation between the Earth crust, mantle, and core. In this work, we summarize various sample digestion and combined preconcentration methods for the determination of gold amounts in rocks. They include fire assay, fire assay combined with Te coprecipitation and instrumental neutron activation analysis (INAA) or laser ablation inductively coupled plasma mass spectrometry, fusion combined with Te coprecipitation and anion exchange resins, dry chlorination, wet acid digestion combined with precipitation, ion exchange resins, solvent extraction, polyurethane foam, extraction chromatography, novel solid adsorbents, and direct determination by INAA. In addition, the faced challenges and future perspectives in this field are discussed.

Keywords: gold; sample preparation; preconcentration; geological samples

1. Introduction

Gold is one of the rare elements and precious metals present in the Earth crust with average concentrations in the igneous, sedimentary, and metamorphic rocks varying between 0.5 and 5 ppb [1]. Proper quantification of the gold abundances in basic rocks is critical for many leading-edge areas of geoscience, such as planetary differentiation [2–10], redistribution of elements during crustal processes [11–17], and ore genesis [18–23]. According to the latest results reported by Brenan and McDonough [10], the metal–silicate partition coefficient of Au is approximately 300, whereas its minimum values measured for Os and Ir in the same experiments are $\sim 10^7$, which differs from the former parameter by at least a factor of 10^4. The authors concluded that not all highly siderophile elements (HSEs) were affected by the core formation in the same way, and that the abundances of elements such as osmium and iridium required the addition of a late veneer. As an illustration, Figure 1 shows the metal-silicate partition coefficients plotted as functions of the oxygen fugacity. Fischer-Gödde et al. analyzed samples of orogenic peridotite massifs and xenoliths, whose Rh and Au contents revealed the presence of HSEs in the primitive mantle (PM), which differed from that of any known group of chondrites and could be explained by the contributions from meteoritic components detected in ancient lunar impact melt rocks [5]. Figure 2 shows the ratio diagrams of Au/Ir vs. ^{187}Os/^{188}Os and Au/Ir vs. Rh/Ir constructed for the HSEs of the PM and different chondrite classes and groups.

In addition, as no explorations conducted during the past 50 years detected deep gold deposits in Earth's crust, the world is thought to be tottering on the precipice of peak gold [24]. As a result, it is very important to discover large deposits of gold and meet the production demand in the field of gold ore genesis. However, the precise determination of ultra-low gold contents in rocks is an extremely difficult task as compared with detecting other trace elements because gold represents a monoisotopic element and exhibits a nugget effect. The latter requires the analysis of relatively large amounts of rocks to obtain meaningful data, which are still characterized by large deviations. Since gold has only one isotope (^{197}Au), it cannot be quantified by an isotope dilution (ID) method that is very precise for trace element determination and, therefore, requires a simple analytical method for its quantitative recovery during pretreatment.

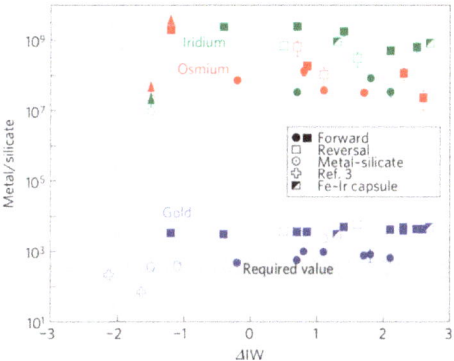

Figure 1. Metal-silicate partition coefficients as functions of the oxygen fugacity. Reprinted with permission from [10].

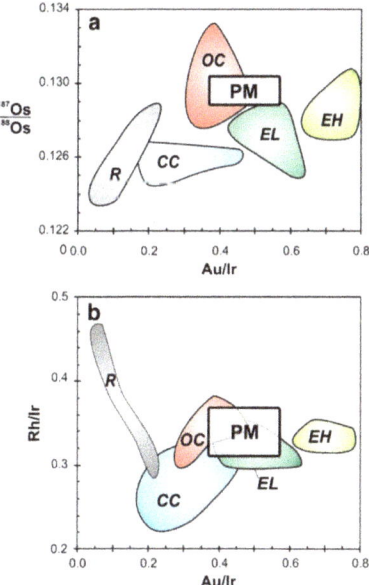

Figure 2. Ratio diagrams of Au/Ir vs. ^{187}Os/^{188}Os and Au/Ir vs. Rh/Ir constructed for the PM and different chondrite classes and groups. The HSE composition determined for the PM model is similar to those of ordinary (OC) and enstatite (EL) chondrites, but differs from the compositions of carbonaceous (CC) and Rumuruti (R) chondrites. Reprinted with permission from [5].

Although modern analytical techniques such as inductively-coupled plasma mass spectrometry (ICP-MS), graphite furnace atomic absorption spectroscopy (GFAAS), and instrumental neutron activation analysis (INAA) are highly sensitive, their use for the direct determination of gold levels in geological samples is complicated because of the low concentrations of gold and interfering effects of matrix components. To increase the reliability of gold quantitation methods, the major matrix components must be separated first, which can be achieved by proper sample preparation and the preconcentration of gold [25–28].

Sample digestion and/or preconcentration technologies used for the determination of gold as well as platinum group metals (PGEs) with ppm concentrations were previously reviewed by Perry, Barefoot, Balcerzak, Pyrzynska, Mokhodoeva, Myasoedova, and their co-authors [25,29–35]; however, methods for the detection of ultra-low gold levels in rocks (which are considerably more complex than ore-grade samples) have not been discussed in detail. This review includes the recent significant contributions to the determination of gold amounts in rocks, especially those focused on sample preparation and the preconcentration of gold prior to its determination. As of today, almost all distributions and parameters of gold in geological samples have been determined by fire assay. To validate them and obtain more data, independent methods such as diisobutyl ketone (DIBK) extraction chromatography and standardization combined with ID ICP-MS cation exchange resin analysis have been developed [5,36]. However, due to the long procedure or expensive spiking, a simple and reliable technique for measuring the contents of gold in rocks must be used. In this work, we describe various sample digestion methods that are often coupled with enrichment techniques and discuss their mutual effects.

2. Dry Digestion Methods

2.1. Fire Assay

Fire assay (FA) and cupellation methods are classical assaying techniques that have been successfully used for the estimation of gold amounts in ores for many centuries [37]. Furthermore, FA has always served as the arbitration method of gold measurement that involves not only the digestion of samples, but also the enrichment of gold and PGEs since it allows the extraction of these precious metals and their separation from base metals in the matrices. The relatively large sample weights used for FA can overcome the nugget effect, which represents its significant advantage. However, the large reagent blanks that result from the large amounts of fluxes introduce significant biases into the determination of ultra-low gold levels. In addition, the uncertainty of the chemical interactions between various flux constituents makes the quality of the obtained results highly dependable on the experience of the analyst.

Lead fire assay (Pb-FA) and nickel sulfide fire assay (NiS-FA) are widely used FA methods. Pb-FA utilized for the collection of Au, Pt, Pd, and Rh is the application of metallurgy in analysis. A simple Pb-FA method has been developed in 1994 for the determination of ultra-low Au, Pt, and Pd contents in rocks by Hall et al. [38]; its limit of detection (LOD) and recovery of gold are 2 ppb and 90%, respectively. The large relative standard deviations (RSD) of 11–55% are likely caused by the sample inhomogeneity rather than by the analytical method. This conclusion is in good agreement with the recoveries of gold ranging from 74–86% at concentrations between 30 and 300 ppb [39]. NiS-FA can be used to collect both PGEs and gold, but its efficiency for gold determination has been very low. Juvonen et al. [40] compared NiS-FA with Pb-FA in terms of their collection efficiencies of Au, Pt, Pd, and Rh and found that the gold recovery of NiS-FA was twice as low as that of Pb-FA at low gold concentrations. Plessen and Erzinger [41] reported that the gold recovery did not exceed 70% when NiS-FA was used to analyze rocks. Although the gold recovery of Pb-FA is better than that of NiS-FA, the environmental pollution and harm to analysts caused by Pb discouraged its further development [42,43]. However, the bottleneck of the analysis of the low gold contents in rocks by NiS-FA consists of the high amounts of reagent blanks and low recovery efficiency of gold.

Many researchers have attempted to solve this problem. In order to lower the reagent blank, Asif and Parry [44] prepared a mini button by reducing the amount of nickel reagent; however, the recovery of gold exhibited a significant reduction. Lu et al. [45] found that FA combined with a Te coprecipitation purification process could lower the NiO blank to 0.24 ppb. In terms of the gold recovery, two specific directions exist: combining NiS-FA with Te coprecipitation to reduce the loss during the acid dissolution process and employing a solid direct determination technology (such as INAA and laser ablation inductively coupled plasma mass spectrometry (LA-ICPMS)) to reduce the loss and possible pollution in each process following the FA step.

2.1.1. NiS-FA + Te Coprecipitation

As large amounts of gold and small PGEs contents are lost in the acid solution during the dissolution of the NiS button, a second enrichment step that would not increase the amount of total dissolved solids must be introduced to improve the gold recovery (especially at low concentrations). Jackson et al. [46] applied the Te coprecipitation method to gold analysis for the first time and increased the recovery of gold from ore samples to 90%. Savard et al. [47] obtained the same results for ore samples, but the recovery remained low for rock samples. In addition, although the sample weight in this method was as large as 15 g, the RSD amounted to 20–50% (the reason for this phenomenon is not clearly understood). Oguri et al. [48] obtained high recoveries (>97%) for low gold concentrations by repeating the NiS-FA procedure under the reduced conditions (produced by graphite powders) and Te coprecipitation at the optimal conditions corresponding to a temperature of 210 °C and time of 75 min. In order to lower the reagent blank and simplify the process, a semi-open NiS dissolution system preventing volatile losses was suggested by Gros et al. [49]. In addition, Sun and Sun [50] proposed a novel NiS-FA method involving Fe, which ensured the self-disintegration of the entire assay button into powder without its mechanical crush. When some particular samples are used (such as black shale or samples containing magnetite), conventional NiS-FA is not applicable. Li [51] ignited black shale samples before the NiS-FA step to eliminate organic matter. Juvonen et al. [52] used potassium tartrate as a reducing agent to successfully prepare an assay button during the analysis of samples containing magnetite.

2.1.2. FA + INAA or LA-ICPMS

After a method that used only 0.5 g of nickel was developed (which decreased the button weight to below 1 g), it enabled the direct analysis of small gold-containing buttons by INAA or LA-ICPMS [44]. Asif et al. [53] proposed a simple method based on NiS-FA and INAA to determine the levels of PGEs and gold in samples. However, the LOD of gold was 2 ppb, which made this technique unsuitable for ultra-low gold contents. Bedard and Barnes [54] compared the capacities of FA-ICP-MS and FA-INAA to determine the gold amounts in geological samples and found that for the specimens rich in gold both methods performed adequately; however, for the low-concentration samples (crustal rocks), ICP-MS was preferable. Jarvis et al. [55] established for the first time a method for gold quantitation based on the combination of NiS-FA with LA-ICP-MS having an LOD of 10 ppb. An ultraviolet (UV) laser ablation ICP-MS was used to directly analyze the gold contents in NiS-FA buttons by Jorge et al. [56]. The obtained LOD of gold was as low as 1.7 ppb, and the RSDs were better than 10%. Later, Resano et al. [57,58] ground NiS buttons to obtain more homogeneous samples and used polyethylene wax as a binder to pelletize the resulting powders (possible interferences were eliminated by utilizing a double-focusing sector field mass spectrometer). Resano et al. [59], Vanhaecke et al. [60], and Compernolle et al. [61] discussed the possibility of combining Pb-FA with LA-ICPMS for the analysis of gold in geological samples. Meanwhile, Compernolle et al. [61] reported the absence of significant differences between the results obtained by the standard addition, internal standard, and external standard methods using Pb-FA with LA-ICPMS for gold determination. Table 1 summarizes various FA digestion methods used in literature studies. Figure 3 illustrates the main stages of the four techniques for the gold determination by the NiS-FA method described in several works. Here, number 1 denotes the method that combines NiS-FA alone with ICP-MS according to Plessen and Erzinger [41]. The polyethylene terephthalate (PET) bottle is used to store the gold solution. Number 2 denotes the technique that combines the NiS-FA method with Te coprecipitation for gold extraction based on the works of Jackson et al. and Savard et al. [46,47]. Simpler methods that combine NiS-FA with INAA or LA-ICP-MS are represented by numbers 3 and 4, respectively. They are based on the approaches developed by Asif et al. and Jorge et al. [53,56].

Table 1. A summary of various FA digestion methods.

Sample Weight/g	Collector/Flux	Separation Technique	Detection Technique	LOD/ppb	Reference
10–30	Pb/Na$_2$CO$_3$, Na$_2$B$_4$O$_7$, SiO$_2$, flour(C)		ICP-MS	2	[38]
50	NiS/Na$_2$CO$_3$, Na$_2$B$_4$O$_7$, CaF$_2$		ICP-MS	0.023	[41]
15	NiS/Na$_2$CO$_3$, Na$_2$B$_4$O$_7$, SiO$_2$	Te coprecipitation	ICP-MS	1.69	[46]
15	NiS/Na$_2$CO$_3$, Na$_2$B$_4$O$_7$, SiO$_2$	Te coprecipitation	ICP-MS	0.484	[47]
20	NiS/Na$_2$CO$_3$, Na$_2$B$_4$O$_7$, SiO$_2$	duplicate NiS-FA and Te coprecipitation	ICP-MS	0.053	[48]
20	NiS/Na$_2$CO$_3$, Li$_2$B$_4$O$_7$, SiO$_2$	Te coprecipitation	ICP-MS	0.33	[49]
10	NiS/Na$_2$CO$_3$, Na$_2$B$_4$O$_7$		INAA	2	[53]
10–15	NiS/Na$_2$CO$_3$, Na$_2$B$_4$O$_7$		UV-LA-ICP-MS	1.7	[54,60]
50	Pb/Na$_2$CO$_3$, NaOH, Na$_2$B$_4$O$_7$, SiO$_2$, C		FS-LA-ICP-MS	4	

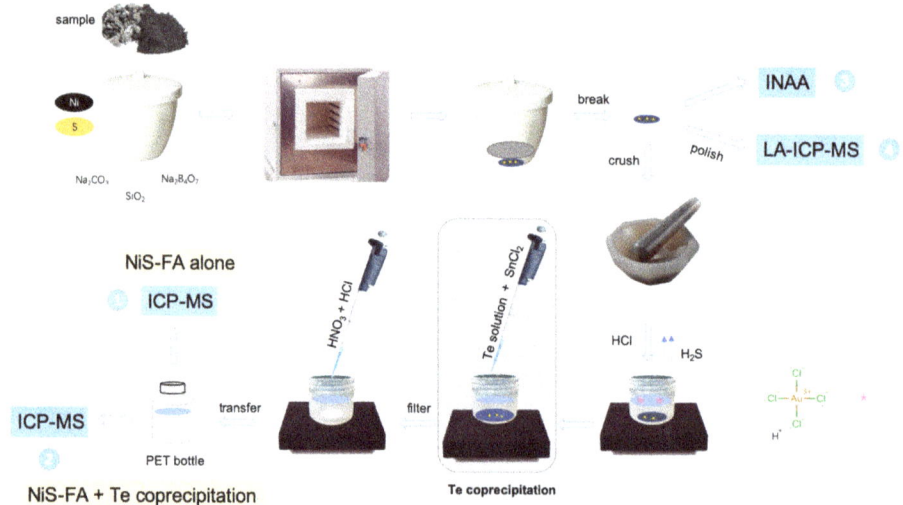

Figure 3. Main stages of the gold determination by the NiS-FA method.

2.2. Fusion

Fusion is an effective sample decomposition method that is mainly used to dissolve acid-insoluble substances. In the process of fusion, insoluble samples are converted into sodium compounds that are soluble in water or acids. Fusion is very different from FA, although both techniques are dry digestion methods. In particular, fusion is just a sample digestion method, whereas FA involves both sample digestion and gold enrichment, and its ingredients and process are more complicated. Fusion is rarely used to decompose samples for the determination of gold contents; for this purpose, a sodium peroxide method is most commonly used. The greatest advantages of this technology include the ability to effectively decompose sulfide and refractory minerals and a large sample mass (up to 20 g). However, large contents of sodium peroxide with impurities are difficult to purify, and the contaminations resulting from the interaction of the flux with the crucible wall can negatively affect the course of analysis [62]. Another disadvantage of this method is the formation of gel-like soluble alkaline silicates during fusion that make the solution difficult to analyze.

Due to the absence of gold enrichment, fusion buttons are not suitable for direct testing (such as FA), but must be dissolved and treated by other enrichment techniques. The most common combination technologies utilize ion exchange resins [63,64] and Te co-precipitation [65–67]. A summary of various fusion digestion methods is presented in Table 2.

Table 2. A summary of various fusion digestion methods.

Sample Weight/g	Crucible	Flux	Separation Technique	Detection Technique	LOD/ppb	Reference
0.5	zirconium	Na_2O_2	Anion resin	GFAAS	-	[63]
1.0	graphite	Na_2O_2	Anion resin	USN-ICP-TMS+NAA	ppt/-	[64]
1–20	zirconium	$Na_2O_2/NaKCO_3/KOH$	Se-Te coprecipitation	ICP-MS	0.58	[65]
1–20	Corundum	Na_2O_2	Te coprecipitation	ICP-MS	0.007	[66,67]
2	Corundum	Na_2O_2	Te coprecipitation	ICP-MS	0.32	

2.2.1. Fusion + Anion Exchange Resin

Enzweiler [63] studied the recovery efficiencies of Pt, Pd, and Au from silicate rocks using a sodium peroxide fusion procedure followed by anion exchange resin separation. The utilized technique was found to be not very efficient: the recovery of gold was as low as 76% due to the formation of hydroxychloro compounds in alkaline solution that were not converted to chloro complexes upon the acidification with HCl that was required for quantitative anion exchange. Later, Dai et al. [64] combined an extra chlorination step with this technique to oxidize PGEs and gold and thus achieve the best retention in the anion resin separation process. However, no valuable data were obtained for gold because of the poor linear regression correlation of the external calibration curve for ICP-MS analysis.

2.2.2. Fusion + Te Coprecipitation

Amosse [65] described a method for the extraction of Pd, Pt, Rh, Ru, Ir, and Au using Se and Te as carriers in the presence of catalyst KI after the fusion with sodium peroxide, sodium potassium carbonate, and potassium hydroxide. The catalyst improved the recovery of Ir from 33% to 97.5%, although it had no effect on the Au recovery efficiency. In order to simply this process, Jin and Zhu [66] used sodium peroxide fusion and Te coprecipitation to measure the PGE and Au levels in geological matrices. The procedural reagent blanks and recovery of Au were 0.044 ppb and 80%, respectively. Qi et al. [67] purified HCl and $SnCl_2$ by Te coprecipitation to further lower the reagent blanks and achieved a good Au recovery of 96.3%.

2.2.3. Fusion of the Residue Formed after Acid Digestion

Another important application of fusion is the ability to recover all gold from the sample residue after acid digestion. Totland et al. [68] used the Na_2O_2 + Na_2CO_3 mixture and pure Na_2O_2 to dissolve the sample residue formed after the HNO_3 + HCl + HF + $HClO_4$ microwave digestion process. Jarvis et al. [69], Tsimbalist et al. [70], and Coedo et al. [71] employed the same method to treat the acid-insoluble residue formed after the HNO_3 + HCl + HF acid digestion. The insoluble residue of black shale samples produced after BrF_3 digestion was melted by $KBrF_4$ [72].

2.3. Dry Chlorination

Chlorine has been the most important gold leaching agent from 1850 to 1900. Afterwards, because of the higher selectivity and lower price of cyanide, the latter gradually replaced chlorine in the gold leaching of sulfide ore. Since that time, little research on the chlorination of gold has been conducted. The advantages of chlorination include very low procedural reagent blanks, high analytical efficiency, and large sample weight. However, the low recovery of gold is a drawback of this method. Nesbitt et al. [73] studied the chlorination mechanism of gold in aqueous solutions and found that the gold particle size was the main factor affecting it. Perry and Van Loon [74] utilized the chlorination process for the determination of ultra-low gold contents in rocks for the first time. A gold recovery of 60% was achieved under the following optimal conditions: 3.5 h, 580 °C, 0.5 g NaCl, and 15 g sample powder. Although the recovery was relatively low, the sensitivity and precision of this technique were comparable or better than those of the FA method. In order to further improve the data precision, Perry et al. [75] attempted to increase the sample weight to 250 g to reduce the nugget effect. Furthermore, chlorination was also used to oxidize PGEs and gold to higher oxidation states that were more strongly absorbed by anion resins [64].

3. Wet Acid Digestion Methods

Wet acid digestion is a widely used sample preparation technique. It is considered an alternative to FA for the extraction of PGEs and gold from geological samples. This method has many advantages, such as simplicity, high speed, low cost, robustness, simple reagent ingredients, low blank, and high degree of sample universality. However, the sample weight used during acid treatment is usually below 10 g, which is smaller than that of the dry digestion methods; as a result, the sampling errors obtained by this technique are large due to the stronger nugget effect. In addition, its efficiency depends on the ratio of the sample weight to the acid volume.

The best acid for gold extraction is aqua regia (aq. reg.) that can dissolve compounds insoluble in HCl or HNO_3 alone. The addition of HF, NH_4F, Br_2, $KClO_3$, or H_2O_2 significantly increases its strength. HF or NH_4F can react with SiO_2 to form SiF_4 and destroy the silicate structure, which facilitates the quantificational extraction of gold from geological samples. In many studies [37,39,76–79], it was found that HSEs could not be quantifiably extracted from rocks without desilication. Br_2, $KClO_3$, or H_2O_2 can oxide HCl to produce more chlorine gas, which increases the gold solubility. Normally, an additional preconcentration step must be performed after acid digestion during the analysis of ultra-low gold contents in rocks. The commonly used enrichment methods use precipitation [26,80–83], ion exchange resins [5,71,81,84–88], solvent extraction [27,89–98], polyurethane foam (PUF) [99–130], extraction chromatography [36,131–133], and solid adsorbents [134–137]. Figure 4 shows the common preconcentration methods utilized with wet acid digestion.

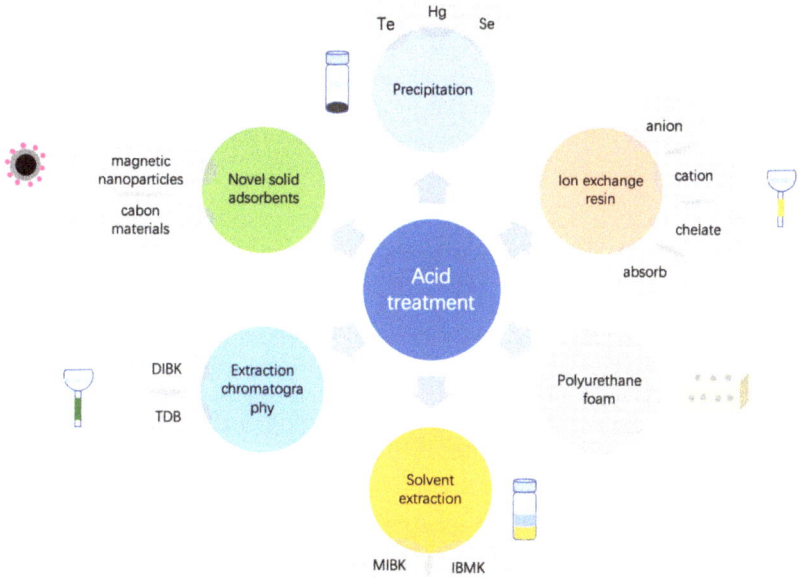

Figure 4. Common preconcentration methods used with wet acid digestion.

3.1. Acid Digestion + Precipitation

A method combining acid digestion with Te coprecipitation was originally used in 1978 to measure the amounts of gold and silver in rocks [80]. Later, Gupta [81] pointed out that Te interfered with the determination of gold and silver during the atomization step; therefore, this method was not suitable for quantifying gold and silver, but could be used to measure the contents of other PGEs instead. In order to prevent the loss of gold during the dissolution of tellurium buttons, acid digestion, Te coprecipitation, and INNA were utilized to determine the gold contents in rocks [82]. The gold recovery and LOD of this method were 96% and 0.7 ppb, respectively. Eller et al. [26] used a polytetrafluoroethylene (PTFE) pressure bomb combined with HNO_3, HF, aqua regia, and $HClO_4$ to dissolve rock samples, while Se coprecipitation was performed to preconcentrate gold. The obtained recovery of Au was greater than 98%. Niskavaara and Kontas [83] combined HF and aqua regia acid digestion with Hg coprecipitation to determine the concentrations of Au, Pd, Pt, Rh, Ag, Te, and Se in geological samples. The observed poor precision of gold in the low-concentration samples was likely caused by sample heterogeneity. Table 3 lists various methods for Au preconcentration.

Table 3. Preconcentration of Au by various precipitation methods.

Sample Weight/g	Digestion	Dissolution	Separation Technique	Detection Technique	LOD/ppb	Reference
2–5	Teflon beaker	HF + aq.reg.	Te coprecipitation	GFAAS	-	[81]
0.3–1.3	Teflon beaker	HNO_3 + HF + $HClO_4$ + HCl	Te coprecipitation	INAA	0.7	[82]
0.5–1.5	PTFE bomb	HNO_3 + HF + aq.reg. + $HClO_4$ + HCl	Se coprecipitation	GFAAS/TXRF	0.2/1.2	[26]
0.5	Borosilicate tube	HCl + HNO_3	Hg coprecipitation	GFAAS	0.3	[83]

3.2. Acid Digestion + Ion Exchange Resin

- Anion exchange resins

Since gold exists in the form of AuCl$_4^-$ complex anions under acidic conditions, when an acidic solution containing AuCl$_4^-$ is passed through an anion exchange resin, AuCl$_4^-$ is adsorbed on the resin and becomes separated from other basic metals. However, it is difficult to elute the AuCl$_4^-$ species strongly adsorbed on the anion resin surface. Barredo and Polo [84] developed a method for the analysis of Au, Ag, and Cd in rock samples, which were digested with HF, aqua regia, and HClO$_4$, preconcentrated with Dowex 1-X8 anion resin, and eluted with ammonia solution.

- Cation exchange resins

In contrast to the anion exchange resins, when an acidic solution containing AuCl$_4^-$ is passed through a cation exchange resin, gold is transported through the resin column in the form of AuCl$_4^-$ complex anions separated from metal cations. Gupta [81,85] compared the enrichment efficiencies of Dowex 50W-X8 cation exchange resin combined with Te coprecipitation for PGEs and gold. The cation exchange resin method is recommended for the determination of µg/g levels of gold. Based on the results reported by Meisel et al. [86] showing that all PGEs exhibit similar chemical behavior in a chromatographic column, they can be used as the ideal internal standards for the calculation of Rh concentrations. Fischer et al. [5] used a Carius tube filled with inverse aqua regia to dissolve rock samples and Eichrom 50W-X8 cation exchange resin to separate Re, Ir, Ru, Pt, Rh, Pd, and Au from the matrix. The concentrations of monoisotopic Rh and Au elements were calculated by the standardization to the ^{101}Ru and ^{193}Ir signal intensities, and the Ru and Ir concentrations were determined by isotope dilution.

- Chelating resins

Chelating resins contain chelating groups that can selectively adsorb gold and separate it from other matrix elements in solution. They combine the ionization exchange and complexation reactions and, therefore, exhibit good selectivity and strong binding energies as compared with those of the ordinary ion exchange resins. Coedo et al. [71] decomposed geological samples by aqua regia and HF and then separated PGEs and gold using tetraethylenepentamine chelating resin. Wu et al. [87] used YPA$_4$ chelating resin as both the solid phase extractant and chemical modifier to determine the Au, Pd, and Pt contents in geological samples by electrothermal vaporization inductively coupled plasma atomic emission spectrometry (ETV-ICP-AES). The elution of gold from the two chelating resins mentioned above was performed by ashing.

- Chelate absorption resins

Chelate absorption resins combine the high selectivity of chelating resins with the high adsorption efficiency of absorption resins. Spheron Thiol 1000 chelate absorption resin was used to extract low gold levels by Medved et al. [88]. Table 4 summarizes the Au extraction methods using ion exchange resins.

Table 4. Various methods for the preconcentration of Au by ion exchange resins.

Sample Weight/g	Digestion	Dissolution	Separation Technique	Detection Technique	LOD/ppb	Reference
2	-	HNO$_3$ + HCl + HClO$_4$ + HF	Anion exchange resin	GFAAS	0.2	[84]
5	Teflon beaker	HF + aq.reg. + HNO$_3$ + HCl	Cation exchange resin	GFAAS	-	[85]
2	Carius tube/HPA-s	HCl + HNO$_3$	Cation exchange resin	ICP-MS	-	[5]
0.25	Microwave digestion	aq.reg. + HF + HClO$_4$	Chelating resin	FI-ICP-MS	1.2	[71]
0.05–1.5	Mild heating	HNO$_3$ + HClO$_4$ + HF	Chelating resin	ETV-ICP-AES	0.075	[87]
5–10	-	aq.reg.	Chelating sorbent	GFAAS	0.5	[88]

3.3. Acid Digestion + Solvent Extraction/Dispersive Liquid–Liquid Microextraction

Solvent extraction (also called liquid-liquid extraction) is a classical method for Au separation and preconcentration characterized by the low reagent blank, high enrichment efficiency, and simple operation. The most widely used solvents for low-level Au determination are methyl isobutyl ketone (MIBK) and isobutyl methyl ketone (IBMK).

Terashima [89] established a simple method utilizing aqua regia for sample digestion and MIBK for gold extraction to determine gold concentrations in 60 geological reference materials. As aqua regia only partially interacted with the samples, Terashima et al. [90] adopted aqua regia and HF for dissolving the entire gold content. The obtained results were very close to those of INAA. Normally, the solvent extraction method is used in conjunction with GFAAS. However, the Fe spectral line of 242.4 nm can interfere with the sensitive Au line of 242.8 nm in the determination of Au by GFAAS. In addition, as the polarities of Au and Fe are close, they are always extracted together by the same solvent. Therefore, the elimination of iron impurities from the organic phase is the key to the accurate determination of low gold levels. A two-stage solvent extraction method (using diethyl ether and MIBK) was proposed by Yokohama et al. to prevent iron interference and effectively concentrate gold [27]. The results obtained for reference materials were in good agreement with the INAA data. Ramesh et al. [91] used the Zeeman background correction technology for GFAAS, washed the MIBK organic phase with a wash solution, and centrifuged it twice for the total removal of iron.

Chattopadhyay and Sahoo [92] used the sequential digestion of HBr-Br_2 and aqua regia, sequential extraction of IBMK and toluene, and Te coprecipitation enrichment method to determine traces of gold in geological samples. Monteiro et al. [93] examined the stability of gold in IBMK extracts and found that closed polypropylene containers (with absorbance measurement changes of less than 3%) were more suitable than both the open polypropylene and closed glass containers over periods of up to 22 h.

As the liquid-liquid extraction method requires the use of large volumes of organic solvents (which are often toxic), Rezaee et al. [94] introduced a dispersive liquid–liquid microextraction (DLLME) method that was highly sensitive, efficient, and powerful for the preconcentration and determination of trace elements. Shamsipur and Ramezani [95] applied the DLLME technology to form an adduct between Victoria Blue dye and $AuCl_4^-$ using acetone as a dispersant and chlorobenzene as an extractant to determine the ultra-trace amounts of gold by GFAAS. Another DLLME method for the determination of gold traces using dicyclohexylamine as the extractant, acetone as the dispersant, and chloroform as the extraction solvent was established by Kagaya et al. [96]. Calle et al. [97] applied the DLLME technology to preconcentrate the ion pairs formed between $AuCl_4^-$ and $[CH_3(CH_2)_3]_4N^+$ in a microliter-range volume of chlorobenzene using acetone as a disperser solvent for the determination of ultra-low gold contents. Fazelirad et al. [98] used benzyldimethyltetradecyl ammonium chloride dihydrate to form an ion pair with $AuCl_4^-$, acetone as the dispersant, and 1-hexyl-3-methylimidazolium hexafluorophosphate ([Hmim][PF_6]) ionic solution as the extractant for gold extraction. Table 5 lists the solvent extraction/dispersive liquid-liquid microextraction methods for Au preconcentration.

Table 5. Methods for the preconcentration of Au by solvent extraction/dispersive liquid-liquid microextraction.

Sample Weight/g	Digestion	Dissolution	Separation Technique	Detection Technique	LOD/ppb	Reference
0.1–2.0	Borosilicate beaker	HNO_3 + HCl	MIBK	GFAAS	-	[89]
0.5–2.0	Teflon beaker	aq.reg. + HF	MIBK	GFAAS	-	[90]
1	PFTE beaker	$HClO_4$ + HF + aq.reg.	diethyl ether + MIBK	GFAAS	0.13	[27]
10	Glass beaker	aq.reg.	MIBK	GFAAS	0.1	[91]
5–10	Borosilicate beaker	HBr + Br_2	IBMK	GFAAS	15	[92]
10	Erlenmeyer flask	HCl + HNO_3	IBMK	GFAAS	0.2	[93]
0.2	-	aq.reg.	DLLME	GFAAS	0.005	[95]
0.003–0.03	Eppednorf vail	HNO_3 + HCl	DLLME	GFAAS	1.5	[97]
0.02	-	HNO_3 + HCl + HF	DLLME	GFAAS	0.002	[98]

3.4. Acid Digestion + PUF

PUF was originally used for the selective adsorption of gold from 0.2 M HCl solution in 1970 by Bowen [99]. After that, this technology has become widely spread, owing to its excellent adsorption selectivity, high enrichment efficiency, simple operation process, and low analysis cost [100,101]. The mechanism of PUF adsorption may be based on physical adsorption, solvent extraction, or ion exchange. However, the majority of studies support the solvent extraction-based mechanism. Gesser [102], Oren [103], Lo and Chow [104], and Jones et al. [105] examined the PUF adsorption of Ga, Fe, Sn, and Rh, and concluded that PUF was a "solid solvent-extractant". In addition, a possible cation chelation mechanism was suggested by Hamon based on the results of his detailed studies on the PUF adsorption of Co thiocyanate and salts of several organic acids [106]. However, as shown by Wang et al. [107], the PUF adsorption of gold involved a reduction reaction, indicating that PUF reduced $AuCl_4^-$ (+3) to Au (0) followed by its deposition on the foam surface (Figure 5). It can be hypothesized that the mechanisms of PUF adsorption are not the same for different compounds.

Figure 5. Scanning electron microscopy images of (**a**) PUF, (**b**) PUF-NH$_2$, and (**c**) PUF-NH$_2$ adsorbed Au. Reprinted with permission from [107].

PUF can be used as a special adsorbent for gold due to its very high selectivity. Schiller [108] compared the adsorption efficiencies of PbS, Fe(OH)$_3$, Al(OH)$_3$, Dowex 1-X8, and PUF toward gold. It was found that quantitative recovery of gold could be achieved within 90 min using PbS and PUF. However, the PUF method is superior to the PbS one in terms of the signal-to-noise ratio during the testing step. With the continuous optimization of various experimental conditions (including those for the pretreatment of PUF [109,110], types and concentrations of the acid digestion reagents and adsorption conditions [111–114], and elution conditions [115,116]), this method has become one of the main techniques for the enrichment of gold in geological samples [117–121]. To further improve the selectivity and adsorption capacity of PUF, many researchers coated it with organic reagents (such as MIBK and TBP) [122–126], bonded chelating ligands to the PUF matrix [127], and functionally modified it by the ligand coupling [128–130] with the PUF skeleton.

3.5. Acid Digestion + Extraction Chromatography

As extraction chromatography represents a chromatographic method, it exhibits the basic characteristics of chromatography. Unlike the commonly used chromatographic techniques, in this method, an inert carrier of organic extractant is supported on the column as the stationary phase for separation, and a solution of various inorganic acids is utilized as the mobile phase. This technology combines the high selectivity of extractants in solvent extraction with the effectiveness of chromatography separation, which significantly reduces the amounts of organic extractants and is less dangerous and easy to operate. In the 1970s, Pohlandt and Steele [131,132] and Bao [133] began to use porous silicon and polytrifluorochloroethylene as an inert carrier for tributyl phosphate extractant to analyze the gold amounts in ore samples and assay grains. Later, Pitcairn and Warwick [36] coated polyacrylamide resin with DIBK to preconcentrate ultra-low gold contents in rocks. The LOD of this method was as low as 0.002 ppb.

3.6. Acid Digestion + Novel Solid Adsorbents

Activated carbon enrichment is a widely used method for gold determination. This technique is very simple, and its separation effect is very strong during pulp application. However, the common elution method involves the direct ashing of activated carbon, which causes the loss of gold and contamination. To solve this problem, a new simple technology was developed by Hassan et al. [134], in which gold was adsorbed on granular activated carbon followed by graphite GFAAS analysis. Other carbon materials such as carbon nanotubes were also used for gold extraction. In order to improve its selectivity and adsorption capacity, Dobrowolski et al. [135] compared the effects of nitric acid, ethylenediamine, and (3-aminopropyl) triethoxysilane on the modification of carbon nanotubes. The efficiency of analysis can also be improved using a hybrid adsorbent. Xue et al. [136] developed an adsorbent composed of cellulose fibers, activated carbon, and anion exchange resin for the preconcentration and separation of Au, Pd, and Pt in geological samples. Furthermore, in recent years, magnetic nanoparticles have been widely used in sample extraction due to their unique magnetic response, large surface area, and chemically modifiable surface. Ye et al. [137] established an on-line method for the Au, Pd, and Pt determination with 4'-aminobenzo-15-crown-5-ether functionalized magnetic nanoparticles. The obtained LOD of gold was 0.16 ppb. Table 6 summarizes the Au preconcentration methods using PUF/extraction chromatography/novel solid absorbents discussed in this work.

Table 6. Methods for the preconcentration of Au by PUF/extraction chromatography/novel solid absorbents.

Sample Weight/g	Digestion	Dissolution	Separation Technique	Detection Technique	LOD/ppb	Reference
10	Polypropylene beaker	aq.reg.	PUF	GFAAS	0.23	[115]
10–20	-	aq.reg.	MIBK-loaded PUF	GFAAS	-	[124]
4	Teflon pot	HNO$_3$ + HF + HCl + aq.reg.	DIBK-loaded CG71 resin	ICP-MS	0.002	[36]
0.2	Microwave vessel	aq.reg.	Single granular carbon	GFAAS	0.9	[134]
0.5	Microwave vessel	aq.reg.	Modified carbon nanotubes	SS-HR-CS-GFAAS	0.002	[135]
10	PFA vessel	aq.reg.	hybrid adsorbent	GFAAS	0.008	[136]
5–10	Hot-plate	aq.reg.	magnetic nanoparticles	FI-column-GFAAS	0.16	[137]

4. Direct Determination Methods

INAA is a reliable multi-element analysis method for the direct quantitative analysis of solid samples. It possesses very high Au sensitivity and is minimally affected by matrix effects. It also allows avoiding losses and contamination during the pre-treatment stage because no digestion or pre-enrichment of the sample is required. Constantin used this technique to determine the gold contents of 82 geochemical reference materials in 2006 and 2009 [28,138] and compared them with the results obtained by other analytical techniques. Table 7 compares various analytical methods commonly used for the gold determination in rocks.

Table 7. Various analytical methods commonly used for the gold determination in rocks.

Sample Weight/g	Dissolution	Separation Technique	Detection Technique	LOD/ppb	Reference
15	NiS/Na$_2$CO$_3$, Na$_2$B$_4$O$_7$, SiO$_2$	Te coprecipitation	ICP-MS	0.484	[47]
1–20	Na$_2$O$_2$	Te coprecipitation	ICP-MS	0.007	[66]
0.5–1.5	HNO$_3$ + HF + aq.reg. + HClO$_4$ + HCl	Se coprecipitation	GFAAS/TXRF	0.2/1.2	[26]
2	HCl + HNO$_3$	Cation exchange resin	ICP-MS	-	[5]
0.02	HNO$_3$ + HCl + HF	DLLME	GFAAS	0.002	[98]
10	aq.reg.	PUF	GFAAS	0.23	[115]
4	HNO$_3$ + HF + HCl + aq.reg.	DIBK-loaded CG71 resin	ICP-MS	0.002	[36]
0.2	aq.reg.	Single granular carbon	GFAAS	0.9	[134]
5–10	aq.reg.	magnetic nanoparticles	FI-column-GFAAS	0.16	[137]
1–3	-	-	INAA	~0.1	[138]

5. Conclusions and Perspectives

In this review, various sample digestion and combined preconcentration methods for the determination of ultra-low gold contents in rocks are summarized. Although some breakthroughs have been achieved in recent years, many important problems remain to be solved, owing to the heterogeneous distribution of gold in rocks, unknown forms of Au in different types of samples, and limited knowledge of the stability of gold compounds. These issues can also create promising opportunities in the study area; thus, future works in this field should focus on the following points:

1. In order to eliminate the nugget effect, the sample weight must be large enough. However, it leads to low digestion efficiency and requires the use of complex operating procedures. Therefore, additional studies must be performed to improve the sample representativeness.
2. Normally, the results obtained by the methods described above exhibit large deviations, which are often attributed to the nugget effect. However, homogeneous reference materials are required to confirm this conclusion.
3. Although FA can attack the entire rock sample, its relatively large reagent blank makes it difficult to determine the low gold content precisely. Wet acid digestion can solve this problem, but aqua regia may partially dissolve the rock samples. The desilication by HF is an effective process; however, it is inconvenient for the use in high-pressure ashers and Carius tubes and may cause a severe interference of ^{181}Ta^{16}O into ^{197}Au determination when ICP-MS is utilized. In addition, the formation of insoluble fluoride compounds may also cause the loss of gold.
4. It should be noted that gold solutions must be analyzed as soon as possible after separation due to the instability of gold in both the HCl and thiourea media. In addition, the memory effect and instrument damage caused by their usage are normally large. Therefore, future research works may focus on the development of suitable media for gold elution and quantitation.

Funding: This research was funded by the National Natural Science Foundation of China (grant nos. 41703021 and 41403021).

Conflicts of Interest: The authors declare no conflict of interest.

References

1. Pitcairn, I.K. Background concentrations of gold in different rock types. *Appl. Earth Sci. (Trans. Inst. Min. Metall. B)* **2011**, *120*, 31–38. [CrossRef]
2. Morgan, J.W.; Wandless, G.A.; Petrie, R.K.; Irving, A.J. Composition of The Earth's Upper Mantle—I. Siderophile Trace Elements in Ultramafic Nodules. *Tectonophysics* **1981**, *75*, 47–67. [CrossRef]
3. Mitchell, R.H.; Keays, R.R. Abundance and distribution of gold, palladium and iridium in some spinel and garnet Iberzolites: Implications for the nature and origin of precious metal-rich intergranular components in the upper mantle. *Geochim. Cosmochim. Acta* **1981**, *45*, 2425–2442. [CrossRef]

4. Pattou, L.; Lorand, J.P.; Gros, M. Non-chondritic platinum-group element ratios in the Earth's mantle. *Nature* **1996**, *379*, 712–715. [CrossRef]
5. Fischer-Gödde, M.; Becker, H.; Wombacher, F. Rhodium, gold and other highly siderophile elements in orogenic peridotites and peridotite xenoliths. *Chem. Geol.* **2011**, *280*, 365–383. [CrossRef]
6. Maier, W.D.; Peltonen, P.; McDonald, I. The concentration of platinum-group elements and gold in southern African and Karelian kimberlite-hosted mantle xenoliths: Implications for the noble metal content of the Earth's mantle. *Chem. Geol.* **2012**, *302–303*, 119–135. [CrossRef]
7. Lorand, J.P.; Pattou, L.; GROS, M. Fractionation of Platinum-group Elements and Gold in the Upper Mantle: A Detailed Study in Pyrenean Orogenic Lherzolites. *J. Petrol.* **1999**, *40*, 957–981. [CrossRef]
8. Lorand, J.P.; Bodinier, J.L.; Dupuy, C. Abundance and distribution of gold in the orogenic-type spinel peridotites from Ariege (Northeastern Pyrenees, France). *Geochim. Cosmochim. Acta* **1989**, *53*, 3085–3090. [CrossRef]
9. Maier, W.D.; Barnes, S.J.; Campbell, I.H. Progressive mixing of meteoritic veneer into the early Earth's deep mantle. *Nature* **2009**, *460*, 620–623. [CrossRef]
10. Brenan, J.M.; McDonough, W.F. Core formation and metal–silicate fractionation of osmium and iridium from gold. *Nat. Geosci.* **2009**, *2*, 798–801. [CrossRef]
11. Cameron, E.M. Archean gold: Relation to granulite formation and redox zoning in the crust. *Geology* **1988**, *16*, 109–112. [CrossRef]
12. Cameron, E.M. Scouring of gold from the lower crust. *Geology* **1989**, *17*, 26–29. [CrossRef]
13. Cameron, E.M. Derivation of gold by oxidative metamorphism of a deep ductile shear zone: Part 2. Evidence from the Bamble Belt, south Norway. *J. Geochem. Explor.* **1989**, *31*, 149–169. [CrossRef]
14. Hofmann, A.; Pitcairn, I.K.; Wilson, A. Gold mobility during Palaeoarchaean submarine alteration. *Earth. Planet. Sc. Lett.* **2017**, *462*, 47–54. [CrossRef]
15. Patten, C.G.C.; Pitcairn, I.K.; Teagle, D.A.H. Hydrothermal mobilisation of Au and other metals in supra-subduction oceanic crust: Insights from the Troodos ophiolite. *Ore. Geol. Rev.* **2017**, *86*, 487–508. [CrossRef]
16. Connors, K.A.; Noble, D.C.; Bussey, S.D. Initial gold contents of silicic volcanic rocks: Bearing on the behavior of gold in magmatic systems. *Geology* **1993**, *21*, 937–940. [CrossRef]
17. Webber, A.P.; Roberts, S.; Taylor, R.N.; Pitcairn, I.K. Golden plumes: Substantial gold enrichment of oceanic crust during ridge-plume interaction. *Geology* **2013**, *41*, 87–90. [CrossRef]
18. Pitcairn, I.K. Sources of Metals and Fluids in Orogenic Gold Deposits: Insights from the Otago and Alpine Schists, New Zealand. *Econ. Geol.* **2006**, *101*, 1525–1546. [CrossRef]
19. Fleet, M.E.; Crocket, J.H.; Liu, M.H. Laboratory partitioning of platinum-group elements (PGE) and gold with application to magmatic sulfide–PGE deposits. *Lithos* **1999**, *47*, 127–142. [CrossRef]
20. Hou, Z.Q.; Zhou, Y.; Wang, R. Recycling of metal-fertilized lower continental crust: Origin of non-arc Au-rich porphyry deposits at cratonic edges. *Geology* **2017**, *45*, 6–9. [CrossRef]
21. Pitcairn, I.K.; Craw, D.; Teagle, D.A.H. Metabasalts as sources of metals in orogenic gold deposits. *Miner. Deposita.* **2015**, *50*, 373–390. [CrossRef]
22. Patten, C.G.C.; Pitcairn, I.K.; Teagle, D.A.H. Michelle Harris, Mobility of Au and related elements during the hydrothermal alteration of the oceanic crust: Implications for the sources of metals in VMS deposits. *Miner. Deposita.* **2016**, *51*, 179–200. [CrossRef]
23. Boskabadi, A.; Pitcairn, I.K.; Broman, C. Carbonate alteration of ophiolitic rocks in the Arabian–Nubian Shield of Egypt: Sources and compositions of the carbonating fluid and implications for the formation of Au deposits. *Int. Geol. Rev.* **2017**, *4*, 391–419. [CrossRef]
24. Kerr, R.A. Is the World Tottering on the Precipice of Peak Gold? *Science* **2012**, *335*, 1038–1039. [CrossRef]
25. Perry, B.J.; Barefoot, R.R.; Van Loon, J.C. Inductively coupled plasma mass spectrometry for the determination of platinum group elements and gold. *Trac-Trend. Anal. Chem.* **1995**, *14*, 388–397.
26. Eller, R.; Alt, F.; Tolg, G. An efficient combined procedure for the extreme trace analysis of gold, platinum, palladium and rhodium with the aid of graphite furnace atomic absorption spectrometry and total-reflection X-ray fluorescence analysis. *Fresenius. Z. Anal. Chem.* **1989**, *334*, 723–739. [CrossRef]
27. Yokoyama, T.; Yokota, T.; Hayashi, S. Determination of trace gold in rock samples by a combination of two-stage solvent extration and graphite furnace atomic absorption spectrometry: The problem of iron interference and its solution. *Geochem. J.* **1996**, *30*, 175–181. [CrossRef]

28. Constantin, M. Determination of Au, Ir and thirty-two other elements in twelve geochemical reference materials by instrumental neutron activation analysis. *J. Radioanal. Nucl. Chem.* **2006**, *267*, 407–414. [CrossRef]
29. Barefoot, R.R. Determination of the precious metals in geological materials by inductively coupled plasma mass spectrometry. *J. Anal. At. Spectrom.* **1998**, *13*, 1077–1084. [CrossRef]
30. Barefoot, R.R.; Van Loon, J.C. Recent advances in the determination of the platinum group elements and gold. *Talanta* **1999**, *49*, 1–14. [CrossRef]
31. Balcerzak, M. Sample Digestion Methods for the Determination of Traces of Precious Metals by Spectrometric Techniques. *Anal. Sci.* **2002**, *18*, 737–750. [CrossRef] [PubMed]
32. Pyrzynska, K. Recent developments in the determination of gold by atomic spectrometry techniques. *Spectrochim. Acta B* **2005**, *60*, 1316–1322. [CrossRef]
33. Mokhodoeva, O.B.; Myasoedova, G.V.; Kubrakova, I.V. Sorption Preconcentration in Combined Methods for the Determination of Noble Metals. *J. Anal. Chem.* **2007**, *62*, 607–622. [CrossRef]
34. Myasoedova, G.V.; Mokhodoeva, O.B.; Kubrakova, I.V. Trends in Sorption Preconcentration Combined with Noble Metal Determination. *Anal. Sci.* **2007**, *23*, 1031–1039. [CrossRef] [PubMed]
35. Pyrzynska, K. Sorbent materials of separation and preconcentration of gold in environmental and geological samples-A review. *Anal. Chim. Acta* **2012**, *741*, 9–14. [CrossRef] [PubMed]
36. Pitcairn, I.K.; Warwick, P.E. Method for Ultra-Low-Level Analysis of Gold in Rocks. *Anal. Chem.* **2006**, *78*, 1290–1295. [CrossRef] [PubMed]
37. Chow, A.; Beamish, F.E. An experimental evaluation of neutron activation, wet assay and fire assay methods of determining gold in ores. *Talanta* **1967**, *14*, 219–231. [CrossRef]
38. Hall, G.E.M.; Pelchat, J.C. Analysis of geological materials for gold, platinum and palladium at low ppb levels by fire assay-ICP mass spectrometry. *Chem. Geol.* **1994**, *115*, 61–72. [CrossRef]
39. Hall, G.E.M.; Vaive, J.E.; Coope, J.A. Bias in the analysis of geological materials for gold using current methods. *J. Geochem. Explor.* **1989**, *34*, 157–171. [CrossRef]
40. Juvonen, M.R.; Bartha, A.; Lakomaa, T.M. Comparison of Recoveries by Lead Fire Assay and Nickel Sulfide Fire Assay in the Determination of Gold, Platinum, Palladium and Rhenium in Sulfide Ore Samples. *Geostand. Geoanal. Res.* **2004**, *28*, 123–130. [CrossRef]
41. Plessen, H.G.; Erzinger, J. Determination of the Platinum-Group Elements and Gold in Twenty Rock Reference Materials by Inductively Coupled Plasma-Mass Spectrometry (ICP-MS) after Pre-Concentration by Nickel Sulfide Fire Assay. *Geostand. Geoanal. Res.* **1998**, *22*, 187–194. [CrossRef]
42. Cerceau, C.; Carvalho, C.F.; Rabelo, C.S. Recovering lead from cupel waste generated in gold analysis by Pb-Fire assay. *J. Environ. Manag.* **2016**, *183*, 771–776. [CrossRef]
43. Porter, K.A.; Kirk, C.; Fearey, D. Elevated Blood Lead Levels Among Fire Assay Workers and Their Children in Alaska, 2010–2011. *Public. Health. Rep.* **2015**, *130*, 440–446. [CrossRef]
44. Asif, M.; Parry, S.J. Elimination of Reagent Blank Problems in the Fire-assay Pre-concentration of the Platinum Group Elements and Gold with a Nickel Sulphide Bead of Less Than One Gram Mass. *Analyst* **1989**, *114*, 1057–1059. [CrossRef]
45. Lu, C.F.; He, H.L.; Zhou, Z.R.; Zhi, X.X.; Li, B.; Zhang, Q. Determination of platinum-group elements and gold in geochemical exploration samples by nickel sulfide fire assay-ICP-MS, II. Reduction of reagent blank. *Rock Miner. Anal.* **2002**, *21*, 7–11.
46. Jackson, S.; Fryer, B.; Gosse, W.; Healey, D.; Longerich, H.; Strong, D. Determination of the precious metals in geological materials by inductively coupled plasma-mass spectrometry (ICP-MS) with nickel sulphide fire-assay collection and tellurium coprecipitation. *Chem. Geol.* **1990**, *83*, 119–132. [CrossRef]
47. Savard, D.; Barnes, S.J.; Meisel, T. Comparison between Nickel-Sulfur Fire Assay Te Co-precipitation and Isotope Dilution with High-Pressure Asher Acid Digestion for the Determination of Platinum-Group Elements, Rhenium and Gold. *Geostand. Geoanal. Res.* **2010**, *34*, 281–291. [CrossRef]
48. Oguri, K.; Shimoda, G.; Tatsumi, Y. Quantitative determination of gold and the platinum-group elements in geological samples using improved NiS fire-assay and tellurium coprecipitation with inductively coupled plasma-mass spectrometry. *Chem. Geol.* **1999**, *157*, 189–197. [CrossRef]
49. Gros, M.; Lorand, J.P.; Luguet, A. Analysis of platinum group elements and gold in geological materials using NiS fire assay and Te coprecipitation; the NiS dissolution step revisited. *Chem. Geol.* **2002**, *185*, 179–190. [CrossRef]

50. Sun, Y.; Sun, M. Nickel sulfide fire assay improved for pre-concentration of platinum-group elements in geological samples: A practical means of ultra-trace analysis combined with inductively coupled plasma-mass spectrometry. *Analyst* **2005**, *130*, 664–669. [CrossRef]
51. Li, C.S.; Chai, C.F.; Li, X.L. Determination of Platinum-Group Elements and Gold in Two Russian Candidate Reference Materials SCHS-1 and SLg-1 by ICP-MS after Nickel Sulfide Fire Assay Preconcentration. *Geostand. Geoanal. Res.* **1998**, *22*, 195–197. [CrossRef]
52. Juvonen, R.; Lakomaa, T.; Soikkeli, L. Determination of gold and the platinum group elements in geological samples by ICP-MS after nickel sulphide fire assay: Difficulties encountered with different types of geological samples. *Talanta* **2002**, *58*, 595–603. [CrossRef]
53. Asif, M.; Parry, S.J.; Malik, H. Instrumental neutron activation analysis of a nickel sulphide fire assay button to determine the platinum-group elements and gold. *Analyst* **1992**, *117*, 1351–1353. [CrossRef]
54. Bédard, L.P.; Barnes, S.J. A comparison of the capacity of FA-ICP-MS and FA-INAA to determine platinum-group elements and gold in geological samples. *J. Radioanal. Nucl. Chem.* **2002**, *254*, 319–329. [CrossRef]
55. Jarvis, K.E.; Williams, J.G.; Parry, S.J.; Bertalan, E. Quantitative determination of the platinum-group elements and gold using NiS fire assay with laser ablation-inductively coupled plasma-mass spectrometry (LA-ICP-MS). *Chem. Geol.* **1995**, *124*, 37–46. [CrossRef]
56. Jorge, A.P.S.; Enzweiler, J.; Shibuya, E.K. Platinum-Group Elements and Gold Determination in NiS Fire Assay Buttons by UV Laser Ablation ICP-MS. *Geostand. Geoanal. Res.* **1998**, *22*, 47–55. [CrossRef]
57. Resano, M.; McIntosh, K.S.; Vanhaecke, F. Laser ablation-inductively coupled plasma-mass spectrometry using a double-focusing sector field mass spectrometer of Mattauch–Herzog geometry and an array detector for the determination of platinum group metals and gold in NiS buttons obtained by fire assay of platiniferous ores. *J. Anal. At. Spectrom.* **2012**, *27*, 165–173.
58. Resano, M.; Ruiz, E.G.; McIntosh, K.S.; Vanhaecke, F. Laser ablation-inductively coupled plasma-dynamic reaction cell-mass spectrometry for the determination of platinum group metals and gold in NiS buttons obtained by fire assay of platiniferous ores. *J. Anal. At. Spectrom.* **2008**, *23*, 1599–1609. [CrossRef]
59. Resano, M.; Ruiz, E.G.; McIntosh, K.S.; Hinrichs, J. Comparison of the solid sampling techniques laser ablation-ICP-MS, glow discharge-MS and spark-OES for the determination of platinum group metals in Pb buttons obtained by fire assay of platiniferous ores. *J. Anal. At. Spectrom.* **2006**, *21*, 899–909. [CrossRef]
60. Vanhaecke, F.; Resano, M.; Koch, J. Femtosecond laser ablation-ICP-mass spectrometry analysis of a heavy metallic matrix: Determination of platinum group metals and gold in lead fire-assay buttons as a case study. *J. Anal. At. Spectrom.* **2010**, *25*, 1259–1267. [CrossRef]
61. Compernolle, S.; Wambeke, D.; Raedt, I.D. Evaluation of a combination of isotope dilution and single standard addition as an alternative calibration method for the determination of precious metals in lead fire assay buttons by laser ablation-inductively coupled plasma-mass spectrometry. *Spectrochim. Acta B* **2012**, *67*, 50–56. [CrossRef]
62. Xue, G. *The Analytical Chemistry of Gold*; Aerospace Press: Beijing, China, 1990; pp. 74–75.
63. Enzweiler, J.; Potts, P.J. The separation of platinum, palladium and gold from silicate rocks by the anion exchange separation of chloro complexes after a sodium peroxide fusion: An investigation of low recoveries. *Talanta* **1995**, *42*, 1411–1418. [CrossRef]
64. Dai, X.X.; Koeberl, C.; Fröschl, H. Determination of platinum group elements in impact breccias using neutron activation analysis and ultrasonic nebulization inductively coupled plasma mass spectrometry after anion exchange preconcentration. *Anal. Chim. Acta* **2001**, *436*, 79–85. [CrossRef]
65. Amossé, J. Determination of Platinum-Group Elements and Gold in Geological Matrices by Inductively Coupled Plasma-Mass Spectrometry (ICP-MS) after Separation with Selenium and Tellurium Carriers. *Geostand. Geoanal. Res.* **1998**, *22*, 93–102. [CrossRef]
66. Jin, X.D.; Zhu, H.P. Determination of platinum group elements and gold in geological samples with ICP-MS using a sodium peroxide fusion and tellurium co-precipitation. *J. Anal. At. Spectrom.* **2000**, *15*, 747–751. [CrossRef]
67. Qi, L.; Gregoire, D.C.; Zhou, M.F. Determination of Pt, Pd, Ru and Ir in geological samples by ID-ICP-MS using sodium peroxide fusion and Te co-precipitation. *Geochem. J.* **2003**, *37*, 557–565. [CrossRef]
68. Totland, M.M.; Jarvis, I.; Jarvis, K.E. Microwave digestion and alkali fusion procedures for the determination of the platinum-group elements and gold in geological materials by ICP-MS. *Chem. Geol.* **1995**, *124*, 21–36.

69. Jarvis, I.; Totland, M.M.; Jarvis, K.E. Determination of the platinum-group elements in geological materials by ICP-MS using microwave digestion, alkali fusion and cation-exchange chromatography. *Chem. Geol.* **1997**, *143*, 27–42. [CrossRef]
70. Tsimbalist, V.G.; Anoshin, G.N.; Mitkin, V.N.; Razvorotneva, L.L.; Golovanova, N.P. Observations on New Approaches for the Determination of Platinum-Group Elements, Gold and Silver in Different Geochemical Samples from Siberia and the Far East. *Geostand. Geoanal. Res.* **2000**, *12*, 171–182. [CrossRef]
71. Coedo, A.G.; Dorado, M.T.; Padilla, I.; Alguacil, F. Preconcentration and matrix separation of precious metals in geological and related materials using metalfix-chelamine resin prior to inductively coupled plasma mass spectrometry. *Anal. Chim. Acta* **1997**, *340*, 31–40. [CrossRef]
72. Mitkin, V.N.; Galizky, A.A.; Korda, T.M. Some Observations on the Determination of Gold and the Platinum-Group Elements in Black Shales. *Geostand. Geoanal. Res.* **2000**, *24*, 227–240. [CrossRef]
73. Nesbitt, C.C.; Milosavljevic, E.B.; Hendrix, J.L. Determination of the Mechanism of the Chlorination of Gold in Aqueous Solutions. *Ind. Eng Chem. Res.* **1990**, *29*, 1696–1700. [CrossRef]
74. Perry, B.J.; Van Loon, J.C. Dry-chlorination Inductively Coupled Plasma Mass Spectrometric Method for the Determination of Platinum Group Elements in Rocks. *J. Anal. At. Spectrom.* **1992**, *7*, 883–888. [CrossRef]
75. Perry, B.J.; Speller, D.V.; Barefoot, R.R.; Van Loon, J.C. A large sample, dry chlorination, ICP-MS analytical method for the determination of platinum group elements and gold in rocks. *Can. J. Appl. Spectrosc.* **1993**, *38*, 131.
76. Ely, J.C.; Neal, C.R.; O'Neill Jr, J.A.; Jain, J.C. Quantifying the platinum group elements (PGEs) and gold in geological samples using cation exchange pretreatment and ultrasonic nebulization inductively coupled plasma-mass spectrometry (USN-ICP-MS). *Chem. Geol.* **1999**, *157*, 219–234. [CrossRef]
77. Dale, C.W.; Burton, K.W.; Pearson, D.G.; Gannoun, A.; Alard, O.; Argles, T.W.; Parkinson, I.J. Highly siderophile element behaviour accompanying subduction of oceanic crust: Whole rock and mineral-scale insights from a high-pressure terrain. *Geochim. Cosmochim. Acta* **2009**, *73*, 1394–1416. [CrossRef]
78. Dale, C.W.; Macpherson, C.G.; Pearson, D.J.; Hammond, S.J.; Arculus, R.J. Inter-element fractionation of highly siderophile elements in the Tonga Arc due to flux melting of a depleted source. *Geochim. Cosmochim. Acta* **2012**, *89*, 202–225. [CrossRef]
79. Li, J.; Zhao, P.P.; Liu, J.G. Reassessment of Hydrofluoric Acid Desilicification in the Carius Tube Digestion Technique for Re–Os Isotopic Determination in Geological Samples. *Geostand. Geoanal. Res.* **2015**, *39*, 17–30. [CrossRef]
80. Fryer, B.J.; Kerrich, R. Determination of precious metals at ppb-levels in rocks by a combined wetchemical and flameless atomic absorption method. *At. Absorpt. Newsletter.* **1978**, *17*, 4–60.
81. Sen Gupta, J.G. Determination of trace and ultra-trace amounts of noble metals in geological and related materials by graphite-furnace atomic-absorption spectrometry after separation by ion-exchange or co-precipitation with tellurium. *Talanta* **1989**, *36*, 651–656. [CrossRef]
82. Elson, C.M.; Chatt, A. Determination of Gold in Silicate Rocks and Ores by Coprecipitation with Tellurium and Neutron Activation. *Anal. Chim. Acta* **1983**, *155*, 305–310. [CrossRef]
83. Niskavaara, H.; Kontas, E. Reductive coprecipitation as a separation method for the determination of gold, palladium, platinum, rhodium, silver, selenium and tellurium in geological samples by graphite furnace atomic absorption spectrometry. *Anal. Chim. Acta* **1990**, *231*, 273–282. [CrossRef]
84. Barredo, F.B.; Polo, C.P. The Simultaneous Determination of Gold, Silver and Cadmium at ppb Levels in Silicate Rocks by Atomic Absorption Spectrometry with Electrothermal Atomization. *Anal. Chim. Acta* **1977**, *94*, 283–287. [CrossRef]
85. Sen Gupta, J.G. Determination of noble metals in silicate rocks, ores and metallurgical samples by simultaneous multi-element graphite furnace atomic absorption spectrometry with Zeeman background correction. *Talanta* **1993**, *40*, 791–797. [CrossRef]
86. Meisel, T.; Fellner, N. Moser, J. A simple procedure for the determination of platinum group elements and rhenium (Ru, Rh, Pd, Re, Os, Ir and Pt) using ID-ICP-MS with an inexpensive on-line matrix separation in geological and environmental materials. *J. Anal. At. Spectrom.* **2003**, *18*, 720–726. [CrossRef]
87. Wu, Y.W.; Jiang, Z.C.; Hu, B.; Duan, J.K. Electrothermal vaporization inductively coupled plasma atomic emission spectrometry determination of gold, palladium, and platinum using chelating resin YPA$_4$ as both extractant and chemical modifier. *Talanta* **2004**, *63*, 585–592. [CrossRef] [PubMed]

88. Medved, J.; Bujdoš, M.; Matúš, P.; Kubová, J. Determination of trace amounts of gold in acid-attacked environmental samples by atomic absorption spectrometry with electrothermal atomization after preconcentration. *Anal. Bioanal. Chem.* **2004**, *379*, 60–65. [CrossRef]
89. Terashima, S. Determination of Gold in Sixty Geochemical Reference Samples by Flameless Atomic Absorption Spectrometry. *Geostand. Geoanal. Res.* **1988**, *12*, 57–60. [CrossRef]
90. Terashima, S.; Itoh, S.; Ando, A. Gold in Twenty-Six Janpanese Geochemical Reference Samples. *Geostand. Geoanal. Res.* **1992**, *16*, 9–10. [CrossRef]
91. Ramesh, S.L.; Sunder Raju, P.V.; Anjaiah, K.V.; Mathur, R. Determination of Gold in Rocks, Ores, and Other Geological Materials by Atomic Absorption Techniques. *Atom. Spectrosc.* **2001**, *22*, 263–269.
92. Chattopadhyay, P.; Sahoo, B.N. Modified Decomposition Procedure for the Determination of Gold in Geological Samples by Atomic Absorption Spectrometry. *Analyst* **1992**, *117*, 1481–1484. [CrossRef]
93. Monteiro, M.I.C.; Lavatori, M.P.A.; de Oliveira, N.M.M. Determination of Gold in Ores by Isobutyl Methyl Ketone Extraction and Electrothermal Atomic Absorption Spectrometry: A Stability Study of the Metal in Organic Media. *Geostand. Geoanal. Res.* **2003**, *27*, 245–249. [CrossRef]
94. Rezaee, M.; Assadi, Y.; Milani Hosseini, M.R.; Aghaee, E.; Ahmadi, F.; Berijani, S. Determination of organic compounds in water using dispersive liquid-liquid microextraction. *J. Chromatogr. A* **2006**, *1116*, 1–9. [CrossRef]
95. Shamsipur, M.; Ramezani, M. Selective determination of ultra trace amounts of gold by graphite furnace atomic absorption spectrometry after dispersive liquid–liquid microextraction. *Talanta* **2008**, *75*, 294–300. [CrossRef]
96. Kagaya, S.; Takata, D.; Yoshimori, T.; Kanbara, T.; Tohda, K. A sensitive and selective method for determination of gold (III) based on electrothermal atomic absorption spectrometry in combination with dispersive liquid–liquid microextraction using dicyclohexylamine. *Talanta* **2010**, *80*, 1364–1370. [CrossRef]
97. De La Calle, I.; Pena-Pereira, F.; Cabaleiro, N.; Lavilla, I.; Bendicho, C. Ion pair-based dispersive liquid–liquid microextraction for gold determination at ppb level in solid samples after ultrasound-assisted extraction and in waters by electrothermal-atomic absorption spectrometry. *Talanta* **2011**, *84*, 109–115. [CrossRef]
98. Fazelirad, H.; Taher, M.A.; Nasiri-Majd, M. GFAAS determination of gold with ionic liquid, ion pair based and ultrasound-assisted dispersive liquid–liquid microextraction. *J. Anal. At. Spectrom.* **2014**, *29*, 2343–2348. [CrossRef]
99. Bowen, H.J.M. Absorption by Polyurethane Foams: New Method of Separation. *J. Chem. Soc. A* **1970**, 1082–1085. [CrossRef]
100. Lemos, V.A.; Santos, M.S.; Santos, E.S. Application of polyurethane foam as a sorbent for trace metal pre-concentration—A review. *Spectrochim. Acta B* **2007**, *62*, 4–12. [CrossRef]
101. Dmitrienko, S.G.; Zolotov, Y.A. Polyurethane foams in chemical analysis: Sorption of various substances and its analytical applications. *Russ. Chem. Rev.* **2002**, *71*, 159–174. [CrossRef]
102. Gesser, H.D. Open-Cell Polyurethane Foam Sponge as a "Solvent Extractor" for Gallium and Iron. *Sep. Sci.* **1976**, *11*, 317–327. [CrossRef]
103. Oren, J.J. The solvent extraction of Fe(III) from acidic chloride solutions by open cell polyurethane foam sponge. *Can. J. Chem.* **1979**, *57*, 2023–2036. [CrossRef]
104. LQ, V.S.K.; Chow, A. Extraction of Tin by Use of Polyurethane Foam. *Talanta* **1981**, *28*, 157–164.
105. Jones, L.; Nel, I.; Koch, K.R. Polyurethane Foams as Selective Sorbents for Noble Metals. *Anal. Chim. Acta* **1986**, *182*, 61–70. [CrossRef]
106. Hamon, R.F. The Cation-Chelation Mechanism of Metal-Ion Sorption by Polyurethanes. *Talanta* **1982**, *29*, 313–326. [CrossRef]
107. Wang, H.Y.; Liu, Y.H.; Xue, D.S. Synthesis of Amino Polyurethane Foam and Its Application in Trace Gold Enrichment in Geological Samples. *Rock. Miner. Anal.* **2016**, *35*, 409–414.
108. Schiller, P. Determination of Trace Amounts of Gold in Natural Sweet Waters by Non-destructive Activation Analysis after Preconcentration. *Anal. Chim. Acta* **1971**, *54*, 364–368. [CrossRef]
109. Han, J.F. Discussion on the Ability of Different Acid - base Concentration for Foam Plastics to Rich Gold. *Anhui. Chem. Ind.* **2017**, *43*, 50–54.
110. Moawed, E.A. Effect on the Chromatographic Behavior of Gold of the Process Used to Acid-Wash Polyurethane Foam. *Chromatographia* **2008**, *67*, 77–84. [CrossRef]

111. Wu, J.; Zhang, M.J.; Xiong, Y.X. Determination of Trace Gold in Geological Samples Combining Foam Adsorption Inductively Coupled Plasma-Mass Spectrometry with Closed Water Bath Dissolution. *J. Central China Normal. Univ.* **2017**, *51*, 626–637.
112. Zheng, L. Determination of Gold in Geochemical Samples by Foam Adsorption-Graphite Furnace Atomic Absorption Spectrometry. *Low Carbon World.* **2016**, *3*, 218–219.
113. Hu, X.C. Determination of Trace Gold in Geochemical Samples by Inductively Coupled Plasma Mass Spectrometry (ICP-MS X-2): Foam Adsorption Separation. *Technol. Outlook* **2015**, *23*, 140.
114. Saeed, M.M.; Ghaffar, A. Adsorption Syntax of Au (III) on Unloaded Polyurethane Foam. *J. Radioanal. Nucl. Chem.* **1998**, *232*, 171–177. [CrossRef]
115. Liu, X.L.; Wen, T.Y.; Sun, W.J. Determination of Au and Pt in Geological Samples by Graphite Furnace Atomic Absorption Spectrometry with Concentrate and Extraction by Foam Plastics and Thiourea. *Rock. Miner. Anal.* **2013**, *32*, 576–580.
116. He, Z.L. Determination of Trace Gold in Geochemical Samples by Graphite Furnace Atomic Absorption Spectrometry with Thiourea Adsorbed by Foam. *Xinjiang Nonferrous Metal.* **2016**, *6*, 49–50.
117. Chen, K. Determination of Gold in Geological Samples by Foam Adsorption Atomic Absorption Spectrometry. *Chem. Manag.* **2018**, *2*, 208–209.
118. Peng, Z.J. Determination of Trace Gold by Graphite Furnace Atomic Absorption Spectrometry with Foam Absorption. *Guangdong Chem.* **2015**, *15*, 222–223.
119. Han, K.Y. Study on Determination of Trace Gold in Carbonate Rocks. *Acad. BBS* **2017**, 272.
120. Xiong, Z.C. Spectrometric Determination of Trace Gold in Rocks and Soils with Polyurethane Foam. *Rock. Miner. Anal.* **1984**, *4*, 364–367.
121. Gu, T.X.; Zhang, Z.; Wang, C.S.; Yan, W.D. Preparation and Certification of High-Grade Gold Ore Reference Materials (GAu 19-22). *Geostand. Geoanal. Res.* **2001**, *25*, 153–158. [CrossRef]
122. Braun, T.; Farag, A.B. Chemical Enrichment and Separation of Gold in the Tributylphosphate-Thiourea-Perchloric acid system. *Anal. Chim. Acta* **1973**, *65*, 115–126. [CrossRef]
123. Braun, T.; Huszar, E. Separation of Trace Amounts of Cobalt from Nickel in the Tri-n-Octylamine-Hydrochloric Acid System. *Anal. Chim. Acta* **1973**, *64*, 77–84. [CrossRef]
124. Yan, M.C.; Wang, C.S.; Cao, Q.X. Eleven Gold Geochemical Reference Samples (GAu 8-18). *Geostand. Geoanal. Res.* **1995**, *19*, 125–133. [CrossRef]
125. El-Shahawi, M.S.; Bashammakh, A.S.; Al-Sibaai, A.A. Solid phase preconcentration and determination of trace concentrations of total gold (I) and/or (III) in sea and wastewater by ion pairing impregnated polyurethane foam packed column prior flame atomic absorption spectrometry. *Int. J. Miner. Process.* **2011**, *100*, 110–115. [CrossRef]
126. Bashammakh, A.S.; Bahhafi, S.O.; Al-Shareef, F.M.; El-Shahawi, M.S. Development of an Analytical Method for Trace Gold in Aqueous Solution Using Polyurethane Foam Sorbents: Kinetic and Thermodynamic Characteristic of Gold (III) Sorption. *Anal. Sci.* **2009**, *25*, 413–418. [CrossRef]
127. El-Shahat, M.F.; Moawed, E.A.; Farag, A.B. Chemical enrichment and separation of uranyl ions in aqueous media using novel polyurethane foam chemically grafted with different basic dyestuff sorbents. *Talanta* **2007**, *71*, 236–241. [CrossRef]
128. Xue, D.S.; Wang, H.Y.; Liu, Y.H. Cytosine-functionalized polyurethane foam and its use as a sorbent for the determination of gold in geological samples. *Anal. Methods.* **2016**, *8*, 29–39. [CrossRef]
129. Moawed, E.A.; Zaid, M.A.A.; El-Shahat, M.F. The chromatographic behavior of group (IIB) metal ions on polyurethane foam functionalized with 8-hydroxyquinoline. *Anal. Bioanal. Chem.* **2004**, *378*, 470–478.
130. Moawed, E.A.; Moawed, M.F. Synthesis, characterization of low density polyhydroxy polyurethane foam and its application for separation and determination of gold in water and ores samples. *Anal. Chim. Acta* **2013**, *788*, 200–207. [CrossRef]
131. Pohlandt, C.; Steele, T.W. Separation of the non-volatile noble metals by reversed-phase extraction chromatography. *Talanta* **1972**, *19*, 839–850. [CrossRef]
132. Pohlandt, C.; Steele, T.W. Chromatographic determination of matte-leach separation and noble metals in residues. *Talanta* **1974**, *21*, 919–925. [CrossRef]
133. Bao, G.M. Determination of Trace Gold in Ores by Extraction Layer-Atom Absorption Method. *Chin. J. Anal. Chem.* **1977**, *6*, 428–431.

134. Hassan, J.; Shamsipur, M.; Karbasi, M.H. Single granular activated carbon microextraction and graphite furnace atomic absorption spectrometry determination for trace amount of gold in aqueous and geological samples. *Microchem. J.* **2011**, *99*, 93–96. [CrossRef]
135. Dobrowolski, R.; Mróz, A.; Dąbrowska, M.; Olszański, P. Solid sampling high-resolution continuum source graphite furnace atomic absorption spectrometry for gold determination in geological samples after preconcentration onto carbon nanotubes. *Spectrochim. Acta B* **2017**, *132*, 13–18. [CrossRef]
136. Xue, D.S.; Wang, H.Y.; Liu, Y.H.; Shen, P. Multicolumn solid phase extraction with hybrid adsorbent and rapid determination of Au, Pd and Pt in geological samples by GF-AAS. *Miner. Eng.* **2015**, *81*, 149–151. [CrossRef]
137. Ye, J.J.; Liu, S.X.; Tian, M.M. Preparation and characterization of magnetic nanoparticles for the on-line determination of gold, palladium, and platinum in mine samples based on flow injection micro-column preconcentration coupled with graphite furnace atomic absorption spectrometry. *Talanta* **2014**, *118*, 231–237. [CrossRef]
138. Constantin, M. Trace Element Data for Gold, Iridium and Silver in Seventy Geochemical Reference Materials. *Geostand. Geoanal. Res.* **2009**, *33*, 115–132. [CrossRef]

© 2019 by the authors. Licensee MDPI, Basel, Switzerland. This article is an open access article distributed under the terms and conditions of the Creative Commons Attribution (CC BY) license (http://creativecommons.org/licenses/by/4.0/).

Review

Modern Methods of Sample Preparation for the Analysis of Oxylipins in Biological Samples

Ivan Liakh [1], Alicja Pakiet [2], Tomasz Sledzinski [1] and Adriana Mika [1,2,*]

[1] Department of Pharmaceutical Biochemistry, Medical University of Gdansk, Debinki 1, 80-211 Gdansk, Poland; liakh_ivan@mail.ru (I.L.); tsledz@gumed.edu.pl (T.S.)
[2] Department of Environmental Analysis, Faculty of Chemistry, University of Gdansk, Wita Stwosza 63, 80-308 Gdansk, Poland; alicjapakiet@gmail.com
* Correspondence: adrianamika@tlen.pl; Tel.: +48-585235190

Academic Editor: Nuno Neng
Received: 26 March 2019; Accepted: 17 April 2019; Published: 25 April 2019

Abstract: Oxylipins are potent lipid mediators derived from polyunsaturated fatty acids, which play important roles in various biological processes. Being important regulators and/or markers of a wide range of normal and pathological processes, oxylipins are becoming a popular subject of research; however, the low stability and often very low concentration of oxylipins in samples are a significant challenge for authors and continuous improvement is required in both the extraction and analysis techniques. In recent years, the study of oxylipins has been directly related to the development of new technological platforms based on mass spectrometry (LC–MS/MS and gas chromatography–mass spectrometry (GC–MS)/MS), as well as the improvement in methods for the extraction of oxylipins from biological samples. In this review, we systematize and compare information on sample preparation procedures, including solid-phase extraction, liquid–liquid extraction from different biological tissues.

Keywords: sample preparation; oxylipins; protein precipitation; liquid–liquid extraction; solid-phase extraction; biological samples

1. Introduction

Oxylipins are biologically important lipid mediators which are formed by the oxidation of polyunsaturated fatty acids (PUFAs) and include hydroperoxy, hydroxy, oxo and epoxy fatty acids [1]. Oxylipins are produced via three enzymatic pathways in a reaction catalyzed by cyclooxygenase (COX), lipoxygenase (LOX), and cytochrome P450 (CYP450) or via non-enzymatic autoxidation [1,2]. Oxylipins formed from PUFAs are octadecanoids derived from linoleic acid (18:2n-6; LA) and α-linolenic acid (18:3n-3;ALA), eicosanoids derived from dihomo-γ-linolenic acid (20:3n-6; DGLA), arachidonic acid (20:4n-6; ARA) and eicosapentaenoic acid (20:5n-3; EPA), as well as docosanoids derived from adrenic acid (22:4n-6; AdA) and docosahexaenoic acid (22:6n-3; DHA) [1]. Two of the most important long-chain PUFAs, which are precursors of oxylipins with strong anti- and pro-inflammatory properties, are EPA and ARA [3]. Most of the pro-inflammatory molecules involved in cell signaling cascades are generated from ARA, whereas the anti-inflammatory ones are derived from EPA [4,5]. Oxylipins derived from EPA are generally less potent or produced less efficiently than the analogous oxylipins derived from ARA [6,7]. ARA and EPA compete with each other for binding to COX-1 and COX-2. EPA is also an inhibitor of ARA oxidation by COX-1 (to a lesser extent by COX-2) to prostaglandin H2 (PGH2), and at the same time, ARA inhibits the conversion of EPA to prostaglandin H3 (PGH3). This close interaction between ARA and EPA modulates the production of thromboxane A2 (TXA2), thromboxane A3 (TXA3), prostaglandin I2 (PGI2), prostaglandin I3 (PGI3), PGH2, PGH3, and may have anti-inflammatory effects caused by the inhibition of the ARA metabolism [1].

Eicosanoids formed from ARA are the most prevalent compounds in the oxylipin family [1]. ARA is a component of membrane phospholipids. It can be released by the phospholipase A2, as well as formed from diacylglycerol by diacylglycerol lipase [8]. ARA can be metabolized to hydroxyeicosatetraenoic acids (HETEs), dihydroxyeicosatetraenoic acids (DiHETEs), epoxyeicosatrienoic acids (EETs), prostaglandins (PGs) and thromboxane (TX), see Figure 1.

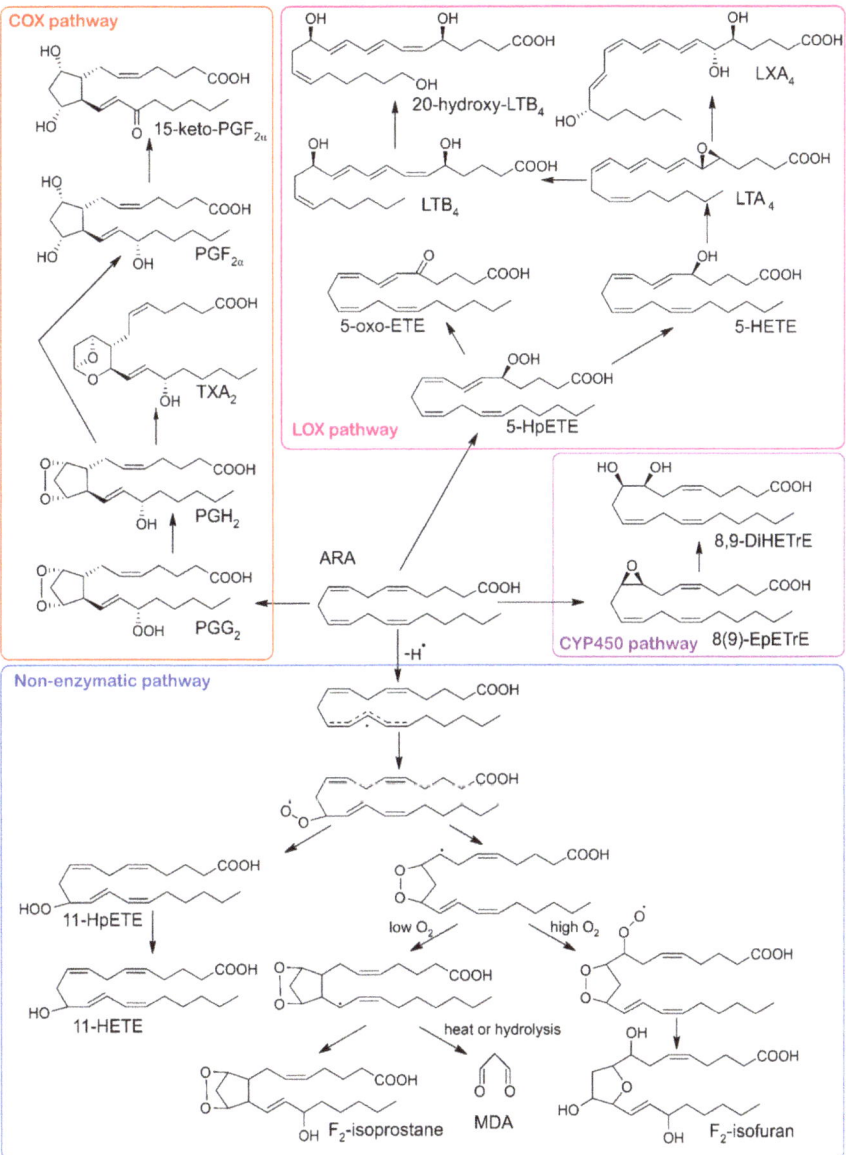

Figure 1. Conversion of ARA into oxylipins by COX, LOX, CYP 450 pathways and by the non-enzymatic pathway. COX, cyclooxygenase; LOX, lipoxygenase; CYP450, cytochrome P450; ARA, arachidonic acid; 5-HpETE, 5-hydroperoxyeicosatetraenoic acid; 5-oxo-ETE, 5-oxo-eicosatetraenoic acid; LTB$_4$, leukotriene B4; 20-hydroxy-LTB$_4$, 20-hydroxy-leukotriene B4; 5-HETE, 5-hydroxyeicosatetraenoic acid;

LTA$_4$, leukotriene A4; LXA$_4$, lipoxin A4; 8(9)-EpETrE, 8,9-epoxyeicosatrienoic acid; 8,9-DiHETrE, 8,9-dihydroxyeicosatrienoic acid; PGG$_2$, prostaglandin G2; PGH$_2$, prostaglandin H2; TXA$_2$, thromboxane A2; PGF$_{2\alpha}$, prostaglandin F2α; 15-keto-PGF$_{2\alpha}$, 15-keto-prostaglandin F2α; 11-HpETE, 11-hydroperoxyeicosatetraenoic acid; 11-HETE, 11-hydroxyeicosatetraenoic acid; MDA, malondialdehyde.

EPA is a precursor of well-known eicosanoids, such as the PG 3-series and TXs (COX) and 5-series leukotrienes (LTs) (LOX) [9], as shown in Figure 2.

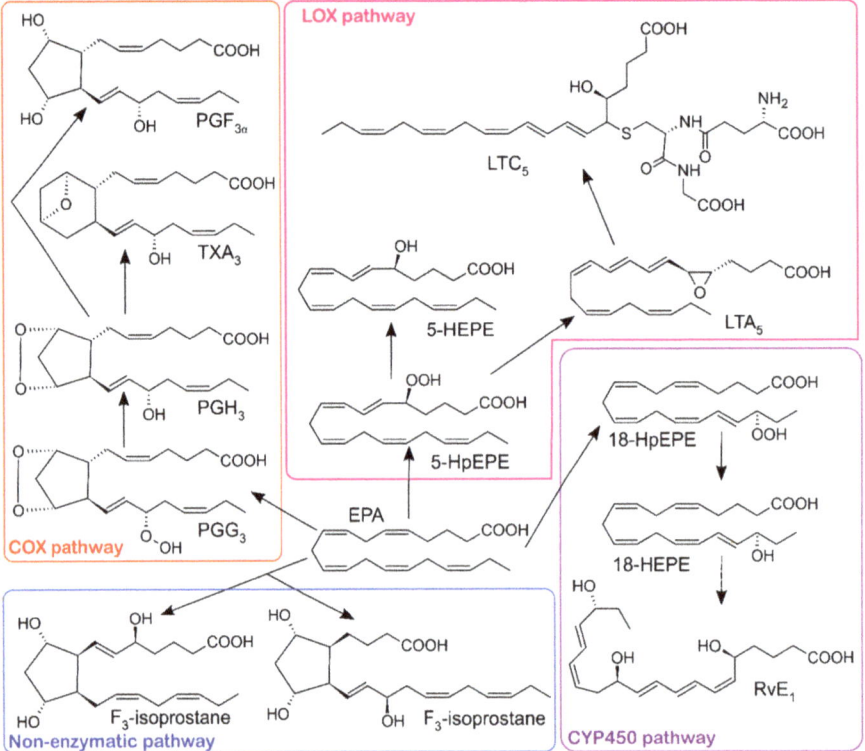

Figure 2. Conversion of EPA into oxylipins by COX, LOX, CYP 450 pathways and by the non-enzymatic pathway. COX, cyclooxygenase; LOX, lipoxygenase; CYP450, cytochrome P450; PGF$_{3\alpha}$, prostaglandin F3α; TXA$_3$, thromboxane A3; PGH$_3$, prostaglandin H3; PGG$_3$, prostaglandin G3; LTC$_5$, leukotriene C5; 5-HEPE, 5-hydroxyeicosapentaenoic acid; LTA$_5$, leukotriene A5; 5-HpEPE, 5-hydroperoxyeicosatetraenoic acid; 18-HpEPE, 18-hydroperoxyeicosatetraenoic acid 18-HEPE, 18-hydroxyeicosapentaenoic acid; RvE$_1$, resolvin E1.

Oxylipins play a major role in regulating inflammatory processes. Depending on the precursor (n-6 or n-3 PUFA), oxylipins can respectively initiate inflammation [10–12] or, on the contrary, be anti-inflammatory agents [13,14]. In many tissues, oxylipins that originate from non-enzymatic autoxidation play the role of oxidative stress markers [15–17]. Additionally, oxylipins have a large influence on a diverse range of processes, such as ovulation, the initiation of labor, bone metabolism, nerve growth and development, wound healing, kidney function, blood-vessel tone, blood coagulation, immune responses [18,19], and play a role in pathological processes, see Table 1.

Table 1. Selected oxylipins and their physiopathological functions in human studies.

Disease	Oxylipin	Precursor	Direction of Change	Function	Ref.
Obesity	5-, 11-, 20-HETE	ARA	↑	promotion of inflammation, blood pressure regulation	[10,20,21]
	15-HETE	ARA	↑	substrates for lipoxins synthesis	
	12,13-DiHOME	LA	↓	brown adipose tissue lipid uptake activation	
	12,13-Di/EpOME	LA	↓	putative markers of adipose lipolysis	
	15-HETrE	DGLA	↑	antiproliferative function	
	5-, 8-, 12-HETE	ARA	↑	associated with low-grade inflammation	[22]
	PGD2	ARA	↑	polarization of adipose tissue macrophage against inflammation	[23]
Metabolic syndrome	20-HETE	ARA	↑	vascular inflammation, angiogenesis	[15,24]
	F2-isoprostanes	ARA	↑	oxidative stress marker	
	LXA4	ARA	↓	promotion of inflammation resolution	[25]
	8-iso-PGF2α	ARA	↑	oxidative stress marker	[26,27]
Type II diabetes	11,12-, 14,15-DiHETrE	ARA	↑	EpETrE's less active metabolites	[28–30]
	13-oxo-ODE	LA	↑	inhibition of inflammation	
	11(12)-, 14(15)-EpETrE	ARA	↑	vasodilation	
	9(10)-EpOME	LA	↑	leukotoxin	
	12(13)-EpOME	LA	↑	putative marker of adipose lipolysis	
	9(10)-EpODE	ALA	↑	putative markers of adipose lipolysis	
Hypothyroidism	PGI2	ARA	↑	platelet activation inhibitor	[31]
	PGE2	ARA	↑	promotion of arterial thrombosis	
	12-HETE	ARA	↓	blood pressure regulation	
Hyperthyroidism	12-HETE	ARA	↑	blood pressure regulation	[31]
	20-HETE	ARA	↑	vasoconstriction	
Sepsis	11-HETE	ARA	↓	promotion of inflammation	[32]
	PGE2	ARA	↓	vasodilation	
	TXB2	ARA	↓	downstream metabolite of TXA2 which is involved in platelet aggregation and vasoconstriction	
Achilles tendopathy	13-HODE	LA	↑	association with pain	[33]
	12,13-DiHOME	LA	↑	association with pain	
Coronary artery disease	9-HETE	ARA	↑	oxidative stress marker	[16]
	F2-isoprostanes	ARA	↑	oxidative stress marker	

Table 1. Cont.

Disease	Oxylipin	Precursor	Direction of Change	Function	Ref.
Myocardial infraction	11-dehydro-TXB2	ARA	↑	oxidative stress marker	[17]
	2,3-dinor-TXB2	ARA	↑	oxidative stress marker	
Atherosclerosis	9-HODE in LDL	LA	↑	lipid peroxidation marker	[34]
Acute respiratory distress syndrome	9(10)-EpOME	LA	↑	leukotoxin	[35]
Asthma	PGE2	ARA	↑	promotion of inflammation	[11,36]
	PGI2	ARA	↑	inhibition of thrombosis and inflammation	
	TXB2	ARA	↑	promotion of inflammation	
	PGF2α	ARA	↓	promotion of inflammation	
	6-keto-PGF1α	ARA	↓	oxidative stress marker	
Cystic fibrosis	LXA4	ARA	→	inhibition of inflammation	[13]
	RvE1	EPA	→	promotion of inflammation resolution, positively associated with better lung function	
Alzheimer's disease	total HODE	LA	↑	in vivo lipid peroxidation marker	[37]
Schizophrenia	8-iso-PGF2α	ARA	↑	oxidative stress marker	[38]
Breast cancer	9-, 13-HODE	LA	↑	PPAR-γ ligand	[14,39–41]
	9-, 13-HOTrE	LA	↑	inhibition of inflammation	
	12-HHTrE	ARA	↑	polymorphonuclear leucocytes (PMN) chemotaxis enhancer	
Colorectal cancer	2,3-dinor-PGF2α	ARA	↑	oxidative stress marker	[42]
	19-HETE	ARA	↑	possible competitive antagonist of 20-HETE	
	12-keto-LTB4	ARA	↑	inactive metabolite of pro-inflammatory LTB4	
	9-HODE	LA	↓	PPAR-γ ligand	
	13-HODE	LA	↓	promotion of apoptosis	
	PGE2	ARA	↑	promotion of inflammation	[12,36]
	PGI2	ARA	↓	inhibition of thrombosis and inflammation	
Non-small cell lung cancer	15S-HETE	ARA	↓	induction of apoptosis	[43]
	13S-HODE	LA	↓	induction of apoptosis	
Prostate cancer	LTB4	ARA	↑	promotion of survival and proliferation of cancerous cells	[44]

LDL, Low density lipoprotein; PMN, Polymorphonuclear neutrophil; PPAR-γ, Peroxisome proliferator-activated receptor γ.

The important functions performed by oxylipins and their constant presence in biological fluids such as blood, urine, and cerebrospinal fluid (CSF), make them potential biomarkers [19]. However, the main problem in the study of oxylipins is their enormous heterogeneity associated with a large number of oxidation pathways. Depending on the type of oxidation, they form many different molecules with similar structures, chemistry and physical properties, which makes it difficult to simultaneously determine them using traditional methods [7,45,46]. In addition, most oxylipins are present at low concentrations and their detection and quantification require methods with high sensitivity [47]. Therefore, this review is aimed at describing the most popular methods of the preparation and extraction of oxylipins for quantitative analysis in various human and animal biofluids, solid tissues and cell cultures.

2. Sample Preparation

2.1. Sample Collection and Storage

Oxylipins are very unstable compounds, and this must be taken into account during the collection of materials for research. Since tissue degradation and free radical oxidation can occur within a few seconds, the material should be procured as quickly as possible: Tissue samples must be quickly frozen in liquid nitrogen, biological fluids collected and stored on ice prior to processing, and cells collected in cold solvents. Considering that non-enzymatic lipid peroxidation can occur even at −20 °C (degradation and loss of analytes have been found for some resolvins and prostanoids derived from DHA and EPA), all samples should be stored at −80 °C, and freeze/thaw cycles should be avoided [48]. To prevent this problem, some antioxidants can be used, and this is described below in Section 2.2. Golovko et al. observed that the storage of brain tissue powder at −80 °C for about four weeks resulted in a two- to four-fold decrease in PG mass [49]. Even the short storage of blood at room temperature before further processing has a huge impact on the concentration of several oxylipins in the plasma: After the storage of whole blood for 60 min, the levels of several oxylipins are greatly reduced (e.g., 15-HETE and 14(15)-EpETrE), whereas other analytes are formed ex vivo (e.g., PGE2) [50]. Leaving samples in the centrifuge for several minutes after centrifuging and prior to collection and freezing, can also lead to a significant decrease in the levels of some oxylipins [50].

When using clinical material, attention should be paid to such a factor as the use of heparin in the treatment of patients, because it leads to heparin-induced phospholipase A2 activity and the elevation of oxylipin levels. In addition, it is advisable to collect the samples at the same time during the day, in order to reduce the potential impact of a circadian rhythm, which can affect several oxylipins, whose concentrations decrease during the day after the morning peak [51]. In serum, coagulation is, in part, mediated by the ARA cascade and causes a massive (ex vivo) formation of several oxylipins (TxB2, 12-HETE), also the detectability in the serum of low-concentration mediators (e.g., resolvins) is higher compared to plasma [50].

In animal experiments, it is also necessary to consider the method used for the euthanasia of the animal and the subsequent processing of the brain sample. It was shown that during decapitation, the level of PGs in the brain of rats was 10–40 times higher than that of rodents euthanized by focused microwave radiation. This difference is primarily the result of the thermal inactivation of enzymes involved in the post-mortem formation of PGs, and to a lesser extent, the capture or destruction of PGs under the action of microwave radiation [49]. Microwave irradiation at a temperature of 70–80 °C prevents postmortem induction in brain eicosanoids and allows the measurement of true levels of eicosanoids, while problems such as significant PG heat-destruction or the trapping of denatured proteins were not detected [49]. In addition, in order to prevent further PG formation after death, the proteins may be heat-denatured in a boiling water bath for 5 min before analysis [52].

Another problem associated with container transfer loss occurs when using, as a surrogate matrix, phosphate buffered saline (PBS)–methanol (MeOH) containing butylated hydroxytoluene (BHT)/ethylenediaminetetraacetic acid (EDTA) (called PMC). Due to the lipophilic nature of the oxylipins and the lack of protein in the PMC matrix, there may be nonspecific binding, which leads to the loss of analytes due to adsorption, especially to the hydrophobic surface of polypropylene materials during transfer with test tubes, bubbles and tips in sample preparation. Usually, these losses during sample preparation are compensated for by the calibration curve; however, with initially low levels of oxylipins in the sample and/or a long transfer time, there can be a significant loss of analytes [53].

Various undesirable components may be present in blood collection tubes available on the market. Silicones can be used as stoppers lubricants or as internal surface coatings. To control the surface wetting density and viscosity surfactants (polyethylene glycols or polyvinylpyrrolidones) and polymeric gels are used. In addition, coagulation activators/inhibitors, polymers and plasticizers presented on the plastic tube walls and rubber stoppers can be added to this list. All the described agents released from plastic containers can be detected after matrix-assisted laser desorption/ionization time-of-flight (MALDI TOF) mass spectrometry and disturb the analysis results. [54]. Furthermore, some substances such as ultraviolet (UV) stabilizers from standard polyvinyl chloride tubes or plasticizers from standard polypropylene microcentrifuge tubes can interfere with liquid chromatography–mass spectrometry (LC–MS) applications [55,56] and analysis by gas chromatography–mass spectrometry (GC–MS) [57]. Nevertheless, in the case of collecting urine samples, polypropylene tubes are used throughout the subsequent process to avoid the binding of eicosanoids to glass surfaces [58].

2.2. Pre-Extraction Additives

2.2.1. Antioxidants

One of the problems that reduce the accuracy of the analysis of biological samples is the formation of oxylipins after sampling. To prevent this process from occurring, COX and soluble epoxide hydrolase (sEH) inhibitors can be used. For example, it is recommended to add 100 mM trans-4-[4-(3-adamantan-1-yl-ureido)cyclohexyloxy]-benzoic acid (t-AUCB) to inhibit the soluble epoxide hydrolase in human plasma. In addition, esterase and protease inhibitors are added to the samples to prevent enzymatic degradation or the formation of oxylipins. [59]. To prevent the formation of ex vivo eicosanoids in urine samples, indomethacin should be added to them immediately [58,60,61]. Furthermore, to prevent the oxidation and breakdown of oxylipins, plasma samples can be stored in MeOH containing Paraoxon—an acetylcholinesterase inhibitor, 12-(3-adamantan-1-yl-ureido)dodecanoic acid (AUDA)—an inhibitor of sEH, and phenylmethylsulfonyl fluoride (PMSF)—a serine protease inhibitor of thrombin [62,63].

Triphenylphosphine (TPP) and radical-scavenging BHT can be added to the tissues during sample collection [47,64]. TPP was used to reduce peroxides to their monoatomic equivalents, and BHT was used to quench radical-catalyzed reactions. Both reagents prevent the conversion of PUFAs to peroxyl radicals [47], and oxylipin degradation or formation (e.g., 11-HETE, 9-HETE, isoprostanes (IsoP)) by autoxidation during sample preparation [50]. In the analysis of lipids, different concentrations of BHT (0.005% to 0.2%) can be used [10,47,65,66]. To assess the need for antioxidants, Golovko et al. analyzed three identical brain samples with 0.1% BHT, 0.005% BHT or without BHT added to the acetone and chloroform used in the extraction. It was found that using only 0.005% BHT prevents the reduction of the mass of 6-oxo-PGF$_1$, whereas 0.1% BHT produces a precipitate in brain samples, which can clog the LC system [49].

Considering the ability of hydroxylated lipids to be converted to glucuronides and other conjugates prior to isolation, Newman et al. incubated urine samples for 3 h at 37 °C with 400 units of Helix pomatia type H-1 glucuronidase to release dihydroxy lipids from their glucuronides [67]. Morgan et al., during the procedure of the synthesis of internal standards, used an addition of 10 µL of SnCl$_2$ (100 mM in water) per mL of sample before extracting lipids from samples with immune cells, for 10 min at room temperature before the extraction to reduce more hydroperoxides unstable to alcohols [68].

2.2.2. Standards

In order to normalize the extraction efficiency and instrument response, the internal standard (IS) is added to the sample before extraction. Deuterated ISs play an important role in the extraction and storage process. Since the deuterated IS is either a similar lipid metabolite or a molecule with similar chemical characteristics, the IS will have the same extraction efficiency and decomposition rate as a lipid metabolite, which will allow the calculation of the amount that is lost as a result of the extraction process or degradation [69]. For oxylipin analysis by mass spectrometry (MS), a mixture of deuterated species is used, in addition to a sufficient number of deuterium atoms (^{2}H), and analytes can be labeled with ^{13}C atoms [70]. Because it is impractical to use an IS for each species analyzed, at least one IS for each lipid class is used, and is selected based on structural similarities [71]. Wang et al. used only 26 deuterated ISs for the analysis of 184 eicosanoids by ultra-high-performance liquid chromatography (UHPLC)/MS [72]. The most commonly used ISs are d4-PGE$_2$, d4-PGD$_2$, d8-12(S)-HETE, d8-5(S)-HETE, d4-PGF$_{2\alpha}$, d4-LTB$_4$, d11-14,15-DiHETrE, d4-9(S)-HODE, d4-12(13)-EpOME, d11-14,15-EET, d11-8,9-EET and d11-11,12-EET [39,73–76]. Although it is important to choose ISs that are not altered during the extraction, Hennebelle et al. found that d4-PGE$_2$ used as the IS underwent degradation during the plasma preparation (hydrolysis process), and, as a consequence, oxylipins that were analyzed with d4-PGE$_2$ cannot be quantified. These included THF-diol, epoxy-keto-octadecenoic acid (EKODE), PGE$_1$, PGD$_1$, PGF$_{2\alpha}$, PGE$_2$, PGD$_2$, PGJ$_2$, PGB$_2$, PGE$_3$, PGD$_3$, 15-deoxy-PGJ$_2$, resolvin E$_1$, 9,12,13-TriHOME, 9,10,13-TriHOME and 11,12-,15-TriHETrE [64].

However, according to the variability of preparation procedures, the use of only one IS is not enough to overcome that problem, and several deuterated ISs can be used [51,63,77]. For example, Yang et al. used deuterated ISs for the extraction of prostaglandins, diols, epoxides and other oxylipins. ISs were added to the samples before extraction (d4-6-keto-PGF$_{1\alpha}$, d4-PGE$_2$, 10,11 DHHep, d6-20-HET-, d4-9-(S)-HODE, d8-5-HETE, d8-11,12-EET). After that, to calculate the recovery rates of each IS, another standard synthetic acid, 1-cyclohexyluriedo-3-dodecanoic acid (CUDA), was added before analysis [47]; 1-phenylurea-3-hexanoic acid (PUHA) may also be added with CUDA as a quality marker for the analysis [73].

2.3. Extraction Methods

It is not always possible to analyze a sample without first isolating the components from the natural matrix. At the same time, in order to reduce the lower limit of detection and increase the sensitivity of determinations, as a rule, it becomes necessary to concentrate them with respect to the matrix components present in the tissue under study. In this case, the separation procedures can significantly simplify the analysis and increase its selectivity, eliminating the influence of interfering impurities.

Considering that oxylipins are present in very low concentrations and that many species are very unstable at room temperature, sample preparation and extraction should be carried out in cold conditions [49,78,79]. Some oxylipins may also be formed during the extraction process. The homogenization stage may activate the synthesis of eicosanoids [49,80]. Due to the high activity of the nonenzymatic and enzymatic processes in blood, urine, solid tissues, or other samples from humans and animals, they often contain very few intact PGs; therefore, sometimes measuring PG metabolite levels is more important [58]. Similarly, more meaningful results might be expected from the determination of oxidized plasma prostanoid metabolites, which are assumed not to be formed as rapidly during the sampling procedure compared to peripheral plasma prostanoid concentrations [81].

Sample preparation includes steps such as adding solvents, acids and antioxidants; extraction; homogenization; centrifugation; the hydrolysis of esterified lipids; the derivatization process. In this review, we will focus on the most commonly used extraction methods that are implemented in the determination of oxylipins.

2.3.1. Protein Precipitation

The extraction of free oxylipins from a biological matrix, such as plasma or tissue, is difficult. Analytes have a wide polarity range and are prone to decomposition during auto-oxidation (all oxylipins) and when treated with a base (PG) or acid (epoxy-FA). If analyte concentrations significantly exceed the limit of quantitation (LOQ) of the instrument, the sample can be directly injected after protein precipitation (PPT) by dilution with organic solvents. However, most analyses require the pre-concentration of the samples using liquid–liquid extraction (LLE) or, most often, solid-phase extraction (SPE); however, in almost all cases, prior PPT is required [50]. In addition to removing the protein, adding an organic solvent to a biological fluid also disrupts the bonds between the metabolites and the proteins present in the solution. As a result, the obtained concentrations of metabolites are total metabolite concentrations equivalent to the sum of the bound and unbound (free) metabolite concentrations [82]. Additionally, PPT with acids can catalyze the hydrolysis of certain conjugates, such as glucuronides and sulfates [83].

Satomi et al. compared PPT using different water-soluble organic solvents such as MeOH, ethanol (EtOH), isopropanol (IPA) or acetonitrile (ACN), followed by methyl tert-butyl ether (MTBE)-based lipid extraction, to determine which would be the best method for sample preparation for LC–MS-based lipidomics analysis. ACN deproteinization is less effective than alcohol, potentially due to insufficient protein denaturation, which causes lipid decomposition through lipase activation. Therefore, protein precipitation by alcohol was evaluated as the best lipid extraction method [84]. MeOH extraction was appropriate for partly hydrophilic lipid species, and IPA was effective for hydrophobic lipids such as triacylglycerols. The best approach to cover a wide range of lipid species using a simple preparation procedure, in practice, is thought to be EtOH extraction [84].

Similar to lipid extraction, when using PPT for the extraction of oxylipins, various types of solvents can be used, their choice of which depends on the type of tissue and the class of the analyte to be determined [39]. PPT by adding two volumes of water-miscible organic solvents, such as MeOH, generally provides a high extraction efficiency (>90%) [85]. However, in the case of the analysis of eicosanoids, this extraction method is not very suitable due to the presence of eicosanoids in the samples at very low concentrations, and precipitation together with the proteins to which they bind unspecifically [85]. Lee et al. used methanol-based protein precipitation, which was followed by the LC–tandem mass spectrometry (MS/MS) analysis of 20 oxylipin levels in the serum of women with endometriosis [71]. Heemskerk et al. performed PPT by the addition of methanol to an adipocyte-conditioned medium or plasma to analyze adipose tissue PUFA synthesis and anti-inflammatory lipid and oxylipin plasma profiles using LC–MS/MS [86]. Although, according to Satomi [84], alcohol precipitation is superior to that of ACN, other authors obtained good results using ACN [87]. Chocholoušková et al. applied ACN for the denaturation during the UHPLC/MS determination of oxylipins in human plasma and found a high efficiency of ACN in protein precipitation [39]. Zein et al. used ACN in PPT for the HPLC-electrospray ionization (ESI)-MS/MS analysis of free fatty acids, eicosanoids and docosanoids in human gingival crevicular fluid (GCF), saliva and serum [88]. In addition, Wang et al. used ACN PPT for the analysis of free arachidonic acid in plasma with the use of LC–MS/MS [89].

For the most efficient protein removal, organic solvents such as MeOH and ACN can be combined with zinc sulfate [90]. Kortz et al. used PPT with a solution consisting of MeOH:zinc sulfate (4:1 v/v) before performing an online SPE-LC-MS/MS analysis of PUFAs and eicosanoids in human plasma [91]. Additionally, Klawitter et al., to measure concentrations of 15-F2t-isoprostane in human plasma and urine samples, used PPT with MeOH/zinc sulfate. Afterwards, the PPT samples were injected into the HPLC system and extracted online. No carry-over and no matrix inferences such as ion suppression or enhancement were observed [92]. Good results can also be achieved with serial PPT using different solvents. Bessonneau et al. compared the performance of the non-lethal in vivo solid-phase microextraction (SPME) sampling method for rat plasma with PPT (acetone/hexane/chloroform) for monitoring the time profile of blood eicosanoids. The results obtained for 12-HETE and ARA were significantly correlated with those obtained using conventional PPT [93].

2.3.2. Liquid–Liquid Extraction

Liquid–liquid extraction (LLE) is a widely used sample preparation method for extracting all major classes of lipids, including phospholipids, ceramides, sphingomyelins, and cholesterol esters [94]. Until recently, the chloroform:MeOH Folch [95] and Bligh and Dyer [96] methods of extraction were the most common methods. In the Folch procedure, a 2:1 chloroform:MeOH mixture is added in a volume 20 times higher than that of the sample. The subsequent addition of saline solution allows a lower layer to be obtained consisting of all lipids, and an upper layer consisting of contaminants. At the final stage, a solution consisting of a chloroform:MeOH:water in a ratio of 8:4:3 is used to affirm lipid separation in the chloroform layer [95]. The adapted procedure of Bligh and Dyer differs in the amount of solvent used (3 mL of MeOH:chloroform mixture 2:1 per gram or mL of the sample). After stirring, 1 mL of chloroform and 1.8 mL of water were added to separate the solution into two phases. A shorter extraction time and the use of chloroform to re-extract tissue improves lipid yields compared to the Folch procedure [97]. However, the main disadvantage of the Folch et al. or Bligh and Dyer protocols is that they lead to the quantitative extraction of most lipids, including those that can be present in very high concentrations in biological tissues (cholesterol, triacylglycerols and phospholipids). High levels of these lipids can interfere with the analysis of oxylipins, which are usually present in very low concentrations [72]. Using the chloroform–MeOH or chloroform–methanol–water method, both hydrophilic and lipophilic substances can be extracted simultaneously; however, because using chloroform is problematic in LC–MS, such an approach requires the further removal of chloroform by lyophilization [82]. Another approach for the extraction of the main classes of lipids is separation using MTBE [98]. A comparison of the extraction efficiency of classes of lipids in human plasma by the MeOH, MeOH:EtOH (1:1) and MTBE methods shows the best result for the separation of lipids using MTBE (3,125 metabolites were detected using LC–TOF with C18 and HILIC columns in positive mode summary) [98]. There is little information about the use of MTBE for the extraction of oxylipins, but it is known that Rund et al. successfully used MTBE extraction for the analysis of IsoP and IsoF formed in HCT116 cells [59].

Biofluids

When determining oxylipins in plasma, the ratio of plasma to precipitant is between 1:1.35 v/v to 1:4 v/v in various studies. The nature of the extraction solvent has a profound effect on the process extraction efficiency [82]. Various types of organic solvents can be used to extract oxylipins. In the Fleming Laboratory (Frankfurt, Germany), double extraction with ethyl acetate (EA) is used to determine levels of fatty acid epoxides in murine plasma or bone marrow extracellular fluid, obtained from flushed-out femurs. In addition, oxylipins were extracted from plasma with sodium acetate, followed by extraction with EA [99]. Using a modification of the Golovko acetone extraction method [49], Pier et al. identified 10 different PGs in human ovarian follicular fluid [100]. For the determination of seven F2-isoP isomers among classes III, IV, and VI in the blood plasma of pregnant women, Larose et al. developed a method including hydrolysis by KOH, double pre-extraction with hexane, and consequent triple extraction with EA:hexane (3:1) [101]. Hall and Murphy used extraction by the Bligh and Dyer method, substituting methylene chloride for chloroform to quantitate production of 5-HETE, 5-HPETE, 5-oxo-ETE in red blood cell (RBC) ghosts [102].

Solid Tissues

Unlike biofluids, the extraction of oxylipins from solid tissues is preceded by a homogenization process, which itself can activate the synthesis of some oxylipins. To prevent this, special additives may be used, see Section 2.2 [79].

The most common LLE method for tissue extraction involving chloroform is the Bligh and Dyer method [78,94]. However, due to the wide range of extracted lipids, matrix effects and the response of analytical equipment, the authors also used other solvents for LLE from tissue. In order to increase the extraction of eicosanoids, reduce chemical background noise and reduce the preparation time, Brose et al. changed the LLE protocol by replacing acetone:chloroform with MeOH. Using a smaller volume of solvents, modified single-stage extraction with MeOH resulted in a much higher (96.7 ± 9.9%) extraction of the internal standard, which may be the result of eliminating analyte loss through transfer/evaporation steps [52]. In another work, Brose et al. used LLE with acetone to extract prostaglandins and isoprostanes (PGE2, PGD2, isoPGE2 such as PGE2, entPGE2, 8-isoPGE2, 11β-PGE2, PGD2, and 15(R)-PGD2) from murine brain [103]. Urban et al. established that for the extraction of PGs from pig brain tissue, the use of an EtOH:10 mM phosphate buffer (85:15) as the extraction solvent, showed better results compared to EtOH:dichloromethane (1:1), MeOH:10 mM phosphate buffer (85:15) and 10 mM phosphate buffer [104].

Cell Cultures

LLE with hexane:EA is very often used for studying endogenous oxylipins from cell cultures. Yang et al. used this method to investigate levels of PGD2, 15-keto-PGE2, 13,14-dihydro-15-keto-PGE2, PGD3, 8-iso-PGE2, 8-iso 15-keto PGF2α, PGF3 α, and 8-iso PGF3 α in human non-small-cell lung cancer cells (A549) and human colon carcinoma cells (DLD-1) [66]. Kempen et al. used hexane:EA (1:1, v/v) LLE for the quantification of ARA metabolites (PGE2, 11-HETE, 5-HETE,12-HETE) from human lung cancer cells H1299 and A549, and a rat leukemia cell line RBL-1 [105]. In addition, using hexane:EA (1:1, v/v) LLE, Schroeder et al. analyzed endogenous levels of eicosanoids (PGE2, 5-HETE, 12-HETE, 15-HETE, and 13-HODE) in cell lysates of squamous cell carcinoma cell lines of the head and neck (HNSCC) [106]. Morgan et al. used the hexane:IPA:acetic acid (HAc) LLE procedure to measure esterified oxylipins generated by immune cells: hydro(pero)xyeicosatetraenoic acids (H(p)ETEs), hydroxyoctadecadienoic acids (HODEs), hydroxydocosahexaenoic acids (HDOHEs) and keto-eicosatetraenoic acids (KETEs), attached to either phosphatidylethanolamine (PE) or phosphatidylcholine (PC). Using this extraction method allows the simultaneous monitoring of up to 23 different oxylipins, with better recoveries of standards and analytes than the classical Bligh and Dyer method [68]. Michaelis et al. used LLE with hexane/EA (1:1, v/v) to determine cytochrome P450 (CYP) epoxygenase-derived epoxyeicosatrienoic acids (EETs) in bovine endothelial cells [107]. PGD2 and PGE2 were measured by Cao et al. in culture supernatants from A549 cells and RAW 264.7 cells after hexane/EA (1:1, v/v) extraction [108]. All extraction steps must be performed under conditions with minimal light levels to reduce the potential for the photodegradation of the eicosanoid metabolites [106].

2.3.3. Solid-Phase Extraction

The basis of the SPE method is the selective separation of analytes between the liquid and the solid phase. The main goal is to remove the compounds that cause matrix effects during the analysis and to concentrate analytes, thus increasing the sensitivity as well as improving the detection limits. By removing interfering compounds and impurities, SPE thereby protects analytical systems and increases efficiency, and when proper solvents are used for elution, tunable selectivity becomes possible [97]. The most commonly used SPE cartridges may be reversed phase (RP) (C18), normal phase (silica) and ion exchange (anion or cation) phase. The basic principle for the use of reverse-phase SPE is that aliphatic fragments in oxylipins can interact with non-polar stationary phases. Silica retains polar compounds, typically used for sample clean-up. Anion-exchange polymer-based resins selectively retain oxylipins based on both hydrophobic and anion-exchange interactions. Polymeric sorbent (containing both lipophilic and hydrophilic functional groups) allows the retention of more lipid metabolites [109], see Figure 3.

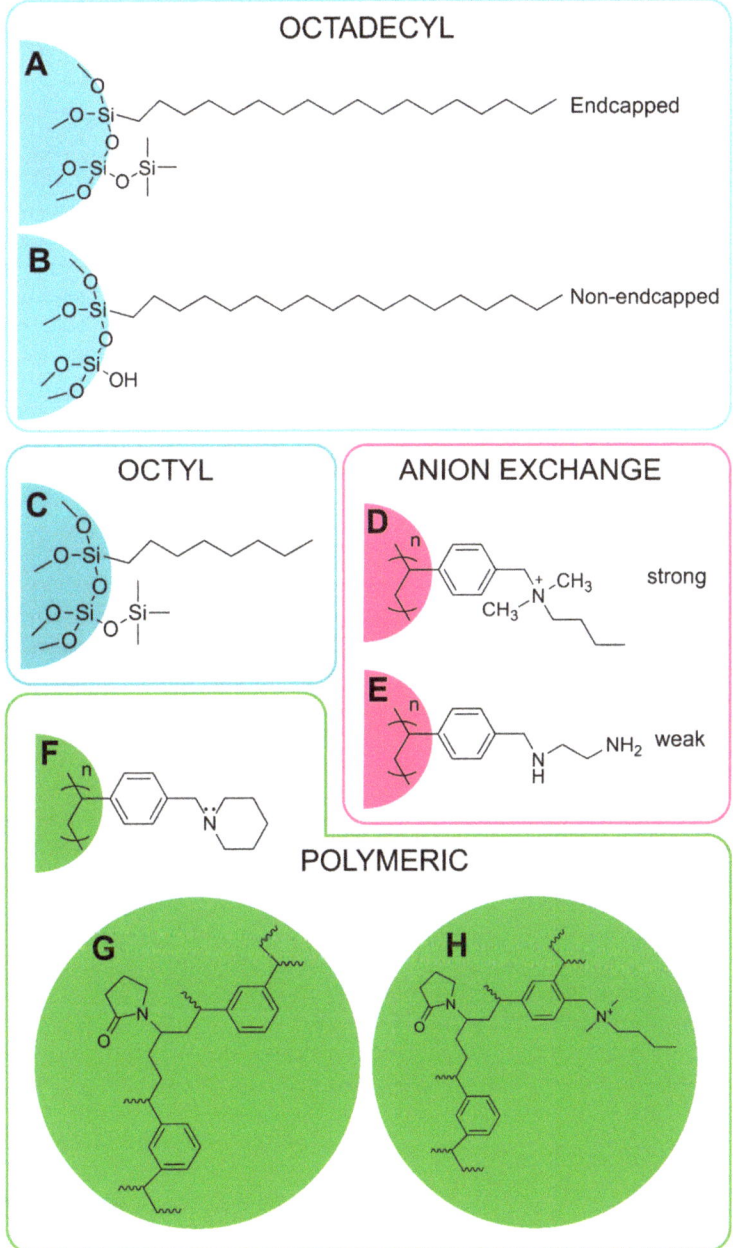

Figure 3. Examples of the most commonly used types of solid-phase extraction (SPE) phases for oxylipin extraction. A: Strata C18-E; B: Chromabond C18; C: Discovery DSC-8; D: Strata-X-A; E: Strata-X-AW; F: Strata-X; G: Oasis HLB; H: Oasis MAX.

Usually, the process consists of several stages. At the first stage, the cartridge containing the solid sorbent is conditioned with a suitable solvent. Then, a tested sample is loaded onto it and subsequently, analytes bind to the sorbent. In the next step, the cartridge is washed to remove unwanted impurities.

At the last stage, the analytes are eluted from the cartridge with a solvent, which is selected taking into account the chemical and physical characteristics with respect to the analyte [110].

Biofluids

Blood/Serum/Plasma

Serum and plasma samples can be directly loaded onto the SPE cartridge [51,76]; however, they usually require pretreatment, since analytes can be bound to proteins, which reduces SPE extraction. To break the bonds between the analyte and the protein, one of the following methods can be used: Changing the pH of the sample to the extreme (pH < 3 or pH > 9) with acids or bases in the concentration range of 0.1 M or more, the above-described protein precipitation using polar solvents (ACN, MeOH or acetone), and treatment with acids or inorganic salts (formic acid (FAc), perchloric acid, trichloroacetic acid, ammonium sulfate, sodium sulfate or zinc sulfate). Less commonly, ultrasonic treatment of biological fluid for 15 min is used [109]. The calcium chelator, EDTA, which can sequester calcium ions and then inhibit phospholipase A2, is used to prevent the ex vivo formation of eicosanoids during the preparation of plasma samples [78]. Oxylipins can either be bound to circulating plasma proteins, such as albumin or can be included in lipoproteins, and may be released during the elution step (e.g., EETs) [111]. This is due to the dissociation of fatty acids from lipoproteins or albumin as a result of the denaturation of proteins by MeOH as an eluent of SPE. It can lead to an overestimation of oxylipin levels, so it is very important to use a sufficient number of IS during the extraction [72].

Considering the fact that there are a wide variety of sample preparation methods for the efficient pre-concentration and extraction of oxylipins from the matrix, Ostermann et al. compared SPE protocols on various cartridges for the analysis of free (non-esterified) oxylipins in human plasma [112]. Classical reverse-phase (RP) material was compared to novel polymeric stationary phases with polar groups, such as Strata X (Phenomenex, Torrance, CA, USA) or Oasis HLB (Waters, Eschborn, Germany). For the extraction of weakly acidic oxylipins from biological samples, materials with anion-exchange properties were used. Compared with other described sample preparation methods, the SepPak tC18 SPE protocol from Serhan lab (Boston, MA, USA [113]) and SPE column (6 mL, 500 mg, 37–55 µm; Waters, Boston, MA, USA), most effectively extracted free oxylipins from plasma, and for many analytes (especially for non-polar epoxides), this protocol gave the highest peak area [112].

However, the extraction efficiency depends not only on the type of column used but also on the SPE conditions (solvents used for column conditioning, washing and elution). Galvão et al. compared their own method of extracting eicosanoids from plasma samples by SPE using Sep-Pak C18 cartridges (500 mg, 2.8 mL) with two previously published methods, in which EtOH, hexane and EA were used for washing and elution. The selection criteria were: A lower consumption of organic solvents and a greater recovery of eicosanoids. Using 2 mL MeOH, 2 mL water/0.1% HAc for column conditioning, 2 mL water/0.1% HAc for the removal of impurities and 2 mL MeOH/0.1% HAc for sample elution, they achieved a better recovery compared with other methods, especially for the extraction of LTs, HETEs, and lipoxin (LX) A4 [114].

Urine

Before target extraction, urine samples can be diluted with a buffer with an appropriate pH or with water. [115,116]. To better dissolve compounds from urine, acid or base hydrolysis is used for basic and acidic compounds, respectively. The urine is heated for 15–20 min after adding a strong base (for example, 10 M KOH) or an acid (usually concentrated HCL), then cooled and diluted with the buffer, and the pH is adjusted accordingly for the SPE procedure. Enzymatic hydrolysis can also be used [109]. For extraction, organic acids are suggested, since mineral acids promote the faster dehydration of urine PGE2 to PGA2 [58].

Sterz et al. compared SPE methods when developing a method for the quantitative determination of seven types of eicosanoids in urine. Each tested SPE cartridge allowed the extraction of all types of eicosanoids (except 12-HETE). C18 RP-SPE (Bond Elut C18) was ideal for extracting LTE4 and 12-HETE. Polymeric SPE (Oasis®HLB, Waters, Eschborn, Germany and Strata X33u, Phenomenex, Aschaffenburg, Germany) was favorable for the extraction of 2,3-dinor-TxB2 and 11-dehydro-TXB [117]. Polymeric RP/strong anion exchange SPE (Oasis®MAX, Waters, Eschborn, Germany) was good for most analytes (optimal for extracting PGG2α species and tetranor PGE-M). Polymeric RP/weak anion exchange SPE (Easy cartridge) showed the worst results for the extraction of eicosanoids compared to all the others [117]. Medina et al. carried out a study of three types of different SPE cartridges: Strata X-AW (100 mg, 3 mL; Phenomenex, Torrance, CA, USA), C18 Sep-Pak classic cartridge, and OasisHLB (both 200 mg, 6 mL; Waters, Milford, MA, USA), in order to establish the most efficient one for the extraction of 13 eicosanoids in human urine [118]. The best recovery was shown for Strata X-AW cartridges (93–107%), when the recovery for Oasis HLB and C18 Sep-Pak cartridges was between 59–71%. Due to the weak ion-exchange interaction of the resin, Strata X-AW provided better reproducibility while the other types of cartridges showed unrepeatable extraction, which depended on the IsoPs nature [118].

Other Biofluids

Practically any biological fluid from which it is necessary to extract oxylipins can be subjected to SPE. Gouveia-Figueira et al. used Waters Oasis HLB SPE cartridges (60 mg, 30 µm), with the elution of 3 mL ACN, 2 mL MeOH, and 1 mL EA to isolate oxylipins in bronchial wash and bronchoalveolar lavage samples [119]. Panthi et al., to develop an optimal method for isolating lipid mediators of inflammation in the tear film, compared different PPT and LLE techniques with one SPE method. Only in the case of SPE (Strata X-AW, 33µ polymeric weak anion; Phenomenex, Torrance, CA), it was possible to isolate six analytes of PG and IsoP in small volumes of tears [120]. Using SPE (Sep-Pak C18, 500 mg, 6 mL; Waters, Boston, MA, USA), Giera et al. extracted several important lipid mediators in human synovial fluid [121]. SPE cartridges (HyperSepRetain PEP; Thermo Fisher Scientific, Waltham, MA, USA) were also useful to exclude impurities in the case of the preparation of platelet-rich plasma samples [122]. Wang et al. used SPE (C18 column, 500 mg; Biotage, Uppsala, Sweden) to isolate oxylipins generated by phagocytes from blood plasma [123].

When examining the levels of oxylipins in human milk samples, the pre-extraction procedure is not necessary. For sample preparation for SPE, only double centrifugation is required to remove the fat layer [124]. Robinson et al. used SPE to determine oxylipin levels in human milk. Using Oasis HLB 96-well plates for SPE (Waters), they quantified eighteen oxylipins: 6-keto- PGF1α, TXB2, PGE2, LXB4, LXA4, LTB4, 15-hydroxyeicosatetraenoic acid (HETE), 12-HETE, 5-HETE, resolvin (Rv) D1, RvD2, 7(S)-maresin (MAR) 1,7(R) MAR1, protectin D1, protectin DX, 18-hydroxyeicosapentaenoic acid (HEPE), 14-hydroxydo-cosahexaenoic acid (HDHA), 17-HDHA [125]. For the same purposes, Wu et al. used Oasis HLB cartridges (60 mg, 30 µm; Waters, Milford, MA, USA), which allowed the extraction and simultaneous detection of 31 oxylipins from human milk [124]. Thus, SPE is a popular method for extracting and concentrating oxylipins from various biological fluids due to the high efficiency it displays with respect to removing impurities from the sample.

Solid Tissues

Solid tissues also need some pretreatment. They are homogenized either in water, in a polar organic solvent (e.g., MeOH or ACN), or in mixtures of water with these solvents, for RP or ion exchange cleanup procedures. Together with the iron chelator EDTA, diethylenetriaminepentaacetic acid (DTPA) may be included in the extraction process before homogenization to limit the formation of eicosanoids [79]. Care must be taken with the choice of pH for extraction. Despite the fact that acidic conditions stabilize the free carboxylic acid form of eicosanoids, and reduce protein binding [126], eicosanoids can be altered by the extraction procedure at extreme pH values [127]. Tissue extracts obtained with mid-polar to non-polar solvents can be processed using normal-phase SPE procedures. After centrifugation or filtration to remove the precipitated proteins and solids, the pH of the sample may need to be adjusted. The analyte may adsorb onto the SPE packing or alternatively, it may simply pass through, free from interferences [109].

Among solid tissues, oxylipin levels are most often examined in brain tissue that has high moisture and fat contents. This must be taken into account when choosing the extraction solvent for such lipophilic oxylipins as eicosanoids because they will bind to fats. Solvents for the SPE of eicosanoids from brain tissue must dissolve the eicosanoids, permeate the matrix of the brain tissue, destroy the tissues, release the eicosanoids and, finally, cause protein precipitation [126]. For the extraction of oxylipins from rat brain, different authors successfully used various methods and types of cartridges: Arnold et al. used Bond Elut Certify II columns to identity EETs in rat brain [9]; Masoodi et al. used C18-E columns (500 mg, 6 mL; Phenomenex, Macclesfield, UK) for the analysis of LTs, resolvins, protectins and related hydroxy-fatty acids in rat brain [113]; Yue et al. used Oasis®HLB (1 cm^3, 30 mg, 30 μm; Waters Corporation, Milford, MA, USA) for the determination of bioactive eicosanoids [126]. In addition, Yue et al. found that MeOH, EtOH or acetone, together with phosphoric acid or FAc, have a similar extraction efficiency, but MeOH and FAc evaporate more easily and are more compatible with SPE (Oasis®HLB, 1 cm^3, 30 mg, 30 μm; Waters), and anhydrous ACN is a stronger mobile phase than MeOH and more efficiently elutes eicosanoids [126].

Blewett et al. used SPE (Oasis SPE cartridge; Waters, Milford, MA) with the HPLC–ESI–MS method for the simultaneous determination of 23 eicosanoids in rat kidney tissue [128]. Le Faouder et al. optimized the sample preparation and the extraction process with an extraction yield of 80%, ranging from 65% to 98%. Using SPE on a C18 cartridge (15 mL, 200 mg; Macherey Nagel) they obtained the separation of 26 PUFA derivatives in colonic tissues of mice [129]. Weylandt et al. used SPE with an anion exchange column (Bond Elut Certify II; Agilent, Santa Clara, CA, USA) to determine the profile of lipid mediators formed from PUFAs (18-HEPE, 17-HDHA,15-HETE) in mouse liver [130]. Jelińska et al. successfully used SPE cartridges (Bakerbond C18, 3 mL, 500 mg, from J.T. Baker. Hampton, NH, USA) to extract eicosanoids (13-HODE, 9-HODE, 15-HETE, 12-HETE, 5-HETE) and PGE2 from 7,12-dimethylbenzanthracene (DMBA)-induced tumors in rats for further LC–MS/MS analysis [131].

Cell Cultures

Cell culture media also require the dilution of the media with water or a buffer at the proper pH to ensure that the analyte is freely dissolved in the sample. If a particulate-laden cell culture medium is difficult to pass through the SPE device, it may need to be vortexed and centrifuged prior to SPE [109]. The direct analysis of oxylipins from the culture medium is complicated by the fact that they can be rapidly metabolized in vitro (e.g., PGE2 and PGD2) [58].

SPE has been widely used for the analysis of oxylipins in cell culture supernatants and lysates and showed greater efficiency both in the number of analytes found and in the values of recovery, lower limits of detection (LLOD) and lower limits of quantitation (LLOQ) [2,85] compared to LLE. Deems et al. developed a procedure for isolating eicosanoids from media and cells from a cell culture. After the medium collection, the eicosanoids were isolated from medium by SPE using Strata-X SPE columns (Phenomenex, Torrance, CA, USA), which allowed over 60 discrete chemical species of eicosanoid to be identified [132]. Wang and DuBois used 6 mL Sep-Pak C18 cartridges (Waters Associates) or a 6 mL octadecyl silica (ODS) column to extract eicosanoids from the cell-free culture medium. More polar materials were removed by subsequent elution with 15% aqueous EtOH. After extraction, the recoveries of 6-keto-PGF1α, TXB2, PGE2, PGF2α, LTB4, LTC4, 5-HETE, 12-HETE, and 15-HETE were 90.5%, 90.6%, 92.5%, 98.1%, 86.1%, 98.3%, 95.3%, 99.8%, and 92.8%, respectively [58]. Le Faouder et al. optimized the sample preparation and the extraction process with extraction using SPE on a C18 cartridge (200 mg, 15 mL; Macherey Nagel), and obtained the separation of 26 oxylipins in human Caco-2 epithelial cells and human primary foam cells [129].

2.3.4. LLE or SPE?

LLE is a widely used method of sample preparation for the extraction of target compounds from aqueous samples; however, when compared with SPE, in samples prepared by LLE a wider peak of phospholipids appeared during the LC–MS analysis [89]. Considering that the SPE method is more suitable for processing a large number of samples, SPE has become a popular method of sample preparation in terms of reproducibility, less use of organic solvents and ease of use. In addition, SPE is very compatible with automatic analysis systems [89]. Although LLE extraction efficiency is usually higher than SPE, many endogenous impurities are extracted using LLE, which can affect separation and quantification [72]. Among the disadvantages of LLE with organic solvents is the poor extraction of hydrophilic compounds, such as PGs and LTs, compared to SPE [53]. SPE has been shown to provide better purification and enrichment than LLE [97]. The disadvantages of SPE are the high cost of SPE cartridges, the fact that it is time-consuming (depending on the procedure used), and the fact that for unstable eicosanoids a long extraction process can be detrimental for further accurate analysis [126].

Often LLE is not used separately, but precedes subsequent SPE to improve the purification of oxylipins [133,134]. Tajima et al. extracted total lipids from the tissues of the right hemisphere of the mouse brain using the Bligh and Dyer method with minor changes and for the further isolation of oxylipins, samples of the aqueous layer were subjected to SPE [135]. However, when comparing the SPE and LLE methods, different authors obtain ambiguous results about how the efficiency of extraction is influenced not only by the choice of solvent for LLE and the cartridge model for SPE. Furthermore, the type of tissue and examined analytes may influence the SPE and LLE efficiency. Sterz et al. compared the LLE and SPE methods for the isolation of eicosanoids from urine and found that LLE allows all analytes to be extracted; however, the extraction efficiency increased with decreasing urine pH (optimal at pH 4). Despite the fact that the signal intensity for some analytes was lower when using LLE than SPE, the best signal-to-noise (S/N) ratios were achieved after LLE over the complete range of analytes [117]. Rago and Fu compared LLE by HAc/IPA/hexane (2:20:30, $v/v/v$) with SPE (Oasis®HLB, Milford, MA, USA) for the extraction of eicosanoids from human and monkey plasma samples. The SPE method showed better recovery and reproducibility than the LLE method [136]. In the case of polar eicosanoids (LTB4, PGD2, PGE2, PGF2α, 13, 14-PGE2 and 8-iso PGF2α), the extraction recovery rate was, on average, 63% enhanced when using SPE [136]. Additionally, when comparing LLE extraction with sodium acetate [99] using six different SPE methods, Ostermann et al. considered LLE as an inappropriate sample preparation for the analysis of oxylipins in plasma [112]. However, Golovko et al. demonstrated that the extraction of PGs from tissue with acetone, followed by LLE, significantly increased the sensitivity level of the LC–MS/MS analysis compared to other extraction methods, due to a significant reduction in the background chemical noise. Dissolving the residue from the lipid extracts of n-hexane:IPA with acetone is less time-consuming and expensive compared to cleaning C18 cartridges. Besides this, PGs are more stable in n-hexane:IPA extracts [49]. Based on the above, in recent years, SPE has more often been chosen for oxylipin analysis than LLE; however, the extraction method finally applied should depend on the studied oxylipin groups.

Most SPE cartridges and extraction procedures used for the purification of oxylipins are presented in Table 2.

Table 2. Representative SPE cartridges and extraction conditions for purification of selected oxylipins.

Sample	Stationary Phase	Sample Preparation	Column Precondition	Sample Wash	Elution	Reference
		OCTADECYL PHASES				
Human plasma (1 mL) Rat Carrageenan-induced air pouch fluid (1 mL)	Bond Elut C18	Pouch fluid: + 1 mL heparinized saline centrifugation at 1000× g, 4 °C, 10 min	2 mL MeOH 2 mL H$_2$O	2 mL MeOH 2 mL petroleum ether	1 mL methyl formate	[137]
Rat brain (~500 mg) rat liver (~500 mg) plasma (500 µL)	C18-E (6 mL, 500 mg)	Tissues: homogenization in H$_2$O, adjusted to 15% MeOH to 3 mL incubation on ice, 30 min centrifugation at 3000 rpm, 5 min + 0.025 mM HCL (pH 3) Plasma: dilution with H$_2$O, adjusted to 15% MeOH to 3 mL incubation on ice, 30 min centrifugation at 3000 rpm, 5 min + 0.025 mM HCl (pH 3)	20 mL MeOH 20 mL H$_2$O	20 mL 15% MeOH 20 mL H$_2$O 10 mL Hex	15 mL methyl formate	[113]
Rat serum (400 µL) rat mammary tumor (~200 mg)	Bakerbond C18 (3 mL, 500 mg)	Serum: + 0.5 mL MeOH + 4 mL MeOH Tumor: homogenization with 2 mL H$_2$O on ice 4 °C, 30 min centrifugation at 3000 rpm, 5 min	10 mL MeOH 10 mL H$_2$O	2 mL H$_2$O 2 mL 10% MeOH	3 × 0.5 mL MeOH	[131]
Human phagocytes culture supernatant	Waters C18	1:10 dilution with H$_2$O (pH ~3.5)	1× MeOH 2× H$_2$O	1× H$_2$O 1× Hex	6 mL methyl formate	[138]
Human plasma (300 µL)	Sep-Pak C18 (2.8 mL, 500 mg)	+ 1.5 mL MeOH/ACN (1/1, v/v) 4 °C, overnight centrifugation at 400× g, 20 min dilution with H$_2$O to 10% MeOH/ACN	2 mL MeOH 2 mL 0.1% HAc 10 mL ethanol 20 mL H$_2$O 10 mL methanol 10 mL H$_2$O	sample + 2 mL 0.1% HAc sample acidified with 0.1 M HCl 10 mL H$_2$O 1 mL 35% ethanol sample acidified with 0.1 M HCl 5 mL 15% MeOH 5 mL H$_2$O 2.5 mL Hex	2 mL 0.1% HAc in MeOH 2 mL ethanol 2 mL Hex	[114]
Mouse brain tissue (20 mg)	Sep-Pak C18	microwave processing homogenization in 3 mL 15% MeOH, 0.005% BHT (pH 3) centrifugation at 2000× g, 4 °C, 10 min		20 mL 15% MeOH 20 mL H$_2$O dried with syringe air	10 mL methyl formate	[49]

Table 2. Cont.

Sample	Stationary Phase	Sample Preparation	Column Precondition	Sample Wash	Elution	Reference
Mouse lung homogenate (50 µL)	Sep-Pak tC18 (6 mL, 500 mg)	+ 0.45 mL MeOH, 10 min vortex 0.5 mL supernatant collected + 4.5 mL HCL in H$_2$O (pH 3.5)	12 mL MeOH 12 mL H$_2$O	12 mL H$_2$O 6 mL Hex	9 mL methyl formate	[139]
	MonoSpin™ C18	+ 0.45 mL MeOH, 10 min vortex centrifugation at 9000× g, 4 °C, 5min evaporation with N$_2$, 40 °C reconstituted in 0.1 mL MeOH + 0.9 mL HCl in H$_2$O (pH 3.5)	0.3 mL MeOH 0.3 mL H$_2$O	centrifugation at 9000× g, 4 °C, 1 min 0.3 mL H$_2$O 0.3 mL Hex centrifugation at 9000× g, 4 °C, 1 min	2 × 0.3 mL methyl formate	[140]
Rat hypothalamus (95 mg)	Hypersep C18 (3 mL, 500 mg)	homogenization in 2 mL water acidified with 1 M HCL to pH 3.0 centrifugation at 7000 rpm	6 mL MeOH 6 mL H$_2$O	2 mL 2% HAc	2 mL MeOH	[140]
OCTYL PHASES						
Human plasma (1 mL)	Discovery DSC-C8		1 mL MeOH 1 mL 0.1% FAc	1 mL 0.1% FAc	1 mL MeOH	[141]
POLYMERIC PHASES						
Human plasma (500 µL)	Bond Elut Certify II (3 mL, 200 mg)	+ 10 µL 0.2 mg/mL BHT, EDTA, 100 µM indomethacin, 100 µM soluble epoxide hydrolase inhibitor trans-4-[4-3-adamantan-1-yl-ureido)-cyclohexyloxy]-benzoic acid in 50% MeOH + 1.4 mL ice cold MeOH, −80 °C, 30 min 10 min, 4 °C, 20,000× g N2 to final volume <1 mL + 2 mL 0.1 M disodium hydrogen phosphate buffer (pH 5.5)	1 × 1% HAc in EA/n-Hex (75/25, v/v) 1× MeOH 1 × 0.1 M disodium hydrogen phosphate buffer (pH 6)	3 mL H$_2$O 3 mL 50% MeOH dried under N$_2$, 1 min	2 mL 1% HAc in EA/n-Hex (75/25, v/v)	[59]
Rat brain/heart/kidney/liver/lung/pancreas/red blood cells (50 mg) plasma (200 µL)	Bond Elut Certify II	+ 0.5 mL distilled H$_2$O + 0.5 mL MeOH + 300 µL (100 µL for plasma) 10 M NaOH, 60 °C, 20 min + 300 µL 60% HAc + 2 mL 1 M sodium acetate buffer (pH 6) adjusted to pH 6 centrifugation	2 mL MeOH 2 mL 5% MeOH, 0.1 M sodium acetate buffer (pH 7)	2 mL MeOH/H$_2$O (1/1, v/v)	2 mL Hex/EA (75/25, v/v)	[9]
Human whole-blood and platelet-rich plasma (250 µL)	HyperSep Retain PEP	dilution with 5% MeOH, 0.1% HAc with 0.009 mM BHT	5% MeOH, 0.1% HAc	2 × 5% MeOH, 0.1% HAc	1 mL EA 1 mL MeOH	[122]

Table 2. Cont.

Sample	Stationary Phase	Sample Preparation	Column Precondition	Sample Wash	Elution	Reference
Human milk (1.5–2 mL)	Oasis HLB (60 mg, 30 μm)	centrifugation + 10 μL 0.2 mg/mL BHT/EDTA	4 mL EA 8 mL MeOH 8 mL 5% MeOH, 0.1% HAc	8 mL 5% MeOH, 0.1% HAc	2 mL EA 3 mL ACN 2 mL EA	[124]
Human plasma (250 μL)	Oasis HLB (60 mg, 30 μm)	+ 0.2 mg BHT/EDTA	0.5 mL MeOH		2 mL EA	[51]
Human plasma (50–200 μL)	Oasis HLB (3 mL, 60 mg)	Free oxylipins: 200 μL plasma + 10 μL 0.2 mg/mL of BHT and TPP and 1 mg/mL EDTA in 50% MeOH + 1 mL 5% MeOH, 0.1% FAc Total oxylipins: 50–100 μL plasma + 10 μL 0.2 mg/mL of BHT and TPP and 1 mg/mL EDTA in 50% MeOH + 0.1 mL 0.1% acetic acid and 0.1% BHT in MeOH 80 °C, overnight + 0.2 mL of 0.25 M sodium carbonate solution, 60 °C, 30 min, constant shaking + 25 μL HAc, 1.575 mL H_2O to pH 4–6	1× EA 1× MeOH 2 × 5% MeOH, 0.1% FAc	2 × 5% MeOH, 0.1% FAc dried under −20 psi, 20 min	0.5 mL MeOH 1.5 mL EA into 6 μL of 30% glycerol in MeOH	[64]
Human plasma (200 μL)	Oasis HLB 96-well plates (30 mg)		1 mL MeOH 1 mL H_2O	1.5 mL 5% MeOH	1.2 mL MeOH	[142]
Human serum/plasma/washed platelets (200 μL)	Oasis HLB (1 mL, 10 mg)	+ 1 mL MeOH centrifugation at 20,000× g, 4 °C, 10 min	0.15 mL MeOH 1 mL 0.03% FAc in H_2O	1 mL 0.03% FAc in H_2O 1 mL 15% ethanol, 0.03% FAc in H_2O 3 mL petroleum ether	0.2 mL MeOH	[143]
Human serum (500 μL) Human sputum (500 μL) Human bronchoalveolar lavage fluid (500 μL)	Oasis HLB 3 mL (60 mg)	+ 1.5 mL 5% HAc	2 mL MeOH 2 mL 0.1% HAc	2 mL 0.1% HAc	2 mL MeOH	[144]
Human serum (950 μL)	Oasis HLB (60 mg)	+ 10 mL 0.2 mg/mL BHT and EDTA	1× EA 1× MeOH	2 × 5% MeOH, 0.1% HAc dried under 0.2 bar	500 μL MeOH 1500 μL EA	[76]

Table 2. Cont.

Sample	Stationary Phase	Sample Preparation	Column Precondition	Sample Wash	Elution	Reference
Human urine	Oasis HLB (3 mL, 60 mg)	dilution with H$_2$O to 2 mL + 0.5 mL 0.5% HAc	3 mL H$_2$O 3 mL 0.1% HAc	3 mL 0.1% HAc dried at −30 kPa, 30 min	3 mL ACN	[116]
	Oasis HLB (30 mg)	dilution with H$_2$O to 2.7 mL + 0.3 mL 1% HAc	3 mL H$_2$O 3 mL 0.1% HAc	1 mL 50% MeOH 3 mL 0.1% HAc dried at −30 kPa, 30 min	1 mL MeOH	
Human cerebrospinal fluid (0.11–1 mL) Rat cortical brain tissue	Oasis HLB (30 mg)	cerebrospinal fluid: dilution with 0.12 M potassium phosphate buffer with 5 mM magnesium chloride, 0.113 mM BHT Tissue: + 0.12 M potassium phosphate buffer with 5 mM magnesium chloride, 0.113 mM BHT centrifugation at 10,000 rpm, 30 min	1 mL MeOH 1 mL H$_2$O	3 × 1 mL 5% MeOH	MeOH	[145]
Mouse serum (1–10 μL) Human lung epithelial cells (5 × 10^4 cells/well) Rat fibroblast cell line culture medium (50 μL)	Oasis HLB (10 mg)	+ 2× MeOH vortex, 10 s dilution with H$_2$O to 10% MeOH	1 mL MeOH 2 × 0.75 mL 5% MeOH	2 × 1 mL 5% MeOH	1 mL MeOH	[146]
Cow heart (~130 mg) cow liver (~320 mg) Pig/elk/cow brain (~80–180 mg) Human plasma (250 μL) Human milk (500 μL) Cell medium 2 mL	Oasis HLB (60 mg, 30 μm)	Tissues: homogenization in 1 mL MeOH with 10 μL 0.2 mg/mL BHT/EDTA in 50% MeOH centrifugation at 2125× g, 10 min diluted to 5% MeOH Plasma, milk, cell medium: + 10 μL 0.2 mg/mL BHT/EDTA in 50% MeOH	2 mL EA 2 × 2 mL MeOH 2 mL 5% MeOH, 0.1% HAc	2 × 2 mL 5% MeOH, 0.1% HAc	3 mL ACN 2 mL MeOH 1 mL EA	[75]
Mouse serum/bronchoalveolar lavage fluid (250 μL)	Oasis HLB (60 mg)	+ 0.2% w/w TPP/BHT	2 mL EA 2 × 2 mL MeOH 2 mL 5% MeOH, 0.1% HAc	1.5 mL 5% MeOH, 0.1% HAc dried under vacuum, 20 min	0.5 mL MeOH 2 mL EA	[47]

Table 2. Cont.

Sample	Stationary Phase	Sample Preparation	Column Precondition	Sample Wash	Elution	Reference
Rat cortical brain tissue (20 mg)	Oasis HLB (30 mg)	homogenization with 0.2 mL MeOH, 0.4 µL FAc on ice centrifugation at 14,000 rpm, 0 °C, 10 min + 1.8 mL H$_2$O	1 mL MeOH 1 mL acetone 2 mL Hex 1 mL acetone 1 mL MeOH 2 mL H$_2$O	3 mL H$_2$O 1 mL 10% MeOH dried under Ar pressure, 10 min	2 mL ACN	[126]
Rat kidney (100 mg)	Oasis	homogenization in 0.2 mL MeOH with 0.01 M BHT and 5 µL FAc centrifugation at 14,000 rpm, 0 °C, 15 min dilution with H$_2$O to 2 mL	2 mL 0.1% FAc 2 mL MeOH 2 mL EA	2 mL 0.1% FAc 2 mL 10% MeOH, 0.1% FAc	1.5 mL 0.1% FAc with 0.01 BHT 0.5 mL MeOH, 0.2% FAc with 0.01 BHT	[128]
Human plasma/urine (1 mL)	Oasis MAX	Urine: + 1 mL 40 mM FAc (pH 2.6) Plasma: + 1 mL 1M KOH in MeOH, 37 °C, 30 min + 1 mL MeOH + 0.2 mL 5 M HCL + 1.7 mL 40 mM FAc to pH 2.6 centrifugation at 10,000× g, 4 °C, 10 min	2 mL MeOH 2 mL 20 mM FAc	2 mL 2% ammonium hydroxide 2 mL MeOH/20 mM FAc (40/60, v/v) 2 mL Hex 2 mL Hex/EA (70/30, v/v)	EA	[147]
Human plasma (200 µL)	Strata-X (3 mL, 60 mg, 33 µm)	+ 0.8 mL H$_2$O, pH 3 acidified with HCL to pH 3	2 mL MeOH 2 mL H$_2$O, pH 3	10% MeOH in H$_2$O pH 3	2 mL MeOH	[148]
Human plasma (500 µL)	Strata-X (3 mL, 200 mg, 33 µm)	+ 1.5 mL 90% MeOH centrifugation at 6000 rpm, 10 min	3 mL MeOH 3 mL H$_2$O	3 mL H$_2$O	3 mL MeOH	[19]
Human plasma (200 µL)	Strata-X (6 mL, 200 mg, 33 µm)	+ 0.5 mL cold MeOH with 20 mg/mL BHT/EDTA −80 °C, 30 min centrifugation at 14,000 rpm, 4 °C, 10 min	6 mL MeOH 6 mL H$_2$O	6 mL 10% MeOH air-dried, 2 min	6 mL MeOH with 0.0004% w/v BHT	[53]
Human whole blood	Strata-X 96-well plates (60 mg/well, 33 µm)	1:1 dilution with RPMI-1640 medium (+ 25 mM Hepes and L-glutamine) + calcium ionophore A23187 (final concentration 30 µM), 37 °C, 30 min centrifugation at 1300 rpm, 10 min + 10% MeOH to 1 mL	MeOH H$_2$O	10% MeOH	1.0 mL MeOH	[149]
Control human plasma (20 µL) Mouse and human tissue: adipose, liver, muscle (2 mg)	Strata-X (3 mL, 60 mg)	Plasma: + 1 mL phosphate salt buffer Tissues: homogenization in 10% MeOH	3 mL MeOH 3 mL H$_2$O	10% MeOH	1 mL MeOH	[72]

228

Table 2. *Cont.*

Sample	Stationary Phase	Sample Preparation	Column Precondition	Sample Wash	Elution	Reference
Cell culture, cell medium (2 mL)	Strata-X	Medium: + 0.1 mL ethanol centrifugation at 3000 rpm, 5 min Cells: + 0.5 mL MeOH + 1 mL phosphate-buffered saline centrifugation at 3000 rpm, 5 min	2 mL MeOH 2 mL H_2O	1 mL 10% MeOH	1 mL MeOH	[132]
ANION-EXCHANGE PHASES						
Zebrafish embryo	Strata-X-A (3 mL, 200 mg, 22 µm)	saponification	3 mL MeOH 3 mL H_2O	4 mL MeOH—cholesterol 3 × 6 mL air 4 mL ACN—α-tocopherol	4 mL FAc/MeOH/ACN (5/47.5/47.5)	[76]
Human urine	Strata-X-AW (3 mL, 100 mg)	+ 200 mM MeOH/HCl centrifugation at 10,000 rpm, 5 min	2 mL MeOH 2 mL H_2O	4 mL H_2O	1 mL MeOH	[56]
VARIOUS PHASES						
	Bond Elut C18 (3 mL, 500 mg)		5 mL MeOH 5 mL H_2O	5 mL H_2O 3 mL 5% MeOH dried under vacuum	4 mL MeOH	
	Oasis HLB (6 mL, 500 mg)	+ 11.25 mL MeOH/CH3Cl (2:1, v/v), room temperature, 1 h + 3.75 mL CH3Cl + 3.75 mL water centrifugation at 2500 rpm, 10 min CH3Cl phase evaporated and reconstituted in 100 µL MeOH	5 mL MeOH 5 mL ACN 5 mL H_2O	3 mL 5% ACN dried under vacuum	4 mL ACN	
Human urine (2 mL)	Strata-X (6 mL, 200 mg)		5 mL MeOH 5 mL H_2O	3 mL 10% MeOH	4 mL MeOH	[117]
	Chromabond Easy (6 mL, 200 mg) Oasis MAX (6 mL, 500 mg)		5 mL 2% FAc in MeOH 5 mL H_2O	5 mL H_2O 3 mL 25% MeOH 3 mL ACN dried under vacuum	2 × 2 mL MeOH	
Mouse colon tissue Human epithelial colorectal adenocarcinoma cells Cell supernatant	Marchery Nagel C18 (15 mL, 200 mg)	Colon tissue: homogenization in 0.5 mL of HBSS + 1 mL MeOH centrifugation at 900× g, 4 °C, 15 min Cells/supernatant: + 1 mL MeOH centrifugation at 900× g, 4 °C, 15 min	10 mL MeOH 10 mL 0.02 M HCl/MeOH (90/10, v/v)	5 mL 0.02 M HCl/MeOH (90/10, v/v) dried under aspiration	5 mL methyl formate	[29]
Foam macrophages supernatant Mouse peritoneal exudate	Oasis HLB 96-well plates	centrifugation at 20,000× g, 4 °C, 20 min	4 mL MeOH 4 mL H_2O	2 mL 5% MeOH 2 mL 10% MeOH 2 mL H_2O dried under aspiration, 15 min centrifugation at 200× g, 2 min	4 mL methyl formate	

Table 2. *Cont.*

Sample	Stationary Phase	Sample Preparation	Column Precondition	Sample Wash	Elution	Reference
Human plasma (500 µL)	Oasis HLB (3 mL, 60 mg, 30 µm)	1:1 dilution with 5% MeOH acidified with 0.1% HAc centrifugation at 20,000× g, 4 °C, 10 min	1× EA 1× MeOH 2 × 5% MeOH, 0.1% HAc	2 × 5% MeOH, 0.1% HAc	0.5 mL MeOH 1.5 mL EA	[112]
		1:1 dilution with 40% MeOH centrifugation at 20,000× g, 4 °C, 10 min	1× EA 1× MeOH 1 × 20% MeOH, 0.1% FAc	1 × 20% MeOH, 0.1% FAc	2.0 mL MeOH	
	SepPak tC18 (6 mL, 500 mg, 37–55 µm)	+ 1.5 mL 20% MeOH centrifugation at 20,000× g, 4 °C, 10 min + 80 µL conc. HAc to pH 3	3× MeOH 3× H$_2$O	10 mL H$_2$O 6 mL Hex	8 mL methyl formate	
	Bond Elut Certify II (3 mL, 200 mg, 47–60 µm)	+ 500 µL 1 M sodium acetate buffer (pH 6) centrifugation at 20,000× g, 4 °C, 10 min	1× MeOH 1 × 0.1 M sodium acetate buffer, 5% MeOH	1× MeOH/H$_2$O (50/50, v/v)	2.0 mL n-Hex/EA (25/75, v/v) 2.0 mL n-Hex/EA (75/25, v/v)	
	Strata-X (3 mL, 100 mg, 33 µm)	1:1 dilution with 20% MeOH centrifugation at 20,000× g, 4 °C, 10 min	3.5 mL MeOH 3.5 mL H$_2$O	3.5 mL 10% MeOH	1.0 mL MeOH	

ACN: Acetonitrile; BHT: Butylated hydroxytoluene; EA: Ethyl acetate; EDTA: Ethylenediaminetetraacetic acid; FAc: Formic acid; HAc: Acetic acid; MeOH: Methanol; TPP: Triphenylphosphine.

2.3.5. New Approaches in Oxylipin Extraction

Although SPE is currently the most widely used method for extracting oxylipins, much attention is paid to the development of solvent-free and miniaturized extraction systems. These new methods include stir-bar-sorptive extraction (SBSE) and liquid-phase microextraction (LPME), but the most popular in oxylipin research is solid-phase microextraction (SPME), used for matrices like blood [93], urine [152] and plasma [153]. The advantages of miniaturization include minimal use of solvents and a small sample volume; however, a very small sample volume can cause problems such as insufficient sensitivity. Typically, SPME and SBSE are used in combination with GC analysis, but they can also be used in combination with LC. LPME can be used with both GC and LC [154,155].

Another new solution in the extraction methods of oxylipins is the semi-automatic microextraction by packed sorbents (MEPS) technique. Unlike conventional SPE, the MEPS sorbent bed is integrated into a liquid handling syringe, which allows work with low sample extraction and washing solvent volumes, and manipulations are performed either manually or in combination with laboratory robotics MEPS [156]. Perestrelo et al. showed good selectivity and sensitivity (LOD 0.37 ng/mL and LOQ 1.22 ng/mL) for the measurement of urinary LTB4 using the MEPS technique with a new digitally controlled syringe (eVols) combined with UHPLC [156]. In this study, they compared the performance of the eight MEPS sorbent materials, and porous graphitic carbon (PGC) sorbent was chosen for the MEPS procedure because it provided better extraction efficiency and reproducibility. The study showed that the developed method is accurate and offers simplicity, reduced sample preparation and analysis time, low cost and minimal consumption of extraction solvent compared with traditional methodologies [156].

A promising alternative to classical SPE and LLE can be deproteinization using ferromagnetic particles, as this is a fast procedure suitable for a large number of samples. In addition, there is no need for centrifugation, a vacuum or pressure. Suhr et al. demonstrated high-efficiency ferromagnetic particle enhanced deproteination in combination with online SPE for the sample clean-up of seven eicosanoids in human plasma samples. Before protein denaturation using ACN, the ferromagnetic bead suspension was added to the sample. High-speed vortexing leads to the binding of the denatured proteins to the surface of the particles, simultaneously forming a pellet after being placed on the magnetic separator [157].

Dried blood spot analysis (DBS) is becoming a popular method. A small drop of whole blood is placed on filter paper and air dried. Then, analytes are removed from the spot by solvent. Benefits include the use of smaller sample volumes (20–25 µL of whole blood), ease of collection, and simplified transport and storage requirements; among the disadvantages are limited stability of the analyzed substances on the filter paper and background interferences from the paper strongly interact with the metabolites [82]. The most common solvent is MeOH. However, the partial pre-addition of water can increase the efficiency of organic extraction by reducing the interaction between the cellulose and the hydroxyl groups of the target analyte with water [158]. Hewawasam et al., during the extraction of DBS with 80% aqueous MeOH, quantified 21 biologically significant oxylipins. This method has a higher LOD compared to other recently described methods, but the LOD value is suitable for a quick and routine analysis of a number of oxylipins in DBS samples [65].

To minimize human intervention and provide high precision, accuracy and throughput in the quantitative determination of eicosanoids, Ferreiro-Vera et al. used an online SPE–LC–MS/MS analysis method based on direct injection of the biofluid into an automated SPE workstation, where sample desalting and deproteinization occur. After this, the resulting eluate is directly injected into the LC–MS analyzer without affecting the electrospray performance [159]. Wagner used the online SPE–LC–MS system equipped with two quaternary pumps and an external Rheodyne MX Series II switching valve, and the UHPLC System and a Strata-X SPE-cartridge (20 × 2.0 mm, 25 µm; Phenomenex, Germany) for the MS detection of 8-iso-PGF2α [160]. Kita et al. used a column-switching reversed-phase LC–MS/MS technique for the quantitation of eicosanoids in murine macrophage-like RAW264.7 cells [161]. Using LC10AD pumps (Shimadzu, Kyoto, Japan), a 3033 autosampler (Shiseido, Tokyo, Japan), and an electrically controlled six-port switching valve, they combined isocratic conditions to achieve three-step gradient separation. They used a tee connector for the online dilution of samples to obtain optimal concentrations of MeOH and FAc, which allowed for the analysis of 14 lipid mediators within 10 min (throughput of 96 samples/24 h) with maximum sensitivity and minimal carryover [161]. Using online SPE–LC–MS/MS analysis, Kortz et al. profiled seven PUFAs and 94 oxidized metabolites in human plasma with a total analysis time of 13 min per sample [91].

Since anion-exchange SPE is not suitable for the extraction of polar lipids such as PG and TX, Sanaki et al. developed a new approach for the analysis of oxidized fatty acids in mouse lung homogenate samples using LC–MS/MS combined with mixed-mode extraction with a spin column [139]. The method is based on the adsorption of oxylipins in neutral conditions on a column (which contained trimethylaminopropyl and octadecyl groups bonded to silica), and all processing procedures (sample loading, washing and elution) were carried out by centrifugation. Mixed-mode SPE allows extraction efficiencies to be obtained of ≥70% for 61 oxylipins and is much more effective compared to RP-SPE. Moreover, the extraction time was reduced from 1 h per six samples to 10 min, and the use of organic solvents was reduced from >20 mL per sample to <1 mL per sample [139].

2.4. Derivatization Process

The derivatization step is determined by the chosen analytical method. For liquid chromatography (LC) using spectrophotometric and fluorimetric detectors, derivatization allows compounds to be obtained which are sensitive to these types of detectors [162]. Derivatization can also be used as a diagnostic tool to determine which functional groups exist in the oxylipin molecule (hydroxyl groups, carbonyl groups, etc.).

HPLC coupled with fluorescent detectors (HPLC–FLD) requires analytes to be derivatized into a complex that fluoresces, as these compounds contain no aromatic or naturally fluorescing systems. A simple derivatization reaction with ADAM (9-anthryl diazomethane) and the fluorescent detection of the resultant product (ADAM can be used to derivatize –COOH groups) was originally used for the analysis of PGs and was later used for the derivatization of eicosanoids before HPLC separation [163]. Nithipatikom et al. used SPE followed by derivatization with 2-(2,3-naphthalimino)ethyl trifluoromethanesulfonate (NE-OTf) to determine 14,15-EET, 11,12-EET, and a mixture of 8,9-EET and 5,6-EET from bovine coronary artery endothelial cells using HPLC–FLD [164]. Yue et al., using derivatization with NE-OTf, developed a method for the HPLC–FLD detection of bioactive eicosanoids including PG, DiHETrE, HETE, EET, and ARA from rat cortical brain tissue [126].

In the case of HPLC equipped with an ultraviolet detector (HPLC–UV), the analytes must have an active chromophore. Some eicosanoids have specific chromophores such as LTs (which contain conjugated triene) but many eicosanoids do not have any active chromophores (i.e., prostanoids) [97]. Chavis et al. used RP–HPLC with a UV detector (237nm) for the quantification of 5-HETE and 12-HETE in human plasma without derivatization [165]. Aghazadeh-Habashi et al. used derivatization with NE-OTf for the simultaneous quantification of eicosanoids in human plasma and rat heart and kidney by the HPLC method using fluorescence detection [166]. Yue et al. coupled the derivatization of eicosanoids with fluorescence detection using RP–HPLC to determine bioactive PGs, EETs, DiHETEs and HETEs in rat brain tissue, and achieved limits of detection (LODs) and LOQs ranging from 2–20 to 20–70 pg on the column, respectively [126].

Derivatization could lead to a better separation efficacy and improved MS detection. Bollinger et al. describe a new derivatization reagent N-(4-aminomethylphenyl)pyridinium (AMPP). The conversion of the carboxylic acid of eicosanoids to a cationic AMPP amide enabled detection in positive-ESI mode, and manifold improvements in the sensitivity of LC–ESI–MS/MS detection [146]. Meckelmann et al. evaluated whether pentafluorobenzyl (PFB) derivatization and electron capture atmospheric pressure chemical ionization (ECAPCI) (−) or AMPP derivatization and ESI (+) could improve the RP–LC–MS/MS analysis of oxylipins in human plasma, and found that PFB derivatization led to a low sensitivity (LOD~10 nM, 100 fmol on the column, 32 pg on the column) and thus was not suitable for detection by the MS instrumentation used. Compared to AMPP/ESI (+), direct ESI (−) yielded a higher sensitivity for several OH-PUFAs, but it did not result in better analytical performance [167]. This discrepancy with the results obtained by Bollinger et al. may be explained by the different MS instrumentation (Waters triple-quadrupole (QqQ) vs. Sciex QqQ) because the Waters QqQ instruments used (Premier and Quattro) are significantly more sensitive in the positive mode than in the negative ESI mode [167].

GC derivatization is associated with obtaining more volatile compounds, reducing the polarity of functional groups, and as a consequence, improving the chromatographic properties of a substance, or obtaining specific products for a particular type of detector [168]. Some examples of derivatization methods are: N-acylation, methoxyamine formation, esterification, and trimethylsilyl (TMS) ether formation [127]. After the derivatization step, many analytes can be detected simultaneously. However, on the other hand, eicosanoid volatility is increased by derivatization, together with polarity and thermal lability; therefore, GC–MS is applicable mainly to eicosanoids, including the catalytic reduction of highly polar, nonvolatile and thermally labile cysteinyl LTs [18]. In contrast to the quantitative analysis of fatty acids, where only one step of derivatization is required (e.g., methyl, trimethylsilyl or PFB esters of fatty acids), one-step derivatization is not suitable for all oxylipins due to the presence of different functional groups [80]. In the GC–MS and GC–MS/MS analysis of eicosanoids, derivatization with fluorine-rich reagents such as PFB bromide is required, because eicosanoids have two or three thermally labile chemical functionalities, i.e., carboxylic, hydroxylic and keto groups [80,137]. ECAPCI–MS analysis of fatty acyls which have been derivatized as PFB esters is more sensitive compared to ESI analysis of underivatized fatty acyls enabling the efficient normal-phase chiral separation of oxylipins [169]. Knott et al. used esterification with PFB bromide, methoxylation and finally, trimethylsilylation with N, O-bis(trimethylsilyl)-trifluoroacetamide (BSTFA) for GC analysis of PGs in human synovial cells and chondrocyte cultures [170]. Additional derivatization of hydroxyl-groups with N-methyl-trimethylsilyl-trifluoroacetamide (MSTFA) to TMS-ethers made the highly polar oxylipins suitable for GC separation and, at the same time, provided further information for structure elucidation [171]. Fulton found that the conversion of 5,6-δ-lactone to 5,6-DHT permitted convenient derivatization with PFB and more sensitive GC–MS without the need for further purification of biological samples obtained from perfused kidney [60]. However, in the case of LTB4, HPLC–UV and LC–MS, in contrast to GC–MS analysis, a previous derivatization procedure to determine LTB4 in biological fluids is not required, thus increasing recovery and reducing the time for sample pretreatment [156]. The main disadvantages of derivatization are the labor intensity, increase in the total procedure time, risk of thermal decomposition, instability of the derivatives and the lack of reproducibility of the yield of the derivatives, and also loss of analyte due to incomplete interaction, contamination with reagents and undesirable side reactions [78,158,172,173].

3. Methods of Oxylipin Analysis

Presently, there are many methods for measuring the levels of oxylipins in human biological samples. However, they all have certain limitations. This is primarily due to the fact that oxylipins are present in extremely low concentrations in biological matrices, have limited stability and are subject to degradation and auto-oxidation [65]. In addition, many oxylipins, especially those derived from the same original fatty acid, have very similar structures [80]. Therefore, their analysis requires rapid, highly-sensitive and accurate analytical methods [112,174], see Figure 4.

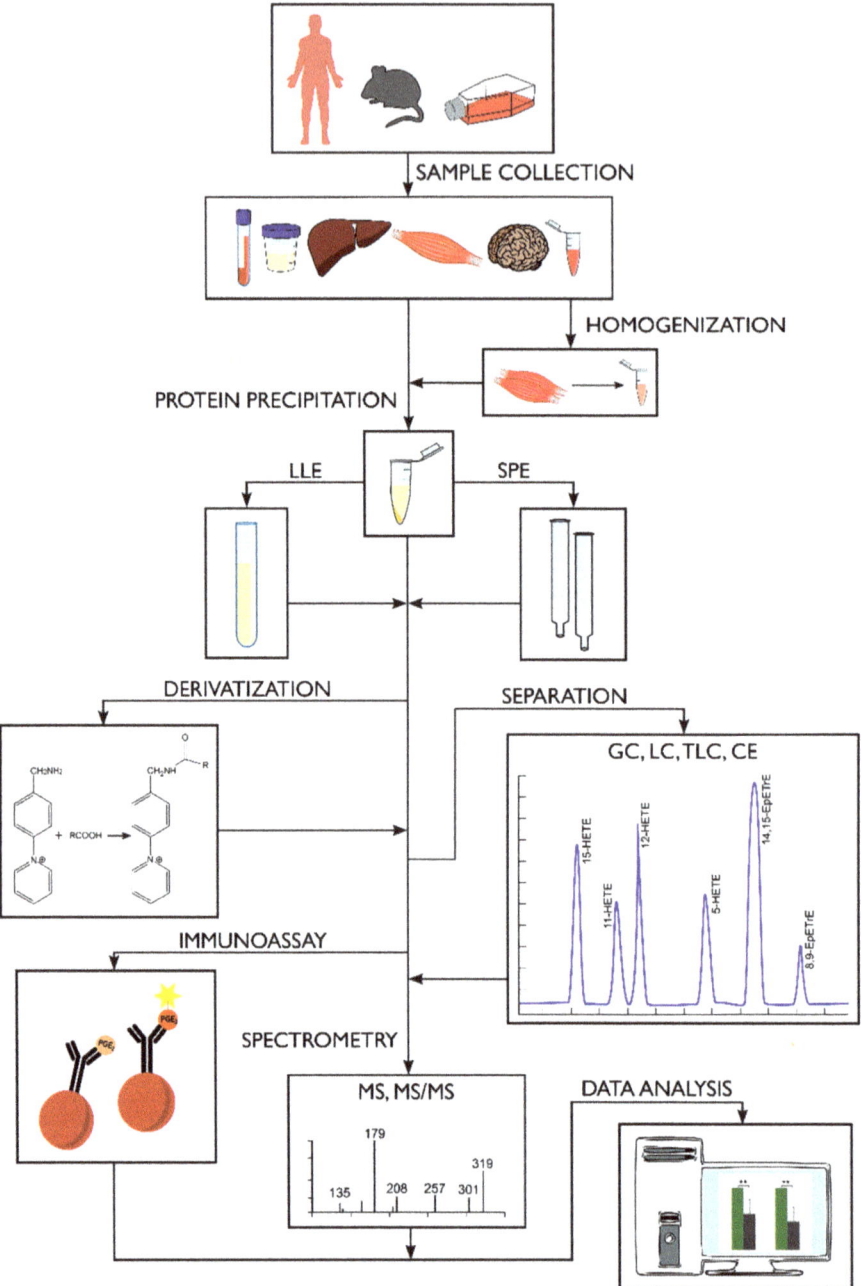

Figure 4. Scheme of the complete analytical procedure of oxylipins.

Various methods, including immunoassay, thin-layer chromatography (TLC), HPLC–UV, HPLC–FLD, GC–MS and LC–MS were used to analyze oxylipins. However, some of these methods are not specific or sensitive enough, and often require derivatization [53]. Currently, more often

GC–MS and LC–MS are used for oxylipin determination. GC–MS has long been a common analytical technique for the quantitative and structural interpretation of eicosanoids; however, due to the cost of equipment and the difficulty in preparing samples, fewer laboratories use GC–MS [127]. GC–MS requires purification steps after derivatization to remove impurities formed during the derivatization process, which can make the analysis laborious, expensive and time-consuming [59]. Among the less frequently used techniques for the analysis of oxylipins, capillary electrophoresis with a photodiode array detector (CE–UV) can be noted, which was sufficiently sensitive to detect and measure EET and DHET enantiomers from murine liver (unlike chiral-phase HPLC) [111,175]. The most popular method for detecting oxylipins is LC–MS. One of the reasons is that MS allows quantitative determination at very low levels in complex matrices. LC-MS gives a better separation of the isomers compared with HPLC–UV and immunoassay [97].

4. Conclusions

Oxylipins play an important role in various biologic processes. A large number of enzymatic and non-enzymatic pathways lead to the formation of hundreds of oxylipins. Effective simultaneous determination of a large number of compounds is possible only during sample preparation by the PP, LLE and SPE methods, both individually and jointly. Modern sample preparation techniques make the analysis more economical and less invasive and time-consuming.

Author Contributions: A.M. conceived and designed the review. I.L. and A.P. studied the literature and wrote the manuscript. T.S. wrote and verified the manuscript. All authors accepted the final version of the review.

Funding: This research was supported by the National Science Centre of Poland (grant no. NCN 2016/21/D/NZ5/00219 and 2016/22/E/NZ4/00665).

Conflicts of Interest: The authors declare no conflict of interest.

References

1. Gabbs, M.; Leng, S.; Devassy, J.G.; Monirujjaman, M.; Aukema, H.M. Advances in Our Understanding of Oxylipins Derived from Dietary PUFAs. *Adv. Nutr. An Int. Rev. J.* **2015**, *6*, 513–540. [CrossRef] [PubMed]
2. Vigor, C.; Bertrand-Michel, J.; Pinot, E.; Oger, C.; Vercauteren, J.; Le Faouder, P.; Galano, J.M.; Lee, J.C.Y.; Durand, T. Non-enzymatic lipid oxidation products in biological systems: ASSESSMENT of the metabolites from polyunsaturated fatty acids. *J. Chromatogr. B Anal. Technol. Biomed. Life Sci.* **2014**, *964*, 65–78. [CrossRef] [PubMed]
3. Shearer, G.C.; Walker, R.E. An overview of the biologic effects of omega-6 oxylipins in humans. *Prostaglandins Leukot. Essent. Fat. Acids* **2018**, *137*, 26–38. [CrossRef]
4. Khan, S.A.; Ali, A.; Khan, S.A.; Zahran, S.A.; Damanhouri, G.; Azhar, E.; Qadri, I. Unraveling the Complex Relationship Triad between Lipids, Obesity, and Inflammation. *Mediators Inflamm.* **2014**, *2014*, 1–16. [CrossRef]
5. Yeung, J.; Hawley, M.; Holinstat, M. The expansive role of oxylipins on platelet biology. *J. Mol. Med.* **2017**, *95*, 575–588. [CrossRef]
6. Christophersen, O.A.; Haug, A. Animal products, diseases and drugs: a plea for better integration between agricultural sciences, human nutrition and human pharmacology. *Lipids Health Dis.* **2011**, *10*, 16. [CrossRef]
7. Spector, A.A.; Kim, H.-Y. Cytochrome P450 epoxygenase pathway of polyunsaturated fatty acid metabolism. *Biochim. Biophys. Acta - Mol. Cell Biol. Lipids* **2015**, *1851*, 356–365. [CrossRef] [PubMed]
8. Yang, J.; Dong, H.; Hammock, B.D. Profiling the regulatory lipids: Another systemic way to unveil the biological mystery. *Curr. Opin. Lipidol.* **2011**, *22*, 197–203. [CrossRef]
9. Arnold, C.; Markovic, M.; Blossey, K.; Wallukat, G.; Fischer, R.; Dechend, R.; Konkel, A.; Von Schacky, C.; Luft, F.C.; Muller, D.N.; et al. Arachidonic acid-metabolizing cytochrome P450 enzymes are targets of ω-3 fatty acids. *J. Biol. Chem.* **2010**, *285*, 32720–32733. [CrossRef]
10. Austin, C.; Sordillo, L.M.; Zhang, C.; Fenton, J.I. Obesity is positively associated with arachidonic acid-derived 5- and 11-hydroxyeicosatetraenoic acid (HETE). *Metabolism* **2017**, *70*, 177–191.

11. Zhou, J.; Chen, L.; Liu, Z.; Sang, L.; Li, Y.; Yuan, D. Changes in erythrocyte polyunsaturated fatty acids and plasma eicosanoids level in patients with asthma. *Lipids Health Dis.* **2018**, *17*, 206. [CrossRef] [PubMed]
12. Rigas, B.; Goldman, I.S.; Levine, L. Altered eicosanoid levels in human colon cancer. *J. Lab. Clin. Med.* **1993**, *122*, 518–523.
13. Yang, J.; Eiserich, J.P.; Cross, C.E.; Morrissey, B.M.; Hammock, B.D. Metabolomic profiling of regulatory lipid mediators in sputum from adult cystic fibrosis patients. *Free Radic. Biol. Med.* **2012**, *53*, 160–171. [CrossRef] [PubMed]
14. Kumar, N.; Gupta, G.; Anilkumar, K.; Fatima, N.; Karnati, R.; Reddy, G.V.; Giri, P.V.; Reddanna, P. 15-Lipoxygenase metabolites of α-linolenic acid, [13-(S)-HPOTrE and 13-(S)-HOTrE], mediate anti-inflammatory effects by inactivating NLRP3 inflammasome. *Sci. Rep.* **2016**, *6*, 31649. [CrossRef] [PubMed]
15. Tsai, I.-J.; Croft, K.D.; Mori, T.A.; Falck, J.R.; Beilin, L.J.; Puddey, I.B.; Barden, A.E. 20-HETE and F2-isoprostanes in the metabolic syndrome: The effect of weight reduction. *Free Radic. Biol. Med.* **2009**, *46*, 263–270. [CrossRef] [PubMed]
16. Shishehbor, M.H.; Zhang, R.; Medina, H.; Brennan, M.-L.; Brennan, D.M.; Ellis, S.G.; Topol, E.J.; Hazen, S.L. Systemic elevations of free radical oxidation products of arachidonic acid are associated with angiographic evidence of coronary artery disease. *Free Radic. Biol. Med.* **2006**, *41*, 1678–1683. [CrossRef] [PubMed]
17. Foegh, M.L.; Zhao, Y.; Madren, L.; Rolnick, M.; Stair, T.O.; Huang, K.S.; Ramwell, P.W. Urinary thromboxane A 2 metabolites in patients presenting in the emergency room with acute chest pain. *J. Intern. Med.* **1994**, *235*, 153–161. [CrossRef]
18. Tsikas, D. Application of gas chromatography-mass spectrometry and gas chromatography-tandem mass spectrometry to assess in vivo synthesis of prostaglandins, thromboxane, leukotrienes, isoprostanes and related compounds in humans. *J. Chromatogr. B Biomed. Appl.* **1998**, *717*, 201–245. [CrossRef]
19. Berkecz, R.; Lísa, M.; Holčapek, M. Analysis of oxylipins in human plasma: Comparison of ultrahigh-performance liquid chromatography and ultrahigh-performance supercritical fluid chromatography coupled to mass spectrometry. *J. Chromatogr. A* **2017**, *1511*, 107–121. [CrossRef]
20. Lynes, M.D.; Leiria, L.O.; Lundh, M.; Bartelt, A.; Shamsi, F.; Huang, T.L.; Takahashi, H.; Hirshman, M.F.; Schlein, C.; Lee, A.; et al. The cold-induced lipokine 12,13-diHOME promotes fatty acid transport into brown adipose tissue. *Nat. Med.* **2017**, *23*, 631–637. [CrossRef]
21. Xi, S.; Pham, H.; Ziboh, V.A. 15-Hydroxyeicosatrienoic acid (15-HETrE) suppresses epidermal hyperproliferation via the modulation of nuclear transcription factor (AP-1) and apoptosis. *Arch. Dermatol. Res.* **2000**, *292*, 397–403. [CrossRef]
22. Möller, K.; Ostermann, A.I.; Rund, K.; Thoms, S.; Blume, C.; Stahl, F.; Hahn, A.; Schebb, N.H.; Schuchardt, J.P. Influence of weight reduction on blood levels of C-reactive protein, tumor necrosis factor-α, interleukin-6, and oxylipins in obese subjects. *Prostaglandins Leukot. Essent. Fat. Acids* **2016**, *106*, 39–49. [CrossRef]
23. Virtue, S.; Masoodi, M.; de Weijer, B.A.M.; van Eijk, M.; Mok, C.Y.L.; Eiden, M.; Dale, M.; Pirraco, A.; Serlie, M.J.; Griffin, J.L.; et al. Prostaglandin profiling reveals a role for haematopoietic prostaglandin D synthase in adipose tissue macrophage polarisation in mice and humans. *Int. J. Obes.* **2015**, *39*, 1151–1160. [CrossRef]
24. Hoopes, S.L.; Garcia, V.; Edin, M.L.; Schwartzman, M.L.; Zeldin, D.C. Vascular actions of 20-HETE. *Prostaglandins Other Lipid Mediat.* **2015**, *120*, 9–16. [CrossRef] [PubMed]
25. Yu, D.; Xu, Z.; Yin, X.; Zheng, F.; Lin, X.; Pan, Q.; Li, H. Inverse Relationship between Serum Lipoxin A4 Level and the Risk of Metabolic Syndrome in a Middle-Aged Chinese Population. *PLoS ONE* **2015**, *10*, e0142848. [CrossRef]
26. Liu, J.-B.; Li, W.-J.; Fu, F.-M.; Zhang, X.-L.; Jiao, L.; Cao, L.-J.; Chen, L. Inverse correlation between serum adiponectin and 8-iso-prostaglandin F2α in newly diagnosed type 2 diabetes patients. *Int. J. Clin. Exp. Med.* **2015**, *8*, 6085–6090.
27. Mukhtar, M.H.; El-Emshaty, H.M.; Alamodi, H.S.; Nasif, W.A. The Activity of Serum 8-Iso-Prostaglandin F2α as Oxidative Stress Marker in Patients with Diabetes Mellitus Type 2 and Associated Dyslipidemic Hyperglycemia. *J. Diabetes Mellit.* **2016**, *6*, 318–332. [CrossRef]
28. Grapov, D.; Adams, S.H.; Pedersen, T.L.; Garvey, W.T.; Newman, J.W. Type 2 Diabetes Associated Changes in the Plasma Non-Esterified Fatty Acids, Oxylipins and Endocannabinoids. *PLoS ONE* **2012**, *7*, e48852. [CrossRef]

29. Altmann, R.; Hausmann, M.; Spöttl, T.; Gruber, M.; Bull, A.W.; Menzel, K.; Vogl, D.; Herfarth, H.; Schölmerich, J.; Falk, W.; et al. 13-Oxo-ODE is an endogenous ligand for PPARγ in human colonic epithelial cells. *Biochem. Pharmacol.* **2007**, *74*, 612–622. [CrossRef]
30. Caligiuri, S.P.B.; Parikh, M.; Stamenkovic, A.; Pierce, G.N.; Aukema, H.M. Dietary modulation of oxylipins in cardiovascular disease and aging. *Am. J. Physiol. Circ. Physiol.* **2017**, *313*, H903–H918. [CrossRef]
31. Yao, X.; Sa, R.; Ye, C.; Zhang, D.; Zhang, S.; Xia, H.; Wang, Y.; Jiang, J.; Yin, H.; Ying, H. Effects of thyroid hormone status on metabolic pathways of arachidonic acid in mice and humans: A targeted metabolomic approach. *Prostaglandins Other Lipid Mediat.* **2015**, *118–119*, 11–18. [CrossRef]
32. Bruegel, M.; Ludwig, U.; Kleinhempel, A.; Petros, S.; Kortz, L.; Ceglarek, U.; Holdt, L.M.; Thiery, J.; Fiedler, G.M. Sepsis-associated changes of the arachidonic acid metabolism and their diagnostic potential in septic patients*. *Crit. Care Med.* **2012**, *40*, 1478–1486. [CrossRef]
33. Gouveia-Figueira, S.; Nording, M.L.; Gaida, J.E.; Forsgren, S.; Alfredson, H.; Fowler, C.J. Serum levels of oxylipins in achilles tendinopathy: An exploratory study. *PLoS ONE* **2015**, *10*, 1–17. [CrossRef]
34. Jira, W.; Spiteller, G.; Carson, W.; Schramm, A. Strong increase in hydroxy fatty acids derived from linoleic acid in human low density lipoproteins of atherosclerotic patients. *Chem. Phys. Lipids* **1998**, *91*, 1–11. [CrossRef]
35. Ozawa, T.; Sugiyama, S.; Hayakawa, M.; Satake, T.; Taki, F.; Iwata, M.; Taki, K. Existence of Leukotoxin 9,10-Epoxy-12-Octadecenoate in Lung Lavages from Rats Breathing Pure Oxygen and from Patients with the Adult Respiratory Distress Syndrome. *Am. Rev. Respir. Dis.* **1988**, *137*, 535–540. [CrossRef] [PubMed]
36. Wang, D.; Dubois, R.N. Prostaglandins and cancer. *Gut* **2006**, *55*, 115–122. [CrossRef]
37. Yoshida, Y.; Yoshikawa, A.; Kinumi, T.; Ogawa, Y.; Saito, Y.; Ohara, K.; Yamamoto, H.; Imai, Y.; Niki, E. Hydroxyoctadecadienoic acid and oxidatively modified peroxiredoxins in the blood of Alzheimer's disease patients and their potential as biomarkers. *Neurobiol. Aging* **2009**, *30*, 174–185. [CrossRef]
38. Dietrich-Muszalska, A.; Olas, B. Isoprostenes as indicators of oxidative stress in schizophrenia. *World J. Biol. Psychiatry* **2009**, *10*, 27–33. [CrossRef] [PubMed]
39. Chocholoušková, M.; Jirásko, R.; Vrána, D.; Gatěk, J.; Melichar, B.; Holčapek, M. Reversed phase UHPLC/ESI-MS determination of oxylipins in human plasma: a case study of female breast cancer. *Anal. Bioanal. Chem.* **2019**, *411*, 1239–1251. [CrossRef]
40. Kapadia, R.; Yi, J.-H.; Vemuganti, R. Mechanisms of anti-inflammatory and neuroprotective actions of PPAR-gamma agonists. *Front. Biosci.* **2008**, *13*, 1813–1826. [CrossRef] [PubMed]
41. Goetzl, E.J.; Gorman, R.R. Chemotactic and chemokinetic stimulation of human eosinophil and neutrophil polymorphonuclear leukocytes by 12-L-hydroxy-5,8,10-heptadecatrienoic acid (HHT). *J. Immunol.* **1978**, *120*, 526–531. [PubMed]
42. Zhang, L.; Chen, B.; Zhang, J.; Li, J.; Yang, Q.; Zhong, Q.; Zhan, S.; Liu, H.; Cai, C. Serum polyunsaturated fatty acid metabolites as useful tool for screening potential biomarker of colorectal cancer. *Prostaglandins Leukot. Essent. Fat. Acids* **2017**, *120*, 25–31. [CrossRef] [PubMed]
43. Yuan, H.; Li, M.-Y.; Ma, L.T.; Hsin, M.K.Y.; Mok, T.S.K.; Underwood, M.J.; Chen, G.G. 15-Lipoxygenases and its metabolites 15(S)-HETE and 13(S)-HODE in the development of non-small cell lung cancer. *Thorax* **2010**, *65*, 321–326. [CrossRef] [PubMed]
44. Larré, S.; Tran, N.; Fan, C.; Hamadeh, H.; Champigneulles, J.; Azzouzi, R.; Cussenot, O.; Mangin, P.; Olivier, J.L. PGE2 and LTB4 tissue levels in benign and cancerous prostates. *Prostaglandins Other Lipid Mediat.* **2008**, *87*, 14–19. [CrossRef] [PubMed]
45. Spickett, C.M.; Pitt, A.R. Oxidative Lipidomics Coming of Age: Advances in Analysis of Oxidized Phospholipids in Physiology and Pathology. *Antioxid. Redox Signal.* **2015**, *22*, 1646–1666. [CrossRef] [PubMed]
46. Lundström, S.L.; Levänen, B.; Nording, M.; Klepczynska-Nyström, A.; Sköld, M.; Haeggström, J.Z.; Grunewald, J.; Svartengren, M.; Hammock, B.D.; Larsson, B.-M.; et al. Asthmatics Exhibit Altered Oxylipin Profiles Compared to Healthy Individuals after Subway Air Exposure. *PLoS ONE* **2011**, *6*, e23864. [CrossRef]
47. Yang, J.; Schmelzer, K.; Georgi, K.; Hammock, B.D. Quantitative Profiling Method for Oxylipin Metabolome by Liquid Chromatography Electrospray Ionization Tandem Mass Spectrometry. *Anal. Chem.* **2009**, *81*, 8085–8093. [CrossRef]

48. Colas, R.A.; Shinohara, M.; Dalli, J.; Chiang, N.; Serhan, C.N. Identification and signature profiles for pro-resolving and inflammatory lipid mediators in human tissue. *AJP Cell Physiol.* **2014**, *307*, C39–C54. [CrossRef]
49. Golovko, M.Y.; Murphy, E.J. An improved LC-MS/MS procedure for brain prostanoid analysis using brain fixation with head-focused microwave irradiation and liquid-liquid extraction. *J. Lipid Res.* **2008**, *49*, 893–902. [CrossRef]
50. Willenberg, I.; Ostermann, A.I.; Schebb, N.H. Targeted metabolomics of the arachidonic acid cascade: current state and challenges of LC-MS analysis of oxylipins. *Anal. Bioanal. Chem.* **2015**, *407*, 2675–2683. [CrossRef]
51. Strassburg, K.; Huijbrechts, A.M.L.; Kortekaas, K.A.; Lindeman, J.H.; Pedersen, T.L.; Dane, A.; Berger, R.; Brenkman, A.; Hankemeier, T.; Van Duynhoven, J.; et al. Quantitative profiling of oxylipins through comprehensive LC-MS/MS analysis: Application in cardiac surgery. *Anal. Bioanal. Chem.* **2012**, *404*, 1413–1426. [CrossRef]
52. Brose, S.A.; Baker, A.G.; Golovko, M.Y. A Fast One-Step Extraction and UPLC–MS/MS Analysis for E2/D2 Series Prostaglandins and Isoprostanes. *Lipids* **2013**, *48*, 411–419. [CrossRef] [PubMed]
53. Yuan, Z.-X.; Majchrzak-Hong, S.; Keyes, G.S.; Iadarola, M.J.; Mannes, A.J.; Ramsden, C.E. Lipidomic profiling of targeted oxylipins with ultra-performance liquid chromatography-tandem mass spectrometry. *Anal. Bioanal. Chem.* **2018**, *410*, 6009–6029. [CrossRef] [PubMed]
54. Drake, S.K.; Bowen, R.A.R.; Remaley, A.T.; Hortin, G.L. Potential Interferences from Blood Collection Tubes in Mass Spectrometric Analyses of Serum Polypeptides. *Clin. Chem.* **2004**, *50*, 2398–2401. [CrossRef]
55. Ito, R.; Miura, N.; Iguchi, H.; Nakamura, H.; Ushiro, M.; Wakui, N.; Nakahashi, K.; Iwasaki, Y.; Saito, K.; Suzuki, T.; et al. Determination of tris(2-ethylhexyl)trimellitate released from PVC tube by LC–MS/MS. *Int. J. Pharm.* **2008**, *360*, 91–95. [CrossRef]
56. Schauer, K.L.; Broccardo, C.J.; Webb, K.M.; Covey, P.A.; Prenni, J.E. Mass Spectrometry Contamination from Tinuvin 770, a Common Additive in Laboratory Plastics. *J. Biomol. Tech.* **2013**, *24*, jbt.13-2402-004. [CrossRef]
57. Haned, Z.; Moulay, S.; Lacorte, S. Migration of plasticizers from poly(vinyl chloride) and multilayer infusion bags using selective extraction and GC–MS. *J. Pharm. Biomed. Anal.* **2018**, *156*, 80–87. [CrossRef] [PubMed]
58. Wang, D.; DuBois, R.N. Measurement of Eicosanoids in Cancer Tissues. *Methods Enzymol.* **2007**, *433*, 27–50. [PubMed]
59. Rund, K.M.; Ostermann, A.I.; Kutzner, L.; Galano, J.M.; Oger, C.; Vigor, C.; Wecklein, S.; Seiwert, N.; Durand, T.; Schebb, N.H. Development of an LC-ESI(-)-MS/MS method for the simultaneous quantification of 35 isoprostanes and isofurans derived from the major n3- and n6-PUFAs. *Anal. Chim. Acta* **2018**, *1037*, 63–74. [CrossRef] [PubMed]
60. Fulton, D.; Falck, J.R.; McGiff, J.C.; Carroll, M.A.; Quilley, J. A method for the determination of 5,6-EET using the lactone as an intermediate in the formation of the diol. *J. Lipid Res.* **1998**, *39*, 1713–1721. [PubMed]
61. Araujo, P.; Mengesha, Z.; Lucena, E.; Grung, B. Development and validation of an extraction method for the determination of pro-inflammatory eicosanoids in human plasma using liquid chromatography-tandem mass spectrometry. *J. Chromatogr. A* **2014**, *1353*, 57–64. [CrossRef]
62. Mueller, M.J.; Mène-Saffrané, L.; Grun, C.; Karg, K.; Farmer, E.E. Oxylipin analysis methods. *Plant J.* **2006**, *45*, 472–489. [CrossRef]
63. Balvers, M.G.J.; Verhoeckx, K.C.M.; Bijlsma, S.; Rubingh, C.M.; Meijerink, J.; Wortelboer, H.M.; Witkamp, R.F. Fish oil and inflammatory status alter the n-3 to n-6 balance of the endocannabinoid and oxylipin metabolomes in mouse plasma and tissues. *Metabolomics* **2012**, *8*, 1130–1147. [CrossRef]
64. Hennebelle, M.; Otoki, Y.; Yang, J.; Hammock, B.D.; Levitt, A.J.; Taha, A.Y.; Swardfager, W. Altered soluble epoxide hydrolase-derived oxylipins in patients with seasonal major depression: An exploratory study. *Psychiatry Res.* **2017**, *252*, 94–101. [CrossRef]
65. Hewawasam, E.; Liu, G.; Jeffery, D.W.; Muhlhausler, B.S.; Gibson, R.A. A stable method for routine analysis of oxylipins from dried blood spots using ultra-high performance liquid chromatography–tandem mass spectrometry. *Prostaglandins Leukot. Essent. Fat. Acids* **2018**, *137*, 12–18. [CrossRef]
66. Yang, P.; Felix, E.; Madden, T.; Fischer, S.M.; Newman, R.A. Quantitative high-performance liquid chromatography/electrospray ionization tandem mass spectrometric analysis of 2- and 3-series prostaglandins in cultured tumor cells. *Anal. Biochem.* **2002**, *308*, 168–177. [CrossRef]
67. Newman, J.W.; Watanabe, T.; Hammock, B.D. The simultaneous quantification of cytochrome P450 dependent linoleate and arachidonate metabolites in urine by HPLC-MS/MS. *J. Lipid Res.* **2002**, *43*, 1563–1578. [CrossRef]

68. Morgan, A.H.; Hammond, V.J.; Morgan, L.; Thomas, C.P.; Tallman, K.A.; Garcia-Diaz, Y.R.; McGuigan, C.; Serpi, M.; Porter, N.A.; Murphy, R.C.; et al. Quantitative assays for esterified oxylipins generated by immune cells. *Nat. Protoc.* **2010**, *5*, 1919–1931. [CrossRef]
69. Dumlao, D.S.; Buczynski, M.W.; Norris, P.C.; Harkewicz, R.; Dennis, E.A. High-throughput lipidomic analysis of fatty acid derived eicosanoids and N-acylethanolamines. *Biochim. Biophys. Acta - Mol. Cell Biol. Lipids* **2011**, *1811*, 724–736. [CrossRef]
70. Mesaros, C.; Lee, S.H.; Blair, I.A. Analysis of epoxyeicosatrienoic acids by chiral liquid chromatography/electron capture atmospheric pressure chemical ionization mass spectrometry using [13C]-analog internal standards. *Rapid Commun. Mass Spectrom.* **2010**, *24*, 3237–3247. [CrossRef] [PubMed]
71. Lee, Y.H.; Cui, L.; Fang, J.; Chern, B.S.M.; Tan, H.H.; Chan, J.K.Y. Limited value of pro-inflammatory oxylipins and cytokines as circulating biomarkers in endometriosis - A targeted 'omics study. *Sci. Rep.* **2016**, *6*, 1–7. [CrossRef] [PubMed]
72. Wang, Y.; Armando, A.M.; Quehenberger, O.; Yan, C.; Dennis, E.A. Comprehensive ultra-performance liquid chromatographic separation and mass spectrometric analysis of eicosanoid metabolites in human samples. *J. Chromatogr. A* **2014**, *1359*, 60–69. [CrossRef] [PubMed]
73. Shearer, G.C.; Harris, W.S.; Pedersen, T.L.; Newman, J.W. Detection of omega-3 oxylipins in human plasma and response to treatment with omega-3 acid ethyl esters. *J. Lipid Res.* **2010**, *51*, 2074–2081. [CrossRef]
74. Gouveia-Figueira, S.; Späth, J.; Zivkovic, A.M.; Nording, M.L. Profiling the oxylipin and endocannabinoid metabolome by UPLC-ESI-MS/MS in human plasma to monitor postprandial inflammation. *PLoS ONE* **2015**, *10*, 1–29. [CrossRef]
75. Gouveia-Figueira, S.; Nording, M.L. Validation of a tandem mass spectrometry method using combined extraction of 37 oxylipins and 14 endocannabinoid-related compounds including prostamides from biological matrices. *Prostaglandins Other Lipid Mediat.* **2015**, *121*, 110–121. [CrossRef]
76. Schuchardt, J.P.; Schmidt, S.; Kressel, G.; Dong, H.; Willenberg, I.; Hammock, B.D.; Hahn, A.; Schebb, N.H. Comparison of free serum oxylipin concentrations in hyper- vs. normolipidemic men. *Prostaglandins Leukot. Essent. Fat. Acids* **2013**, *89*, 19–29. [CrossRef] [PubMed]
77. Hellström, F.; Gouveia-Figueira, S.; Nording, M.L.; Björklund, M.; Fowler, C.J. Association between plasma concentrations of linoleic acid-derived oxylipins and the perceived pain scores in an exploratory study in women with chronic neck pain. *BMC Musculoskelet. Disord.* **2016**, *17*, 103. [CrossRef] [PubMed]
78. Astarita, G.; Kendall, A.C.; Dennis, E.A.; Nicolaou, A. Targeted lipidomic strategies for oxygenated metabolites of polyunsaturated fatty acids. *Biochim. Biophys. Acta - Mol. Cell Biol. Lipids* **2015**, *1851*, 456–468. [CrossRef] [PubMed]
79. Maskrey, B.H.; O'Donnell, V.B. Analysis of eicosanoids and related lipid mediators using mass spectrometry. *Biochem. Soc. Trans.* **2008**, *36*, 1055–1059. [CrossRef] [PubMed]
80. Tsikas, D.; Zoerner, A.A. Analysis of eicosanoids by LC-MS/MS and GC-MS/MS: A historical retrospect and a discussion. *J. Chromatogr. B Anal. Technol. Biomed. Life Sci.* **2014**, *964*, 79–88. [CrossRef] [PubMed]
81. Schweer, H.; Kammer, J.; Kühl, P.G.; Seyberth, H.W. Determination of peripheral plasma prostanoid concentration: an unreliable index of "in vivo" prostanoid activity. *Eur. J. Clin. Pharmacol.* **1986**, *31*, 303–305. [CrossRef]
82. Vuckovic, D. Current trends and challenges in sample preparation for global metabolomics using liquid chromatography-mass spectrometry. *Anal. Bioanal. Chem.* **2012**, *403*, 1523–1548. [CrossRef]
83. Tan, Z.-R.; Ouyang, D.-S.; Zhou, G.; Wang, L.-S.; Li, Z.; Wang, D.; Zhou, H.-H. Sensitive bioassay for the simultaneous determination of pseudoephedrine and cetirizine in human plasma by liquid-chromatography–ion trap spectrometry. *J. Pharm. Biomed. Anal.* **2006**, *42*, 207–212. [CrossRef]
84. Satomi, Y.; Hirayama, M.; Kobayashi, H. One-step lipid extraction for plasma lipidomics analysis by liquid chromatography mass spectrometry. *J. Chromatogr. B* **2017**, *1063*, 93–100. [CrossRef]
85. Martin-Venegas, R.; Jáuregui, O.; Moreno, J.J. Liquid chromatography-tandem mass spectrometry analysis of eicosanoids and related compounds in cell models. *J. Chromatogr. B* **2014**, *964*, 41–49. [CrossRef]
86. Heemskerk, M.M.; Dharuri, H.K.; van den Berg, S.A.A.; Jónsdóttir, H.S.; Kloos, D.-P.; Giera, M.; van Dijk, K.W.; van Harmelen, V. Prolonged niacin treatment leads to increased adipose tissue PUFA synthesis and anti-inflammatory lipid and oxylipin plasma profile. *J. Lipid Res.* **2014**, *55*, 2532–2540. [CrossRef]

87. Polson, C.; Sarkar, P.; Incledon, B.; Raguvaran, V.; Grant, R. Optimization of protein precipitation based upon effectiveness of protein removal and ionization effect in liquid chromatography-tandem mass spectrometry. *J. Chromatogr. B. Analyt. Technol. Biomed. Life Sci.* **2003**, *785*, 263–275. [CrossRef]
88. Zein Elabdeen, H.R.; Mustafa, M.; Szklenar, M.; Rühl, R.; Ali, R.; Bolstad, A.I. Ratio of Pro-Resolving and Pro-Inflammatory Lipid Mediator Precursors as Potential Markers for Aggressive Periodontitis. *PLoS ONE* **2013**, *8*, e70838. [CrossRef]
89. Wang, W.; Qin, S.; Li, L.; Chen, X.; Wang, Q.; Wei, J. An Optimized High Throughput Clean-Up Method Using Mixed-Mode SPE Plate for the Analysis of Free Arachidonic Acid in Plasma by LC-MS/MS. *Int. J. Anal. Chem.* **2015**, *2015*, 1–6. [CrossRef]
90. Kortz, L.; Helmschrodt, C.; Ceglarek, U. Fast liquid chromatography combined with mass spectrometry for the analysis of metabolites and proteins in human body fluids. *Anal. Bioanal. Chem.* **2011**, *399*, 2635–2644. [CrossRef]
91. Kortz, L.; Dorow, J.; Becker, S.; Thiery, J.; Ceglarek, U. Fast liquid chromatography-quadrupole linear ion trap-mass spectrometry analysis of polyunsaturated fatty acids and eicosanoids in human plasma. *J. Chromatogr. B Anal. Technol. Biomed. Life Sci.* **2013**, *927*, 209–213. [CrossRef]
92. Klawitter, J.; Haschke, M.; Shokati, T.; Klawitter, J.; Christians, U. Quantification of 15-F2t-isoprostane in human plasma and urine: results from enzyme-linked immunoassay and liquid chromatography/tandem mass spectrometry cannot be compared. *Rapid Commun. Mass Spectrom.* **2011**, *25*, 463–468. [CrossRef]
93. Bessonneau, V.; Zhan, Y.; De Lannoy, I.A.M.; Saldivia, V.; Pawliszyn, J. In vivo solid-phase microextraction liquid chromatography-tandem mass spectrometry for monitoring blood eicosanoids time profile after lipopolysaccharide-induced inflammation in Sprague-Dawley rats. *J. Chromatogr. A* **2015**, *1424*, 134–138. [CrossRef]
94. Tumanov, S.; Kamphorst, J.J. Recent advances in expanding the coverage of the lipidome. *Curr. Opin. Biotechnol.* **2017**, *43*, 127–133. [CrossRef]
95. Folch, J.; Lees, M.; Sloane Stanley, G.H. A simple method for the isolation and purification of total lipides from animal tissues. *J. Biol. Chem.* **1957**, *226*, 497–509.
96. Bligh, E.G.; Dyer, W.J. A Rapid Method of Total Lipid Extraction and Purification. *Can. J. Physiol. Pharmacol.* **1959**, *37*, 911–917.
97. Puppolo, M.; Varma, D.; Jansen, S.A. A review of analytical methods for eicosanoids in brain tissue. *J. Chromatogr. B Anal. Technol. Biomed. Life Sci.* **2014**, *964*, 50–64. [CrossRef]
98. Yang, Y.; Cruickshank, C.; Armstrong, M.; Mahaffey, S.; Reisdorph, R.; Reisdorph, N. New sample preparation approach for mass spectrometry-based profiling of plasma results in improved coverage of metabolome. *J. Chromatogr. A* **2013**, *1300*, 217–226. [CrossRef]
99. Fromel, T.; Jungblut, B.; Hu, J.; Trouvain, C.; Barbosa-Sicard, E.; Popp, R.; Liebner, S.; Dimmeler, S.; Hammock, B.D.; Fleming, I. Soluble epoxide hydrolase regulates hematopoietic progenitor cell function via generation of fatty acid diols. *Proc. Natl. Acad. Sci.* **2012**, *109*, 9995–10000. [CrossRef]
100. Pier, B.; Edmonds, J.W.; Wilson, L.; Arabshahi, A.; Moore, R.; Bates, G.W.; Prasain, J.K.; Miller, M.A. Comprehensive profiling of prostaglandins in human ovarian follicular fluid using mass spectrometry. *Prostaglandins Other Lipid Mediat.* **2018**, *134*, 7–15. [CrossRef]
101. Larose, J.; Julien, P.; Bilodeau, J.-F. Analysis of F 2 -isoprostanes in plasma of pregnant women by HPLC-MS/MS using a column packed with core-shell particles. *J. Lipid Res.* **2013**, *54*, 1505–1511. [CrossRef]
102. Hall, L.M.; Murphy, R.C. Electrospray mass spectrometric analysis of 5-hydroperoxy and 5-hydroxyeicosatetraenoic acids generated by lipid peroxidation of red blood cell ghost phospholipids. *J. Am. Soc. Mass Spectrom.* **1998**, *9*, 527–532. [CrossRef]
103. Brose, S.A.; Thuen, B.T.; Golovko, M.Y. LC/MS/MS method for analysis of E 2 series prostaglandins and isoprostanes. *J. Lipid Res.* **2011**, *52*, 850–859. [CrossRef] [PubMed]
104. Urban, M.; Enot, D.P.; Dallmann, G.; Körner, L.; Forcher, V.; Enoh, P.; Koal, T.; Keller, M.; Deigner, H.P. Complexity and pitfalls of mass spectrometry-based targeted metabolomics in brain research. *Anal. Biochem.* **2010**, *406*, 124–131. [CrossRef] [PubMed]
105. Kempen, E.C.; Yang, P.; Felix, E.; Madden, T.; Newman, R.A. Simultaneous Quantification of Arachidonic Acid Metabolites in Cultured Tumor Cells Using High-Performance Liquid Chromatography/Electrospray Ionization Tandem Mass Spectrometry. *Anal. Biochem.* **2001**, *297*, 183–190. [CrossRef]

106. Schroeder, C.P.; Yang, P.; Newman, R.A.; Lotan, R. Eicosanoid metabolism in squamous cell carcinoma cell lines derived from primary and metastatic head and neck cancer and its modulation by celecoxib. *Cancer Biol. Ther.* **2004**, *3*, 847–852. [CrossRef]
107. Michaelis, U.R.; Xia, N.; Barbosa-Sicard, E.; Falck, J.R.; Fleming, I. Role of cytochrome P450 2C epoxygenases in hypoxia-induced cell migration and angiogenesis in retinal endothelial cells. *Investig. Ophthalmol. Vis. Sci.* **2008**, *49*, 1242–1247. [CrossRef] [PubMed]
108. Cao, H.; Xiao, L.; Park, G.; Wang, X.; Azim, A.C.; Christman, J.W.; van Breemen, R.B. An improved LC–MS/MS method for the quantification of prostaglandins E2 and D2 production in biological fluids. *Anal. Biochem.* **2008**, *372*, 41–51. [CrossRef]
109. Guide to Solid Phase Extraction. *SUPELCO Bull. 910* **1998**, 1–12.
110. Späth, J. Oxylipins in human plasma – method development and dietary effects on levels. Master's Thesis, Umea Universitet, Umea, Sweden, 2014.
111. VanRollins, M.; VanderNoot, V.A. Simultaneous resolution of underivatized regioisomers and stereoisomers of arachidonate epoxides by capillary electrophoresis. *Anal. Biochem.* **2003**, *313*, 106–116. [CrossRef]
112. Ostermann, A.I.; Willenberg, I.; Schebb, N.H. Comparison of sample preparation methods for the quantitative analysis of eicosanoids and other oxylipins in plasma by means of LC-MS/MS. *Anal. Bioanal. Chem.* **2015**, *407*, 1403–1414. [CrossRef] [PubMed]
113. Masoodi, M.; Mir, A.A.; Petasis, N.A.; Serhan, C.N.; Nicolaou, A. Simultaneous lipidomic analysis of three families of bioactive lipid mediators leukotrienes, resolvins, protectins and related hydroxy-fatty acids by liquid chromatography/electrospray ionisation tandem mass spectrometry. *Rapid Commun. Mass Spectrom.* **2008**, *22*, 75–83. [CrossRef]
114. Galvão, A.F.; Petta, T.; Flamand, N.; Bollela, V.R.; Silva, C.L.; Jarduli, L.R.; Malmegrim, K.C.R.; Simões, B.P.; de Moraes, L.A.B.; Faccioli, L.H. Plasma eicosanoid profiles determined by high-performance liquid chromatography coupled with tandem mass spectrometry in stimulated peripheral blood from healthy individuals and sickle cell anemia patients in treatment. *Anal. Bioanal. Chem.* **2016**, *408*, 3613–3623. [CrossRef]
115. Okemoto, K.; Maekawa, K.; Tajima, Y.; Tohkin, M.; Saito, Y. Cross-Classification of Human Urinary Lipidome by Sex, Age, and Body Mass Index. *PLoS ONE* **2016**, *11*, e0168188. [CrossRef] [PubMed]
116. Balgoma, D.; Larsson, J.; Rokach, J.; Lawson, J.A.; Daham, K.; Dahlén, B.; Dahlén, S.-E.; Wheelock, C.E. Quantification of Lipid Mediator Metabolites in Human Urine from Asthma Patients by Electrospray Ionization Mass Spectrometry: Controlling Matrix Effects. *Anal. Chem.* **2013**, *85*, 7866–7874. [CrossRef] [PubMed]
117. Sterz, K.; Scherer, G.; Ecker, J. A simple and robust UPLC-SRM/MS method to quantify urinary eicosanoids. *J. Lipid Res.* **2012**, *53*, 1026–1036. [CrossRef] [PubMed]
118. Medina, S.; Domínguez-Perles, R.; Gil, J.I.; Ferreres, F.; García-Viguera, C.; Martínez-Sanz, J.M.; Gil-Izquierdo, A. A ultra-pressure liquid chromatography/triple quadrupole tandem mass spectrometry method for the analysis of 13 eicosanoids in human urine and quantitative 24 hour values in healthy volunteers in a controlled constant diet. *Rapid Commun. Mass Spectrom.* **2012**, *26*, 1249–1257. [CrossRef]
119. Gouveia-Figueira, S.; Karimpour, M.; Bosson, J.A.; Blomberg, A.; Unosson, J.; Pourazar, J.; Sandström, T.; Behndig, A.F.; Nording, M.L. Mass spectrometry profiling of oxylipins, endocannabinoids, and N-acylethanolamines in human lung lavage fluids reveals responsiveness of prostaglandin E2 and associated lipid metabolites to biodiesel exhaust exposure. *Anal. Bioanal. Chem.* **2017**, *409*, 2967–2980. [CrossRef] [PubMed]
120. Panthi, S.; Chen, J.; Wilson, L.; Nichols, J.J. Detection of Lipid Mediators of Inflammation in the Human Tear Film. *Eye Contact Lens Sci. Clin. Pract.* **2018**, *0*, 1. [CrossRef] [PubMed]
121. Giera, M.; Ioan-Facsinay, A.; Toes, R.; Gao, F.; Dalli, J.; Deelder, A.M.; Serhan, C.N.; Mayboroda, O.A. Lipid and lipid mediator profiling of human synovial fluid in rheumatoid arthritis patients by means of LC–MS/MS. *Biochim. Biophys. Acta - Mol. Cell Biol. Lipids* **2012**, *1821*, 1415–1424. [CrossRef] [PubMed]
122. Rauzi, F.; Kirkby, N.S.; Edin, M.L.; Whiteford, J.; Zeldin, D.C.; Mitchell, J.A.; Warner, T.D. Aspirin inhibits the production of proangiogenic 15(S)-HETE by platelet cyclooxygenase-1. *FASEB J.* **2016**, *30*, 4256–4266. [CrossRef]
123. Wang, C.; Colas, R.A.; Dalli, J.; Arnardottir, H.H.; Nguyen, D.; Hasturk, H.; Chiang, N.; Van Dyke, T.E.; Serhan, C.N. Maresin 1 Biosynthesis and Proresolving Anti-infective Functions with Human-Localized Aggressive Periodontitis Leukocytes. *Infect. Immun.* **2016**, *84*, 658–665. [CrossRef] [PubMed]

124. Wu, J.; Gouveia-Figueira, S.; Domellöf, M.; Zivkovic, A.M.; Nording, M.L. Oxylipins, endocannabinoids, and related compounds in human milk: Levels and effects of storage conditions. *Prostaglandins Other Lipid Mediat.* **2016**, *122*, 28–36. [CrossRef] [PubMed]
125. Robinson, D.T.; Palac, H.L.; Baillif, V.; Van Goethem, E.; Dubourdeau, M.; Van Horn, L.; Martin, C.R. Long chain fatty acids and related pro-inflammatory, specialized pro-resolving lipid mediators and their intermediates in preterm human milk during the first month of lactation. *Prostaglandins Leukot. Essent. Fat. Acids* **2017**, *121*, 1–6. [CrossRef]
126. Yue, H.; Strauss, K.I.; Borenstein, M.R.; Barbe, M.F.; Rossi, L.J.; Jansen, S.A. Determination of bioactive eicosanoids in brain tissue by a sensitive reversed-phase liquid chromatographic method with fluorescence detection. *J. Chromatogr. B Anal. Technol. Biomed. Life Sci.* **2004**, *803*, 267–277. [CrossRef] [PubMed]
127. O'Donnell, V.B.; Maskrey, B.; Taylor, G.W. Eicosanoids: Generation and detection in mammalian cells. *Methods Mol Biol* **2009**, *462*, 5–23.
128. Blewett, A.J.; Varma, D.; Gilles, T.; Libonati, J.R.; Jansen, S.A. Development and validation of a high-performance liquid chromatography-electrospray mass spectrometry method for the simultaneous determination of 23 eicosanoids. *J. Pharm. Biomed. Anal.* **2008**, *46*, 653–662. [CrossRef] [PubMed]
129. Le Faouder, P.; Baillif, V.; Spreadbury, I.; Motta, J.-P.; Rousset, P.; Chêne, G.; Guigné, C.; Tercé, F.; Vanner, S.; Vergnolle, N.; et al. LC–MS/MS method for rapid and concomitant quantification of pro-inflammatory and pro-resolving polyunsaturated fatty acid metabolites. *J. Chromatogr. B* **2013**, *932*, 123–133. [CrossRef] [PubMed]
130. Weylandt, K.H.; Krause, L.F.; Gomolka, B.; Chiu, C.Y.; Bilal, S.; Nadolny, A.; Waechter, S.F.; Fischer, A.; Rothe, M.; Kang, J.X. Suppressed liver tumorigenesis in fat-1 mice with elevated omega-3 fatty acids is associated with increased omega-3 derived lipid mediators and reduced TNF-α. *Carcinogenesis* **2011**, *32*, 897–903. [CrossRef] [PubMed]
131. Jelińska, M.; Białek, A.; Mojska, H.; Gielecińska, I.; Tokarz, A. Effect of conjugated linoleic acid mixture supplemented daily after carcinogen application on linoleic and arachidonic acid metabolites in rat serum and induced tumours. *Biochim. Biophys. Acta - Mol. Basis Dis.* **2014**, *1842*, 2230–2236. [CrossRef]
132. Deems, R.; Buczynski, M.W.; Bowers-Gentry, R.; Harkewicz, R.; Dennis, E.A. Detection and Quantitation of Eicosanoids via High Performance Liquid Chromatography-Electrospray Ionization-Mass Spectrometry. In *Methods in Enzymology*; Academic Press: New York, NY, USA, 2007; ISBN 9780123738950.
133. Takabatake, M.; Hishinuma, T.; Suzuki, N.; Chiba, S.; Tsukamoto, H.; Nakamura, H.; Saga, T.; Tomioka, Y.; Kurose, A.; Sawai, T.; et al. Simultaneous quantification of prostaglandins in human synovial cell-cultured medium using liquid chromatography/tandem mass spectrometry. *Prostaglandins Leukot. Essent. Fat. Acids* **2002**, *67*, 51–56. [CrossRef]
134. Masoodi, M.; Nicolaou, A. Lipidomic analysis of twenty-seven prostanoids and isoprostanes by liquid chromatography/electrospray tandem mass spectrometry. *Rapid Commun. Mass Spectrom.* **2006**, *20*, 3023–3029. [CrossRef] [PubMed]
135. Tajima, Y.; Ishikawa, M.; Maekawa, K.; Murayama, M.; Senoo, Y.; Nishimaki-Mogami, T.; Nakanishi, H.; Ikeda, K.; Arita, M.; Taguchi, R.; et al. Lipidomic analysis of brain tissues and plasma in a mouse model expressing mutated human amyloid precursor protein/tau for Alzheimer's disease. *Lipids Health Dis.* **2013**, *12*, 68. [CrossRef] [PubMed]
136. Rago, B.; Fu, C. Development of a high-throughput ultra performance liquid chromatography–mass spectrometry assay to profile 18 eicosanoids as exploratory biomarkers for atherosclerotic diseases. *J. Chromatogr. B* **2013**, *936*, 25–32. [CrossRef] [PubMed]
137. Margalit, A.; Duffin, K.L.; Isakson, P.C. Rapid Quantitation of a Large Scope of Eicosanoids in Two Models of Inflammation: Development of an Electrospray and Tandem Mass Spectrometry Method and Application to Biological Studies. *Anal. Biochem.* **1996**, *235*, 73–81. [CrossRef] [PubMed]
138. Dalli, J.; Serhan, C.N. Specific lipid mediator signatures of human phagocytes: microparticles stimulate macrophage efferocytosis and pro-resolving mediators. *Blood* **2012**, *120*, e60–e72. [CrossRef] [PubMed]
139. Sanaki, T.; Fujihara, T.; Iwamoto, R.; Yoshioka, T.; Higashino, K.; Nakano, T.; Numata, Y. Improvements in the High-Performance Liquid Chromatography and Extraction Conditions for the Analysis of Oxidized Fatty Acids Using a Mixed-Mode Spin Column. *Mod. Chem. Appl.* **2015**, *3*, 1000161.
140. Petta, T.; Moraes, L.A.B.; Faccioli, L.H. Versatility of tandem mass spectrometry for focused analysis of oxylipids. *J. Mass Spectrom.* **2015**, *50*, 879–890. [CrossRef]

141. Shinde, D.D.; Kim, K.-B.; Oh, K.-S.; Abdalla, N.; Liu, K.-H.; Bae, S.K.; Shon, J.-H.; Kim, H.-S.; Kim, D.-H.; Shin, J.G. LC–MS/MS for the simultaneous analysis of arachidonic acid and 32 related metabolites in human plasma: Basal plasma concentrations and aspirin-induced changes of eicosanoids. *J. Chromatogr. B* **2012**, *911*, 113–121. [CrossRef]
142. Chen, G.; Zhang, Q. Comprehensive analysis of oxylipins in human plasma using reversed-phase liquid chromatography-triple quadrupole mass spectrometry with heatmap-assisted selection of transitions. *Anal. Bioanal. Chem.* **2019**, *411*, 367–385. [CrossRef]
143. Yasumoto, A.; Tokuoka, S.M.; Kita, Y.; Shimizu, T.; Yatomi, Y. Multiplex quantitative analysis of eicosanoid mediators in human plasma and serum: Possible introduction into clinical testing. *J. Chromatogr. B* **2017**, *1068–1069*, 98–104. [CrossRef]
144. Thakare, R.; Chhonker, Y.S.; Gautam, N.; Nelson, A.; Casaburi, R.; Criner, G.; Dransfield, M.T.; Make, B.; Schmid, K.K.; Rennard, S.I.; et al. Simultaneous LC-MS/MS analysis of eicosanoids and related metabolites in human serum, sputum and BALF. *Biomed. Chromatogr.* **2018**, *32*, e4102. [CrossRef] [PubMed]
145. Miller, T.M.; Donnelly, M.K.; Crago, E.A.; Roman, D.M.; Sherwood, P.R.; Horowitz, M.B.; Poloyac, S.M. Rapid, simultaneous quantitation of mono and dioxygenated metabolites of arachidonic acid in human CSF and rat brain. *J. Chromatogr. B Anal. Technol. Biomed. Life Sci.* **2009**, *877*, 3991–4000. [CrossRef]
146. Bollinger, J.G.; Thompson, W.; Lai, Y.; Oslund, R.C.; Hallstrand, T.S.; Sadilek, M.; Turecek, F.; Gelb, M.H. Improved Sensitivity Mass Spectrometric Detection of Eicosanoids by Charge Reversal Derivatization. *Anal. Chem.* **2010**, *82*, 6790–6796. [CrossRef]
147. Lee, C.-Y.J.; Jenner, A.; Halliwell, B. Rapid preparation of human urine and plasma samples for analysis of F2-isoprostanes by gas chromatography-mass spectrometry. *Biochem. Biophys. Res. Commun.* **2004**, *320*, 696–702. [CrossRef]
148. Caligiuri, S.P.B.; Aukema, H.M.; Ravandi, A.; Guzman, R.; Dibrov, E.; Pierce, G.N. Flaxseed consumption reduces blood pressure in patients with hypertension by altering circulating oxylipins via an α-linolenic acid-induced inhibition of soluble epoxide hydrolase. *Hypertension* **2014**, *64*, 53–59. [CrossRef] [PubMed]
149. Song, J.; Liu, X.; Wu, J.; Meehan, M.J.; Blevitt, J.M.; Dorrestein, P.C.; Milla, M.E. A highly efficient, high-throughput lipidomics platform for the quantitative detection of eicosanoids in human whole blood. *Anal. Biochem.* **2013**, *433*, 181–188. [CrossRef]
150. Lebold, K.M.; Kirkwood, J.S.; Taylor, A.W.; Choi, J.; Barton, C.L.; Miller, G.W.; La Du, J.; Jump, D.B.; Stevens, J.F.; Tanguay, R.L.; et al. Novel liquid chromatography–mass spectrometry method shows that vitamin E deficiency depletes arachidonic and docosahexaenoic acids in zebrafish (Danio rerio) embryos. *Redox Biol.* **2014**, *2*, 105–113. [CrossRef] [PubMed]
151. García-Flores, L.A.; Medina, S.; Gómez, C.; Wheelock, C.E.; Cejuela, R.; Martínez-Sanz, J.M.; Oger, C.; Galano, J.M.; Durand, T.; Hernández-Sáez, Á.; et al. Aronia - Citrus juice (polyphenol-rich juice) intake and elite triathlon training: A lipidomic approach using representative oxylipins in urine. *Food Funct.* **2018**, *9*, 463–475. [CrossRef]
152. Mizuno, K.; Kataoka, H. Analysis of urinary 8-isoprostane as an oxidative stress biomarker by stable isotope dilution using automated online in-tube solid-phase microextraction coupled with liquid chromatography-tandem mass spectrometry. *J. Pharm. Biomed. Anal.* **2015**, *112*, 36–42. [CrossRef]
153. Rodríguez Patiño, G.; Castillo Rodríguez, M.A.; Ramírez Bribiesca, J.E.; Ramírez Noguera, P.; Gonsebatt Bonaparte, M.E.; López-Arellano, R. Development of a method for the determination of 8-iso-PGF2α in sheep and goat plasma using solid-phase microextraction and ultra-performance liquid chromatography/tandem mass spectrometry. *Rapid Commun. Mass Spectrom.* **2018**, *32*, 1675–1682. [CrossRef]
154. Hyötyläinen, T. Critical evaluation of sample pretreatment techniques. *Anal. Bioanal. Chem.* **2009**, *394*, 743–758. [CrossRef]
155. Prosen, H. Applications of liquid-phase microextraction in the sample preparation of environmental solid samples. *Molecules* **2014**, *19*, 6776–6808. [CrossRef] [PubMed]
156. Perestrelo, R.; Silva, C.L.; Câmara, J.S. Determination of urinary levels of leukotriene B4 using ad highly specific and sensitive methodology based on automatic MEPS combined with UHPLC-PDA analysis. *Talanta* **2015**, *144*, 382–389. [CrossRef]

157. Suhr, A.C.; Bruegel, M.; Maier, B.; Holdt, L.M.; Kleinhempel, A.; Teupser, D.; Grimm, S.H.; Vogeser, M. Ferromagnetic particles as a rapid and robust sample preparation for the absolute quantification of seven eicosanoids in human plasma by UHPLC-MS/MS. *J. Chromatogr. B Anal. Technol. Biomed. Life Sci.* **2016**, *1022*, 173–182. [CrossRef] [PubMed]
158. Balashova, E.E.; Trifonova, O.P.; Maslov, D.L.; Lokhov, P.G. Application of dried blood spot for analysis of low molecular weight fraction (metabolome) of blood. *Heal. Prim. Care* **2018**, *2*, 1–11.
159. Ferreiro-Vera, C.; Mata-Granados, J.M.; Priego-Capote, F.; Quesada-Gómez, J.M.; Luque de Castro, M.D. Automated targeting analysis of eicosanoid inflammation biomarkers in human serum and in the exometabolome of stem cells by SPE–LC–MS/MS. *Anal. Bioanal. Chem.* **2011**, *399*, 1093–1103. [CrossRef]
160. Wagner, B.M. Entwicklung Eines Multidimensionalen bioanalytischen Modellsystems für die Gesicherte Identifikation der Auswirkungen von Oxidativen Stressfaktoren auf den Organismus. Ph.D. Thesis, Medical University of Graz, Graz, Austria, 2014.
161. Kita, Y.; Takahashi, T.; Uozumi, N.; Shimizu, T. A multiplex quantitation method for eicosanoids and platelet-activating factor using column-switching reversed-phase liquid chromatography–tandem mass spectrometry. *Anal. Biochem.* **2005**, *342*, 134–143. [CrossRef]
162. Parkinson, D.R. Analytical Derivatization Techniques. In *Comprehensive Sampling and Sample Preparation*; Elsevier Science: Saint Louis, MO, USA, 2012; ISBN 9780123813749.
163. Moraes, L.A.; Giner, R.M.; Paul-Clark, M.J.; Perretti, M.; Perrett, D. An isocratic HPLC method for the quantitation of eicosanoids in human platelets. *Biomed. Chromatogr.* **2004**, *18*, 64–68. [CrossRef] [PubMed]
164. Nithipatikom, K.; Pratt, P.F.; Campbell, W.B. Determination of EETs using microbore liquid chromatography with fluorescence detection. *Am. J. Physiol. Circ. Physiol.* **2000**, *279*, H857–H862. [CrossRef]
165. Chavis, C.; Fraissinet, L.; Chanez, P.; Thomas, E.; Bousquet, J. A Method for the Measurement of Plasma Hydroxyeicosatetraenoic Acid Levels. *Anal. Biochem.* **1999**, *271*, 105–108. [CrossRef] [PubMed]
166. Aghazadeh-Habashi, A.; Asghar, W.; Jamali, F. Simultaneous determination of selected eicosanoids by reversed-phase HPLC method using fluorescence detection and application to rat and human plasma, and rat heart and kidney samples. *J. Pharm. Biomed. Anal.* **2015**, *110*, 12–19. [CrossRef]
167. Meckelmann, S.W.; Hellhake, S.; Steuck, M.; Krohn, M.; Schebb, N.H. Comparison of derivatization/ionization techniques for liquid chromatography tandem mass spectrometry analysis of oxylipins. *Prostaglandins Other Lipid Mediat.* **2017**, *130*, 8–15. [CrossRef] [PubMed]
168. Schulze, B. Oxylipins and their involvement in plant response to biotic and abiotic stress. Ph.D. Thesis, Friedrich-Schiller-Universität Jena, Jena, Germany, 18 October 2005.
169. Mesaros, C.; Blair, I.A. Targeted Chiral Analysis of Bioactive Arachidonic Acid Metabolites Using Liquid-Chromatography-Mass Spectrometry. *Metabolites* **2012**, *2*, 337–365. [CrossRef]
170. Knott, I.; Dieu, M.; Burton, M.; Lecomte, V.; Remacle, J.; Raes, M. Differential effects of interleukin-1α and β on the arachidonic acid cascade in human synovial cells and chondrocytes in culture. *Agents Actions* **1993**, *39*, 126–131. [CrossRef]
171. Jira, W.; Spitellera, G.; Richter, A. Increased Levels of Lipid Oxidation Products in Rheumatically Destructed Bones of Patients Suffering from Rheumatoid Arthritis. *Zeitschrift Naturforsch. Sect. C J. Biosci.* **1998**, *53*, 1061–1071. [CrossRef]
172. Molnár-Perl, I. AMINO ACIDS | Liquid Chromatography. In *Encyclopedia of Separation Science*; Wilson, I., Poole, C., Cooke, M., Eds.; Elsevier: Oxford, 2000; ISBN 978-0-12-226770-3.
173. Prinsen, H.C.M.T.; Schiebergen-Bronkhorst, B.G.M.; Roeleveld, M.W.; Jans, J.J.M.; de Sain-van der Velden, M.G.M.; Visser, G.; van Hasselt, P.M.; Verhoeven-Duif, N.M. Rapid quantification of underivatized amino acids in plasma by hydrophilic interaction liquid chromatography (HILIC) coupled with tandem mass-spectrometry. *J. Inherit. Metab. Dis.* **2016**, *39*, 651–660. [CrossRef]

174. Quehenberger, O.; Armando, A.M.; Brown, A.H.; Milne, S.B.; Myers, D.S.; Merrill, A.H.; Bandyopadhyay, S.; Jones, K.N.; Kelly, S.; Shaner, R.L.; et al. Lipidomics reveals a remarkable diversity of lipids in human plasma. *J. Lipid Res.* **2010**, *51*, 3299–3305. [CrossRef]
175. VanderNoot, V.A.; VanRollins, M. Capillary Electrophoresis of Cytochrome P-450 Epoxygenase Metabolites of Arachidonic Acid. 2. Resolution of Stereoisomers. *Anal. Chem.* **2002**, *74*, 5866–5870. [CrossRef]

© 2019 by the authors. Licensee MDPI, Basel, Switzerland. This article is an open access article distributed under the terms and conditions of the Creative Commons Attribution (CC BY) license (http://creativecommons.org/licenses/by/4.0/).

Review

A Review of the Extraction and Determination Methods of Thirteen Essential Vitamins to the Human Body: An Update from 2010

Yuan Zhang [1], Wei-e Zhou [2], Jia-qing Yan [1], Min Liu [1], Yu Zhou [1], Xin Shen [1], Ying-lin Ma [1], Xue-song Feng [3], Jun Yang [1] and Guo-hui Li [1,*]

1. Department of Pharmacy, National Cancer Center/National Clinical Research Center for Cancer/Chinese Academy of Medical Sciences and Peking Union Medical College, Beijing 100021, China; zhangyuan@cicams.ac.cn (Y.Z.); yanjiaqing5566@126.com (J.-q.Y.); liumin26081@163.com (M.L.); yz0908hospital@163.com (Y.Z.); hmuegw@126.com (X.S.); ylmacmu@163.com (Y.-l.M.); yangjun_99@126.com (J.Y.)
2. Graduate School of Peking Union Medical College, Peking Union Medical College, Chinese Academy of Medical Sciences, Beijing 100032, China; zhouweiexby@163.com
3. School of Pharmacy, China Medical University, Shenyang 110013, China; voncedar@126.com
* Correspondence: lgh0603@cicams.ac.cn; Tel.: +86-10-8778-8573

Academic Editor: Derek J. McPhee
Received: 11 May 2018; Accepted: 17 June 2018; Published: 19 June 2018

Abstract: Vitamins are a class of essential nutrients in the body; thus, they play important roles in human health. The chemicals are involved in many physiological functions and both their lack and excess can put health at risk. Therefore, the establishment of methods for monitoring vitamin concentrations in different matrices is necessary. In this review, an updated overview of the main pretreatments and determination methods that have been used since 2010 is given. Ultrasonic assisted extraction, liquid–liquid extraction, solid phase extraction and dispersive liquid–liquid microextraction are the most common pretreatment methods, while the determination methods involve chromatography methods, electrophoretic methods, microbiological assays, immunoassays, biosensors and several other methods. Different pretreatments and determination methods are discussed.

Keywords: vitamins; extraction; determination; review

1. Introduction

As one of the seven major nutrients, vitamins play important roles in the body. Vitamins are involved in the processes of normal metabolism and cell regulation, and they are necessary for growth and development; thus, they are chemicals that we all need to stay healthy [1,2]. There are thirteen vitamins that are recognized as playing roles in human nutrition [3]. Based on their solubility, these vitamins can be divided into fat-soluble vitamins and water-soluble vitamins. The former contains vitamin A, D, E and K, while the latter group includes the B-complex and C vitamins.

A number of biological functions in the body have been associated with the fat-soluble vitamins [4–12]. Once the amount of vitamins cannot meet the body's needs, the vitamins must be supplied from the diet. The functions and dietary sources of these fat-soluble vitamins are represented in Table 1 [13].

Table 1. List of fat-soluble vitamins [13].

Vitamin Name	Function	Dietary Sources
Vitamin A	Helps with (1) healthy mucous membranes; (2) skin, vision, tooth and bone growth; (3) health of the immune system.	From animal sources (retinol): liver, eggs, fortified margarine, butter, cream, cheese, fortified milk.
		From plant sources (beta-carotene): dark orange vegetables (pumpkin, sweet potatoes, winter squash, carrots), fruits (cantaloupe, apricots), dark green leafy vegetables.
Vitamin K	Required for correct blood clotting.	Vegetables from the cabbage family, leafy green vegetables, milk; it is also produced in the intestinal tract by the bacteria.
Vitamin E	Helps to protect the cell walls.	Nuts and seeds, egg yolks, liver, wholegrain products, wheat germ, leafy green vegetables and polyunsaturated plant oils.
Vitamin D	Required to properly absorb calcium.	Fortified margarine, fortified milk, fatty fish, liver, egg yolks; the skin can also produce vitamin D when it is exposed to sunlight.

B-complex and C vitamins are water-soluble vitamins. The B-group is a big family, which contains B_1 (thiamine), B_2 (riboflavin), B_3 (niacin), B_5 (pantothenic acid), B_6 (pyridoxine), B_8 (biotin), B_9 (folic acid), B_{12} (cyanocobalamine) and related substances. In metabolic processes, several B-group vitamins act mainly as coenzymes to produce energy and play important roles [14]. Vitamin C is one of the most important vitamins which is also indispensable for life and is involved in many important physiological processes, such as iron absorption, the immune response and so on [3]. The functions and dietary sources of these water-soluble vitamins are represented in Table 2 [13].

Table 2. List of water-soluble vitamins [13].

Vitamin Name	Benefits	Dietary Sources
Ascorbic Acid (Vitamin C)	Ascorbic acid is an antioxidant, and it is a portion of an enzyme that is required for protein metabolism. It also helps with iron absorption and is important for the health of the immune system.	Found in vegetables and fruits, especially: kiwifruit, mangoes, papayas, lettuce, potatoes, tomatoes, peppers, strawberries, cantaloupe and so on.
Thiamine (Vitamin B_1)	Thiamine is a portion of an enzyme that is required for energy metabolism, and it is important for nerve function.	Found in moderate amounts in all nutritious foods: nuts and seeds, legumes, wholegrain/enriched cereals and breads, pork.
Riboflavin (Vitamin B_2)	Riboflavin is a portion of an enzyme that is required for energy metabolism. It is also important for skin health and normal vision.	Enriched, wholegrain cereals and breads, leafy green vegetables, milk products.
Niacin (Vitamin B_3)	Niacin is a portion of an enzyme that is required for energy metabolism. It is also important for skin health as well as the digestive and nervous systems.	Peanut butter, vegetables (particularly leafy green vegetables, asparagus and mushrooms), enriched or wholegrain cereals and breads, fish, poultry and meat.
Pantothenic Acid (Vitamin B_5)	Pantothenic acid is a portion of an enzyme that is required for energy metabolism.	It is widespread in foods.
Pyridoxine (Vitamin B_6)	Pyridoxine is a portion of an enzyme that is required for protein metabolism. It also helps with the production of red blood cells.	Fruits, vegetables, poultry, fish, meat.
Folic Acid (Vitamin B_9)	Folic acid is a portion of an enzyme that is required for creating new cells and DNA.	Liver, orange juice, seeds, legumes, leafy green vegetables. It is now added to many refined grains.
Cobalamin (Vitamin B_{12})	Cobalamin is a portion of an enzyme required for the production of new cells, and it is important to the function of nerves.	Milk, milk products, eggs, seafood, fish, poultry, meat. It is not present in plant foods.
Biotin (Vitamin H)	Biotin is a portion of any enzyme that is required for energy metabolism.	It is widespread in foods and can be produced by bacteria in the intestinal tract.

Different vitamins are necessary for the body to maintain normal health, as reported by the US National Institute of Health [15]. In recent years, the essential roles of vitamins in human health have received extensive attention. For people who are at risk of vitamin deficiencies, vitamin supplementation is regarded as an effective treatment (e.g., intake of multivitamin tablets). However, an overdose of vitamins can be toxic in nature [16–20]. In addition, interactions of vitamins and other drugs are often reported [21]. Consequently, in order to use vitamins reasonably, it is essential to develop rapid, accurate, reliable and efficient methods for the simultaneous separation and quantitation of multiple vitamins in different matrices.

Recently, research on vitamins has attracted widespread interest. The number of publications involving vitamins has increased significantly which demonstrates that these issues are becoming more and more popular. A lot of pretreatment and determination methods of vitamins were developed before 2010 [22–32]. At that time, general pretreatment techniques included liquid–liquid extraction (LLE), solid-phase extraction (SPE) and so on, while the determination methods include chromatography methods, electrophoretic methods and others. As we know, great progress in analytical chemistry has been achieved and some new analytical instruments have been developed since 2010. Considering few comprehensive reviews of pretreatment and determination of vitamins has been published systematically, in this paper, we presented a review of the most common sample preparation methods, including ultrasonic assisted extraction (UAE), supercritical fluid extraction (SFE), SPE, LLE, dispersive liquid–liquid microextraction (DLLME) and different analysis methods, including chromatography methods, electrophoretic methods, microbiological assays, immunoassays, biosensors and others which have been reported and used to analyze vitamins since 2010. The pretreatment methods used are summarized in Table 3 [33–102].

Table 3. Pretreatment methods, sample matrices and targets of the recent articles.

Pretreatments	Determination Methods	Sample Matrix	Analytes	Ref.
liquid–liquid extraction (LLE)	liquid chromatography-ultraviolet detection (LC-UV)	Human serum	Vitamins A (retinol, retinyl esters), E (α- and γ-tocopherol) and D (25-OH vitamin D)	[33]
ultrasonic assisted extraction (UAE), filtration	LC-UV	Multivitamin capsule	Benfotiamine (B_1). Pyridoxine hydrochloride (B_6), mecobalamin (B_{12})	[34]
LLE	LC-UV	Milk, fruit juice and vegetable beverage	Vitamins E (a-, c- and d-tocopherol) and D (cholecalciferol and ergocalciferol)	[35]
Dilute and shoot	LC-UV	Honey	Vitamin B_2, riboflavin; vitamin B_3, nicotinic acid; vitamin B_5, pantothenic acid; vitamin B_9, folic acid; and vitamin C, ascorbic acid	[36]
UAE, filtration	LC-UV	Mineral tablets	Thiamine, riboflavin, niacinamide, pantothenic acid, pyridoxine, folic acid and ascorbic acid	[37]
Protein precipitation, centrifugation and filtration	LC-UV	Rat plasma	Vitamins D_3 and K_1	[38]
supercritical fluid extraction (SPE)	LC-UV	Combined feed, premixes, and biologically active supplements	Ascorbic acid (C), nicotinic acid (B_3 or PP), nicotinamide (B_3 or PP), pyridoxine hydrochloride (B_6), riboflavin (B_2) and thiamine hydrochloride (B_1)	[39]
Protein precipitation, filtration	LC-FLD	Plasma	Vitamin B_2 (riboflavin)	[40]
LLE	LC-UV	Human serum	All-trans-retinol, retinyl acetate, a-tocopherol, a-tocopheryl acetate	[41]
UAE	LC-UV	Vitamin tablets	10 vitamins (7 water-soluble and 3 fat-soluble)	[42]
UAE	LC-UV	Energy drinks	Caffeine and water-soluble vitamins	[43]
LLE	LC-UV	Pharmaceutical formulations	Fat-soluble vitamins	[44]
Dilute and shoot	LC-UV	Red bull and other energy drinks	Caffeine and vitamin B_6	[45]
UAE	LC-UV	Vitamin premixes, bioactive dietary supplements and pharmaceutical preparations	12 water-soluble vitamins	[46]
UAE	LC-UV	Food samples, human plasma and human adipose tissue	Retinol, tocopherols, coenzyme Q_{10} and carotenoids	[47]
Filtration, dilute and shoot	LC-UV	A.marmelos fruit juice	Vitamin C, polyphenols, organic acids and sugars	[48]
SPE	LC-UV	Meat products	Vitamin B_{12}	[49]

Table 3. Cont.

Pretreatments	Determination Methods	Sample Matrix	Analytes	Ref.
Extraction, filtration	LC-UV	Leafy vegetables	Riboflavin (vitamin B_2), niacin (vitamin B_3), pantothenic acid (vitamin B_5) and pyridoxine (vitamin B_6)	[50]
SPE	LC-UV	Leaves of Suaeda vermiculata	Vitamin B group	[51]
UAE	LC-UV	12 multivitamin/multimineral pharmaceuticals and preparations and human serum	B-complex vitamins and vitamin C	[52]
Extraction, filtration	LC-UV	Fruits and vegetables	L-ascorbic and dehydroascorbic acids	[53]
LLE	LC-FLD	Human serum	Vitamin K	[54]
Dilute and shoot	LC-UV	Fruit beverages and in pharmaceutical preparations	Vitamin C	[55]
Dilute and shoot	LC-UV	Pharmaceutical solid dosage	B-group vitamins and atorvastatin	[56]
LLE	LC-UV	Urine	Vitamin B_{12}	[57]
SPE	LC-UV	Cereal samples	Tocopherols, tocotrienols and carotenoids	[58]
Extraction, filtration	Flow-Injection MS/MS	Nutritional supplements	B vitamins	[59]
dispersive liquid–liquid microextraction (DLLME)	LC-DAD-MS	Infant foods and several green vegetables	Vitamins D and K	[60]
LLE	LC-MS	Bovine milk	Vitamins A, E and b-carotene	[61]
LLE	LC-MS	Serum	25-Hydroxyvitamin D_3 and 25-hydroxyvitamin D_2	[62]
LLE	LC-MS	Infant formula and adult nutritionals	Vitamins D_2 and D_3	[63]
LLE	LC-MS	Human plasma	Vitamins A, D and E	[64]
Centrifugation and filtration	LC-MS	Vegetables and fruits	Ascorbic and dehydroascorbic acids	[65]
UAE	LC-DAD-MS	Green leafy vegetables	Fat and water-soluble vitamins	[66]
LLE	LC-MS	Infant formula and adult nutritionals	Vitamins D_2 and D_3	[67]
LLE	LC-DAD-MS	Milk	Carotenoids and fat-soluble vitamins	[68]
UAE	LC-MS	Nutritional formulations	Fat- and water-soluble vitamins	[69]
Centrifugation and filtration	LC-MS	Rice	Folates	[70]
SPE	LC-MS	Neonatal dried blood spots	25-Hydroxyvitamin D_3	[71]
LLE	LC-MS	Blood	Vitamin K_1	[72]
LLE	LC-MS	Serum	25(OH) Vitamin D_3 and D_2	[73]
Centrifugation and filtration	LC-MS	SRM 1849 infant/adult nutritional formula powder	Water-soluble vitamins	[74]
LLE	LC-MS	Milk	Vitamin D_3	[75]
UAE	LC-corona-charged aerosol detector	Infant milk and dietary supplement	Water-soluble vitamins	[76]
UAE	MEKC	Food supplements	Water-soluble vitamins	[77]
UAE	MEKC	Commercial multivitamin pharmaceutical formulation	Water- and fat-soluble vitamins	[78]
UAE	MEKC	Multivitamin formulation	Water- and fat-soluble vitamins	[79]
LLE	MEKC	Multivitamin tablets and vitamin E soft capsules	Fat-soluble vitamins	[80]
Dilute and shoot	HPTLC	Standard stock solutions	Vitamins B_1, B_2, B_6 and B_{12}	[81]
Extraction, dilute and shoot	PCR, PLS and TLC	Pharmaceutical formulations	Vitamins B_1, B_6 and B_{12}	[82]
SPE	Spectrophotometry	Energy drinks	Caffeine and B vitamins	[83]
Extraction, dilute and shoot	Spectrofluorimetry	Pharmaceuticals	Water-soluble vitamins	[84]
DLLME	Spectrofluorimetry	Tablets and urine samples	Vitamin B_1	[85]
Filtration	Spectrofluorimetry	Corn steep liquor	Vitamins B_2, B_3, B_6 and B_7	[86]
Dilute and shoot	Spectrofluorimetry	Multivitamin drugs, food additives and energy drinks	Fat- and water-soluble vitamins	[87]
Dilute and shoot	Spectrofluorimetry	Dosage forms	Water-soluble vitamins	[88]

Table 3. *Cont.*

Pretreatments	Determination Methods	Sample Matrix	Analytes	Ref.
Centrifugation	Voltammetry	Fruit juices and wine	Ascorbic acid content	[89]
Centrifugation, UAE	Voltammetric Sensor	Food samples	Vitamin C and vitamin B_6	[90]
Dilution	Electrode	Human plasma	Vitamins B_2, B_9 and C	[91]
No previous preparation	Electrode	Pharmaceutical samples and fruit juices	Vitamins C, B_1 and B_2	[92]
Dilution	Electrode	Orange juice samples	Vitamins B_2, B_6 and C	[93]
Centrifugation, filtration	Nanosensor	Food samples	Vitamins B_9	[94]
No previous preparation	Sensor	Aqueous solutions	Vitamins B_1, amino acids and drug substances	[95]
Dilution	Sensors	Polymerization samples	Vitamin B_3	[96]
Dilution	Nanocomposites	Honey samples	Vitamins B_2, B_6 and C	[97]
Dilution	SFC-MS/MS	Standard stock solutions	Water- and fat-soluble vitamins	[98]
Centrifugation	HPLC, ELISA	Serum	Vitamins A, C and D	[99]
SPE	Microbiological assays	Infant formula	B group vitamins	[100]
UAE	LC-UV	Rice bran powder	Vitamins B_1, B_2, B_3, B_6 and B_{12}	[101]
SPE	LC-UV	Plasma	Retinol and α-tocopherol	[102]

2. Sample Pretreatment Methods

Sample extraction and purification is vitally important. In the pretreatment processes of vitamins, different substances can be separated and preconcentrated which can improve the analytical performance significantly (e.g., selectivity, sensitivity, accuracy) [30,31]. In the past, soxhlet extraction and heating under reflux have been the most commonly used methods; both these two methods have limitations due to the high costs of time and organic solvents. With the development of analysis technics, many eco-friendly and effective sample preparation technologies have emerged and are becoming more and more popular. These new sample preparation technologies can significantly decrease the personnel costs and time consumed [49,51,58,83,100,102].

The applied pretreatment methods heavily depend on the type of matrix used. A sample type can be divided into two categories based on its state during the determination of vitamins. Liquid samples and solid samples are most common for liquid samples such as serum, juice, milk and so on. There is no need for grinding, and homogeneity, sonicating and heating reflux are often employed for extraction. Then, LLE is frequently used. For solid samples, such as capsules, tablets, meat and so on, grinding and homogeneity are necessary. After the targets have been extracted, different pretreatment methods are needed. In brief, for different samples, different pretreatment methods have been established, e.g., for most solid samples, homogenization, drying and sieving through a screen were adopted, while for serum, the samples can be processed with deproteinization only.

2.1. Protein Precipitation, Centrifugation and Filtration

Samples such as whole blood, serum, plasma or urine are a complex mixture of biologically active compounds that can bond to the target analyte, interfere with determination or have a negative impact on the stability of the observed compound. To increase compatibility with the detection and separation techniques, some sample preparation procedures are worth considering, such as protein precipitation, centrifugation and filtration [36,38,40,42,45,48,55,59].

Modern detection tools are far more sensitive and some samples are relatively simple to collect and require only a few steps to purify. Methodology using a "dilute and shoot" technique is gaining popularity and is now frequently used. Ciulu et al. [36] published a method for honey samples that used a diluted mixture of 1 mL of 2 M NaOH with 12.5 mL of 1 M phosphate buffer followed by filtration through a PVDF membrane filter. Dabre et al. [42] published a method for vitamin tablets that used a solved and diluted solution followed by filtration through an 0.2 µm PTFE membrane filter.

Leacock et al. [45] developed a method for energy drinks that used a diluted mixture of 60:40 phosphate buffer/methanol solution and then delivered the sample to the detecting instrument. Different diluents were used for different targets; among the diluents, phosphate buffer was the most commonly used diluent [36,45]. Buffer can reduce the fluctuation of pH effectively and reduce the effect of pH on retention time.

Gershkovich et al. [38] described a protein precipitation method for rat plasma centrifuged at 10,000 rpm for 10 min at 5 °C followed by the addition of ice cold acetonitrile. After being filtered using PVDF 0.22 mm centrifugal filters, the supernatant was transferred into autosampler vials and directly measured by HPLC. Brian et al. [40] described a protein precipitation method for plasma that had been centrifuged after the addition of TCA using a 96-well protein precipitation plate filtration system.

Considering that some precipitation reagents used were not directly compatible with the chromatographic equipment, a complete set of protein precipitation reagents were proposed and documented. Still, the most abundantly used reagents include methanol, acetonitrile and a mixture of deionized water with different concentrations of acids.

2.2. Ultrasonic Assisted Extraction

As a high efficiency pretreatment method, UAE can save time and increase the yield and the quality of an extract dramatically [101]. The extraction efficiency can be enhanced by ultrasonic energy through induced cavitation. Since 2010, UAE has gained popularity and is now frequently used [34,42,43,47,52,66,69,76–79,101]. The information block diagram of UAE is shown in Figure 1.

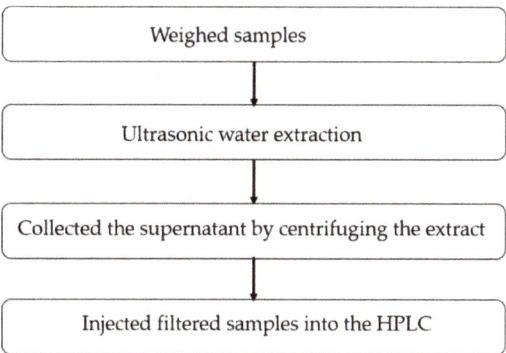

Figure 1. Information block diagram of UAE.

There are many commercial instruments to choose; thus, the UAE instruments are easy to use. After ultrasonic, we can collect the solvents by filtration or centrifugation easily. In general, after extraction, centrifugation or filtration is inevitable. Chen et al. [101] investigated the effects of different factors on the extraction efficiency of vitamins B from rice bran powder. Different factors (e.g., the extraction time, the ratio of the solvent to solid ratio) were optimized by a two-factor center composite response surface method. The extraction time affects the efficiency of extraction significantly. When the solvent to solid ratio is fixed, the increase in extraction time can significantly improve the vitamin B content in the defatted rice bran extract. Good purification (purification factor is 4.55) and recovery (recovery rate is 92.8%) can be obtained using at 323 K at a solvent to solid ratio of 10.0 for 1.5 h in dried, defatted rice bran extract. Unlike most drugs, most vitamins are sensitive to oxidation and ultraviolet light; thus, some antioxidant measures are often carried out. Francisco et al. [35] established an HPLC method for the analysis of vitamins D and E in milk, fruit juice and a vegetable beverage. During the extraction process, saponification was adopted in a nitrogen atmosphere and darkness. Further, the addition of ascorbic acid or BHT as an antioxidant was employed. Muhammad et al. [41] optimized and validated a liquid chromatography (LC) method for fat-soluble vitamins in human serum. Ascorbic acid and BHT were used as stabilizers in order to prevent oxidation during processing and storage.

Compared with the other pretreatment methods (e.g., heating reflux), the UAE methods mentioned above have shorter extraction times with good recovery. However, the methods still consume large volume of solvents.

2.3. Liquid–Liquid Extraction

Another complementary method for the extraction and purification of vitamins is LLE [33,35,41,44,54,57,61–64,67,68,72,73,75,80]. Although it is now experiencing a renaissance, the use of highly toxic volatile compounds makes this procedure experimental and not suitable for clinical and routine use for high throughput analysis.

Fariborz et al. [44] described an LC method for fat-soluble vitamins using the LLE process. Multivitamin syrup was transferred to a test tube, and then ascorbic acid, n-hexane–DEE solution (9:1, v/v) and DMSO were added. After that, vortex and centrifugation were carried out, and then the upper organic layer was gathered. After undergoing extraction four times, the organic phases were joined, evaporated to dryness and the residue was redissolved before determination. This can be taken as the normal procedure of LLE. Francisco et al. [35] compared the extraction efficiencies of different regents. Extraction with diethyl ether and hexane (2 × 50 and 1 × 100 mL) was tested, and the best results were obtained using hexane for the extraction of vitamins D and E. Muhammad et al. [41] optimized and validated an LC method for several fat-soluble vitamins in human serum. During the extraction process, n-hexane, chloroform, diethyl ether, ethyl acetate, and dichloromethane were compared in terms of their extraction efficiency on the targets. The best recovery of the targets was obtained using a two-step extraction process (n-hexane followed by dichloromethane). The above pretreatment methods are mainly used to purify fat-soluble vitamins. For water-soluble vitamins, LLE methods have also been developed. Cloud-point extraction is an alternative method. Berton et al. [57] developed an LC method for vitamin B_{12} using the ionic liquid-based aqueous two-phase system extraction process. The urine sample was transferred to a vial after being centrifuged, and then the ionic liquid (0.2 g of [C_6 mim] [Cl]) was added and fully dissolved into the pretreated sample. Then, K_2HPO_4 (3 g) was added after being vortexed, the homogeneous solution became cloudy and VB_{12} was extracted into the ionic liquid (IL) phase. After five min of stirring (without vortex assistance), two well-defined phases were formed. VB_{12} was gathered in the upper IL-enriched phase which can be directly injected in to the detecting instruments. During the pretreatment process, variables such as pH, temperature and the composition of aqueous two-phase system (ATPS), which can affect the IL-based ATPS approach, were optimized. The average extraction efficiency was over 95% under optimum conditions.

2.4. Dispersive Liquid–Liquid Micro-Extraction

Different miniaturized pre-treatment techniques based on LLE were developed prior to 2010, including SDME, HF-LPME and so on. Since 2010, DLLME has become a very popular environmentally benign sample preparation technique, because it has a lot of advantages, such as low solvent cost and high enrichment factor [60,85].

In the work of Viñas et al. [60], DLLME with HPLC-PDA detection and a comparison with MS/MS detection for vitamins D and K in foods were combined. For the DLLME procedure, the targets were extracted with acetonitrile (3 mL) which was also used as dispersive solvent. Then, an extractant solvent (carbon tetrachloride, 150 µL) was added, the mixture was injected into water directly using a micropipette, and after being shaken and centrifuged, the demented phase was collected and evaporated to dryness. The residue was reconstituted and injected into the LC. This method eliminates interfering compounds in the matrix, is sensitive and has an improved limit of detection (LOD) compared to other methods. Zeeb et al. [85] established a simple and accurate technique for the determination of vitamin B_1 using the DLLME procedure to purify interfering substance in tablets and urine samples. Under optimum conditions for DLLME, 10.0 mL of sample solution containing the analytes (pH = 13) was transferred into a glass test tube with a conic bottom. One milliliter of ferricyanide solution (0.01 mol·L^{-1}) was added to sample and mixed completely. Then, 0.50 mL of acetone (disperser solvent) containing 122 µL of chloroform (micro-extraction solvent) was injected rapidly into the sample solution. Then, the cloudy solution was formed. After that, the dispersed fine droplets of chloroform were sedimented, and the sedimented phase was removed and injected into the analyzer. Subsequently, the samples were determined by spectrofluorimeter. The results showed that this method has a good recovery and limit of quantitation.

Compared with traditional LLE methods, the abundant contact surface of fine droplets and analytes speeds up the mass transferring processes of analytes from the aquatic phase to the organic phase in a DLLME process, which not only greatly enhances the extraction efficiency but also overcomes the time-consumption problem [56,57]. However, the recoveries obtained by the DLLME method are usually not high enough compared with those of other methods. This may be caused by the use of a dispersive solvent which usually decreases the partition coefficients of analytes into the extraction solvents [58,59].

2.5. Solid Phase Extraction

SPE is one of the most common methods to pretreat samples and has been applied to analyze vitamins in different matrices [39,49,51,58,71,83,102]. For liquid samples, SPE is generally directly used to treat real samples [83,102]. However, for solid samples, the analytes are extracted from the sample matrices using organic solvents, like acetonitrile, in advance, and then the SPE procedure is performed to the extract [39,49,51,58]. Generally, SPE cartridge columns are activated before successive washing with different agents during the SPE process. Then, the samples are passed through the cartridges at settled flow rates. The cartridges are then dried, and analytes are eluted from the cartridges.

In the SPE process, solid phase materials, which are useful for extraction, concentration and clean-up, are available in a wide variety of chemistries, adsorbents and sizes. The sorbent selected in SPE controls analytical parameters such as selectivity, affinity and capacity. For this reason, different SPE materials have been used. Because of the different chemical properties of water-soluble and fat-soluble vitamins, SPE methods using different columns have been established.

An LC method was proposed to measure water-soluble vitamin B and C using SPE pretreatment methods by Rudenko et al. [39]. The best results in terms of purification efficiency and recovery rate were obtained with the reversed-phase adsorbent Sep-Pak C_{18} column. The adsorbent was preliminary washed with methanol (5 mL) and distilled water (5 mL). Then, 3 mL of the extract of combined feed was passed through the column. After that, the adsorbent was washed with 1 mL of distilled water. Vitamins were desorbed with 3 mL of methanol. The eluate obtained was analyzed after several treatments. Guggisberg et al. [49] established a purification method for vitamin B_{12} in meat samples using an immunoaffinity column. Fifteen millilitres of the supernatant of the homogenized sample was filtered and loaded onto an immunoaffinity column. Ten millilitres of purified water was used to remove impurities, and then 3 mL of methanol was used to elute the targets by complete denaturation of the antibody. The eluate was concentrated to dryness and reconstituted in the mobile phase before analysis. The recoveries and limits of detection obtained were satisfactory.

Irakli et al. [58] developed and validated a HPLC method for the simultaneous determination of vitamin E and carotenoids in cereals after SPE. As we know, vitamin E is a kind of fat-soluble vitamin; thus, in addition to the SPE columns mentioned above, new columns were adopted to purify the targets. In the experiment, three columns were compared in regard to their purification effects. OASIS cartridges with CH_2Cl_2 as the elution solvent were selected for the extraction of studied targets from cereal samples due to having better recoveries than others. Prior to the extraction of SPE, cartridges were conditioned with 3 mL of methanol and 3 mL of water. Subsequently, the above extracts were applied after the addition of 2 mL water to decrease the percentage of ethanol content in the supernatants and allowed to pass through the bed without suction. After washing with 2 mL of water, the retained constituents were eluted with 2 mL dichloromethane, followed by evaporation to dryness. The residue was reconstituted with 200 µL of methanolic solution of a-tocopherol acetate (IS, 50 µg/mL), and aliquots of 20 µL were injected into an HPLC column. Higashi et al. [71] also established a specific LC-MS/MS method for the determination of vitamin D. During the pretreatment process, the methanolic extracts were combined, diluted with water (400 mL) and purified using an Oasis HLBs cartridge. After successive washing with water (1 mL) and methanol–water (7:3, v/v, 1 mL), the vitamin D metabolites and internal standard (IS) were eluted with ethyl acetate (500 mL). This method was reproducible (intra- and inter-assay relative standard deviations (RSDs), <6.9%) and accurate (analytical recovery, 95.2–102.7%), and the limit of quantification (LOQ) was 3.0 ng/mL. The developed method enabled specific quantification of vitamin D and its metabolites.

SPE methods have shown good behavior in the process of purification producing the desired clean-up effect and achieving automation. With more and more sorbents been developed, more and more SPE columns are becoming optional. Liu et al. [102] developed a novel packed-fiber solid phase extraction procedure based on electrospun nanofibers in human serum. The parameters affecting extraction efficiency were optimized. The LOD for retinol was 0.01 µg/mL and it was 0.3 µg/mL for α-tocopherol. The relative recovery was >90%, which meets the requirements of the analyses of retinol and α-tocopherol in human plasma with satisfactory results.

SPE has many advantages compared with other extraction methods, e.g., complete phase separation, high recovery and low consumption of organic solvents. However, during the process, breakthrough problems may occur. With the development of pretreatment technology, online SPE has been developed to purify some targets. We researched the literature published after 2010, but found no papers that used automatic sample pretreatment techniques to purify vitamins.

2.6. Supercritical Fluid Extraction

During the process of SFE, supercritical fluids (usually CO_2) are adopted as extraction media. Supercritical CO_2 has the advantages of high diffusivity and low viscosity which helps it to diffuse through solid materials easily. The characteristics of supercritical fluids allow faster extraction compared with traditional LC. Chen et al. [101] studied the recovery of B vitamins from rice bran powder. Before the UAE process, the rice bran was degrased using carbon dioxide and the degreasing effect was good.

2.7. Brief Summary

Among all the sample preparation methods, reflux, UAE and SFE are preferred for solid samples, while for liquid samples, LLE, SPE and DLLME are preferred. Reflux extraction methods are traditional methods involving the consumption of large amounts of organic solvents and extraction time. A high extraction efficiency can be obtained with SFE, but expensive instruments are required compared with UAE. Considering the column passing operation, methods like SPE can be complicated. However, multiple samples can be prepared simultaneously by SPE; thus, the total time required can be greatly saved. Moreover, for SPE, it can be coupled with LC to achieve online analysis.

In the study of sample preparation methods, parameter optimization is very important. If the optimization of a method is very complete, it is still a drawback. To find the conditions to allow fast and efficient extraction or clean-up of the target compounds from the sample matrix, the fractional factorial design has been used in some studies to investigate the influences of the extraction conditions [101]. Moreover, the selected design allows the interpretation of results using statistical tests and graphic tools to determine which factors have statistically significant effects as well as to determine which interactions are significant between factors.

3. Analysis Methods

Analytical methods are often divided into three groups—screening, quantitative and confirmatory—according to different purposes. Screening methods are specifically designed to avoid untrue results. They usually offer semi-quantitative results and determine analytes at the level of interest. Quantitative methods are often used for the quantification of targets based on different detectors. The ability to reliably identify a compound is the most important feature of confirmatory methods.

The data from the determination methods of vitamins in foods [34,36,39,43,45,46,48–50,53,58–61, 63,65–70,74–77,86,87,89,90,93–97,100,101], drugs [35,37,39,42,44,46,51,52,56,78–80,82,84,85,88,92] and biological fluids [33,38–41,47,55,57,62,64,71–73,91,99,102] using chromatography methods [33–76], immunochemical methods [99], capillary electrophoresis [77–81] and so on [82–102] have been summarized. Since 2010, HPLC has become the most common method for the determination of vitamins. Considering the rapid development of HPLC and MS detectors, this technology will play an even greater role.

3.1. Liquid Chromatography

Due to its high sensitivity and broad linear range, HPLC has been widely used. At present, RP-HPLC are the most widely used analytical methods for vitamins. With the development of new HPLC equipment, new systems that can tolerate ultra-high pressures have been developed. The new system, which is called ultra-high performance liquid chromatography (UHPLC), uses sub-2-µm-particle columns and improves chromatographic performance significantly, for example, in terms of sensitivity, speed and resolution.

3.1.1. LC Coupled with MS and Multiclass Analyses

Table 4 represents a selection of analytical methods used for the detection of vitamins by RP-HPLC and UHPLC with MS detection reported since 2010.

Table 4. Examples of HPLC-MS/MS methods for the detection of vitamins.

Analysis Time (min)	Instrument Analysis Methods	Column	Mobile Phase	Limit of Detection (LOD)	Limit of Quantification (LOQ)	Ref
45 min	Flow-Injection Tandem Mass Spectrometry (Linear Ion-Trap Mass Spectrometer)	Cadenza CD-C_{18} column (4.6 × 250 mm, 3 µm)	A: 20 mM aqueous ammonium formate (pH 4); B: methanol. Gradient	0.04–48.2 ng/g	0.13–160.6 ng/g	[59]
15 min	HPLC-APCI-MS/MS	Zorbax Eclipse ODS (4.6 × 250 mm, 5 µm)	Acetonitrile, isopropanol and water. Gradient	0.2–0.6 ng/mL	0.8–2 ng/mL	[60]
26 min	High Performance Liquid Chromatography–Ion Trap Mass Spectrometry (HPLC–Msn)	Polaris C_{18} column (2.1 × 150 mm, 5 µm)	A: water; B: methanol. Gradient	no report	0.1 µg/100 mL for all trans-retinol and α-tocopherol and 1 µg/100 mL for β-carotene	[61]
15 min	LC-MS/MS	Zorbax SB-CN column (4.6 × 250 mm, 5 µm)	34% water and 66% methanol. Isocratic	~0.15 ng/g	no report	[62]
3 min	Ultra-Pressure Liquid Chromatography with Tandem Mass Spectrometry Detection (UPLC-MS/MS)	UPLC HSS C_{18} column (2.1 × 100 mm, 1.8 µm)	A: 2 mM NH_4COOH; B: 2 mM NH_4COOH: MeOH. Gradient	The LODs for vitamin D_2 were reported as 0.20 and 0.61 µg/100 g,	The reported LOQ values for vitamin D_3 were 0.47 and 1.44 µg/100 g	[63]
6 min	Liquid Chromatography/Tandem Mass Spectrometry	Ascentis Express C_{18} column (4.6 × 50 mm, 2.7 µm)	A: Ammonium formate in MeOH; B: H_2O. Gradient	0.1 µM for all-trans retinol, 3.3 nM for 25-OH VD_2 and 25-OH VD_3	no report	[64]
5 min	Liquid Chromatography with Tandem-Mass Spectrometry	Prontosil C_{18} analytical column (3 × 250 mm, 3 µm)	0.2% (v/v) formic acid. Isocratic	13 ng/mL for AA and 11 ng/mL for DHAA	44 ng/mL for ascorbic acid (AA) and 38 ng/mL for dehydroascorbic acid (DHAA)	[65]

Table 4. Cont.

Analysis Time (min)	Instrument Analysis Methods	Column	Mobile Phase	Limit of Detection (LOD)	Limit of Quantification (LOQ)	Ref
30 min	HPLC-MS/MS	ACE-100 C_{18} (2.1 × 100 mm, 3 μm)	A: 10 mM ammonium acetate solution (pH 4.5); B: MeOH with 0.1% acetic acid; C: MeOH with 0.3% acetic acid. Gradient	0.07–170 ng/mL	0.2–520 ng/mL	[66]
12 min	HPLC-MS/MS	HSS T_3 (2.1 × 150 mm, 1.7 μm)	A: 2 mM ammonium formate in water; B: 2 mM ammonium formate in methanol. Gradient	0.02 μg/100 g	0.12 μg/100 g	[67]
30 min	LC-DAD-MS/MS	a Supelcosil C_{18} (4.6 mm × 50 mm, 5 μm) and an Alltima C_{18} (4.6 mm × 250 mm, 5 μm) for fat-soluble vitamins, ProntoSIL C_{30} column (4.6 × 250 mm, 3 μm) for carotenoids	A: Methanol; B: isopropanol/hexane (50:50, v/v). Gradient	0.9–15.6 μg/L	2.7–46.8 μg/L	[68]
45 min	LC-MS/MS	Cadenza CD-C_{18} stationary phase (4.6 × 250 mm, 3 μm particles)	A: 20 mM ammonium formate (pH 4.0); B: methanol. Gradient	no report	no report	[69]
8 min	UPLC-MS/MS	HSS T_3 column (2.1 × 150 mm, 1.8 μm)	A: 0.1% of formic acid in water; B: 0.1% of formic acid in acetonitrile. Gradient	0.06–0.45 μg/100 g	0.12–0.91 μg/100 g	[70]
12 min	HPLC-MS/MS	Pro C_{18} RS column (2.0 × 150 mm, 5 μm)	Methanol–10 mM ammonium formate containing 5 mM methylamine. Isocratic	1.5 ng/mL	3 ng/mL	[71]
24 min	UPLC-MS/MS	Alltima C_{18} column (2.1 × 150 mm, 3 μm)	Methanol acidified with 0.1% formic acid. Isocratic	14 ng/L	36 ng/L	[72]
4 min	LC-MS/MS	MAX-RP (2.0 × 50 mm, 4 μm) column	A: 85% methanol, B: 15% ammonium acetate. Isocratic	10 nmol/L	no report	[73]
27 min	LC-MS/MS	HydroRP (2.0 × 250 mm, 4 um) column	A: 0.1% formic acid in water; B: 0.1% formic acid in acetonitrile. Gradient	no report	no report	[74]
24 min	LC-LIT-MS	Polaris C_{18} column (2.1 × 150 mm, 5 μm)	A: Methanol: B: water containing 5 mM ammonium (92:8 v/v). Isocratic	no report	0.01 μg/100 g	[75]

Among these methods, HPLC-MS/MS, which can also be considered to be a confirmatory method, has become the main analytical technique used for the identification of vitamins due to its higher selectivity and sensitivity than other instrumental methods. Being a confirmatory method, MS detection is used to identify and quantify a substance and can be used to confirm a compound's molecular structure. The basic principle of this detection technique is measurement of the mass-to-charge (m/z) ratios of ionized molecules. HPLC-MS/MS is often applied using a triple quadrupole analyzer and a selected reaction monitoring mode. This mode allows for the confirmation of the composition of compounds and provides structural information. In MS/MS, the most intensive ionic fragment from a precursor ion is used for quantification. A less sensitive secondary transition is used as the second criterion for confirmation purposes. This mode also improves the precision and sensitivity of the analysis but does not collect the full scan data. This can limit the availability of the full scan data which could otherwise be used to both identify target analytes and detect additional unknown compounds. The choice of one or another MS approach to monitor certain substances and residues in live animals and animal products can be referred to the European Union Commission Decision 2002/657/EC which established performance criteria and other requirements for analytical methods with different types of detection, including MS. The first step in tandem MS detection is the selection of a precursor ion. The HPLC-MS/MS analysis of vitamins is usually performed with an electrospray ionization (ESI) source operated in positive ionization mode. The protonated molecule [M + H]$^+$ was chosen as a precursor ion for quantitation in all developed methods.

One of the advantages of MS/MS is the fact that complete HPLC separation of the target analytes is not necessary for selective detection. However, it is always advisable to have good chromatographic separation in order to reduce matrix effects that typically result in the suppression or, less frequently, in the enhancement of analyte signals. Therefore, short HPLC columns are generally used, considerably speeding up the analysis. As indicated in Table 4, C_{18} reversed phase based columns are widely used for HPLC multiresidue analytical methods.

Because MS detection is incompatible with most mobile phases, volatile organic modifiers should be used when HPLC is coupled to MS. Thus, formic and acetic acid or their ammonium salts are added to acetonitrile–water or methanol–water mixtures. The typical concentrations of modifiers range from 2 to 20 mmol/L. It has been observed that the higher concentrations lead to reduced signal intensities.

HPLC-MS/MS methods have been applied to quantify vitamins in different matrices successfully [59–75]. Midttun et al. [64] simultaneously determined three vitamins in a small amount of human plasma. Mass spectrometric parameters were optimized before analysis. The LOD for trans-retinols were 0.10 µM and 3.3 nM for 25-OH D_2 and 25-OH D_3, respectively. Thus, the method is able to meet the requirements for determination and can be applied to biological samples. Besides the fat-soluble vitamins, water-soluble vitamins can also be quantified by HPLC-MS/MS and a good level of sensitivity can be obtained. Fenoll et al. [65] established a HPLC-MS/MS method for measuring ascorbic and dehydroascorbic acids in several fruits (e.g., pepper, tomato, orange and lemon). MS/MS transitions of m/z 173→143, 71 were used for ascorbic acid (AA) while m/z 175→115, 87 were used for dehydroascorbic acid (DHAA). The negative ion mode of ESI was chosen. The method was successfully applied for the determination of AA and DHAA without derivatization or oxidation/reduction processes. The major advantages of the method include its simplicity (little sample preparation), speed (analysis time is no more than 5 min) and great sensitivity (LODs were 13 ng/mL for AA and 11 ng/mL for DHAA, respectively). The advanced instrument can also be used for the determination of fat- and water-soluble vitamins simultaneously. Santos et al. [66] described a HPLC-DAD-MS/MS method to determine both fat-soluble and water-soluble vitamins in green leafy vegetables simultaneously. The LOD and LOQ were 0.07–170 ng/mL and 0.2–520 ng/mL, respectively.

A defining feature of HPLC-MS/MS methods is the high cost of the equipment and the large consumption of organic reagents. With the invention of UHPLC in 2004, UHPLC-MS/MS has had a wide range of applications in recent years. Stevens [63] determined the quantities of vitamin D_2 and D_3 in infant formula and adult nutritionals using a HSS C_{18} column (2.1 × 100 mm, 1.8 µm), and the analysis time was less than 3 min which allowed dramatic administrative time savings. Brouwer et al. [70] used an HSS T_3 column (2.1 × 150 mm, 1.8 µm) to separate vitamin B groups. With 1.8 µm particles, the analytes can be separated in less than 8 min. In general, with a decrease in the number of particles in the stationary phase, the analysis time reduces significantly.

Another inevitable drawback of HPLC-MS/MS is the occurrence of abundant matrix effects, which compromise the quantitative aspects and selectivity of the methods. Extracts from different matrices usually have high contents of organic components, such as lipids, protein, etc. These interfering compounds compete with the analytes to reach the droplet surface positions which affects the maximum evaporation efficiency and hampers ionization of the analytes. These components also increase the viscosity of the sample and the surface tension of the droplets generated from the ESI source, hindering the evaporation of the analytes. Therefore, before analysis using HPLC-MS/MS, matrix effects should be examined. In order to reduce the effects of the matrix effects, different approaches have been developed. Different factors can affect the matrix effects in theory. Firstly, better separation of the matrix compounds from the analytes can reduce the matrix effects; thus, researchers have compared and optimized the columns in order to achieve better separation. Secondly, dilution of the sample extracts and the use of internal standards or matrix-matched calibration are also frequently used [65]. Thirdly, an internal standard can be applied to the matrix effects due to its nearly identical chemical and physical properties. Brouwer et al. [70] investigated the matrix effect of the method that they established. They found that ion suppression or enhancement occurs, with suppression being most pronounced for 5-MTHF and 5, 10-CH+THF; hence, the isotopically labelled standards compensated the ion suppression or enhancement, rendering the matrix effect for all compounds between 85.4 and 103.9. Docros et al. [72] also studied the matrix effect of the quantitative method that they created. The peak area ratio of vitamin K_1 to its internal standard in plasma was used to calculate the matrix effect. They found that the matrix has a week ion suppression effect on vitamin K_1 but they seemed to be dependent on each other.

In recent years, great progress has been made in mass detectors. Different mass detectors have been developed, e.g., hybrid triple quadrupole-linear ion trap (QqLIT) instruments. QqLIT is a powerful technique that is used for large-scale screening of targets in real samples with the advantages of the new equipment, such as high resolution and extract confirmatory results. The method is based on a QqQ with the third quadrupole (Q_3) which can be used as either a conventional quadrupole mass filter or a linear ion trap that combines the advantages of the classical QqQ scanning functionality and the possibility of additional sensitive ion trap scans to allow structural analysis within the same operating platform. Due to its high ion accumulation capacity, this method has improved the full-spectrum sensitivity and provides very promising modes, such as enhanced full mass scan and enhanced product-ion and multi-stage scans. All of these features make the technique very powerful for the identification of unknown or suspected analytes, even those with poor fragmentation and at low concentrations. Another attractive capability of QqLIT for semi-targeted analysis is its information dependent acquisition that can automatically combine a survey scan with the dependent (enhanced trap) scan during a single experiment. Trenerry et al. [75] described robust methods using HPLC-QqLIT method and measured the serum vitamin D_3 in different matrices. The level of vitamin D_3 in fresh bovine milk (0.05 µg/100 mL), commercial (natural and fortified) milk samples (0.01–2 µg/100 mL) and a dairy based infant formula (8 µg/100 mL) was obtained without the need for extensive clean-up procedures. The LOQs were 0.01 µg/100 mL and 0.02 µg/100 mL for LC-MSn and LC-MS/MS, respectively.

3.1.2. Liquid Chromatography Coupled with Other Techniques

Classical reversed-phase HPLC with ultraviolet (UV), photodiode array (PDA) and fluorescence detectors [33–58] is still widely used for the routine quantification of vitamins in different types of samples. All of these approaches are quantitative but not confirmatory, as they cannot provide direct evidence of the structure or composition of a substance. UV detection is the most affordable and versatile method, but the least selective and sensitive, while FL detection is much more sensitive and selective.

The vast majority of chromatographic separations of vitamins have been performed with conventional silica-based reversed phased columns (mainly C_{18}) with spherical sorbent particles, 3–5 µm in diameter. The speed of the analysis can be increased through the use of a high temperature or ultra-high pressure system [33–55]. Considering the instability of some vitamins, such as vitamin C, the high column temperature is rarely used. Momenbeik et al. [44] established an HPLC method for the determination of vitamins A, D_3, E and K. A Zorbax-eclipse XDB-C_8 column (150 × 4.6 mm, 5 µm) was employed in the experiment, with the wavelength set at 285 nm. This method has been validated and found to be applicable for routine analyses, but the analysis is too long—30 min. Lorencio et al. [33] assessed the suitability of UHPLC for the simultaneous determination of vitamins A, E and D. The HSS T3 cloumn (2.1 × 100 mm, 1.8 µm) was employed. With the decrease in particles in the stationary phases, the analysis time was much shorter. The method consumes no more than 4 min which improves the efficiency significantly. Klimcazk et al. [55] made a comparison of the UPLC and HPLC methods for the determination of vitamin C. The two methods are both applicable for the determination of vitamin C in routine analyses, while UPLC is faster, more sensitive and more environmentally friendly.

Methanol–water and acetonitrile–water are the most common mobile phases. In addition, three-component mixtures have been reported researchers: water, methanol and acetonitrile [58,72]. In most cases, the mobile phase is modified with acetic, formic acid, acetate acid and so on. In addition, the application of an elution gradient is often adopted.

A simple and rapid method for the simultaneous determination of seven water-soluble vitamins in infant milk and dietary supplement uses LC coupled to a corona-charged aerosol detector. The detection limits range from 0.17 to 0.62 mg/L for dietary supplements and 1.7 to 6.5 mg/L for infant milk. The method is more sensitive than the common UV or DAD methods.

3.1.3. Summary

Multiclass analytical methods have been developed due to progress in chromatography and mass spectrometry methods. This has led to significant trends in the detection of different vitamins in complex samples. The progress in chromatography and mass-spectrometry methods has resulted in the development of multiclass analytical methods which are currently a significant trend in the detection of different vitamins in food and biological samples that can be successfully applied for both quantification and screening purposes. These methods are able to detect both fat- and water-soluble compounds and are of great interest to analytical laboratories due to their simplicity, high sample throughput and cost-effectiveness.

3.2. Electrophoretic Methods

Capillary electrophoresis (CE) is another good quantitative analytical approach that is mainly used when only small amounts of a sample are available. It is a highly efficient, fast and lower solvent-consuming technique in which sample components are separated according to their sizes and charges. Some advantages of CE are its high separation efficiency, ability to analyze several samples simultaneously in multicapillary systems and low consumption of reagents and accessories (packaged columns are not required). Several CE methods have been published and reviewed for the analysis of vitamins since 2010 [77–80].

For the past 6 years, CE has been used for the determination of vitamins in different samples [77–80]. Phosphate and borate buffers which sometimes contain additional organic modifiers, such as sodium polystyrene sulfonate, have been used as running buffers.

Micellar electrokinetic chromatography (MEKC) is the most popular mode of CE. It allows the separation of both ionic and neutral analytes. Applications of the MEKC method as determination methods for the analysis of vitamins have been reported [77–80]. In the work of Danielle et al. [77], to optimize the electrophoretic method, the composition and concentration of the buffer solution, SDS and organic modifier (ethanol) concentrations, pH, temperature, injection time, injection pressure and the inner diameter of the capillary were evaluated. In this work, a good separation of ten water-soluble vitamins was obtained in only 18 min. This analytical procedure is precise (RSD < 6%), accurate (better than 9%), selective, sensitive, robust and simple. Yin et al. [78] examined different conditions, such as microemulsion composition (effect of surfactant, co-surfactants, oil phases, organic solvents, pH and concentration of the buffer.) and the effects of voltage and temperature. After being optimized, a novel microemulsion system consisting of 1.2% (w/w) sodium lauryl sulphate (SDS), 21% (v/v) 1-butanol, 18% (v/v) acetonitrile, 0.8% (w/w) n-hexane and 20 mM borax buffer (pH 8.7) was applied. Aurora-Prado et al. [79] described a rapid determination method for water- and fat-soluble vitamins simultaneously in commercial formulations by MEKC. The final selected buffer contained SDS (surfactant), butan-1-ol (co-surfactant), ethyl acetate (oil) and pH 9.2 tetraborate buffer, modified with 15% (v/v) 2-propanol. The UV detection was set at 214 nm which gave adequate sensitivity without interference from sample excipients. Under the optimized conditions, the vitamins were baseline separated in less than 7 min, and acceptable limits of quantification between 8.40 and 16.23 µg/mL were obtained. The method was considered appropriate for rapid and routine analyses.

3.3. Microbiological Assays

Based on the fact that specific vitamins are necessary for the growth of specific bacteria, microbiological inhibition assays were established. Because they offer biological responses to vitamin activity and allow the determination of different vitamins, microbiological assays have been recognized by international official institutions as the gold standard for many years. However, the traditional microbiological method still suffers from poor precision and accuracy—a relative measurement uncertainty of ±20% is commonly observed. In the past, this microbiological method has often been used as a screening approach, but with the development of technology in recent years, quantitative methods have been established and the precision and accuracy have improved a lot.

Zhang et al. [100] established a "three-in-one" pretreatment method to determine several water-soluble vitamins using VitaFast kits. The VitaFast kits are fast and friendly to users. The experimental results of the assays which have employed "three-in-one" sample preparation methods are in good agreement with those obtained from conventional VitaFast extraction methods. The proposed new sample preparation method will significantly improve the efficiency of infant formulae inspection.

3.4. Biosensors

Since 2010, it has been demonstrated that optical biosensors are excellent tools for detecting vitamins in different matrices [89–97]. Their main advantages are their technical simplicity, low cost and the possibility of being used in field analyses.

Electrochemical sensors based on the use of receptors fabricated through different imprinting approaches have been developed for the detection of vitamins [89–97]. Trace quantities of vitamins B_1, B_2 and C were successfully detected in a microfluidic device by employing electrokinetic separation and electrochemical detection using silver liquid amalgam film–modified silver solid amalgam annular band electrodes (AgLAF–AgSAE). The method is based on the adsorptive accumulation of analytes at the AgLAF–AgSAE in a phosphate buffer (VB_1), a phosphate buffer with Triton X-100 (VB_2) and an alkaline borate buffer with Triton X-100 (VC). The analytical parameters and procedure of electrode activation were optimized. The calibration graphs obtained for vitamins C, B_1 and B_2 were linear, respectively, for the concentration ranges 0.05–12, 0.01–0.1 and 0.05–3 mg/L. The detection limits were calculated and equaled 0.02, 0.003 and 0.009 mg/L, while the repeatability of the peak current was 2%, 1% and 3%, respectively [92]. On the basis of conventional electrodes, modified electrodes were developed. Revin et al. [91] simultaneously determined several vitamins using a heterocyclic conducting polymer modified electrode. The research compared a bare GC electrode and the proposed 3-amino-5-mercapto-1,2,4-triazole modified glassy carbon (p-AMTa) electrode. The former failed to show stable voltammetric signals for the targets while the p-AMTa electrode showed stable voltammetric signals for the vitamins in a mixture with potential differences of 670 mV and 530 mV between riboflavin (RB)-ascorbic acid (AA) and AA-folic acid (FA), respectively. Nie et al. [93] used electroactive species-doped poly(3,4-ethylenedioxythiophene) films to enhance the sensitivity for the electrochemical simultaneous detection of vitamins B_2, B_6 and C. The functionalized PEDOT films were prepared by incorporation of two electroactive species: ferrocene carboxylic acid (Fc-) and ferricyanide (Fe(CN)64-). After comparison, the authors reported that the oxidation peak currents of vitamins obtained at the glassy carbon electrodes (GCEs) modified with electroactive species-doped PEDOT films were much higher than those at the ClO_4-doped PEDOT films and bare GCEs.

Pisoschi et al. [89] developed a determination method for vitamin C in fruit samples by differential pulse voltammetry. Four hundred and seventy millivolts on the carbon paste working electrode and 530 mV on the Pt strip working electrode were used for the determination of ascorbic acid. The influence of the operational parameters on the analytical signal was investigated. The obtained calibration graph showed a linear dependence between the peak height and the ascorbic acid concentration within the range of 0.31–20 mM with a Pt working electrode and within the range 0.07–20 mM with a carbon paste working electrode. The developed method was applied for the determination of vitamin C in wine and juice samples. A quantity of 6.83 mg/100 mL vitamin C was detected using this method.

One of the prospective trends in this field is the development of nanoparticles, quantum dots and nanocomposites. For example, a facile one-pot strategy for the electrochemical synthesis of poly (3,4-ethylenedioxythiophene)/Zirconia nanocomposite has been applied to analyze vitamins B_2, B_6 and C [97]. The obtained PEDOT/ZrO_2NPs nanocomposite film showed a large specific area, high conductivity, rapid redox properties and the presence of encapsulated structures, which make it an excellent sensing platform for sensitive determination of the targets. Detection limits of 0.012 µM, 0.20 µM and 0.45 µM were obtained for vitamin B_2 (VB_2), vitamin B_6 (VB_6) and vitamin C (VC), respectively. Jamali et al. [94] described a novel nanosensor based on Pt:Co nanoalloy ionic liquid carbon paste electrode for the voltammetric determination of vitamin B_9 in food samples. The sensor exhibited an enhanced effectiveness for the electro-oxidation of vitamin B_9 in aqueous solution. The detection limit was 4×10^{-8} M. The proposed modified electrode has several advantages, such as being simple, having high stability, high sensitivity, and excellent catalytic activity, long-term stability and remarkable voltammetric reproducibility for the eletro-oxidation of vitamin B_9. Baghizadeh et al. [90] determined a voltammetric sensor for simultaneous determination of vitamin C and B_6 in food samples using a ZrO_2 nanoparticle/ionic liquids carbon paste electrode. At an optimum condition (pH 7.0), the two peaks separated into ca. 0.44 and 0.82 V for AA and vitamin B_6, respectively. The detection limits for AA and vitamin B_6 were 0.009 and 0.1 µM, respectively. The modified electrode has been successfully applied for assays of AA and vitamin B_6.

Another interesting approach is the application of new potentiometric sensors with Donnan potential as analytics. The Donnan potential at the interface between the ion-exchange polymer and the solution of an electrolyte represents the difference between the Galvani potentials in the ion–exchanger phase and solution phase. Its value can be estimated if one measures the electromotive force (EMF) of the electrochemical circuit. Bobreshova et al. [95] determined the quantities of amino acids, vitamins and drug substances in aqueous solutions using new potentiometric sensors with Donnan potential. Certain regularities of the Donnan potential formation have been studied in systems with polymers of different structures and solutions containing inorganic ions and organic electrolytes in different ionic forms. The developed sensor was introduced as a cross-sensitive electrode into the array of multisensor systems for multicomponent quantitative analysis and the measurement error of electrolytes in aqueous solutions did not exceed 10%.

3.5. Spectrometry

Fluorescence spectrometry can be used as a screening method to detect vitamins in different samples [84,85]. In a previous paper [85], DLLME combined with spectrofluorimetry was applied to the extraction, pre-concentration and analysis of thiamine (vitamin B_1). The method proposed to detect vitamins was based on the oxidation of thiamine with ferricyanide to form fluorescent thiochrome (TC). The excitation wavelengths and emission wavelengths were 375 and 438 nm, respectively. After the optimization, the detection limit reached 0.06 ng/mL. The method was successfully applied to pharmaceutical formulations and human urine. Mallboud et al. [84] described two spectrofluorimetric methods for the detection of some water-soluble vitamins. The first proposed method depends on the oxidation of thiamine to fluorescent thiochrome using iodine/NaOH, while the second method depends on using an acetate buffer of pH 6 for the simultaneous determination of riboflavin and pyridoxine HCl using their native fluorescence levels. Different variables that affect the fluorescence intensity related to the two methods were optimized. The excitation wavelengths and emission wavelengths of thiochrome were 375 and 438 nm, while pyridoxine and riboflavin exhibited intrinsic native fluorescence with excitation and emission maxima at 325, 457 and 415, 527 nm, respectively.

Good recoveries and sensitivities were obtained, and the proposed methods were applied to the analysis of the investigated vitamins in their laboratory-prepared mixtures and pharmaceutical dosage forms.

A simple, sensitive and accurate UV derivative spectrophotometric method for the detection of caffeine and vitamin B groups in energy drinks has been developed [83]. Caffeine was determined in a mixture with vitamin B_2 with the zero-crossing technique from the I derivative spectra (λ = 266.8 nm), and vitamin B_3 in mixture with vitamin B_6 vitamin from the II derivative spectra (λ = 280.1 nm). Vitamin B_{12} was also been determined in a three-component mixture with vitamins B_3 and B_6. Mohamed et al. [88] also described the use of two spectrophotometric methods—derivative and multivariate methods—for the determination of several water-soluble vitamins, the described methods were successfully applied for the determination of vitamin combinations in synthetic mixtures and dosage forms from different manufacturers. The methods are sensitive enough for routine analysis. Monakhova et al. [87] established a chemometrics-assisted spectrophotometric method for the simultaneous detection of vitamins in complex mixtures. The main contribution of the method was the advanced independent component analysis algorithm. The key features of the proposed method are its simplicity, accuracy and reliability. Through comparison with some other established methods (MCR-ALS, SIMPLISMA, other ICA techniques), the proposed method was shown to be comparable or even outperformed other chemometrics methods.

Near-infrared spectroscopy (NIR) methods have been applied for the determination of the vitamin B group [86]. The NIR spectra of the samples were acquired from 1350 nm to 1800 nm. The multivariate regression models obtained by NIR spectroscopy and the PLS method showed a low rate of predictive errors and good correlation coefficients.

3.6. Other Methods

Besides the methods mentioned above, several effective methods have also been developed, e.g., immunoassays [99], high performance thin-layer chromatography [81] and supercritical fluid chromatography [98]. The use of these methods is justified when it is necessary to carry out routine quality control of relatively simple sample compositions. The advantages of these methods include their simplicity, compactness and relatively low cost of the analysis.

Immunoassays are characterised by their high specificity, high sensitivity, simplicity and cost effectiveness which makes them particularly useful for routine uses. These assays are based on a specific reaction between an antibody and an antigen, and they are capable of detecting the low concentration of residues in a short period of time and often do not require laborious extraction or clean-up steps. Enzyme-linked immunosorbent assays (ELISA) are the most widely used immunoassays due to their high sample throughput. These methods can drastically reduce the number of analyses required to detect vitamins in different samples. Martin et al. [99] used the 25-OH vitamin D ELISA Assay kit to determine the quantity of vitamin D; the method was successfully applied into serum samples.

Recently, TLC has been improved to incorporate HPTLC grade stationary phases, automated sample application devices, a controlled development environment, automated development, forced-flow techniques, computer-controlled densitometry, quantitation and fully validated procedures. HPTLC is becoming a routine analytical technique because of its advantages of low operating cost, high sample throughput, simplicity, speed, need for minimum sample clean up, high reproducibility, accuracy, reliability and robustness [81]. Panahi et al. [81] isolated and quantified vitamins B_1, B_2, B_6 and B_{12} using HPTLC. The precoated aluminum-backed silica gel G60 F254 HPTLC plate was developed with nearly 30 different solvent systems. The ethanol–chloroform–acetonitrile–toluene–ammonia–water (7:4:4.5:0.5:1:1) mixture was selected as the mobile phase, and the wavelength was set at 254 nm. The retention factors of vitamins B_1, B_2, B_6 and B_{12} were 0.36, 0.6, 0.85 and 0.46, respectively, the LOQ were 141.72, 42.41, 100.31 and 11.5 ng and the LOD were 42.52, 12.72, 30.09 and 3.45 ng, which is applicable for routine analysis.

SFC is a complementary separation technique to GC and LC. This technique employs supercritical carbon dioxide or subcritical carbon dioxide as the mobile phase. As mentioned above, SFC could be applied into purify the vitamins in rice [101], SFC could also be used to separate the vitamins. Taguchi et al. [98] simultaneously separated and quantified water- and fat-soluble vitamins using a single chromatography technique to unify SFC and LC. In this method, the phase state was continuously changed in the following order: supercritical, subcritical and liquid. The gradient of the mobile phase starting at almost 100% CO_2 was replaced with 100% methanol at the end. As a result, this approach achieved further extension of the polarity range of the mobile phase in a single run and successfully enabled the simultaneous analysis of fat- and water-soluble vitamins with a wide log P range of -2.11 to 10.12. Furthermore, the seventeen vitamins were exceptionally separated in 4 min. The results indicated that the use of dense CO_2 and the replacement of CO_2 by methanol are practical approaches in unified chromatography covering diverse compounds. In conclusion, the SFE method has significant advantages, including (1) the method is environmentally friendly, because the CO_2 is non-flammable and has no negative impact on human health; (2) The super fluid is flexible in adjusting its dissolving power by adding different co-solvents; (3) SFE can inherently eliminate organic solvents and provide cleaner extracts at the same time. The only serious drawback of SFE is its higher investment costs compared to traditional atmospheric pressure extraction techniques.

3.7. Summary of Analysis Methods

Among all the analytical methods, LC is the most popular due to its advantages. LC methods can meet the requirements of the qualitative and quantitative analysis of vitamins in different matrices, such as foods and biological samples, especially when it is combined with MS. However, UV and FLD detectors suffer overlapping peaks when dealing with complex samples, while matrix effects and high costs are necessary when using LC-MS. CE is alternative method for the determination of vitamins which is highly efficient, low solvent-consuming and fast, but its separation reproducibility needs to be enhanced. The main advantages of biosensors include their low cost, technical simplicity and the possibility of being used in field analyses, while their relatively short lifespan restricts the development of technology. Spectrometry is cheap and easy to promote, but its sensitivity does not quite meet requirements sometimes. In a word, with the development of equipment of HPLC and MS, this technology will surely be broadly used, while other technologies, such as electrophoresis and spectrometry, are seen as supplementary methods to be used when necessary.

4. Conclusions

Since 2010, different methods for determination of vitamins in various types of samples have been proposed. The key roles of these methods are played by sample preparation techniques, and the main efforts in this field have been focused on the optimization of the preparation, extraction and clean-up steps and on the enhancement of the environmental safety of these procedures. The method with the most promise in achieving these goals is SPE. The main advantages of this approach are its good compatibility with high throughput multiresidue analytical procedures and its relatively low cost. Therefore, this technique is expected to have the most pronounced development in the future.

The currently proposed analytical approaches for the detection of vitamins are mainly based on HPLC–MS or HPLC-MS/MS. Great advances in HPLC-MS/MS have made it a key technique for the determination of not only vitamins but also other targets. The main trend in this field is the combination of MS detectors with modern chromatographic approaches such as UHPLC and the application of the powerful QqTOF and Orbitrap instruments. These hybrid approaches have made a great contribution to the analysis of trace organic contaminants, including vitamins, and have contributed to the development of multianalyte techniques for the detection of a wide range of substances in a single analytical run. These methods seem poised to be the most frequently used techniques for the purposes of analysis in the future. The main disadvantages of these methods are their complex equipment and high cost. There is currently great interest in the development of screening methods based on microbiological, immunoassays and biosensors, which have the main advantages of being low cost, having short analysis times and the possibility of their onsite use. The clear trend in this field is the miniaturization of screening systems (chips, microarrays, microtiter plates) as well as their automation. We think that these features will maintain the sustainable progress of these methods in the near future.

Author Contributions: Conceived and designed the review: G.-h.L., Y.Z., W.-e.Z., J.-Q.Y., M.L., Y.Z., X.S., Y.-l.M., X.-s.F., J.Y. All authors read and approved the final manuscript.

Acknowledgments: This work was supported by CAMS Innovation Fund for Medical Sciences (CIFMS) (No. 2016-I2M-1-001, 2017-I2M-1-003 and 2017-I2M-1-005), Liaoning planning program of philosophy and social science (No. L17BGL034), Research Program on the Reform of Undergraduate Teaching in General Higher Schools in Liaoning (No. 2016-346), Key Program of the Natural Science Foundation of Liaoning Province of China (No. 20170541027) and Medical Education Scientific Research about "12th Five-Year Plan" (Fifth Batch of Subjects) Founded by China Medical University (No. YDJK2015005).

Conflicts of Interest: The authors declare no conflict of interest.

Abbreviation

LLE	Liquid–Liquid Extraction
SPE	Solid-Phase Extraction
UAE	Ultrasonic Assisted Extraction
SFE	Supercritical Fluid Extraction
DLLME	Dispersive Liquid–Liquid Microextraction
SDME	Single-Drop Microextraction
HF-LPME	Hollow Fibre Liquid Phase Microextraction
QqLIT	Hybrid Triple Quadrupole-Linear Ion Trap
UPLC or UHPLC	Ultra-High Performance Liquid Chromatography
HPLC	High Performance Liquid Chromatography
SFC	Supercritical Fluid Chromatography
HPLC-MS/MS	High-Performance Liquid Chromatography-Tandem Mass Spectrometry
UV	Ultraviolet
PDA	Photodiode Array
DAD	Diode Array Detector
AA	Ascorbic Acid
ELISA	Enzyme-Linked Immunosorbent Assays
TLC	Thin-Layer Chromatography
CE	Capillary Electrophoresis
TOF	Time of Flight Mass Spectrometry
MEKC	Micellar Electrokinetic Chromatography

References

1. Khayat, S.; Fanaei, H.; Ghanbarzehi, A. Minerals in Pregnancy and Lactation: A Review Article. *J. Clin. Diagn. Res.* **2017**, *11*, QE01–QE05. [CrossRef] [PubMed]
2. Glavinic, U.; Stankovic, B.; Draskovic, V.; Stevanovic, J.; Petrovic, T.; Lakic, N.; Stanimirovic, Z. Dietary amino acid and vitamin complex protects honey bee from immunosuppression caused by Nosema ceranae. *PLoS ONE* **2017**, *12*, e0187726. [CrossRef] [PubMed]
3. Eggersdorfer, M.; Laudert, D.; Létinois, U.; McClymont, T.; Medlock, J.; Netscher, T.; Bonrath, W. One hundred years of vitamins—A success story of the natural sciences. *Angew. Chem. Int. Ed. Engl.* **2012**, *51*, 12960–12990. [CrossRef] [PubMed]
4. Lounder, D.T.; Khandelwal, P.; Dandoy, C.E.; Jodele, S.; Grimley, M.S.; Wallace, G.; Lane, A.; Taggart, C.; Teusink-Cross, A.C.; Lake, K.E.; et al. Lower levels of vitamin A are associated with increased gastrointestinal graft-versus-host disease in children. *Blood* **2017**, *129*, 2801–2807. [CrossRef] [PubMed]

5. Ghanbari, A.A.; Shabani, K.; Mohammad Nejad, D. Protective effects of vitamin E consumption against 3MT electromagnetic field effects on oxidative parameters in substantia nigra in rats. *Basic Clin. Neurosci.* **2016**, *7*, 315–322. [CrossRef] [PubMed]
6. Amundson, L.A.; Hernandez, L.L.; Laporta, J.; Crenshaw, T.D. Maternal dietary vitamin D carry-over alters offspring growth, skeletal mineralisation and tissue mRNA expressions of genes related to vitamin D, calcium and phosphorus homoeostasis in swine. *Br. J. Nutr.* **2016**, *116*, 774–787. [CrossRef] [PubMed]
7. Riva, N.; Vella, K.; Meli, S.; Hickey, K.; Zammit, D.; Calamatta, C.; Makris, M.; Kitchen, S.; Ageno, W.; Gatt, A. A comparative study using thrombin generation and three different INR methods in patients on Vitamin K antagonist treatment. *Int. J. Lab. Hematol.* **2017**, *39*, 482–488. [CrossRef] [PubMed]
8. Zhao, Y.; Monahan, F.J.; McNulty, B.A.; Gibney, M.J.; Gibney, E. Effect of vitamin E intake from food and supplement sources on plasma α-and γ-tocopherol concentrations in a healthy Irish adult population. *Br. J. Nutr.* **2014**, *112*, 1575–1585. [CrossRef] [PubMed]
9. Clugston, R.D.; Blaner, W.S. Vitamin A (retinoid) metabolism and actions: What we know and what we need to know about amphibians. *Zoo Biol.* **2014**, *33*, 527–535. [CrossRef] [PubMed]
10. Makarova, A.; Wang, G.; Dolorito, J.A.; Kc, S.; Libove, E.; Epstein, E.H., Jr. Vitamin D3 Produced by Skin Exposure to UVR Inhibits Murine Basal Cell Carcinoma Carcinogenesis. *J. Investig. Dermatol.* **2017**, *137*, 2613–2619. [CrossRef] [PubMed]
11. Shearer, M.J. Vitamin K in parenteral nutrition. *Gastroenterology* **2009**, *137*, S105–S118. [CrossRef] [PubMed]
12. Duffy, M.J.; Murray, A.; Synnott, N.C.; O'Donovan, N.; Crown, J. Vitamin D analogues: Potential use in cancer treatment. *Crit. Rev. Oncol. Hematol.* **2017**, *112*, 190–197. [CrossRef] [PubMed]
13. WebMD Medical Reference from Healthwise. Available online: https://www.webmd.com/food-recipes/vitamin-mineral-sources (accessed on 19 June 2018).
14. Helliwel, K.E. The roles of B vitamins in phytoplankton nutrition: New perspectives and prospects. *New Phytol.* **2017**, *216*, 62–68. [CrossRef] [PubMed]
15. Vitamins. Available online: https://medlineplus.gov/vitamins.html (accessed on 19 June 2018).
16. Hmami, F.; Oulmaati, A.; Amarti, A.; Kottler, M.L.; Bouharrou, A. Overdose or hypersensitivity to vitamin D? *Arch. Pediatr.* **2014**, *21*, 1115–1119. [CrossRef] [PubMed]
17. Kaur, P.; Mishra, S.K.; Mithal, A. Vitamin D toxicity resulting from overzealous correction of vitamin D deficiency. *Clin. Endocrinol.* **2015**, *83*, 327–331. [CrossRef] [PubMed]
18. Hussein, A.M.; Saleh, H.A.; Mustafa, H.N. Effect of sodium selenite and vitamin E on the renal cortex in rats: An ultrastructure study. *Tissue Cell* **2014**, *46*, 170–177. [CrossRef] [PubMed]
19. Weiss, W.P. A 100-Year Review: From ascorbic acid to zinc—Mineral and vitamin nutrition of dairy cows. *J. Dairy Sci.* **2017**, *100*, 10045–10060. [CrossRef] [PubMed]
20. Dia, K.; Sarr, S.A.; Mboup, M.C.; Ba, D.M.; Fall, P.D. Overdose in Vitamin K antagonists administration in Dakar: Epidemiological, clinical and evolutionary aspects. *Pan Afr. Med. J.* **2016**, *46*, 186. [CrossRef]
21. Péter, S.; Navis, G.; de Borst, M.H.; von Schacky, C.; van Orten-Luiten, A.C.B.; Zhernakova, A.; Witkamp, R.F.; Janse, A.; Weber, P.; Bakker, S.J.L.; et al. Public health relevance of drug–nutrition interactions. *Eur. J. Nutr.* **2017**, *56*, 23–36. [CrossRef] [PubMed]
22. Kartsova, L.A.; Koroleva, O.A. Simultaneous determination of water-and fat-soluble vitamins by high-performance thin-layer chromatography using an aqueous micellar mobile phase. *J. Anal. Chem.* **2007**, *62*, 255–259. [CrossRef]
23. Viñas, P.; Balsalobre, N.; López-Erroz, C.; Hernández-Córdoba, M. Liquid chromatographic analysis of riboflavin vitamers in foods using fluorescence detection. *J. Agric. Food. Chem.* **2004**, *52*, 1789–1794. [CrossRef] [PubMed]
24. Cimpoiu, C.; Casoni, D.; Hosu, A.; Miclaus, V.; Hodisan, T.; Damian, G. Separation and identification of eight hydrophilic vitamins using a new TLC method andRaman spectroscopy. *J. Liq. Chromatogr. Relat. Technol.* **2005**, *28*, 2551–2559. [CrossRef]
25. Perveen, S.; Yasmin, A.; Khan, K.M. Quantitative simultaneous estimation of water soluble vitamins, Riboflavin, Pyridoxine, Cyanocobalamin and Folic Acid in neutraceutical products by HPLC. *Open Anal. Chem. J.* **2009**, *3*, 1–5. [CrossRef]
26. Gao, Y.L.; Guo, F.; Gokavi, S.; Chow, A.; Sheng, H.A.; Guo, M.R. Quantification of water-soluble vitamins in milk-based infant formulae using biosensor-based assays. *Food Chem.* **2008**, *110*, 769–776. [CrossRef]

27. Ashraf-Khorassani, M.; Ude, M.; Doane-Weideman, T.; Tomczak, J.; Taylor, L.T. Comparison of gravimetry and hydrolysis/derivatization/gas chromatography-mass spectrometry for quantitative analysis of fat from standard reference infant formula powder using supercritical fluid extraction. *J. Agric. Food. Chem.* **2002**, *50*, 1822–1826. [CrossRef] [PubMed]
28. Lang, R.; Yagar, E.F.; Eggers, R.; Hofmann, T. Quantitative investigation of trigonelline, nicotinic acid, and nicotinamide in foods, urine, and plasma by means of LC-MS/MS and stable isotope dilution analysis. *J. Agric. Food. Chem.* **2008**, *56*, 11114–11121. [CrossRef] [PubMed]
29. Nelson, B.C.; Sharpless, K.E.; Sander, L.C. Improved liquid chromatography methods for the separation and quantification of biotin in NIST standard reference material 3280: Multivitamin/multielement tablets. *J. Agric. Food. Chem.* **2006**, *54*, 8710–8716. [CrossRef] [PubMed]
30. Kostarnoi, A.V.; Golubitskii, G.B.; Basova, E.M.; Budko, E.V.; Ivanov, V.M. High-performance liquid chromatography in the analysis of multicomponent pharmaceutical preparations. *J. Anal. Chem.* **2008**, *63*, 516–529. [CrossRef]
31. Ball, G.F.M. *Vitamins in Foods: Analysis, Bioavailability, and Stability*; CRC Press: Boca Raton, FL, USA, 2005.
32. Park, Y.J.; Jang, J.H.; Park, H.K.; Koo, Y.E.; Hwang, I.K.; Kim, D.B. Determination of Vitamin B12 (Cyanocobalamin) in Fortified Foods by HPLC. *Prev. Nutr. Food Sci.* **2003**, *8*, 301–305. [CrossRef]
33. Granado-Lorencio, F.; Herrero-Barbudo, C.; Blanco-Navarro, I.; Pérez-Sacristán, B. Suitability of ultra-high performance liquid chromatography for the determination of fat-soluble nutritional status (vitamins A, E, D, and individual carotenoids). *Anal. Bioanal. Chem.* **2010**, *397*, 1389–1393. [CrossRef] [PubMed]
34. Poongothai, S.; Ilavarasan, R.; Karrunakaran, C.M. Simultaneous and accurate determination of vitamins B1, B6, B12 and alpha-lipoic acid in multivitamin capsule by reverse-phase high performance liquid chromatographic method. *Int. J. Pharm. Pharm. Sci.* **2010**, *2*, 133–139.
35. Barba, F.J.; Esteve, M.J.; Frígola, A. Determination of vitamins E (α-, γ- and δ-tocopherol) and D (cholecalciferol and ergocalciferol) by liquid chromatography in milk, fruit juice and vegetable beverage. *Eur. Food Res. Technol.* **2011**, *232*, 829–836. [CrossRef]
36. Ciulu, M.; Solinas, S.; Floris, I.; Panzanelli, A.; Pilo, M.I.; Piu, P.C.; Spano, N.; Sanna, G. RP-HPLC determination of water-soluble vitamins in honey. *Talanta* **2011**, *83*, 924–929. [CrossRef] [PubMed]
37. Jin, P.; Xia, L.; Li, Z.; Che, N.; Zou, D.; Hu, X. Rapid determination of thiamine, riboflavin, niacinamide, pantothenic acid, pyridoxine, folic acid and ascorbic acid in Vitamins with Minerals Tablets by high-performance liquid chromatography with diode array detector. *J. Pharm. Biomed. Anal.* **2012**, *70*, 151–157. [CrossRef] [PubMed]
38. Gershkovich, P.; Ibrahim, F.; Sivak, O.; Darlington, J.W.; Wasan, K.M. A simple and sensitive method for determination of vitamins D3 and K1 in rat plasma: Application for an in vivo pharmacokinetic study. *Drug Dev. Ind. Pharm.* **2014**, *40*, 338–344. [CrossRef] [PubMed]
39. Rudenko, A.O.; Kartsova, L.A. Determination of water-soluble vitamin B and vitamin C in combined feed, premixes, and biologically active supplements by reversed-phase HPLC. *J. Anal. Chem.* **2010**, *65*, 71–76. [CrossRef]
40. Petteys, B.J.; Frank, E.L. Rapid determination of vitamin B2 (riboflavin) in plasma by HPLC. *Clin. Chim. Acta* **2011**, *421*, 38–43. [CrossRef] [PubMed]
41. Khan, M.I.; Khan, A.; Iqbal, Z.; Ahmad, L.; Shah, Y. Optimization and validation of RP-LC/UV–VIS detection method for simultaneous determination of fat-soluble anti-oxidant vitamins, all-trans-retinol and α-tocopherol in human serum: Effect of experimental parameters. *Chromatographia* **2010**, *71*, 577–586. [CrossRef]
42. Dabre, R.; Azad, N.; Schwämmle, A.; Lämmerhofer, M.; Lindner, W. Simultaneous separation and analysis of water-and fat-soluble vitamins on multi-modal reversed-phase weak anion exchange material by HPLC-UV. *J. Sep. Sci.* **2011**, *34*, 761–772. [CrossRef] [PubMed]
43. Gliszczyńska-Świgło, A.; Rybicka, I. Simultaneous determination of caffeine and water-soluble vitamins in energy drinks by HPLC with photodiode array and fluorescence detection. *Food Anal. Methods* **2015**, *8*, 139–146. [CrossRef]
44. Momenbeik, F.; Roosta, M.; Nikoukar, A.A. Simultaneous microemulsion liquid chromatographic analysis of fat-soluble vitamins in pharmaceutical formulations: Optimization using genetic algorithm. *J. Chromatogr. A* **2010**, *24*, 3770–3773. [CrossRef] [PubMed]

45. Leacock, R.E.; Stankus, J.J.; Davis, J.M. Simultaneous determination of caffeine and vitamin B6 in energy drinks by high-performance liquid chromatography (HPLC). *J. Chem. Educ.* **2010**, *88*, 232–234. [CrossRef]
46. Bendryshev, A.A.; Pashkova, E.B.; Pirogov, A.V.; Shpigun, O.A. Determination of water-soluble vitamins in vitamin premixes, bioacitve dietary supplements, and pharmaceutical preparations using high-efficiency liquid chromatography with gradient elution. *Moscow Univ. Chem. Bull.* **2010**, *65*, 260–268. [CrossRef]
47. Gleize, B.; Steib, M.; André, M.; Reboul, E. Simple and fast HPLC method for simultaneous determination of retinol, tocopherols, coenzyme Q10 and carotenoids in complex samples. *Food Chem.* **2012**, *134*, 2560–2564. [CrossRef] [PubMed]
48. Yadav, N.; Tyagi, G.; Jangir, D.K.; Mehrotra, R. Rapid determination of polyphenol, vitamins, organic acids and sugars in Aegle marmelos using reverse phase-high performance liquid chromatography. *J. Pharm. Res.* **2011**, *4*, 717–719.
49. Guggisberg, D.; Risse, M.C.; Hadorn, R. Determination of vitamin B12 in meat products by RP-HPLC after enrichment and purification on an immunoaffinity column. *Meat Sci.* **2012**, *90*, 279–283. [CrossRef] [PubMed]
50. Hasan, M.N.; Akhtaruzzaman, M.; Sultan, M.Z. Estimation of vitamins B-complex (B2, B3, B5 and B6) of some leafy vegetables indigenous to Bangladesh by HPLC method. *J. Anal. Sci. Methods Instrum.* **2013**, *3*, 24–29.
51. Chamkouri, N. SPE-HPLC-UV for simultaneous determination of vitamins B group concentrations in Suaeda vermiculata. *Tech. J. Eng. Appl. Sci.* **2014**, *4*, 439–443.
52. Patil, S.S.; Srivastava, A.K. Development and validation of rapid ion-pair RPLC method for simultaneous determination of certain B-complex vitamins along with vitamin C. *J AOAC Int.* **2012**, *95*, 74–83. [CrossRef] [PubMed]
53. Spínola, V.; Mendes, B.; Câmara, J.S.; Castilho, P.C. An improved and fast UHPLC-PDA methodology for determination of L-ascorbic and dehydroascorbic acids in fruits and vegetables. Evaluation of degradation rate during storage. *Anal. Bioanal. Chem.* **2012**, *403*, 1049–1058. [CrossRef] [PubMed]
54. Marinova, M.; Lütjohann, D.; Westhofen, P.; Watzka, M.; Breuer, O.; Oldenburg, J. A Validated HPLC Method for the Determination of Vitamin K in Human Serum–First Application in a Pharmacological Study. *Open Clin. Chem. J.* **2011**, *4*, 17–27. [CrossRef]
55. Klimczak, I.; Gliszczyńska-Świgło, A. Comparison of UPLC and HPLC methods for determination of vitamin C. *Food Chem.* **2015**, *175*, 100–105. [CrossRef] [PubMed]
56. Reema, K.; Itishree, V.; Shantaram, N.; Jagdish, G. Method development and validation for the simultaneous estimation of b-group vitamins and atorvastatin in pharmaceutical solid dosage form by RP-HPLC. *Int. J. Pharm. Chem. Biol. Sci.* **2013**, *3*, 330–335.
57. Berton, P.; Monasterio, R.P.; Wuilloud, R.G. Selective extraction and determination of vitamin B12 in urine by ionic liquid-based aqueous two-phase system prior to high-performance liquid chromatography. *Talanta* **2012**, *97*, 521–526. [CrossRef] [PubMed]
58. Irakli, M.N.; Samanidou, V.F.; Papadoyannis, I.N. Development and validation of an HPLC method for the simultaneous determination of tocopherols, tocotrienols and carotenoids in cereals after solid-phase extraction. *J. Sep. Sci.* **2011**, *34*, 1375–1382. [CrossRef] [PubMed]
59. Bhandari, D.; Van Berkel, G.J. Evaluation of flow-injection tandem mass spectrometry for rapid and high-throughput quantitative determination of B vitamins in nutritional supplements. *J. Agric. Food Chem.* **2012**, *60*, 8356–8362. [CrossRef] [PubMed]
60. Viñas, P.; Bravo-Bravo, M.; López-García, I.; Hernández-Córdoba, M. Dispersive liquid–liquid microextraction for the determination of vitamins D and K in foods by liquid chromatography with diode-array and atmospheric pressure chemical ionization-mass spectrometry detection. *Talanta* **2013**, *115*, 806–813. [CrossRef] [PubMed]
61. Plozza, T.; Trenerry, V.C.; Caridi, D. The simultaneous determination of vitamins A, E and β-carotene in bovine milk by high performance liquid chromatography–ion trap mass spectrometry (HPLC–MSn). *Food Chem.* **2012**, *134*, 559–563. [CrossRef]
62. Tai, S.S.; Bedner, M.; Phinney, K.W. Development of a candidate reference measurement procedure for the determination of 25-hydroxyvitamin D3 and 25-hydroxyvitamin D2 in human serum using isotope-dilution liquid chromatography-tandem mass spectrometry. *Anal Chem.* **2010**, *82*, 1942–1948. [CrossRef] [PubMed]

63. Stevens, J.; Dowell, D. Determination of vitamins D2 and D3 in infant formula and adult nutritionals by ultra-pressure liquid chromatography with tandem mass spectrometry detection (UPLC-MS/MS): First Action 2011.12. *J. AOAC Int.* **2012**, *95*, 577–582. [CrossRef] [PubMed]
64. Midttun, Ø.; Ueland, P.M. Determination of vitamins A, D and E in a small volume of human plasma by a high-throughput method based on liquid chromatography/tandem mass spectrometry. *Rapid Commun. Mass Spectrom.* **2011**, *25*, 1942–1948. [CrossRef] [PubMed]
65. Fenoll, J.; Martínez, A.; Hellín, P.; Flores, P. Simultaneous determination of ascorbic and dehydroascorbic acids in vegetables and fruits by liquid chromatography with tandem-mass spectrometry. *Food Chem.* **2011**, *127*, 340–344. [CrossRef]
66. Santos, J.; Mendiola, J.A.; Oliveira, M.B.; Ibáñez, E.; Herrero, M. Sequential determination of fat-and water-soluble vitamins in green leafy vegetables during storage. *J. Chromatogr. A* **2012**, *1261*, 179–188. [CrossRef] [PubMed]
67. Gilliland, D.L.; Black, C.K.; Denison, J.E.; Seipelt, C.T.; Baugh, S. Simultaneous determination of vitamins D2 and D3 by electrospray ionization LC/MS/MS in infant formula and adult nutritionals: First action 2012.11. *J. AOAC Int.* **2013**, *96*, 1387–1395. [CrossRef] [PubMed]
68. Gentili, A.; Caretti, F.; Bellante, S.; Ventura, S.; Canepari, S.; Curini, R. Comprehensive profiling of carotenoids and fat-soluble vitamins in milk from different animal species by LC-DAD-MS/MS hyphenation. *J. Agric. Food Chem.* **2013**, *61*, 1628–1639. [CrossRef] [PubMed]
69. Phinney, K.W.; Rimmer, C.A.; Thomas, J.B.; Sander, L.C.; Sharpless, K.E.; Wise, S.A. Isotope dilution liquid chromatography-mass spectrometry methods for fat-and water-soluble vitamins in nutritional formulations. *Anal. Chem.* **2011**, *83*, 92–98. [CrossRef] [PubMed]
70. De Brouwer, V.; Storozhenko, S.; Stove, C.P.; Van Daele, J.; Van der Straeten, D.; Lambert, W.E. Ultra-performance liquid chromatography–tandem mass spectrometry (UPLC–MS/MS) for the sensitive determination of folates in rice. *J. Chromatogr. B Anal. Technol. Biomed. Life Sci.* **2010**, *878*, 509–513. [CrossRef] [PubMed]
71. Higashi, T.; Suzuki, M.; Hanai, J.; Inagaki, S.; Min, J.Z.; Shimada, K.; Toyo'oka, T. A specific LC/ESI-MS/MS method for determination of 25-hydroxyvitamin D3 in neonatal dried blood spots containing a potential interfering metabolite, 3-epi-25-hydroxyvitamin D3. *J. Sep. Sci.* **2011**, *34*, 725–732. [CrossRef] [PubMed]
72. Ducros, V.; Pollicand, M.; Laporte, F.; Favier, A. Quantitative determination of plasma vitamin K1 by high-performance liquid chromatography coupled to isotope dilution tandem mass spectrometry. *Anal. Biochem.* **2010**, *401*, 7–14. [CrossRef] [PubMed]
73. Højskov, C.S.; Heickendorff, L.; Møller, H.J. High-throughput liquid–liquid extraction and LCMSMS assay for determination of circulating 25 (OH) vitamin D3 and D2 in the routine clinical laboratory. *Clin. Chim. Acta* **2010**, *411*, 114–116. [CrossRef] [PubMed]
74. Goldschmidt, R.J.; Wolf, W.R. Simultaneous determination of water-soluble vitamins in SRM 1849 Infant/Adult Nutritional Formula powder by liquid chromatography–isotope dilution mass spectrometry. *Anal. Bioanal. Chem.* **2010**, *397*, 471–481. [CrossRef] [PubMed]
75. Trenerry, V.C.; Plozza, T.; Caridi, D.; Murphy, S. The determination of vitamin D3 in bovine milk by liquid chromatography mass spectrometry. *Food Chem.* **2011**, *125*, 1314–1319. [CrossRef]
76. Márquez-Sillero, I.; Cárdenas, S.; Valcárcel, M. Determination of water-soluble vitamins in infant milk and dietary supplement using a liquid chromatography on-line coupled to a corona-charged aerosol detector. *J. Chromatogr. A* **2013**, *1313*, 253–258. [CrossRef] [PubMed]
77. Da Silva, D.C.; Visentainer, J.V.; de Souza, N.E.; Oliveira, C.C. Micellar electrokinetic chromatography method for determination of the ten water-soluble vitamins in food supplements. *Food Anal. Methods* **2013**, *6*, 1592–1606. [CrossRef]
78. Yin, C.; Cao, Y.; Ding, S.; Wang, Y. Rapid determination of water-and fat-soluble vitamins with microemulsion electrokinetic chromatography. *J. Chromatogr. A* **2008**, *1193*, 172–177. [CrossRef] [PubMed]
79. Aurora-Prado, M.S.; Silva, C.A.; Tavares, M.F.M.; Altria, K.D. Rapid determination of water-soluble and fat-soluble vitamins in commercial formulations by MEEKC. *Chromatographia* **2010**, *72*, 687–694. [CrossRef]
80. Liu, Q.; Jia, L.; Hu, C. On-line concentration methods for analysis of fat-soluble vitamins by MEKC. *Chromatographia* **2010**, *72*, 95–100. [CrossRef]
81. Panahi, H.A.; Kalal, H.S.; Rahimi, A.; Moniri, E. Isolation and quantitative analysis of B1, B2, B6 AND B12 vitamins using high-performance thin-layer chromatography. *Pharm. Chem. J.* **2011**, *45*, 125–129. [CrossRef]

82. Elzanfaly, E.S.; Nebsen, M.; Ramadan, N.K. Development and validation of PCR, PLS, and TLC densitometric methods for the simultaneous determination of vitamins b1, b6 and b12 in pharmaceutical formulations. *Pak. J. Pharm. Sci.* **2010**, *23*, 409–415. [PubMed]
83. Pieszko, C.; Baranowska, I.; Flores, A. Determination of energizers in energy drinks. *J. Anal. Chem.* **2010**, *65*, 1228–1234. [CrossRef]
84. Mohamed, A.M.I.; Mohamed, H.A.; Abdel-Latif, N.M.; Mohamed, M.R.A. Spectrofluorimetric determination of some water-soluble vitamins. *J. AOAC Int.* **2011**, *94*, 1758–1769. [CrossRef] [PubMed]
85. Zeeb, M.; Ganjali, M.R.; Norouzi, P. Dispersive liquid-liquid microextraction followed by spectrofluorimetry as a simple and accurate technique for determination of thiamine (vitamin B1). *Microchim. Acta* **2010**, *168*, 317–324. [CrossRef]
86. Xiao, X.; Hou, Y.Y.; Du, J.; Sun, D.; Bai, G.; Luo, G.A. Determination of vitamins B2, B3, B6 and B7 in corn steep liquor by NIR and PLSR. *Trans. Tianjin Univ.* **2012**, *18*, 372–377. [CrossRef]
87. Monakhova, Y.B.; Mushtakova, S.P.; Kolesnikova, S.S.; Astakhov, S.A. Chemometrics-assisted spectrophotometric method for simultaneous determination of vitamins in complex mixtures. *Anal. Bioanal Chem.* **2010**, *397*, 1297–1306. [CrossRef] [PubMed]
88. Mohamed, A.M.; Mohamed, H.A.; Mohamed, N.A.; El-Zahery, M.R. Chemometric methods for the simultaneous determination of some water-soluble vitamins. *J. AOAC Int.* **2011**, *94*, 467–481. [CrossRef] [PubMed]
89. Pisoschi, A.M.; Pop, A.; Negulescu, G.P.; Pisoschi, A. Determination of ascorbic acid content of some fruit juices and wine by voltammetry performed at Pt and carbon paste electrodes. *Molecules* **2011**, *16*, 1349–1365. [CrossRef] [PubMed]
90. Baghizadeh, A.; Karimi-Maleh, H.; Khoshnama, Z.; Hassankhani, A.; Abbasghorbani, M. A voltammetric sensor for simultaneous determination of vitamin C and vitamin B_6 in food samples using ZrO_2 nanoparticle/ionic liquids carbon paste electrode. *Food Anal. Methods* **2015**, *8*, 549–557. [CrossRef]
91. Revin, S.B.; John, S.A. Simultaneous determination of vitamins B_2, B_9 and C using a heterocyclic conducting polymer modified electrode. *Electrochim. Acta* **2012**, *75*, 35–41. [CrossRef]
92. Baś, B.; Jakubowska, M.; Górski, Ł. Application of renewable silver amalgam annular band electrode to voltammetric determination of vitamins C, B1 and B2. *Talanta* **2011**, *84*, 1032–1037. [CrossRef] [PubMed]
93. Nie, T.; Xu, J.K.; Lu, L.M.; Zhang, K.X.; Bai, L.; Wen, Y.P. Electroactive species-doped poly (3,4-ethylenedioxythiophene) films: Enhanced sensitivity for electrochemical simultaneous determination of vitamins B2, B6 and C. *Biosens. Bioelectron.* **2013**, *50*, 244–250. [CrossRef] [PubMed]
94. Jamali, T.; Karimi-Maleh, H.; Khalilzadeh, M.A. A novel nanosensor based on Pt: Co nanoalloy ionic liquid carbon paste electrode for voltammetric determination of vitamin B9 in food samples. *LWT-Food Sci. Technol.* **2014**, *57*, 679–685. [CrossRef]
95. Bobreshova, O.V.; Parshina, A.V.; Agupova, M.V.; Polumestnaya, K.A. Determination of amino acids, vitamins, and drug substances in aqueous solutions using new potentiometric sensors with Donnan potential as analytical signal. *Russ. J. Electrochem.* **2010**, *46*, 1252–1262. [CrossRef]
96. Verma, R.; Gupta, B.D. Fiber optic SPR sensor for the detection of 3-pyridinecarboxamide (vitamin B3) using molecularly imprinted hydrogel. *Sens. Actuators B* **2013**, *177*, 279–285. [CrossRef]
97. Nie, T.; Zhang, K.X.; Xu, J.K.; Lu, L.M.; Bai, L. A facile one-pot strategy for the electrochemical synthesis of poly (3,4-ethylenedioxythiophene)/Zirconia nanocomposite as an effective sensing platform for vitamins B2, B6 and C. *J. Electroanal. Chem.* **2014**, *717–718*, 1–9. [CrossRef]
98. Taguchi, K.; Fukusaki, E.; Bamba, T. Simultaneous analysis for water-and fat-soluble vitamins by a novel single chromatography technique unifying supercritical fluid chromatography and liquid chromatography. *J. Chromatogr. A* **2014**, *1362*, 270–277. [CrossRef] [PubMed]
99. Martin, H.; Petrucka, P.; Buza, J. Determination of Vitamins A, C and D Status Using Serum Markers and a 24-Hour Dietary Recall among Maasai Women of Reproductive Age. *Open Access Library J.* **2014**, *1*, 1–7. [CrossRef]
100. Zhang, H.; Lan, F.; Shi, Y.; Wan, Z.G.; Yue, Z.F.; Fan, F.; Lin, Y.K.; Tang, M.J.; Lv, J.Z.; Xiao, T.; et al. A "three-in-one" sample preparation method for simultaneous determination of B-group water-soluble vitamins in infant formula using VitaFast® kits. *Food Chem.* **2014**, *153*, 371–377. [CrossRef] [PubMed]

101. Chen, C.H.; Yang, Y.H.; Shen, C.T.; Lai, S.M.; Chang, C.M.J.; Shieh, C.J. Recovery of vitamins B from supercritical carbon dioxide-defatted rice bran powder using ultrasound water extraction. *J. Taiwan Inst. Chem. E* **2011**, *42*, 124–128. [CrossRef]
102. Liu, Z.; Kang, X.; Fang, F. Solid phase extraction with electrospun nanofibers for determination of retinol and α-tocopherol in plasma. *Microchim. Acta* **2010**, *168*, 59–64. [CrossRef]

© 2018 by the authors. Licensee MDPI, Basel, Switzerland. This article is an open access article distributed under the terms and conditions of the Creative Commons Attribution (CC BY) license (http://creativecommons.org/licenses/by/4.0/).

MDPI
St. Alban-Anlage 66
4052 Basel
Switzerland
Tel. +41 61 683 77 34
Fax +41 61 302 89 18
www.mdpi.com

Molecules Editorial Office
E-mail: molecules@mdpi.com
www.mdpi.com/journal/molecules

www.ingramcontent.com/pod-product-compliance
Lightning Source LLC
LaVergne TN
LVHW071939080526
838202LV00064B/6639